The G Protein-Coupled Receptors Handbook

Contemporary Clinical Neuroscience

Series Editors:
Ralph Lydic and
Helen A. Baghdoyan

The G Protein-Coupled Receptors Handbook
edited by *Lakshmi A. Devi*, 2005
Attention Deficit Hyperactivity Disorder: *From Genes to Patients*,
edited by *David Gozal and Dennis L. Molfese*, 2005
Genetics and Genomics of Neurobehavioral Disorders
edited by *Gene S. Fisch*, 2003
**Sedation and Analgesia for Diagnostic
and Therapeutic Procedures**
edited by *Shobha Malviya, Norah N. Naughton,
and Kevin K. Tremper*, 2003
Neural Mechanisms of Anesthesia,
edited by *Joseph F. Antognini, Earl E. Carstens,
and Douglas E. Raines*, 2002
Glutamate and Addiction
edited by *Barbara Herman*, 2002
Molecular Mechanisms of Neurodegenerative Diseases
edited by *Marie-Françoise Chesselet*, 2000

Contemporary Clinical Neuroscience

THE G PROTEIN-COUPLED RECEPTORS HANDBOOK

Edited by

LAKSHMI A. DEVI, PhD

*Department of Pharmacology and Biological Chemistry,
Mount Sinai School of Medicine,
New York, NY*

Foreword by

Andreas Engel, PhD

*M.E. Müller Institute, Biozentrum,
University of Basel,
Basel, Switzerland*

Krzysztof Palczewski, PhD

*Department of Ophthalmology,
University of Washington School of Medicine,
Seattle, WA*

HUMANA PRESS ✷ **TOTOWA, NEW JERSEY**

© 2005 Humana Press Inc.
999 Riverview Drive, Suite 208
Totowa, New Jersey 07512
www.humanapress.com

All rights reserved. No part of this book may be reproduced, stored in a retrieval system, or transmitted in any form or by any means, electronic, mechanical, photocopying, microfilming, recording, or otherwise without written permission from the Publisher.

All authored papers, comments, opinions, conclusions, or recommendations are those of the author(s), and do not necessarily reflect the views of the publisher.

The content and opinions expressed in this book are the sole work of the authors and editors, who have warranted due diligence in the creation and issuance of their work. The publisher, editors, and authors are not responsible for errors or omissions or for any consequences arising from the information or opinions presented in this book and make no warranty, express or implied, with respect to its contents.

Production Editor: Tracy Catanese

Cover design by Patricia F. Cleary

Cover illustration: Atomic force microscopy (AFM) of native mouse disk membranes imaged in buffer solution (Fotiadis, D., et al. [2003] Nature 421:127). Based on the AFM data and crystal structure of rhodopsin (Palczewski, K., et al. (2000). Science 289, 739), a 3-D model for the packing arrangement of rhodopsin molecules within the paracrystalline arrays in native disk membranes as computed and 6 rhodopsin molecules are displayed (Liang, Y. et al. [2003] J Biol Chem 278:21,655). Artwork created by and courtesy of Kris Palczewski, Andreas Engel, and Slawek Filipek.

For additional copies, pricing for bulk purchases, and/or information about other Humana titles, contact Humana at the above address or at any of the following numbers: Tel.: 973-256-1699; Fax: 973-256-8341; E-mail: orders@humanapr.com or visit our website at www.humanapress.com

The opinions expressed herein are the views of the authors and may not necessarily reflect the official policy of the National Institute on Drug Abuse or any other parts of the US Department of Health and Human Services. The US Government does not endorse or favor any specific commercial product or company. Trade, proprietary, or company names appearing in this publication are used only because they are considered essential in the context of the studies reported herein.

This publication is printed on acid-free paper. ∞
ANSI Z39.48-1984 (American National Standards Institute) Permanence of Paper for Printed Library Materials.

Photocopy Authorization Policy:
Authorization to photocopy items for internal or personal use, or the internal or personal use of specific clients, is granted by Humana Press Inc., provided that the base fee of US $30.00 per copy is paid directly to the Copyright Clearance Center at 222 Rosewood Drive, Danvers, MA 01923. For those organizations that have been granted a photocopy license from the CCC, a separate system of payment has been arranged and is acceptable to Humana Press Inc. The fee code for users of the Transactional Reporting Service is: [1-58829-365-3/05 $30.00].

Printed in the United States of America. 10 9 8 7 6 5 4 3 2 1

eISBN: 1-59259-919-2

Library of Congress Cataloging-in-Publication Data

G protein-coupled receptors handbook / edited by Lakshmi A. Devi.
     p. cm. -- (Contemporary clinical neuroscience)
  Includes bibliographical references and index.
  ISBN 1-58829-365-3 (alk. paper)
  1. G proteins--Receptors--Handbooks, manuals, et. I. Devi, Lakshmi A. II. Series.
  QP552.G16.G174 2005
  611'.0181--dc22

2004024043

FOREWORD

Comprising the largest class of membrane-bound receptors, the G protein-coupled receptors (GPCRs) also represent one of the most prevalent gene families. These receptors mediate the biological effects of numerous hormones, neurotransmitters, chemokines, odorants, and other sensory stimuli. These in turn control such diverse physiological processes as neurotransmission, cellular metabolism, secretion, cellular differentiation and growth, and inflammatory and immune responses. In short, GPCRs are involved in a myriad of processes in the human body relevant to health and disease. Consequently, the GPCRs are targets of approx 70% of pharmacological therapeutics and provide further important opportunities for the development of new drug candidates with potential applications in all clinical fields. Recent progress in GPCR research has proliferated at a remarkable rate.

Structurally, GPCRs are characterized by a seven-transmembrane a-helical (7TM) configuration of more than 25% homology, but detailed structural knowledge is sparse. Of the three distant families of vertebrate GPCRs, family A is by far the largest group, and includes rhodopsin, adrenergic receptors, and the olfactory subgroups. The receptors for the gastrointestinal peptide hormone family belong to family B, whereas family C includes the metabotropic glutamate/pheromone receptors. The recent availability of the structure of rhodopsin has given a basis to better understand structure–function relationships in other GPCRs. Subsequently, sequence-based predictions and molecular modeling incorporating a multitude of results from biochemical and biophysical analyses can now be scrutinized, demonstrating some degree of success for these methods. Despite the progress made in predicting the critical residues engaged in ligand binding, particularly within the large family A, detailed structural knowledge is still required for understanding the process of signal transduction at a mechanistic level. Current work focuses on determining the structure of other GPCRs, on elucidation of their interactions with ligands, and on conformational changes during their activation process. There is significant hope that additional breakthroughs will occur in the near future.

The classical view that GPCRs function as monomeric entities has been jarred by the emerging concept of GPCR dimerization. Examples of GPCRs that can be biochemically detected in homo- or heteromeric complexes are being reported at an accelerated rate. These findings have not only indicated that many GPCRs exist as homodimers and heterodimers, but also that their oligomeric assemblies could have important functional roles. The important observation of GPCR dimerization came through the direct visualization of rhodopsin dimers in native disk membranes by atomic force microscopy. The ability of GPCRs to specifically oligomerize may provide some insight into how different receptor pathways influence each other. The general acceptance of the existence of GPCR dimers is now likely to have important implications for the development and screening of a new class of drugs.

The G Protein-Coupled Receptors Handbook gives a broad overview on the most recent progress in the rapidly evolving field of GPCR research. It comes at a timely period because of the significant advances that have been made in the last few years in the understanding of the structure and function of GPCRs.

Andreas Engel, PhD
M.E. Müller Institute, Biozentrum
University of Basel
Basel, Switzerland

Krzysztof Palczewski, PhD
Department of Ophthalmology
University of Washington School of Medicine
Seattle, WA

Preface

The intent of *The G Protein-Coupled Receptors Handbook* is to provide a comprehensive overview of recent advances in the G protein-coupled receptor (GPCR) field. From the basics of GPCR structure to dimerization and drug discovery, this book reviews much of the recent advances and current knowledge regarding GPCRs.

The first few chapters focus on the fundamentals of GPCR structure and function. GPCR function is now known to be regulated by a number of mechanisms: ligand-induced conformational changes, stabilizing intramolecular interactions, pharmacological chaperones, and membrane trafficking all play a role in regulating GPCRs. Specific ligand binding causes changes in GPCR conformation, which ultimately result in the activation of intracellular signaling cascades. Meanwhile, the inactive state of the receptor is maintained by stabilizing intramolecular interactions; disruption of these interactions is necessary for receptor activation. Pharmacological chaperones play a role in GPCR folding and maturation, and appear to be involved in a number of human genetic diseases. Finally, membrane trafficking of GPCRs in endocytic and biosynthetic pathways also contribute to the physiological regulation of GPCRs.

GPCRs are present in every cell and interact with a multitude of downstream effectors: heterotrimeric G proteins, regulators of G protein signaling (RGS), arrestins, G protein-coupled receptor kinases (GRKs), and many other GPCR interacting proteins. Heterotrimeric G proteins are among the most important signaling transducers involved in GPCR activity, directly coupling to the receptor and transmitting its information about activation/inactivation to the cell. RGS proteins are involved in the regulation and termination of the signaling process. GRKs catalyze GPCR phosphorylation, promoting receptor desensitization and internalization. Arrestins mediate the desensitization and uncoupling of GPCRs from their G proteins, and may also function as signal transducers. In addition, β-arrestin regulates the sequestration, intracellular trafficking, degradation, and recycling of most GPCRs. More than 50 other GPCR interacting proteins have been identified that function as modulators of GPCR function at various stages of signaling.

The next section of this book explores our current understanding of GPCR dimerization. The emerging concept of dimerization has modified our views of GPCR structure, function, and regulation tremendously. The existence of GPCR dimers has been demonstrated using biochemical methods, such as co-immunoprecipitation, and biophysical approaches, such as fluorescence (FRET) and bioluminescence resonance energy transfer (BRET). Potential domains of GPCR dimerization have been described using computational and experimental approaches. Functional complementation studies have been used to analyze the basis, selectivity, and mechanisms of dimerization. It is now evident that dimerization plays a role in receptor maturation, as many GPCRs have been shown to dimerize prior to their trafficking to the cell surface. There is also some evidence suggesting that dimerization alters the endocytotic and postendocytotic trafficking properties of GPCRs. More importantly, heterodimerization has been shown to modify the pharmacological properties of GPCRs; a finding that could have an enormous impact on the future of drug design.

The final chapters of this book describe some of the most recent developments in the GPCR field, leading to advances in drug discovery. It is now thought that a number of GPCRs functionally interact as heterodimers to mediate analgesic responses. Elucidating the role of GPCRs in mediating pain is also crucial to the development of superior analgesic drugs. Thus, a new wave of drugs specifically targeting heterodimeric receptor complexes may be on the horizon. Another important area of current research consists of investigating the structural plasticity of receptor activation by examining the conserved motifs contributing to the overall receptor structure (and variability among subtypes); this would confer ligand-binding specificity and, thus, could lead to the development of receptor-type selective drugs. Finally, the last chapter describes the identification of natural ligands of orphan GPCRs, i.e., deorphanization. Orphan receptors may represent an untapped drug target. Understanding the evolutionary diversity in GPCR ligand recognition is fundamental to understanding the potential of GPCRs as therapeutic targets.

I thank all the authors for their timely and insightful contributions, the series editors Helen Baghdoyan and Ralph Lydic for suggesting this book as a part of their series, and Ms. Elyse O'Grady at Humana Press for keeping things moving along. I also thank Noura Abul-Husn, Fabien DeCaillot, and José Morón for their extensive input into the chapters. Finally, I am grateful to Dr. Ivone Gomes for her excellent assistance throughout all of the editing and formatting stages. From planning the list of chapters/authors to the realization of this book, it has been a rewarding experience; I hope this book will serve as a helpful guide for those who are interested in learning more about the function and regulation of GPCRs.

Lakshmi A. Devi, PhD

Contents

Foreword v
Preface vii
Contributors xi
Color Plates xv

PART I GPCRs: STRUCTURE AND FUNCTION

1   Structure–Function Relationships in G Protein-Coupled
    Receptors: *Ligand Binding and Receptor Activation* 3
    Dominique Massotte and Brigitte L. Kieffer

2   Molecular Mechanisms Involved in the Activation
    of Rhodopsin-Like Seven-Transmembrane Receptors 33
    Peng Huang and Lee-Yuan Liu-Chen

3   GPCR Folding and Maturation: *The Effect of Pharmacological
    Chaperones* 71
    Ulla E. Petäjä-Repo and Michel Bouvier

4   Regulated Membrane Trafficking and Proteolysis of GPCRs 95
    James N. Hislop and Mark von Zastrow

PART II GPCR ACTIVITY AND ITS REGULATORS

5   Heterotrimeric G Proteins and Their Effector Pathways 109
    Tracy Nguyen Hwangpo and Ravi Iyengar

6   RGS Proteins: *Orchestration of Multiple Signaling Pathways* 135
    Ryan W. Richman and Mariá A. Diversé-Pierluissi

7   G Protein-Coupled Receptor Kinases 149
    Lan Ma, Jingxia Gao, and Xiaoqing Chen

8   Regulators of GPCR Activity: *The Arrestins* 159
    Louis M. Luttrell

9   GPCR Interacting Proteins: *Classes, Assembly, and Functions* 199
    Hongyan Wang, Catherine B. Willmore, and Jia Bei Wang

PART III GPCR DIMERIZATION/OLIGOMERIZATION

10 Biophysical and Biochemical Methods
    to Study GPCR Oligomerization 217
    Karen M. Kroeger, Kevin D. G. Pfleger, and Karin A. Eidne

11 Oligomerization Domains of G Protein-Coupled Receptors:
    Insights Into the Structural Basis of GPCR Association 243
    **Marta Filizola, Wen Guo, Jonathan A. Javitch,
    and Harel Weinstein**

12 Functional Complementation and the Analysis of GPCR
    Dimerization 267
    Graeme Milligan, Juan J. Carrillo, and Geraldine Pascal

13 The Role of Oligomerization in G Protein-Coupled Receptor
    Maturation 287
    **Michael M. C. Kong, Christopher H. So, Brian F. O'Dowd,
    and Susan R. George**

14 Receptor Oligomerization and Trafficking 309
    Selena E. Barlett and Jennifer L. Whistler

15 Modulation of Receptor Pharmacology
    by G Protein-Coupled Receptor Dimerization 323
    **Noura S. Abul-Husn, Achla Gupta, Lakshmi A. Devi,
    and Ivone Gomes**

PART IV RECENT DEVELOPMENTS IN DRUG DISCOVERY

16 Role of Heteromeric GPCR Interactions in Pain/Analgesia 349
    Andrew P. Smith and Nancy M. Lee

17 Conformational Plasticity of GPCR Binding Sites:
    *Structural Basis for Evolutionary Diversity
    in Ligand Recognition* 363
    **Xavier Deupi, Cedric Govaerts, Lei Shi, Jonathan A. Javitch,
    Leonardo Pardo, and Juan Ballesteros**

18 De-Orphanizing GPCRs and Drug Development 389
    Rainer K. Reinscheid and Olivier Civelli

Index 403

CONTRIBUTORS

NOURA S. ABUL-HUSN, MSc • *Department of Pharmacology and Biological Chemistry, Mount Sinai School of Medicine, New York, NY*
JUAN BALLESTEROS, PhD • *Chief Scientific Officer, Novasite Pharmaceuticals Inc., San Diego, CA*
SELENA E. BARTLETT, PhD • *Ernest Gallo Clinic and Research Center, University of California San Francisco, Emeryville, CA*
MICHEL BOUVIER, PhD • *Département de Biochimie, Faculté de Médicine, Université de Montréal, Montréal, Quebec, Canada*
JUAN J. CARRILLO, PhD • *AstraZeneca R & D, Charnwood, Discovery Bioscience, Loughborough, Leicestershire, UK*
XIAOQING CHEN, PhD • *Pharmacology Research Center and Department of Pharmacology, Shanghai Medical College, Fudan University, Shanghai, China*
OLIVIER CIVELLI, PhD • *Department of Pharmacology, University of California Irvine, Irvine, CA*
XAVIER DEUPI, PhD • *Department of Molecular & Cellular Physiology, Stanford University, Palo Alto, California*
LAKSHMI A. DEVI, PhD • *Department of Pharmacology and Biological Chemistry, Mount Sinai School of Medicine, New York, NY*
MARÍA A. DIVERSÉ-PIERLUISSI, PhD • *Department of Pharmacology and Biological Chemistry, Mount Sinai School of Medicine, New York, NY*
KARIN A. EIDNE, PhD, MBA • *Molecular Endocrinology Research Group/ 7TM Receptor Laboratory, Western Australian Institute for Medical Research (WAIMR), Centre for Medical Research, University of Western Australia, Perth, Australia*
ANDREA S. ENGEL, PhD • *M.E. Müller Institute, Biozentrum, University of Basel, Basel, Switzerland*
MARTA FILIZOLA, PhD • *Department of Physiology and Biophysics, Weill Medical College of Cornell University, New York, NY*
JINGXIA GAO, PhD • *Pharmacology Research Center and Department of Pharmacology, Shanghai Medical College, Fudan University, Shanghai, China*
SUSAN R. GEORGE, MD •*Department of Pharmacology, University of Toronto, Toronto, Canada*
IVONE GOMES, PhD • *Department of Pharmacology and Biological Chemistry, Mount Sinai School of Medicine, New York, NY*

CEDRIC GOVAERTS, PhD • *Departments of Cellular and Molecular Pharmacology, University of California, San Francisco, California*
WEN GUO, PhD • *Center for Molecular Recognition, Columbia University College of Physicians & Surgeons, New York, NY*
ACHLA GUPTA, PhD • *Department of Pharmacology and Biological Chemistry, Mount Sinai School of Medicine, New York, NY*
JAMES N. HISLOP • *Departments of Psychiatry and Pharmacology, University of California at San Francisco, San Francisco, CA*
PENG HUANG, PhD • *Department of Pharmacology, Temple University School of Medicine, Philadelphia, PA*
TRACY NGUYEN HWANGPO, BS • *Department of Pharmacology and Biological Chemistry, Mount Sinai School of Medicine, New York, NY*
RAVI IYENGAR, PhD • *Department of Pharmacology and Biological Chemistry, Mount Sinai School of Medicine, New York, NY*
JONATHAN A. JAVITCH, MD, PhD • *Center for Molecular Recognition, Columbia University College of Physicians & Surgeons, New York, NY*
BRIGITTE L. KIEFFER, PhD • *IGBMC/CNRS/INSERM/ULP, Illkirch Cedex, France*
MICHAEL M. C. KONG, BSc • *Department of Pharmacology, Faculty of Medicine, University of Toronto, Canada*
KAREN M. KROEGER, PhD • *Molecular Endocrinology Research Group/ 7TM Receptor Laboratory, Western Australian Institute for Medical Research (WAIMR), Centre for Medical Research, University of Western Australia, Perth, Australia*
NANCY M. LEE, PhD • *California Pacific Medical Center Research Institute, San Francisco, CA*
LEE-YUAN LIU-CHEN, PhD • *Department of Pharmacology, Temple University School of Medicine, Philadelphia, PA*
LOUIS M. LUTTRELL, MD, PhD • *Division of Endocrinology, Diabetes & Medical Genetics, Department of Medicine, Medical University of South Carolina, Charleston, SC*
LAN MA, PhD • *Pharmacology Research Center and Department of Pharmacology, Shanghai Medical College, Fudan University, Shanghai, China*
DOMINIQUE MASSOTTE, PhD • *IGBMC/CNRS/INSERM/ULP, Illkirch Cedex, France*
GRAEME MILLIGAN, PhD • *University of Glasgow, Glasgow, Scotland, UK*
BRIAN F. O'DOWD, PhD • *Department of Pharmacology, Faculty of Medicine, University of Toronto, Toronto, Canada*

Contributors

KRZYSZTOF PALCZEWSKI, PhD • *Department of Ophthalmology, University of Washington, School of Medicine, Seattle, WA*
LEONARDO PARDO, PhD • *Laboratori de Medicina Computacional, Unitat de Bioestadística and Institut de Neurociències, Universitat Autònoma de Barcelona, Bellaterra, Catalunya, Spain*
GERALDINE PASCAL, PhD • *Molecular Pharmacology Group, Division of Biochemistry and Molecular Biology, Institute of Biomedical and Life Sciences, University of Glasgow, Scotland, UK*
ULLA E. PETÄJÄ-REPO, PhD • *Biocenter Oulu and Department of Anatomy and Cell Biology, University of Oulu, Oulu, Finland*
KEVIN D. G. PFLEGER, MA, PhD • *Molecular Endocrinology Research Group/7TM Receptor Laboratory, Western Australian Institute for Medical Research (WAIMR), Centre for Medical Research, University of Western Australia, Perth, Australia*
RAINER REINSCHEID, PhD • *Department of Pharmacology, University of California Irvine, Irvine, CA*
RYAN W. RICHMAN, BA • *Department of Pharmacology and Biological Chemistry, Mount Sinai School of Medicine, New York, NY*
LEI SHI, PhD • *Institute for Computational Biomedicine, Weill Medical College of Cornell University, New York, NY*
ANDREW P. SMITH, PhD • *California Pacific Medical Center Research Institute, San Francisco, CA*
CHRISTOPHER H. SO, BSc • *Department of Pharmacology, Faculty of Medicine, University of Toronto, Toronto, Canada*
MARK VON ZASTROW, MD, PhD • *Departments of Psychiatry and Pharmacology, University of California at San Francisco, San Francisco, CA*
HONGYAN WANG, PhD • *Neuroapoptosis Laboratory, Department of Neurosurgery, Brigham and Women's Hospital, Harvard Medical School, Massachusetts*
JIA BEI WANG, MD, PhD • *Department of Pharmaceutical Sciences, School of Pharmacy, University of Maryland, Baltimore, Maryland*
HAREL WEINSTEIN, DSc • *Department of Physiology and Biophysics, Weill Medical College of Cornell University, New York, NY*
JENNIFER L. WHISTLER, PhD • *Ernest Gallo Clinic and Research Center, University of California San Francisco, Emeryville, California*
CATHERINE B. WILLMORE, PhD • *Department of Pharmaceutical Sciences, School of Pharmacy, University of Maryland, Baltimore, MD*

COLOR PLATES

Color plates 1–7 appear in an insert following p. 368.

Plate 1	Fig. 2 from Chapter 17; for full caption *see* p. 369.
Plate 2	Fig. 3 from Chapter 17; for full caption *see* p. 373.
Plate 3	Fig. 4 from Chapter 17; for full caption *see* p. 374.
Plate 4	Fig. 7 from Chapter 17; for full caption *see* p. 378.
Plate 5	Fig. 8 from Chapter 17; for full caption *see* p. 379.
Plate 6	Fig. 9 from Chapter 17; for full caption *see* p. 382.
Plate 7	Fig. 10 from Chapter 17; for full caption *see* p. 383.

I
GPCRs: Structure and Function

1
Structure–Function Relationships in G Protein-Coupled Receptors

Ligand Binding and Receptor Activation

Dominique Massotte and Brigitte L. Kieffer

1. INTRODUCTION

G protein-coupled receptors (GPCRs) are integral membrane proteins that form the fourth largest superfamily in the human genome, with more than 800 genes identified to date *(1,2)*. Many of these receptors play key physiological roles, and several pathologies have been associated with receptor functional abnormalities *(3,4)*. Therefore, GPCRs represent important targets for drug design within pharmaceutical companies *(5)*. Indeed, GPCRs mediate the effect of numerous ligands, including neurotransmitters, chemo-attractants, hormones, cytokines, and sensory stimuli such as photons and odorants.

GPCRs were named for their common ability to associate with heterotrimeric G proteins (Gαβγ). Binding of extracellular ligands with agonistic properties initiates the signal transduction cascade by triggering conformational changes in the receptor that promote heterotrimeric G protein activation *(6,7)*. Following nucleotide exchange (guanosine diphosphate [GDP] replacement by guanosine triphosphate [GTP]), the tightly associated Gα and Gβγ-subunits separate from each other and from the receptor. Both components are then free to interact and modulate the activity of downstream elements of the signaling cascades, such as adenylyl cyclase, phospholipases, mitogen-activated protein kinases (MAPKs), or calcium and potassium ion channels. Signal transduction is tightly regulated by receptor posttranslational modifications. Among them, receptor phosphorylation by GPCR-specific and -nonspecific kinases modulates subsequent interactions with several intracellular proteins involved in receptor internalization and downregulation *(8,9)* or promoting growth factor receptor transactivation

From: *Contemporary Clinical Neuroscience: The G Protein-Coupled Receptors Handbook*
Edited by: L. A. Devi © Humana Press Inc., Totowa, NJ

(10). Additional regulatory mechanisms ensue from the interplay of G protein subunits with regulators of G protein signaling (RGS) *(11)*.

Tremendous progress has been accomplished within the past few years in dissecting GPCR-mediated signal transduction pathways, but the molecular mechanisms underlying ligand recognition and signal transduction through the membrane are restrained by the lack of detailed receptor structures. To date, only the three-dimensional (3D) structure of rhodopsin has been solved at high resolution *(12)* because of the difficulty in producing large amounts of concentrated integral membrane proteins, even in heterologous expression systems *(13)*. Moreover, purification of GPCRs retaining structural integrity requires defined compositions and ratios of lipids and detergents. Additionally, GPCR size is fairly large (from approx 40 kDa to 200 kDa), which further hampers their study. Altogether, these distinctive features have prevented the acquisition of 3D structural information by means of crystallography as well as nuclear magnetic resonance (NMR) techniques. Thus, most of the structural information gathered to date derives from mutagenesis studies or biochemical and biophysical approaches, to which models based on the rhodopsin structure are now added. Our view depicts GPCRs as a bundle of seven-transmembrane α-helices alternatively connected through intracellular and extracellular loops. The N-terminal part of the receptor is located on the extracellular side of the cytoplasmic membrane, whereas its C-terminal counterpart faces the cytoplasm.

2. GPCR CLASSIFICATION

Few sequences are conserved among the GPCR superfamily, which is often divided into six classes (*see* GPCR Database available at http://gpcr.org/). Distinctive structural elements that characterize the three main GPCR families (A, B, and C) are summarized in Fig. 1.

Class A receptors, also called rhodopsin-like receptors, comprise the largest family of GPCRs. This class of receptors binds ligands from various types, including small molecules such as biogenic amines as well as peptides (*see* Subheading 4.). The overall homology among all class A receptors is restricted to a limited number of highly conserved key residues in the transmembrane regions, suggesting a critical role in the structural or functional integrity of the receptor. Ligand binding to class A receptors is discussed in detail in Subheading 4.

Class B receptors, also called secretin-like receptors, include about 20 different receptors for various hormones and neuropeptides *(2)*. Ligand binding involves both the N-terminus and extracellular loops of the receptor, and to date, no evidence has been obtained regarding interactions occurring within the transmembrane region of these receptors.

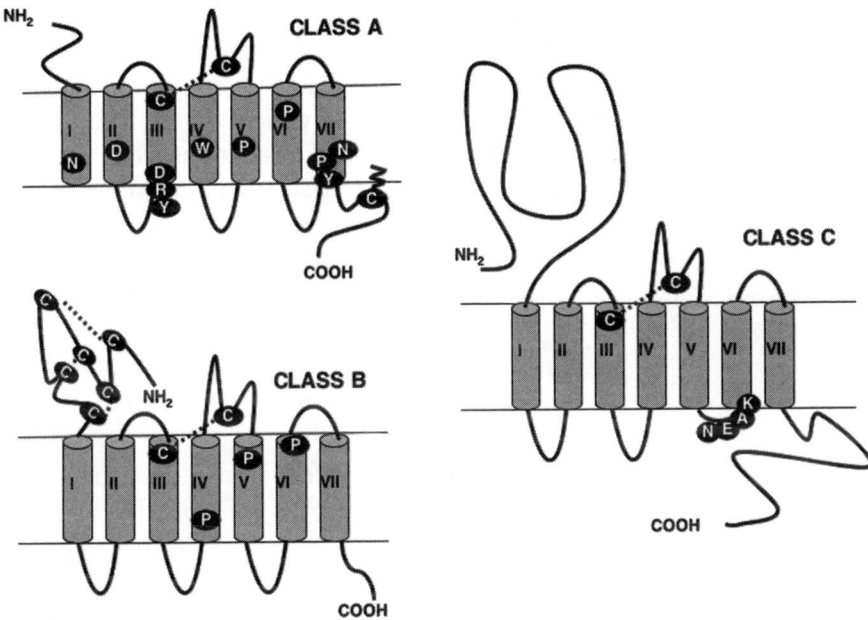

Fig. 1. The G protein-coupled receptor (GPCR) main families. A schematic representation is shown for the three main GPCR classes (A, B, and C) with common structural motifs to each family. The only common motif to class A, B, or C GPCRs is a conserved cysteine residue on helix III and another in the second extracellular loop 2. These cysteines are believed to be connected via a disulfide bridge. For class A receptors (rhodopsin-like family), the most conserved amino acid of each helix is indicated. A putative palmitoylation site is represented in the proximal part of the C-terminus (≩). The DRY motif on helix III and the NPXXY motif on helix VII are conserved among class A GPCRs (see details in Fig. 2A). Class B GPCRs (secretin-like family) share a large amino terminus with conserved cysteine residues and disulfide bridges. Some proline residues are also conserved within the helical bundle, but those residues are different from class A conserved prolines. Class C receptors (metabotropic glutamate family) are characterized by a very large extracellular domain that binds the ligands. The highly conserved motif NEAK (NDSK in the case of the GABA family) on the very short intracellular loop 3 is indicated.

In class C receptors, also known as metabotropic glutamate receptors, ligand recognition is achieved through their very large extracellular domain (300–600 residues). It is organized into two distinct lobes separated by a cavity that binds the ligand in a "Venus flytrap" manner *(14)*.

Classes D and E constitute two minor families that are present in fungi and recently the frizzled/smoothened receptor family was added to the world of GPCRs (15).

3. OVERALL TOPOGRAPHY OF CLASS A RECEPTORS

Despite limited sequence homology, class A receptors exhibit identical structural organization, and their overall topography can be subdivided into three main regions (Fig. 2A).

On the extracellular side, the N-terminal region is involved in ligand binding (see Subheading 4.) and possibly receptor activation (see Subheading 5.2.), whereas the extracellular loops represent important key elements for peptide binding and play a role in receptor selectivity toward ligands (see Subheading 4.3.).

The transmembrane core is comprised of a bundle of seven α-helices that provide a hydrophobic environment critical for nonpeptide as well as small-peptide ligand binding (see Subheadings 4.1. and 4.2.; Fig. 2B). It relays the conformational changes induced upon ligand binding on the extracellular side of the receptor to the intracellular architectural determinants that regulate activation of the signaling cascade (see Subheading 5.2.).

On the intracellular side, the loop regions contain key elements for either direct or scaffolding–protein-dependent interactions with intracellular effectors (see Subheadings 5.4. and 5.5.). Additionally, posttranslational modifications present in the C-terminal are likely to modulate both receptor activation state and G protein coupling (see Subheading 5.3.) as well as to participate in the regulation of receptor internalization and desensitization.

4. LIGAND BINDING IN CLASS A RECEPTORS

Countless studies have been performed on individual receptors that now allow us to draw a fairly consistent picture of the precepts that govern ligand binding to class A receptors. Information on critical determinants has been experimentally obtained using site-directed as well as random mutagenesis, receptor chimeras, and biochemical and biophysical methods. Such experimental data were combined with computer modeling and were used to refine the proposed models. This approach led to an improved template that has been used for ligand-docking studies (3,16). One major outcome was the notion that both the size and nature of the ligand drastically influence the modalities and location of its binding. Hence, only some commonalities may be extracted that are general among ligands or receptors.

4.1. Binding of Small Ligands Within the Receptor Transmembrane Regions

On the basis of the crystallographical information obtained for rhodopsin, Ballesteros et al. *(17)* performed a detailed structural comparison of the D2 dopamine receptor with rhodopsin and concluded that the rhodopsin and biogenic amine receptors may be very similar, despite structural divergence in the transmembrane helical bundle. Indeed, helix kinks at proline (Pro) residues or helix binding or twisting at cysteine, serine, or threonine residues may slightly modify the shape of the ligand-binding pocket and introduce the subtle differences required for class A receptors to bind a structurally diverse collection of ligands. Conserved Pro-kinks in helices V,VI, and VII could adopt different conformations that could significantly change the binding sites of different GPCRs. Nonconserved Pro residues in helices II and IV or nonconserved cysteine/threonine/serine residues in helix III and other helices are another source of potential structural divergence in the binding-site crevice. The authors postulated that GPCRs have evolved in a way that maintains their overall fold by means of alternative molecular mechanisms (structural mimicry) that enable localized variations within their binding sites suitable for recognizing a wide variety of ligands. As a consequence, if the crystal structure of rhodopsin can be used as a template for class A receptor modeling, the particular conformation of the binding site of a given receptor may require substantial refinement to be accurately described at the molecular level.

However, some structural elements represent a very specific signature for a receptor family. Catecholamines and related biogenic amines bind primarily within the transmembrane region of their receptors. The identified binding crevice is outlined by residues from helices III, V,VI, and VII. This binding pocket is common to both agonists and antagonists that likely establish a salt bridge with a conserved aspartate residue on helix III at a position analogous to D113 (3.08)[1] in the β2-adrenergic receptor (AR). Additional key interactions have also been identified that differ between agonists and antagonists *(6,18)*.

---

[1]Amino acids will be referred throughout the text according to the one-letter code. Residue numbering in parentheses corresponds to the nomenclature introduced by Ballesteros and Weinstein for amino acids located in the transmembrane region of the receptor. The first number refers to the helix on which the residue is located. The second number indicates the position of the residue relative to the most conserved amino acid on this helix to which an arbitrary value of 50 is assigned. Residue 3.44 for example is located on helix III six amino acids before the conserved arginine.

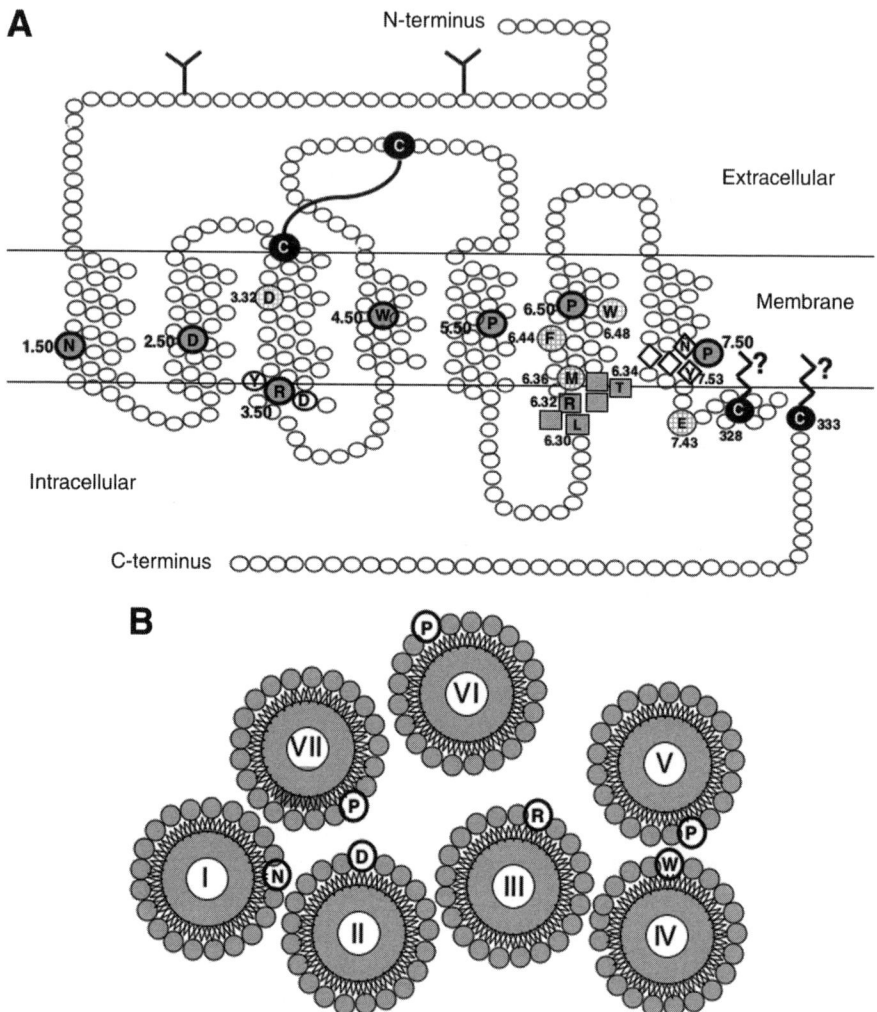

Fig. 2. Schematic representation of class A receptors. (**A**) General organization of the δ-opioid receptor as a prototype for class A receptors. The amino terminus is located at the extracellular side of the membrane and the carboxy-terminus at the cytoplasmic side. *N*-glycosylation consensus sequences are present (Y) on the N-terminus. The conserved cysteines forming a disulfide bridge between helix III and the extracellular loop 2 are indicated. Potential palmitoylation sites at the proximal part of the receptor are represented (⧚). Residues in the transmembrane region are numbered according to the nomenclature of Ballesteros–Weinstein. Highly conserved residues among class A G protein-coupled receptors (GPCRs) are indicated by a grey circle with a thick black border. Important motifs are also located. Residues of the E/DRY motif on helix III are shown in white circles with a thick border. Residues forming the basic $X_1BBX_2X_3B$ motif on helix VI are indicated by gray

Structure–Function Relationships in GPCRs

Based on the mutagenesis studies performed on GPCRs that bind cationic amine neurotransmitters, the aspartate residue in helix III has become a systematic anchorage point for amine ligands in modeling studies. However, opposite effects were observed when this assumption was tested on the µ- and δ-opioid receptors, leading to the conclusion that the aspartate residue in helix III is required for high-affinity binding of agonists to the µ-opioid receptor, but not the δ-opioid receptor *(19,20)*. Moreover, extensive mutagenesis studies performed on the δ-opioid receptors involving peptides and alkaloids acting as agonists as well as antagonists emphasized that the determinants of the opioid receptor binding pocket differ among ligands, despite the presence of a common subset *(21)*. These data again underscore that, although ligand binding in two closely related receptors shares considerable similarities, it displays (on a very fine scale) many subtle differences that preclude direct extrapolation from one set of data to another.

4.2. Ligand Binding to the Receptor Extracellular Regions

Unlike small ligands (photon, biogenic amines, nucleosides, eicosanoids, lysophosphatidic acid, and sphingosine 1-phosphate), peptides bind to the extracellular domains of the receptor. Determinants have been identified in the receptor N-terminus that are essential for recognition of various peptides (90 amino acids), including oxytocin, vasopressin, endothelin, opioids, or substance P. Moreover, the long extracellular N-terminus of the target receptor constitutes the primary high-affinity binding site of large glycoprotein hormones (30 kDa) such as lutotropin/choriogonadotropin (LH/CG), thyrotropin-stimulating hormone (TSH) or follitropin-stimulating hormone (FSH). Upon ligand binding, this domain may undergo a conformational change that allows secondary contacts with extracellular loop regions and eventually leads to receptor activation (reviewed in refs. *6* and *22*). Note-

---

Fig. 2 *(From opposite page)* squares. Residues forming the NPXXY motif on helix VII are indicated by white diamond symbols. Several other important residues for ligand binding and receptor activation mentioned in the text are also indicated in the figure.

(B) Organization of the transmembrane helices as seen from the extracellular side of the membrane. The helices are positioned according to the projection maps of rhodopsin and are believed to be organized sequentially in a counterclockwise manner. Conserved amino acids through class A receptors are indicated.

worthy peptides of rather small size (40 amino acids) show a mixed binding profile through additional transmembrane anchoring in addition to their primary interaction with the extracellular loops *(3,5,23)*.

Studies using point mutants and receptor chimeras clearly showed that the extracellular loops are also involved in receptor selectivity. Extracellular loop 2 appears to be a critical determinant to discriminate among α_1-AR subtypes *(24)*. In the case of µ-, κ-, and δ-opioid receptor types, extracellular loop 1 contains critical elements for µ-selectivity *(25)*, and extracellular loop 2 contains critical elements for κ-selectivity *(26)*. Extracellular loop 3, together with the external parts of helices VI and VII, is involved in δ-ligand selectivity by enhancing the affinity of the receptor for δ-ligands *(25,27–29)*. Additionally, this region also contains important determinants for µ-agonist selectivity *(30)* and for κ-selective alkaloids *(28)*.

4.3. Role of Extracellular Loop 2 and Conserved Disulfide Bridge in Small-Ligand Binding

Although binding of small ligands within the transmembrane core of the receptor is widely acknowledged, a possible involvement of the second extracellular loop has also been proposed for small ligands; however, this involvement is still under debate. Interestingly, two antagonists of the a_{1a}-AR, phentolamine and WB4101, exhibit unusual binding features in which three amino acid residues localized in extracellular loop 2 appear to be critical. This observation suggests a binding profile involving extracellular regions of the receptor that is more similar to what has been described for peptide hormone receptors *(24)*. In the high-resolution bovine rhodopsin structure, the second extracellular loop folds down into the binding-site crevice to form a lid over retinal *(12)*, and one may postulate a similar extracellular loop 2 structure exists in at least some class A receptors. Using the substituted-cysteine accessibility method (SCAM), Shi et al. *(31)* concluded that this may indeed be the case for the D2 dopamine receptor. Another argument in favor of a critical role for extracellular loop 2 comes from the observation that several antibodies directed to extracellular loop 2 induced AR and bradykinin receptor (BR) activation *(32)*. Nearly all class A receptors show the presence of two conserved cysteine residues that are believed to form a disulfide bridge connecting helix III and extracellular loop 2. However, the actual presence of the bridge has been established only in a limited number of cases, including rhodopsin *(12)*, µ-opioid *(33)*, leukotriene LTB4 *(34)*, muscarinic m1 *(35)*, platelet thromboxane *(36)*, TSH *(37)*, or gonadotropin-releasing hormone (GnRH) *(38)* receptors. This disulfide bond may be crucial for both the structural integrity and function of many GPCRs. Its removal by mutagenesis severely disrupted ligand binding to muscarinic acetylcholine *(35)*, opioid *(33)*, and angiotensin (AT)1 *(39)* receptors and destabilized the high-affinity state of the β_2-AR *(40)*. The disulfide bridge likely dictates a relatively rigid architecture by constraining the extracellular loop. This, in turn, shapes the ligand-binding site, rather than contributing directly to ligand binding.

4.4. Relative Distribution of Agonist- and Antagonist-Binding Sites

Interestingly, antagonists are small molecules that invariably bind within the transmembrane region of class A GPCRs. They prevent agonist binding and subsequent receptor activation whether the agonist is a peptide or a small molecule. The generic antagonist binding pocket is located in a region flanked by helices III, V, VI, and VII, in which residues establish the main side-chain interactions with the ligand *(6,17,18)*. Receptor contacts with peptide agonists and nonpeptide antagonists do not substantially overlap at atomic levels in the tachykinin receptor NK1 *(18,41)*, AT1 *(3)*, or opioid receptors *(21)*. Therefore, competitive antagonism would primarily arise from a steric exclusion mechanism *(3,18)*.

5. RECEPTOR ACTIVATION UPON AGONIST BINDING

This section briefly reviews the models currently applied to describe GPCRs modus operandi. Attempts are made to draw a picture of the molecular events that occur upon agonist binding that lead to G protein activation. The role of palmitoylation is discussed. Finally, modulation of the interactions with intracellular partners is envisaged in the light of the receptor susceptibility to adopt multiconformational states.

5.1. Ternary Complex Models of Receptor Activation

In the original ternary complex model (TCM) described by De Lean et al. *(42)*, an agonist-bound activated receptor forms a complex with a G protein, resulting in its activation. This corresponds to a simple example of a receptor isomerization mechanism in which ligand-binding (A) promotes a conformation of receptor (R) that couples to and activates a G protein (G). The next level of progression toward present GPCR models involved the incorporation of different receptor conformations into the scheme. The demonstration of constitutive GPCR activity by Costa and Herz *(43)* indicated that receptors could couple to and activate G proteins in the absence of ligand. This required modification of the original TCM, which did not enable spontaneous formation of the R*G species; this modification resulted in the extended TCM (ETC) (ref. *44*; Fig. 3A). According to the ETC, the receptor

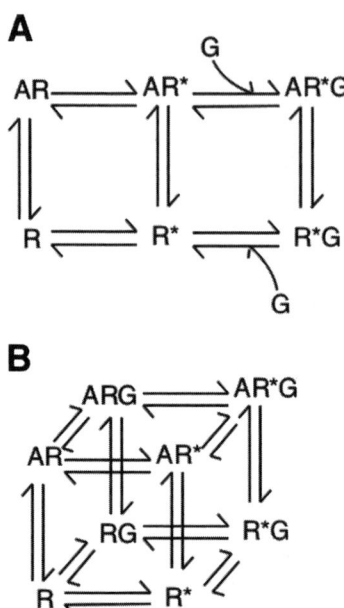

Fig. 3. Ternary models of G protein-coupled receptor (GPCR) activation. (**A**) Extended ternary complex model (ETC) proposed by Samama et al. (64). According to this model, the receptor can spontaneously adopt either an inactive (R) or an active (R*) conformation. Only the activated form (R*) of the receptor can interact with the G protein (G) in the presence or the absence (constitutive activity) of an agonist (A). (**B**) Cubic ternary complex model (CTC) proposed by Weiss et al. (65). In this more thermodynamically complete representation of GPCR activation, both the inactive state (R) and the active state (R*) of the receptor are allowed to interact with the G protein (G).

exists in an equilibrium between an inactive conformation (R) and an active conformation (R*). In absence of agonist, the inactive form R prevails, but a certain fraction of receptors spontaneously assume the R* state because of the low-energy barrier separating the two conformations. Agonists are predicted to bind with highest affinity to R* and to shift the equilibrium to a larger proportion of receptors under the active conformation. Conversely, inverse agonists that have the ability to inhibit agonist-independent activity (also called constitutive activity) stabilize the inactive conformation R, thereby shifting the equilibrium away from R*. On the other hand, neutral antagonists do not influence the equilibrium between R and R*.

In 1996, Weiss et al. *(45)* proposed a more thermodynamically complete model called the cubic TCM (CTC; Fig. 3B). In this model, both the active R* and the inactive R conformations of the receptor are allowed to interact with the G protein, whereas in the ETC model only the active R* receptor state could interact with the G protein. It is presently unclear which of these models better predicts and describes experimental findings with GPCRs. On the practical side, the ETC model has fewer parameters and is simpler to use, whereas the CTC model is more comprehensive but has a greater number of nonestimatable parameters. The choice for the appropriate model may be dictated by the importance of the inactive agonist–receptor–G protein (ARG) state: GPCR systems in which the ARG state is negligible can be accurately described by the ETC model, whereas other systems in which the ARG species plays a role (e.g., cannabinoid receptors *[46]*) require use of the CTC model (ref. *47*; Fig. 3).

Increasing evidence points to the existence of multiple conformational states for GPCRs (*see* Subheadings 5.3. and 5.4.). Additionally, experimental data indicate that neither the ETC nor the CTC model accurately describes the complex behavior of GPCRs. In an attempt to embrace the multiplicity of receptor conformations, multistate models in which the receptor spontaneously alternates between multiple active and inactive states have been proposed *(48,49)*.

5.2. What Do Constitutively Active Mutants and Rhodopsin-Based Models Tell Us About Activation Mechanisms in Class A Receptors?

Some mutations appear to enhance basal activities of GPCRs and, therefore, are believed to mimic the agonist activity and to favor the active state of the receptor. This, in turn, facilitates productive interaction with intracellular G proteins. These mutant receptors are currently called constitutively active mutants (CAMs). The δ-opioid receptor was the first GPCR described as able to modulate second messengers in the absence of agonist *(43)*. A fairly large number of CAMs were incidentally identified from mutagenesis studies on many different GPCRs. These CAMs contributed massively to the set of data that helps explain the mechanisms of receptor activation. The current hypothesis states that CAMs release the conformational constraints of the GPCR inactive state. This was first postulated for the α_{1B}-AR. Mutation of alanine 293 (A 6.34) and replacement by any of the 19 other amino acids generated a CAM, suggesting that the gain of function resulted from the loss of an intramolecular constraint *(50)*. Indeed, the current belief is that agonist binding to a wild-type receptor introduces new molecular contacts that replace the intramolecular interactions constraining the receptor in an inactive conformation. This results in a conformational switch and subsequent receptor activation. However, many CAMs are likely activated by simple disruption of interactions that exist within the receptor inactive conformation, rather than by formation of new intramolecular bonds. Therefore, it should be remembered that the actual structure adopted by CAMs is only an approximation of the real active conformation of the receptor (for a review, *see* ref. *51*).

The crystal structure obtained for rhodopsin corresponds to the inactive form in which 11-*cis* retinal is bound, and this serves as a template to postulate movement of helices III, VI, and VII upon light activation. Class A GPCRs share a good number of conserved structural determinants with rhodopsin. Therefore, the high-resolution structure of rhodopsin has been used as template for GPCR modeling of the transmembrane domains, and the helix movement model has been extended to class A receptors as a common mechanism of activation. According to this hypothesis, ligands activate GPCRs by disrupting the networks of intramolecular contacts that stabilize the ground state. This modifies the conformation of the receptor so that it optimally exposes epitopes that bind and stabilize a conformation of the G protein close to the transition state for GDP–GTP exchange and G protein activation.

Despite the availability of a high-resolution structure of rhodopsin at 2.8 Å, the actual mechanism used to disrupt stabilizing intramolecular interactions remains elusive. Evidence for movements of helix VI relative to helix III have been essentially provided by several different approaches that were mostly applied to rhodopsin. Biophysical studies included Fourier transformed infrared resonance spectroscopy (FTIR), surface plasmon resonance (SPR), tryptophan ultraviolet (UV)-absorbance spectroscopy, and electron paramagnetic resonance spectroscopy (EPR) (reviewed in ref. *52*). Spectral changes were also measured upon *N*,*N'*-dimethyl-*N*(iodoacetyl)-*N'*-(7-nitrobenz-2-oxa-1,3-diazol-4-yl)ethylene-diamine (IANBD) binding to cysteine residues in the β_2-AR *(52,53)*. Additionally, several indirect strategies were used, including generation of *bis*-histidine metal ion-binding sites between cytoplasmic extensions of helices III and VI in rhodopsin receptors *(54)*, β_2-ARs *(55)*, and NK1 receptors *(41)*. Cysteine accessibility was also determined in a β_2-AR CAM *(56)* and random mutagenesis was performed on muscarinic m5 *(57)*, δ-opioid *(58)*, AT_{1A} *(59)*, and C5A chemo-attractant *(60)* receptors.

In rhodopsin and biogenic amine receptors, one key event in the activation process may involve arginine (R3.50) in the highly conserved E/DRY motif at the cytoplasmic side of helix III (Fig. 1). Protonation of this residue would disrupt the ionic interaction with a glutamic acid (E6.30) at position X_1 of a basic "$X_1BB\ X_2\ X_3B$" motif (where B is a basic amino acid and X is a nonbasic amino acid) located at the junction region between intracelullar loop 3 and helix VI (Fig. 2A). Mutagenesis studies have established this mechanism for 5-HT$_{2A}$ receptors *(61)*, H$_2$ histamine receptors *(62)*, α_{1B}-ARs *(63)*, and β_2-ARs *(64)*. Mutagenesis of residues clustered at the junction between helix 3 and intracellular loop 2 in the muscarinic m5 receptor suggested that some of the amino acids adjacent to the E/DRY motif are involved in maintaining the receptor in an inactive state but also alternate with residues required for G protein coupling *(65)*. A similar role in G protein activation was postulated for the N-terminus of intracellular loop 2 in rhodopsin *(66)* and more recently in the V$_{1A}$ vasopressin receptor *(67)*. ^{1}H NMR analysis established a similar structure for the vasopressin and rhodopsin intracellular 2 loops *(67)* but was distinct from the α_{2A}-AR intracellular loop 2 conformation *(68)*. This is of particular interest, because unlike the other two, the α_{2A}-AR is not activated by mutation of the aspartate in the DRY motif and therefore diverges from the consensus model described earlier *(69)*.

In addition to the R3.50–E6.30 salt bridge, the residue X_3 (6.34) of the basic motif is hydrogen-bonded to the arginine R3.50 in rhodopsin *(12)*. Introduction of a lysine at position X_3 revealed that the residue at position 6.34 is also involved in constraining biogenic amine receptors in an inactive form in the α_{1B}-, α_{2A}-, β_1-, and β_2-ARs and in the 5-HT$_{1B}$-, 5-HT$_{2A}$-, and 5-HT$_{2C}$-receptors (refs. *61* and *70* and references therein). However, this strategy may not generalize across all receptors. In the case of opioid receptors, the ionic interaction postulated earlier between E6.30 (X_1 residue of the basic motif) and R3.50 (in the DRY motif) cannot occur, because the glutamate residue E6.30 on helix VI is replaced by a leucine. Moreover, mutation of T6.34 into a lysine does not activate the µ-opioid receptor *(70)*. These data show that the actual interactions depend on the residues and local environments at the intracellular ends of helices III, V, and VI and that sequence differences in this region are likely to support locally different forms of activation mechanisms *(71)*. Interestingly, in the δ-opioid receptor R258 (6.32), the second basic residue of the "$X_1BB\ X_2\ X_3B$" motif would be involved in an ionic bridge with E323 (7.43) on helix VII *(58)*.

A group of mutations comprising tryptophan W173 (4.50) that is strictly conserved in all rhodopsin-like GPCRs (despite its location on the most variable helix IV) induced constitutive activation of the δ-opioid receptor *(58)*. This cluster of mutations could either directly or indirectly affect the orientation of W173, which would play a central role at the helix II–helix IV interface in controlling the orientation and outward motion of helix III during the activation process. W173 is also involved in opioid ligand binding *(21)* and has been located within the binding crevice in the D2 dopamine receptor *(72)*. Because of its high conservation, W173 may represent a key switch for helix III movements in most GPCRs.

Chen et al. *(73)* reported that a phenylalanine F303 (6.44) on helix VI is a key residue involved in α_{1B}-AR transmembrane movement that leads to G protein activation. This residue is highly conserved among GPCRs and is located several residues below those identified as being important for ligand interaction and receptor activation in many GPCRs. A similar role has been assigned to the equivalent phenylalanine residue in chemo-attractant C5A *(60)*, muscarinic m5 *(57)*, and cholecystokinin receptors *(74)*. In the muscarinic m1 receptor, the conserved F374 (6.44) in helix VI is part of a network of interactions involving a leucine residue L116 (3.43) in helix III and the asparagine N414 (7.49) of the NPXXY motif on helix VII *(7, 75)*. Additionally, an important and specific interaction occurs in rhodopsin between the NPXXY motif and the methionine M257 (6.40) on helix VI *(76)*. In the δ-opioid receptor, mutation of the tyrosine Y318 (7.53) of the NPXXY motif into a histidine or replacement of methionine M262 (6.36) in helix VI by a threonine led to constitutive activity *(58)*. Interestingly, a residue equivalent to M262 is highly conserved among the peptide receptor family, and its mutation in the LH receptor is associated with precocious puberty in humans *(77)*. These data support the view that the conserved NPXXY motif plays a central role in the conformational switch that leads to receptor activation and underscore the importance of networks of hydrophobic interactions in maintaining GPCRs in the inactive state. Following agonist binding, these networks of Van der Waals interactions may be disrupted, resulting in the removal of the hydrophobic latch between helices III, VI, and VII. This, in turn, may induce a rotation of helices VI and VII relative to helix III. From the previous examples, it can also be concluded that although activation of class A GPCRs may be associated with similar conformational changes, different receptors may employ specialized sets of intramolecular interactions to produce these changes.

A whole-receptor random mutagenesis strategy applied to the δ-opioid receptor identified 30 mutations distributed throughout the receptor sequence and allowed researchers to draw a general picture of the events leading to receptor activation *(58)*. The N-terminus, extracellular loop 3, and upper portions of helices VI and VII constitute an outward platform that responds to extracellular ligands and initiates transmembrane signaling.

Movement of at least helices VI and VII throughout the transmembrane core then follows, in addition to local re-arrangement of the helices III, VI, and VII which are proposed for rhodopsin and several biogenic amine receptors. Again, a common structural switch might involve the cytoplasmic ends of helices III and VI identified in several class A receptors (histamine H_2 receptors, µ-opioid receptors, ARs, and muscarinic receptors).

Notably, this study identified five amino acid modifications in the N-terminal domain that enhanced spontaneous activity of the δ-opioid receptor (Q12L, D21G, P28L, A30D, and R41Q) *(58)*. Each mutation substantially modified the chemical nature of the amino acid side-chain, introducing or deleting ionic charges or modifying hydrophobicity and structural constraints. This suggests that the N-terminal portion of the receptor is folded as a domain whose structure and spatial orientation influences receptor function. This hypothesis is consistent with the rhodopsin structure, in which the N-terminal domain is folded as a β-sheet and covers the helical bundle like a lid *(12)*. Presently, functional activity of the N-terminal region has been investigated only in glycoprotein hormone GPCRs. For example, the N-terminal tail of the TSH receptor has been proposed to bind spontaneously to the empty receptor and act as an inverse agonist favoring the off-state *(78)*. The present data suggest that the short N-terminal domain of some class A GPCRs may also modulate the on–off transition.

5.3. Palmitoylation: A Modulator of Receptor Activity

Palmitoylation is a posttranslational modification that results in the attachment of a 16-carbon-long saturated acyl chain to a cysteine residue. Unlike other acyl chain additions, palmitoylation is a dynamic process. Several studies have suggested that dynamic palmitoylation could modulate receptor activity by influencing the coupling to G proteins as well as the receptor phosphorylation state.

Mutations of C-terminal cysteine residues have been reported for several GPCRs, and a variety of receptor functions were perturbed following these mutations *(79–81)*. These cysteine residues are often believed to be palmitoylated and, therefore, are involved in the formation of a fourth intracellular loop. Dynamic modulation of the local hydrophobicity through palmitoylation may uncover or mask receptor domains that govern interactions with intracellular effectors such as heterotrimeric G proteins or receptor kinases. For example, depalmitoylation of rhodopsin increased its ability to activate $G_t\alpha$-light-dependent GTPase activity *(82)*. Crystallographical data suggest that helix VIII serves in rhodopsin as a membrane-dependent conformational switch that may adopt a helical structure in the inactive state or a looplike conformation upon rhodopsin activation. Thus, one can speculate that palmitoylation of the two rhodopsin cysteine residues might modify the orientation of helix VIII on the membrane surface *(83)*.

Presently, the impact of C-terminal cysteine mutations on the interactions between receptors and G proteins appears to be receptor type-dependent. Coupling to G proteins was either nonaffected or decreased (ref. *79* and references therein; refs. *81* and *84*). The lack of palmitoylation subsequent to cysteine replacement was established only in some cases. Therefore, some of the effects observed following cysteine mutagenesis may result either from the loss of the cysteine residues themselves or from the modification they carry.

Substitution of the conserved palmitoylated cysteine residues 328/329 into serines resulted in constitutive activation of the 5-HT$_{(4a)}$ receptor, a receptor coupled to Gα_s. More recently, mutation of cysteine C328 into an arginine residue in the δ-opioid receptor as well as replacement of cysteines 348/353 by alanine residues in the μ-opioid receptor conferred agonist-independent activity to both G$\alpha_{i/o}$-coupled receptors *(58,85)*. Although a definite link to the receptor palmitoylation state is still needed for opioid receptors, the data suggest a common role of these residues in the control of receptor activation.

Interestingly, in some cases the lack of palmitoylation appears to have differential effects on the various signaling pathways engaged by a given receptor. A palmitoyl-deficient mutant of the human endothelin (ET)$_A$ receptor was reconstituted in phospholipid vesicles together with various G proteins. This mutant was less effective in stimulating G$_i\alpha$ and G$_q\alpha$ than the wild-type counterpart, but its ability to stimulate G$_o\alpha$ was not affected *(86)*. Similarly, Horstmeyer et al. *(87)* showed that the unpalmitoylated ET$_A$ receptor was still able to couple to Gα_s, but no longer to Gα_q, in Chinese hamster ovary (CHO) cells. On the other hand, an unpalmitoylated triple cysteine mutant in positions 402, 403, and 405 of the receptor ET$_B$ was unable to couple to Gα_i or Gα_q proteins. However, the presence of the palmitoylated cysteine 402 was sufficient to promote coupling to Gα_q but not Gα_i. In the latter case, additional downstream carboxy-terminal elements appear to be required *(88)*.

Phosphorylation by numerous kinases, including protein kinase A (PKA) and GPCR kinases (GRKs), initiates a cascade of events that leads to receptor desensitization (*see* Chapter 7). In addition to a role in receptor–G protein coupling, palmitoylation has been proposed as a key determinant in receptor desensitization. Increased palmitate turnover rates upon agonist stimulation has been reported for numerous receptors, including the β$_2$-ARs

(89,90), α_{2A}-ARs *(91)*, muscarinic m2 *(92)*, and 5-HT$_{4a}$ receptors *(84)*. Several studies linked a lack of palmitoylation to an increased level of receptor phosphorylation. Mutation of the palmitoylated cysteine residue improved PKA phosphorylation of the β_2-AR *(93)* and GRK phosphorylation of the adenosine A3 receptor *(94)*. Conversely, introduction by mutagenesis of a PKA phosphorylation consensus motif at a palmitoylation site of the D1 dopamine receptor gave rise to a palmitoylation-deficient mutant that was constitutively desensitized *(79)*. Moreover, palmitoylation at cysteine C356 and phosphorylation at tyrosine Y352 appear mutually exclusive in the bradykinin B$_2$ receptor, maybe as a result of the close vicinity of the two residues *(95)*. Therefore, palmitoylation could be seen as a molecular switch regulating the accessibility of phosphorylation sites involved in receptor downregulation. As mentioned previously, agonist stimulation increases palmitate turnover, thus promoting receptor depalmitoylation *(89,90)*. This, in turn, would unmask phosphorylation sites that render them readily accessible for phosphorylation and would provide a link to receptor internalization and desensitization. Interestingly, substitution of the carboxy-terminal cysteines by glycine residues decreased both the basal- and agonist-induced level of phosphorylation of the V$_{1a}$ vasopressin receptor that was nonetheless internalized at a faster rate, suggesting receptor-specific effects *(81)*.

Notably, additional palmitoylation sites have been postulated in the intracellular loops for the rat µ-opioid receptor *(96)* and the V$_{1a}$ vasopressin receptors *(81)* besides those identified in their C-terminal region; however, no functional role has been assigned to them yet.

5.4. Receptor Multiple Conformations and G Protein Coupling

It was long-believed that a given GPCR interacts with a particular G protein or a given family of G proteins. However, accumulating evidence has now clearly indicated that several GPCRs can simultaneously interact with G proteins that belong to different families and can activate different signaling cascades—some of which exert opposing effects (for a review, *see* ref. 97). The efficacy of coupling to the various G proteins may then vary according to the receptor type and the interacting G protein but also depends on the agonist.

To date, 23 Gα-subunits have been identified that are classically divided into four different families: G$\alpha_{i/o}$, Gα_s, G$\alpha_{q/11}$, and Gα_{12}. Six β- and 11 γ-subunits that are differentially expressed have also been isolated (e.g., $\beta_1\gamma_2$ are ubiquitous, whereas $\beta_1\gamma_1$ are restricted to visual cells *[98]*). Consequently, the heterotrimeric combinations that can be observed are dependent on the expression pattern of each of the three components. This implies that the actual coupling of a GPCR to a given heterotrimeric G protein may vary among cells, because it is highly dependent on both the availability of the subunits in a given cell as well as their location in close vicinity to the receptor.

Availability of α,β and γ-subunits is not the only factor that influences the type of coupling that will occur between a given receptor and heterotrimeric G proteins. Indeed, determinants that govern the choice of the interacting partners must be present on the receptor. A very large number of studies based on point mutations, chimeras, and synthetic mimetic peptides have pointed to intracellular loops 2 and 3 and the proximal region of the C-terminus as key regions for interaction with and activation of G proteins. Despite a plethora of data, no consensus motifs could be identified as a signature that reflects the interaction of the receptor with one of the four families of Gα proteins *(97,99)*. Fidelity of coupling to a single G protein seems to require a combination of distinct intracellular regions on both intracellular loops 2 and 3 *(65,100)*, but the exact molecular determinants that allow the receptor to distinguish among the various G protein subunits remain unclear.

Recently, Slessareva et al. *(101)* showed that closely related GPCRs achieve selective coupling through multiple and distinct domains located on the G protein α-subunits. This suggests that coupling selectivity ultimately involves subtle and cooperative interactions among various domains on both the G protein and the associated receptor; therefore, multiple conformational states likely exist for a given receptor. Presumably, the various conformations adopted by the receptor are also directly linked to the nature of the agonist. Multiple G protein-coupled states of the β_2-AR were evidenced using various guanylyl nucleotide analogs *(102)*. Moreover, changes in the fluorescence of a reporter group born by a purified β_2-AR were monitored following binding of agonists; these revealed that the extent of changes depended on agonist efficacy *(103)*. Therefore, a given agonist induces particular structural modifications within the receptor that will ultimately contribute to the selectivity of the coupling with the G protein. Experimental evidence was obtained when activation profiles of a Gα_{i1}- or a Gα_{i2}-subunit in fusion with the μ-opioid receptor were compared upon binding of different agonists. The activation profile of the fused Gα_{i1}-subunit closely resembled that observed for the wild-type receptor, which interacts freely with the pool of G proteins present in the cell, whereas activation of the fused Gα_{i2}-subunit was only promoted by a very limited number of the agonists tested *(104)*. Similarly, differential activation of Gα_{o1} and Gα_{i1} was observed with the δ-opioid agonist DADLE *(105)*. Recently, plasmon–waveguide resonance spectroscopy experiments using the δ-opioid receptor confirmed that the affinity of the receptor toward the G protein depends on the agonist, antagonist, or inverse agonist nature of the ligand prebound to the receptor. Moreover, the selectivity of the coupling toward a given Gα-subtype within the Gα$_{i/o}$-family was demonstrated to depend on the agonist DPDPE *(106)*. Upon catecholamine binding, the β$_2$-AR undergoes transitions to two kinetically distinguishable conformational states that were correlated with biological responses in cellular assays. These results support a mechanistic model for GPCR activation in which contacts between the receptor and structural determinants of the agonist stabilize a succession of conformational states with distinct cellular functions *(107)*. The response evoked by the tachykinin NK2 receptor also differed if the receptor bound the complete form of NKA or the naturally occurring truncated NKA 4–10. NKA elevated intracellular calcium level and stimulated cyclic adenosine monophosphate (cAMP) production, whereas NKA 4–10 only affected calcium concentrations. The authors also demonstrated that PKA activation diminished cAMP production, whereas protein kinase C (PKC) activation facilitated the switch from calcium response to cAMP production. To account for these observations, multiple active and desensitized conformations with low, intermediate, or high affinities and with distinct signaling specificities were assumed for the NK2 receptor *(108)*. Similarly, PKA-mediated phosphorylation of the β$_2$-AR switched coupling from stimulatory Gα$_s$ to inhibitory Gα$_{i/o}$ protein *(109)*.

Other mechanisms have been proposed to explain coupling to different pathways. Alternative splicing at the C-terminal region of the receptor may be one additional determinant of receptor–G protein selectivity. In some cases, it also modifies coupling specificity. The strict coupling to Gα$_s$ observed in the case of the 5-HT$_{4a}$ receptor was enlarged to the Gα$_{i/o}$ family in the 5-HT$_{4b}$ variant *(110)*. Similarly, the prostaglandin receptors *(111,112)* distinguish themselves by their affinity for different G protein families. RNA editing in intracellular loop 3 of the 5-HT$_{2c}$ receptor also affected coupling selectivity and efficiency *(113)*. Additionally, G protein coupling selectivity was reported to be modified by μ- and δ-opioid receptors *(114,115)*, AT1 receptors and B$_2$Rs *(116)*, or CCR2/CCR5 *(117)* heterodimerization (*see* Part III).

5.5. Interaction With Intracellular Effectors Other Than G Proteins

Within the scope of proteomics, an increasing number of proteins were identified that interact with intracellular loops or with the carboxy-terminal tail of GPCRs (*see* Part II). Most of the partner candidates have been extracted from yeast two-hybrid screens, glutathione-*S*-transferase (GST) pulldown assays, or gel overlays. Some possess enzymatic properties, including receptor specific (GRK) or nonspecific (PKA, PKC) kinases, nitric oxide synthase, calmodulin, or small G proteins such as Arf or RhoA. Others are scaffolding proteins that act as adaptors, including β-arrestin 1/2, MUPP-1, AKAP 79/250, NHERF 1/2; these possess several important functions. They participate in the targeting of GPCRs to specific subcellular compartments but are also responsible for the clustering of these receptors with various effectors. Finally, interacting proteins can regulate GPCR functions in an allosteric manner (for recent reviews, see refs. *10, 118,* and *119*). In several cases, receptor interactions with these types of molecules were shown to be ligand-dependent *(118,120)*.

6. CONCLUSIONS

Our knowledge of GPCR structure and function has greatly improved over recent years. The structure at high resolution obtained for rhodopsin has confirmed many of the former assumptions based on mutagenesis, biochemical, and biophysical studies. The overall structure of GPCRs—especially those from class A—definitely appears to match that of rhodopsin fairly well. However, the structure that has been obtained for the visual pigment corresponds to the inactive state of the protein, and many aspects of the receptor function remain unsolved. In particular, modeling combined to structure–activity studies of class A GPCRs has revealed that despite a common overall structure, the fine-tuning of ligand binding, receptor activation, and interaction with intracellular partners is fully receptor-dependent. Furthermore, the nature of the ligand appears to be a crucial element as well. Altogether, this suggests that a thorough understanding of the mechanisms underlying receptor activation and subsequent signal cascade initiation corresponds to unique equations in which both the ligand and the receptor type are critical factors. Moreover, the structure of rhodopsin brings insights into only the transmembrane region of the receptor. This leaves a wide gap in the case of GPCRs, where ligand binding is largely assumed to be at the extracellular domains of the receptor and underscores the need for a detailed picture of each receptor of interest. This will require production of large amounts of receptors purified to homogeneity. Although 3D crystals will provide static pictures of the receptor, biochemical and biophysical studies will bring dynamic information that is needed to accurately describe and explain GPCR activation. Therefore, despite the tremendous progress made over recent years, much remains to be deciphered, and, to date, a full understanding of GPCR plasticity remains beyond our reach.

REFERENCES

1. Consortium IHGS. Initial sequensing and analysis of the human genome. Nature 2001;409:860–922.
2. Fredriksson R, Lagerström MC, Lundin L-G, Schiöth HB. The G-protein-coupled receptors in the human genome form five main families. Phylogenetic analysis, paralogon groups, and fingerprints. Mol Pharmacol 2003;63:1256–1272.
3. Flower DR. Modelling G-protein-coupled receptors for drug design. Biochim Biophys Acta 1999;1422:207–234.
4. Seifert R, Wenzel-Seifert K. Constitutive activity of G-protein-coupled receptors: cause of disease and common property of wild-type receptors. Naunyn Schmiedebergs Arch Pharmakol 2002;366:381–416.
5. Klabunde T, Hessler G. Drug design strategies for targeting G-protein-coupled receptors. Chem Bio Chem2002;3:928–944.
6. Gether U. Uncovering molecular mechanisms involved in activation of G protein-coupled receptors. Endocr Rev 2000;21:90–113.
7. Lu Z-L, Saldanha JW, Hulme EC. Seven-transmembrane receptors: crystals clarify. Trends Pharmacol Sci 2002;23:140–146.
8. Tsao P, Cao T, von Zastrow M. Role of endocytosis in mediating downregulation of G-protein-coupled receptors. Trends Pharmacol Sci 2001;22:91–96.
9. Kohout TA, Lefkowitz RJ. Regulation of G protein-coupled receptor kinases and arrestins during receptor desensitization. Mol Pharmacol 2003;63:9–18.
10. Hur E-M, Kim K-T. G protein-coupled receptors signalling and cross-talk: achieving rapidity and specificity. Cell Signal 2002;14:397–405.
11. Ishii M, Kurachi Y. Physiological actions of regulators of G-protein signaling (RGS) proteins. Life Sci 2003;74:163–171.
12. Palczewski K, Kumasaka T, Hori T, et al. Crystal structure of rhodopsin: a G protein-coupled receptor. Science 2000;289:739–745.
13. Massotte D. G protein-coupled receptor overexpression with the baculovirus-insect cell system: a tool for structural and functional studies. Biochim Biophys Acta 2003;1610:77–89.
14. Kunishima N, Shimada Y, Tsuji Y, et al. Structural basis of glutamate recognition by a dimeric metabotropic glutamate receptor. Nature 2000;407:971–977.
15. Malbon CC, Wang H, Moon RT. Wnt signaling and heterotrimeric G-proteins: strange bedfellows or a classic romance? Biochem Biophys Res Commun 2001;287:589–593.
16. Vaidehi N, Floriano WB, Trabanino R, et al. Prediction of structure and function of G protein-coupled receptors. Proc Natl Acad Sci USA 2002;99: 12,622–12,627.
17. Ballesteros JA, Shi L, Javitch JA. Structural mimicry in G protein-coupled receptors: implications of the high-resolution structure of rhodopsin for structure–function analysis of rhodopsin-like receptors. Mol Pharmacol 2001;60:1–19.

18. Strader CD, Fong TM, Graziano MP, Tota MR. The family of G-protein-coupled receptors. FASEB J 1995;9:745–754.
19. Surratt CK, Johnson PS, Moriwaki A, et al. mu opiate receptor. Charged transmembrane domain amino acids are critical for agonist recognition and intrinsic activity. J Biol Chem 1994;269:20,548–20,553.
20. Befort K, Tabbara L, Bausch S, Chavkin C, Evans C, Kieffer BL. The conserved aspartate residue in the third putative transmembrane domain of the δ opioid receptor is not the anionic counterpart for cationic opiate binding but is a constituent of the receptor binding site. Mol Pharmacol 1996;49:216–223.
21. Befort K, Tabbara L, Kling D, Maigret B, Kieffer BL. Role of aromatic transmembrane residues of the δ-opioid receptor in ligand recognition. J Biol Chem 1996;271:10,161–10,168.
22. Ji TH, Grossmann M, Ji I. G protein-coupled receptors. I. Diversity of receptor–ligand interactions. J Biol Chem 1998;273:17,299–17,302.
23. Marshall GR. Peptide interactions with G-protein coupled receptors. Biopolymers 2001;60:246–277.
24. Zhao MM, Hwa J, Perez DM. Identification of critical extracellular loop residues involved in alpha 1-adrenergic receptor subtype-selective antagonist binding. Mol Pharmacol 1996;50:1118–1126.
25. Wang WW, Shahrestanifar M, Jin J, Howells RD. Studies on mu and delta opioid receptor selectivity utilizing chimeric and site-mutagenized receptors. Proc Natl Acad Sci USA 1995;92:12,436–12,440.
26. Wang JB, Johnson PS, Wu JM, Wang WF, Uhl GR. Human kappa opiate receptor second extracellular loop elevates dynorphin's affinity for human mu/kappa chimeras. J Biol Chem 1994;269:25,966–25,969.
27. Varga EV, Li X, Stropova D, et al. The third extracellular loop of the human δ-opioid receptor determines the selectivity of the δ-opioid agonists. Mol Pharmacol 1997;50:1619–1624.
28. Law PY, Wong YH, Loh HH. Mutational analysis of the structure and function of opioid receptors. Biopolymers 1999;51:440–455.
29. Pepin MC, Yue SY, Roberts E, Wahlestedt C, Walker P. Novel "restoration of function" mutagenesis strategy to identify amino acids of the delta-opioid receptor involved in ligand binding. J Biol Chem 1997;272:9260–9267.
30. Xue J-C, Chen C, Zhu J, et al. The third extracellular loop of the μ opioid receptor is important for agonist selectivity. J Biol Chem 1995;270:12,977–12,979.
31. Shi L, Javitch JA. The second extracellular loop of the dopamine D2 receptor lines the binding-site crevice. Proc Natl Acad Sci USA 2004;101:440–445.
32. Schwartz TW, Rosenkilde MM. Is there a 'lock' for all agonist 'keys' in 7TM receptors? Trends Pharmacol Sci 1996;17:213–216.
33. Zhang P, Johnson PS, Zoellner C, et al. Mutation of human μ opioid receptor extracellular "disulfide cysteine" residues alters ligand binding but does not prevent receptor targeting to the cell plasma membrane. Mol Brain Res 1999;72:195–204.

34. Baneres J-L, Martin A, Hullot P, Girard J-P, Rossi J-C, Parello J. Structure-based analysis of GPCR function: conformational adaptation of both agonist and receptor upon leukotriene B4 binding to recombinant BLT1. J Mol Biol 2003;329:801–814.
35. Savarese TM, Wang CD, Fraser CM. Site-directed mutagenesis of the rat m1 muscarinic acetylcholine receptor. Role of conserved cysteines in receptor function. J Biol Chem 1992;267:11,439–11,448.
36. D'Angelo DD, Eubank JJ, Davis MG, Dorn GW 2nd. Mutagenic analysis of platelet thromboxane receptor cysteines. Roles in ligand binding and receptor–effector coupling J Biol Chem 1996;271:6233–6240.
37. Perlman JH, Wang W, Nussenzveig DR, Gershengorn MC. A disulfide bond between conserved extracellular cysteines in the thyrotropin-releasing hormone receptor is critical for binding. J Biol Chem 1995;270:24,682–24,685.
38. Cook JV, Eidne KA. An intramolecular disulfide bond between conserved extracellular cysteines in the gonadotropin-releasing hormone receptor is essential for binding and activation Endocrinology 1997;138:2800–2806.
39. Yamano Y, Ohyama K, Chaki S, Guo DF, Inagami T. Identification of amino acid residues of rat angiotensin II receptor for ligand binding by site directed mutagenesis. Biochem Biophys Res Commun 1992;187:1426–1431.
40. Noda K, Saad Y, Graham RM, Karnik SS. The high affinity state of the beta 2-adrenergic receptor requires unique interaction between conserved and non-conserved extracellular loop cysteines. J Biol Chem 1994;269:6743–6752.
41. Elling CE, Nielsen SM, Schwartz TW. Conversion of antagonist-binding site to metal-ion site in the tachykinin NK-1 receptor. Nature 1995;374:74–77.
42. De Lean A, Stadel JM, Lefkowitz RJ. A ternary complex model explains the agonist-specific binding properties of the adenylate cyclase-coupled beta-adrenergic receptor. J Biol Chem 1980;255:7108–7117.
43. Costa T, Herz A. Antagonists with negative intrinsic activity at δ opioid receptors coupled to GTP-binding proteins. Proc Natl Acad Sci USA 1989;86:7321–7325.
44. Samama P, Cotecchia S, Costa T, Lefkowitz RJ. A mutation-induced activated state of the beta 2-adrenergic receptor. Extending the ternary complex model. J Biol Chem 1993;268:4625–4636.
45. Weiss JM, Morgan PH, Lutz MW, Kenakin TP. The cubic ternary complex receptor-occupancy model. I. model description. J Theor Biol 1996;178:151–167.
46. Bouaboula M, Perrachon S, Milligan L, et al. A selective inverse agonist for central cannabinoid receptor inhibits mitogen-activated protein kinase activation stimulated by insulin or insulin-like growth factor 1. J Biol Chem 1997;272:22,330–22,339.
47. Christopoulos A, Kenakin T. G protein-coupled receptor allosterism and complexing. Pharmacol Rev 2002;54:323–374.
48. Zuscik MJ, Porter JE, Gaivin R, Perez DM. Identification of a conserved switch residue responsible for selective constitutive activation of the beta2-adrenergic receptor. J Biol Chem 1998;273:3401–3407.

49. Kenakin T. Ligand-selective receptor conformations revisited: the promise and the problem. Trends Pharmacol Sci 2003;24:346–354.
50. Kjelsberg MA, Cotecchia S, Ostrowski J, Caron MG, Lefkowitz RJ. Constitutive activation of the α_{1B}-adrenergic receptor by all amino acid substitution at a single site. J Biol Chem 1992;267:1430–1433.
51. Parnot C, Miserey-Lenkei S, Bardin S, Corvol P, Clauser E. Lessons from constitutively active mutants of G protein-coupled receptors. Trends Endocrinol Metabol 2002;13:336–343.
52. Gether U, Asmar F, Meinild AK, Rasmussen SG. Structural basis for activation of G-protein-coupled receptors. Pharmacol Toxicol 2002;91:304–312.
53. Gether U, Ballesteros JA, Seifert R, Sansers-Bush E, Weinstein H, Kobilka BK. Structural instability of a constitutively active G protein-coupled receptor. Agonist-independent activation due to conformational flexibility. J Biol Chem 1997;272:2587–2590.
54. Sheikh SP, Zvyaga TA, Lichtarge O, Sakmar TP, Bourne HR. Rhodopsin activation blocked by metal-ion-binding sites linking transmembrane helices C and F. Nature 1996;383:347–350.
55. Sheikh SP, Vilardarga JP, Baranski TJ, et al. Similar structures and shared switch mechanisms of the beta2-adrenoceptor and the parathyroid hormone receptor. Zn(II) bridges between helices III and VI block activation. J Biol Chem 1999;274:17,033–17,041.
56. Javitch JA, Fu D, Liapakis G, Chen J. Constitutive activation of the beta2 adrenergic receptor alters the orientation of its sixth membrane-spanning segment. J Biol Chem 1997;272:18,546–18,549.
57. Spalding TA, Burstein ES, Henderson SC, Ducote KR, Brann MR. Identification of a ligand-dependent switch within a muscarinic receptor. J Biol Chem 1998;273:21,563–21,568.
58. Décaillot FM, Befort K, Filliol D, Yue SY, Walker P, Kieffer BL. Opioid receptor random mutagenesis reveals how a G protein-coupled receptor turns on. Nat Struct Biol 2003;10:629–636.
59. Parnot C, Bardin S, Miserey-Lenkei S, Guedin D, Corvol P, Clauser E. Systematic identification of mutations that constitutively activate the angiotensin II type 1A receptor by screening a randomly mutated cDNA library with an original pharmacological bioassay. Proc Natl Acad Sci USA 2000;97:7615–7620.
60. Baranski TJ, Herzmark P, Lichtarge O, et al. C5a receptor activation. Genetic identification of critical residues in four transmembrane helices. J Biol Chem 1999;274:15,757–15,765.
61. Shapiro DA, Kristiansen K, Weiner DM, Kroeze WK, Roth BL. Evidence for a model of agonist-induced activation of 5-hydroxytryptamine 2A serotonin receptors that involves the disruption of a strong ionic interaction between helices 3 and 6. J Biol Chem 2002;277:11,441–11,449.
62. Alewijnse AE, Timmerman H, Jacobs EH, et al. The effect of mutations in the DRY motif on the constitutive activity and structural instability of the histamine H(2) receptor. Mol Pharmacol 2000;57:890–898.

63. Sheer A, Cotecchia S. Constitutively active G protein-coupled receptors: potential mechanisms of receptor activation. J Recept Signal Transduct Res 1997;17:57–73.
64. Ballesteros JA, Jensen AD, Liapakis G, et al. Activation of the beta 2-adrenergic receptor involves disruption of an ionic lock between the cytoplasmic ends of transmembrane segments 3 and 6. J Biol Chem 2001;276:29,171–29,177.
65. Burstein ES, Spalding TA, Brann MR. The second intracellular loop of the m5 muscarinic receptor is the switch which enables G-protein coupling. J Biol Chem 1998;273:24,322–24,327.
66. Yamashita T, Terakita A, Shichida Y. Distinct roles of the second and third cytoplasmic loops of bovine rhodopsin in G protein activation. J Biol Chem 2000;275:34,272–34,279.
67. Demene H, Granier S, Muller D, et al. Active peptidic mimics of the second intracellular loop of the V(1A) vasopressin receptor are structurally related to the second intracellular rhodopsin loop: a combined 1H NMR and biochemical study. Biochemistry 2003;42:8204–8213.
68. Chung DA, Zuiderweg ER, Fowler CB, Soyer OS, Mosberg HI, Neubig RR. NMR structure of the second intracellular loop of the alpha 2A adrenergic receptor: evidence for a novel cytoplasmic helix. Biochemistry 2002;41:3596–3604.
69. Chung DA, Wade SM, Fowler CB, et al. Mutagenesis and peptide analysis of the DRY motif in the alpha2A adrenergic receptor: evidence for alternate mechanisms in G protein-coupled receptors. Biochem Biophys Res Commun 2002;293:1233–1241.
70. Huang P, Li J, Chen C, Visiers I, Weinstein H, Liu-Chen LY. Functional role of a conserved motif in TM6 of the rat μ opioid receptor: constitutively active and inactive receptors result from substitutions of Thr6.34(279) with Lys and Asp. Biochemistry 2001;40:13,501–13,509.
71. Huang P, Visiers I, Weinstein H, Liu-Chen L-Y. The local environment at the cytoplasmic end of TM6 of the m opioid receptor differs from those of rhodopsin and monoamine receptors: introduction of an ionic lock between the cytoplasmic ends of helices 3 and 6 by a L6.30(275)E mutation inactivates the μ opioid receptor and reduces the constitutive activity of its T6.34(279)K mutant. Biochemistry 2002;41:11,972–11,980.
72. Javitch JA, Shi L, Simpson MM, et al. The fourth transmembrane segment of the dopamine D2 receptor: accessibility in the binding-site crevice and position in the transmembrane bundle. Biochemistry 2000;39:12,190–12,199.
73. Chen S, Lin F, Xu M, Graham RM. Phe(303) in TMVI of the alpha(1B)-adrenergic receptor is a key residue coupling TM helical movements to G-protein activation. Biochemistry 2002;41:588–596.
74. Jagerschmidt A, Guillaume N, Roques BP, Noble F. Binding sites and transduction process of the cholecystokinin B receptor: involvement of highly conserved aromatic residues of the transmembrane domains evidenced by site-directed mutagenesis. Mol Pharmacol 1998;53:878–885.

75. Lu ZL, Hulme EC. A network of conserved intramolecular contacts defines the off-state of the transmembrane switch mechanism in a seven-transmembrane receptor. J Biol Chem 2000;275:5682–5686.
76. Han M, Smith SO, Sakmar TP. Constitutive activation of opsin by mutation of methionine 257 on transmembrane helix 6. Biochemistry 1998;37:8253–8261.
77. Laue L, Chan WY, Hsueh AJ, et al. Genetic heterogeneity of constitutively activating mutations of the human luteinizing hormone receptor in familial male-limited precocious puberty. Proc Natl Acad Sci USA 1995;92:1906–1910.
78. Vlaeminck-Guillem V, Ho SC, Rodien P, Vassart G, Costagliola S. Activation of the cAMP pathway by the TSH receptor involves switching of the ectodomain from a tethered inverse agonist to an agonist. Mol Endocrinol 2002;16:736–746.
79. Jin H, Xie Z, George SR, O'Dowd BF. Palmitoylation occurs at cysteine 347 and cysteine 351 of the dopamine D1 receptor. Eur J Pharmacol 1999;386:305–312
80. Ponimaskin EG, Heine M, Joubert L, et al. The 5-hydroxytryptamine(4a) receptor is palmitoylated at two different sites and acylation is critically involved n regulation of receptor constitutive activity. J Biol Chem 2002;277:2534–2546.
81. Hawtin SR, Tobin AB, Patel S, Wheatley M. Palmitoylation of the vasopressin V_{1a} receptor reveals different conformational requirements for signaling, agonist-induced receptor phosphorylation, and sequestration. J Biol Chem 2001;276:38,139–38,146.
82. Morrison DF, O'Brien PJ, Pepperbeg DR. Depalmitoylation with hydroxylamine alters the functional properties of rhodopsin. J Biol Chem 1991;266:20,118–20,123.
83. Krishna AG, Menon ST, Terry TJ, Sakmar TP. Evidence that helix 8 of rhodopsin acts as a membrane-dependent conformational switch. Biochemistry 2002;41:8298–8309.
84. Ponimaskin EG, Schmidt MF, Heine M, Bickmeyer U, Richter DW. 5-Hydroxytryptamine 4(a) receptor expressed in Sf9 cells is palmitoylated in an agonist-dependent manner. Biochem J 2001;353:627–634.
85. Brillet K, Kieffer BL, Massotte D. Enhanced spontaneous activity of the mu opioid receptor by cysteine mutations: characterization of a tool for inverse agonist screening. BMC Pharmacol 2003;3:14.
86. Doi T, Sugimoto H, Arimoto I, Hiroaki Y, Fujiyoshi Y. Interactions of endothelin receptor subtypes A and B with Gi, Go, and Gq in reconstituted phospholipid vesicles. Biochemistry 1999;38:3090–3099.
87. Horstmeyer A, Cramer H, Sauer T, Mueller-Esterl W, Schroeder C. Palmitoylation of endothelin A. Differential modulation of signal transduction activity by post-translational modification. J Biol Chem 1996;271:20,811–20,819.
88. Okamoto Y, Ninomiya H, Tanioka M, Sakamoto A, Miwa S, Masaki T. Palmitoylation of human endothelin $_B$. Its critical role in G protein coupling

and a differential requirement for the cytoplasmic tail by G protein subtypes. J Biol Chem 1997;272:21,589–21,596.
89. Mouillac B, Caron M, Dennis M, Bouvier M. Agonist-modulated palmitoylation of β2-adrenergic receptor in Sf9 cells. J Biochem (Tokyo) 1992;267:21,733–21,737.
90. Loisel TP, Ansanay H, Adams L, et al. Activation of the β$_2$-adrenergic receptor–Gα$_s$ complex leads to rapid depalmitoylation and inhibition of repalmitoylation of both the receptor and Gα$_s$. J Biol Chem 1999;274:31,014–31,019.
91. Kennedy M, Limbird L. Mutations of the alpha 2A-adrenergic receptor that eliminate detectable palmitoylation do not perturb receptor–G-protein coupling. J Biol Chem 1993;268:8003–8011.
92. Hayashi M, Haga T. Palmitoylation of muscarinic acetylcholine receptor m2 subtypes: reduction in their ability to activate G proteins by mutation of a putative palmitoylation site, cysteine 457, in the carboxy-terminal tail. Arch Biochem Biophys 1997;340:376–382.
93. Moffet S, Adam L, Bonin H, Loisel TP, Bouvier M, Mouillac B. Palmitoylated cysteine 341 modulates phosphorylation of the β$_2$-adrenergic receptor by the cAMP-dependent protein kinase. J Biol Chem 1996;271:21,491–21,497.
94. Palmer TM, Stiles GL. Identification of threonine residues controlling the agonist-dependent phosphorylation and desensitization of the rat A(3) adenosine receptor. Mol Pharmacol 2000;57:539–545.
95. Soskic V, Nyakatura E, Roos M, Muller-Esterl W, Godovac-Zimmermann J. Correlations in palmitoylation and multiple phosphorylation of rat bradykinin B2 receptor in Chinese hamster ovary cells. J Biol Chem 1999;274:8539–8545.
96. Chen C, Shahabi V, Xu W, Liu-Chen L-Y. Palmitoylation of the rat μ opioid receptor. FEBS Lett 1998;441:148–152.
97. Hermans E. Biochemical and pharmacological control of the multiplicity of coupling at G-protein-coupled receptors. Pharmacol Ther 2003;99:25–44.
98. Schwindinger WF, Robishaw JD. Heterotrimeric G-protein betagammadimers in growth and differentiation. Oncogene 2001;20:1653–1660.
99. Wong SK. G protein selectivity is regulated by multiple intracellular regions of GPCRs. Neurosignals 2003;12:1–12.
100. Abadji V, Lucas-Lenard JM, Chin C, Kendall DA. Involvement of the carboxyl terminus of the third intracellular loop of the cannabinoid CB1 receptor in constitutive activation of G$_s$. J Neurochem 1999;72:2032–2038.
101. Slessareva JE, Ma H, Depree KM, et al. Closely related G-protein-coupled receptors use multiple and distinct domains on G-protein alpha-subunits for selective coupling. J Biol Chem 2003;278:50,530–50,536.
102. Seifert R, Gether U, Wenzel-Seifert K, Kobilka BK. Effects of guanine, inosine, and xanthine nucleotides on beta(2)-adrenergic receptor/G(s) interactions: evidence for multiple receptor conformations. Mol Pharmacol 1999;56:348–358.

103. Ghanouni P, Steenhuis JJ, Farrens DL, Kobilka BK. Agonist-induced conformational changes in the G-protein-coupling domain of the beta 2 adrenergic receptor. Proc Natl Acad Sci USA 2001;98:5997–6002.
104. Massotte D, Brillet K, Kieffer BL, Milligan G. Agonists activate $G_{i1}\alpha$ or $G_{i2}\alpha$ fused to the human mu opioid receptor differently. J Neurochem 2002;81:1372–1382.
105. Moon H-E, Cavalli A, Bahia DS, Hoffmann M, Massotte D, Milligan G. The human δ opioid receptor activates $Gi1\alpha$ more efficiently than $Go1\alpha$. J Neurochem 2001;76:1–10.
106. Alves ID, Salamon Z, Varga E, Yamamura HI, Tollin G, Hruby VJ. Direct observation of G-protein binding to the human delta-opioid receptor using plasmon-waveguide resonance spectroscopy. J Biol Chem 2003;278:48,890–48,897.
107. Swaminath G, Xiang Y, Lee TW, Steenhuis J, Parnot C, Kobilka BK. Sequential binding of agonists to the beta2 adrenoceptor. Kinetic evidence for intermediate conformational states. J Biol Chem 2004;279:686–691.
108. Palanche T, Ilien B, Zoffmann S, et al. The neurokinin A receptor activates calcium and cAMP responses through distinct conformational states. J Biol Chem 2001;276:34,853–34,861.
109. Zamah AM, Delahunty M, Luttrell LM, Lefkowitz RJ. Protein kinase A-mediated phosphorylation of the beta 2-adrenergic receptor regulates its coupling to Gs and Gi. Demonstration in a reconstituted system. J Biol Chem 2002;277:31,249–31,256.
110. Pindon A, van Hecke G, van Gompel P, Lesage AS, Leysen JE, Jurzak M. Differences in signal transduction of two 5-HT4 receptor splice variants: compound specificity and dual coupling with $G\alpha_s$- and $G\alpha_{i/o}$-proteins. Mol Pharmacol 2002;61:85–96.
111. Namba T, Sugimoto Y, Negishi M, et al. Alternative splicing of C-terminal tail of prostaglandin E receptor subtype EP3 determines G-protein specificity. Nature 1993;365:166–170.
112. Negishi M, Namba T, Sugimoto Y, et al. Opposite coupling of prostaglandin E receptor EP3C with Gs and G(o). Stimulation of Gs and inhibition of G(o). J Biol Chem 1993;268:26,067–26,070.
113. Berg KA, Cropper JD, Niswender CM, Sanders-Bush E, Emeson RB, Clarke WP. RNA-editing of the 5-HT(2C) receptor alters agonist-receptor-effector coupling specificity. Br J Pharmacol 2001;134:386–392.
114. Charles AC, Mostovskaya N, Asass K, Evans CJ, Dankovich ML, Hales TG. Coexpression of delta-opioid receptors with mu receptors in GH3 cells changes the functional response to mu agonists from inhibitory to excitatory. Mol Pharmacol 2003;63:89–95.
115. George SR, Fan T, Xie Z, et al. Oligomerization of mu- and delta-opioid receptors. Generation of novel functional properties. J Biol Chem 2000;275:26,128–26,135.
116. AbdAlla S, Lother H, Quitterer U. AT1-receptor heterodimers show enhanced G-protein activation and altered receptor sequestration. Nature 2000;407:94–98.

117. Mellado M, Rodriguez-Frade JM, Vila-Coro AJ, et al. Chemokine receptor homo- or heterodimerization activates distinct signaling pathways. EMBO J 2001;20:2497–2507.
118. Hall RA, Lefkowitz RJ. Regulation of G protein-coupled receptor signaling by scaffold proteins. Circ Res 2002;91:672–680.
119. Bockaert J, Marin P, Dumuis A, Fagni L. The 'magic tail' of G protein-coupled receptors: an anchorage for functional protein networks. FEBS Lett 2003;546:65–72.
120. Wang D, Sadee W, Quillan JM. Calmodulin binding to G protein-coupling domain of opioid receptors. J Biol Chem 1999;274:22,081–22,088.

2
Molecular Mechanisms Involved in the Activation of Rhodopsin-Like Seven-Transmembrane Receptors

Peng Huang and Lee-Yuan Liu-Chen

1. INTRODUCTION

1.1. Seven-Transmembrane Receptors

Seven-transmembrane receptors (7TMRs) comprise a large family of membrane-bound proteins that share a unifying signal transduction mechanism (i.e., upon activation, these receptors signal through G proteins). These receptors are involved in a vast variety of physiological functions, including neurotransmission, function of exocrine and endocrine glands, smell, taste, vision, chemotaxis, embryogenesis, development, human immunodeficiency virus (HIV) infection, oncogenesis, cell growth, and differentiation. More recent studies indicate that these receptors are also associated with and signal through other molecules (1). Therefore, it is more appropriate to use the term 7TMR than G protein-coupled receptors (GPCRs).

To date, rhodopsin is the only receptor of the superfamily for which the high-resolution structure has been determined (2,3). In 1993, Shertler et al. (2) published a project map of the bovine rhodopsin at 9 Å resolution in two dimensions. In 2000, Palczewski et al. reported the structure of bovine rhodopsin in ground (inactive) states at 2.8 Å resolution in three dimensions (3,4). Both reports showed the seven helices of rhodopsin traversing the plane of the membranes in a nonparallel manner, with some transmembranes (TMs) being tilted, an extracellular N-terminal domain, an intracellular C-terminal domain, and three extracellular loops and three intracellular loops connecting the helices (Fig. 1). In the proximal region of the C-terminal

From: *Contemporary Clinical Neuroscience: The G Protein-Coupled Receptors Handbook*
Edited by: L. A. Devi © Humana Press Inc., Totowa, NJ

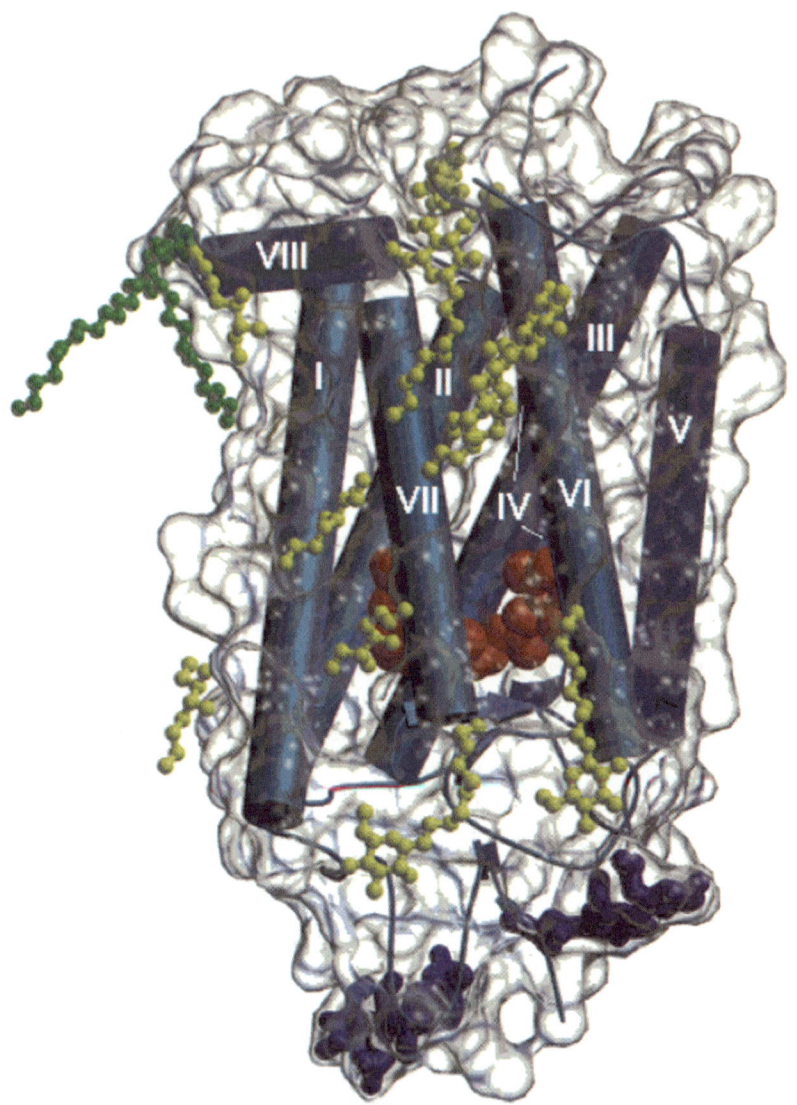

Fig. 1. Three-dimensional crystal structure of rhodopsin with bound detergent and amphiphile molecules. Helical portions of the protein, including the seven TMs, are shown as blue rods, and β-strands are shown as blue arrows. The polypeptide connecting the helices appears as blue coils. A transparent envelope around the protein represents the molecular surface. The dark blue ball-and-stick groups at the bottom of the figure denote carbohydrate groups attached to the protein. Two palmitoyl groups covalently attached to the protein are shown in green. Nonylglucoside and heptanetriol molecules located near the hydrophobic surface of the protein are shown in yellow. (Reprinted from ref. *4* with permission of the American Chemical Society, copyright 2001.)

domain, there is a short helix, helix 8 (H8), parallel to the plane of the plasma membranes (Fig. 1). It is generally accepted that 7TMRs share the structure of the 7-TM bundle *(5,6)*.

In the human genome, there are three major families of 7TMRs (*see* refs. *7* and *8* for a classification scheme): (a) rhodopsin and rhodopsin-like receptors (approx 200) and odorant and taste receptors (several hundreds); (b) glucagons/vasoactive intestinal polypeptide/calcitonin receptors (approx 25); and (c) metabotropic glutamate receptors, γ-aminobutyric acid $(GABA)_B$ receptors and chemosensors (approx 20). Within each family, there is at least 25% homology within the 7-TMs and a distinctive set of highly conserved residues and motifs. The rhodopsin family of 7TMRs, which constitute approx 90% of all 7TMRs *(4)*, are the most extensively studied.

1.2. Numbering Schemes for Rhodopsin-Like 7TMR Sequences

Two residue-numbering schemes are used throughout this chapter: the generic numbering scheme of Ballesteros and Weinstein *(9)* and the residue numbers in the amino acid sequence of the particular receptor being discussed. According to the generic nomenclature, amino acid residues in TMs are assigned two numbers (N1 and N2). N1 refers to the TM number. For N2, the most conserved residue in each TM is assigned 50, and the other residues are numbered in relation to this conserved residue, with numbers decreasing toward the N-terminus and increasing toward the C-terminus. The receptors of the rhodopsin family are characterized by the presence of highly conserved "fingerprint" residues *(8,10)*, including N1.50 in the TM1; D2.50 in the TM2; the DRY(3.49–3.51) motif in the TM3; W4.50 in the TM4; and P5.50, P6.50, and NP7.50XXY motif in TMs5, -6, and -7.

For example, D3.49(134) in bovine rhodopsin is located in TM3 and one amino acid N-terminal to the most conserved R3.50; it is the 134th amino acid residue from the N-terminus. The generic numbering scheme allows easy comparisons of the same residues among different receptors in the rhodopsin family.

1.3. Structural Studies on Rhodopsin

The structure–function relationship of rhodopsin has been extensively studied (reviewed in refs. *11–13*) with low-resolution electron density maps *(2,14,15)* and biochemical approaches, including crosslinking (by genetically engineered disulfide bridges and Zn^{2+} binding sites), site-directed spin labeling, scanning accessibility determinations (reviewed in ref. *11*), and analysis of retinal movement by photo-affinity labeling *(16)*. By X-ray crystallography, Palczewski et al. reported a high-resolution structure of the bovine rhodopsin in ground (inactive) states at 2.8 Å resolution *(3,4)*, which

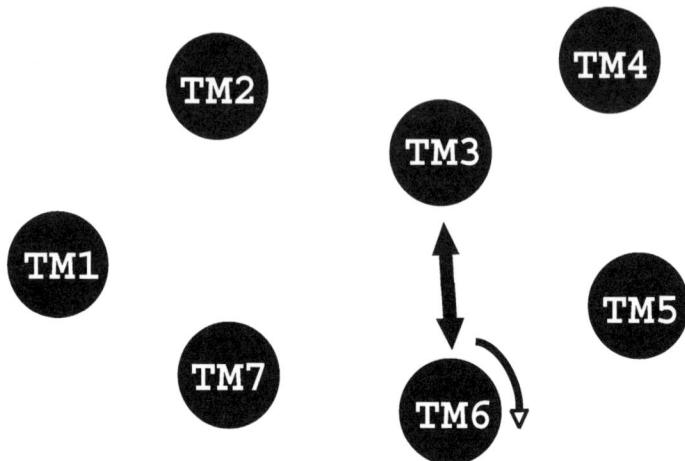

Fig. 2. A diagram to show the conformational changes of TMs during 7TMR activation. The figure is based on the structure of rhodopsin and is viewed from the cytoplasmic side. The circles represent the cytoplasmic ends of each helix. The activation of all 7TMRs probably involves a separation of TM3 and TM6 (the double arrow) and a clockwise rotation in TM6.

currently is the only 7TMR with known structural details (Fig. 1). This report provided much new information and confirmed many detailed structural characteristics inferred from experimental results in previous publications. Rhodopsin has been an excellent model for the other members in the rhodopsin 7TMR family regarding the mechanism by which the receptor may be constrained in inactive states and which molecular interactions may be changed to induce conformational changes from inactive states to active states, thus exposing receptor interaction sites for heterotrimeric G proteins *(10,17)*. Upon activation by light, retinal isomerization triggers outward movement of the helices of rhodopsin, thereby opening a cleft at the cytoplasmic ends of the helices. Additionally, there is an increased distance between TM3 and TM6 (Fig. 2), exposing more inner faces of TM2, -3, –6, and -7, which are believed to be involved in interaction with the G protein transducin. In contrast, the cytoplasmic ends of TM4 and TM5 become less exposed. Overall movement of TM6 exceeds that of TM3, which is more constrained by its central position in the helix bundle *(11)* (Fig. 2).

1.4. Scope of This Chapter

Experimental data support the notion that the helical movements involved in rhodopsin activation likely occur in other 7TMRs of the rhodopsin family

Conformational Changes in 7TMRs Activation 37

(6,18,19; this is summarized under Subheading 2). Additonally, detailed molecular events underlying the movements in TMs during receptor activation (thereby determining the conversion from inactive to activated states) have been partially elucidated experimentally. Section 3 focuses on the mechanisms that are likely to be common to many, if not all, 7TMRs of this family. Structural motifs located in the intracellular ends of TM3 and TM6 are important components of a network of intra- and interhelical interactions stabilizing the ground (inactive) state of the receptors. Disruptions of the interactions appear to play an important role in receptor activation. The interpretations of the mechanisms will be mainly based on substantial structure–function studies of constitutively active mutants (CAMs) of various 7TMRs and the molecular modeling work that integrated the experimental data using rhodopsin-based models. Because it is a vast field, it is not possible to cite all the work related to the topic. We apologize to the authors whose works are not mentioned in the chapter.

2. THE RELATIVE POSITIONS OF TMs ARE CHANGED DURING THE ACTIVATION OF 7TMRs

2.1. Crosslinking of TM3 and TM6 Stabilizes Inactive Conformations

Before the high-resolution crystal structure of rhodopsin was known, cysteine or histidine residues were engineered into cytoplasmic ends of different TM helices of rhodopsin to form disulfide bonds or Zn^{2+} binding sites to assess proximity between residues and to determine effects of structural constraints on receptor activation.

2.1.1. Disulfide Bonds

Cysteine residues engineered into the cytoplasmic ends of TM3 and TM6 of rhodopsin were shown to form Cys–Cys disulfide links after mild oxidation between V3.54(139)C in TM3 and each of the five positions 6.30(247)–6.34(250), but not 6.35(252), in the TM6 *(20,21)*. Additionally, a disulfide bond formed between R3.50(135)C at the end of TM3 and V6.33(250)C at the end of TM6 *(21)*. These results indicate that the cytoplasmic ends of TM3 and TM6 are in close proximity. Formation of any of the six disulfide bonds prevented the activation of rhodopsin, demonstrating the importance of movement of the cytoplasmic ends of TM3 and TM6 in rhodopsin activation *(20,21)*.

Formation of a disulfide bond between A6.29(246)C at the end of TM6 and Q312C in the H8 abolished both transducin activation and phosphorylation by rhodopsin kinase *(21)*, indicating that movement of TM6 relative to

H8 is important for rhodopsin activation. A disulfide bond between K6.28(245)C at the end of TM6 and S338C in the C-terminal domain enhanced transducin activation but abolished phosphorylation by rhodopsin kinase, suggesting that the structure recognized by transducin was stabilized in this mutant *(21)*.

2.1.2. Zn^{2+} Binding Sites

Two properly spaced substituted histidines can bind Zn^{2+}, which in turn restricts movement of the two histidines. When histidines were introduced in the cytoplasmic ends of TM3 and TM6 at V3.53(138) and T6.34(251), binding of Zn^{2+} inhibited rhodopsin activation *(22)*. The inhibitory effects of Zn^{2+} bridges connecting TM3 and TM6 were also demonstrated in other 7TMRs. For example, when His residues were introduced into A3.53(134)H at the end of TM3 and one of the three positions E6.30(268)H, H6.31(269), and L6.34(272)H at the end of TM6 *(23)* in the β_2-adrenergic receptor (AR), activation of the receptor was greatly inhibited by the Zn^{2+} ion. These results indicate that relative movements of TM3 and TM6 are required for 7TMR activation.

2.2. Activation of 7TMRs Involves Rearrangement of the Positions of TM3, TM6, and TM5

Several different strategies were used to examine agonist-dependent dynamic structural changes in TMs of a 7TMR, including site-directed spin labeling studies, fluorescence spectroscopic analysis, and *in situ* disulfide crosslinking.

2.2.1. Site-Direct Spin Labeling and Electron Paramagnetic Resonance Spectroscopy

The general approach has been to individually substitute residues at the cytoplasmic ends of TMs of rhodopsin with cysteines. These cysteines allow covalent incorporation of nitroxide spin labels, of which unpaired electrons can be probed with electron paramagnetic resonance (EPR) spectroscopy *(24)*. These site-directed spin labels allow determination of the environment of a side-chain (aqueous, hydrophobic, buried within membranes) as well as measurement of approximate distances between pairs of spin labels in proteins of less than 25 Å *(24)*.

EPR spectral changes suggest that upon light exposure, rhodopsin activation causes a rigid body tilt or rotation of TM6, moving its cytoplasmic end out from the bundle *(25)* with a rotation of TM6 about its axis (clockwise as viewed from the cytoplasm; Fig. 2), simultaneously increasing the exposure of the cytoplasmic end of TM3 because of a rigid body movement relative to the other TMs (ref. *26*; Fig. 2) and decreasing the exposure of some positions near the end of TM5 *(25)*. Additionally, the distances between V3.54(139)C and K6.31(248)C, T6.34(251)C, or R6.35(252)C at the cytoplasmic ends of TM3 and TM6, respectively, were demonstrated to be increased *(20)*. The data were consistent with the separation of the cytoplasmic ends of TM3 and TM6 and rigid body motion of TM6 relative to TM3 (Fig. 2). The movement of TM3 was interpreted to be relatively small *(20)*.

2.2.2. Fluorescence Spectroscopic Analysis

The use of fluorescence spectroscopic analysis for β_2-AR allowed the first direct structural analysis of conformational changes in a diffusible ligand-activated 7TMR *(27,28*; reviewed in refs. *18* and *29)*. The approach is to label the cysteine(s) with a fluorescent probe that is sensitive to the polarity of the local environment. Gether et al. *(28)* demonstrated that agonists induced a decrease in fluorescence, whereas antagonists caused an increase, and there was a linear correlation between biological efficacy and the change in fluorescence. Subsequently, they mutated all but one, two, or three of the cysteine residues in the β_2-AR and showed that agonist binding caused conformational changes around Cys3.44(125) in TM3 and around Cys6.47(285) in TM6; this was explained by either a clockwise rotation of TM6 (when viewed from the intracellular side; Fig. 2) and/or a tilting of TM6 toward TM5 *(27)*. Therefore, these conformational changes are similar to those in rhodopsin, indicating a shared mechanism of 7TMR activation. However, the relatively slow kinetics of the conformational changes in the β_2-AR is a notable difference.

2.2.3. In Situ Disulfide Crosslinking

"*In situ* disulfide cross-linking," a recent application used by Wess and colleagues *(30)* regarding the disulfide crosslinking strategy to the m3 muscarinic receptor,, allowed examination of agonist-dependent dynamic structural changes of a 7TMR present in its native membrane-bound environment. Mutant receptors are generated that contain the Y5.62(254)C mutation and a second cysteine substitution within the segment K6.29(484)–S6.38(493) at the intracellular ends of TM5 and TM6, respectively. Formation of disulfide bonds during receptor activation was observed between Y5.62(254)C and each residue within the A6.34(489)C–L6.37(492)C segment, indicating that the cytoplasmic ends of TM5 and TM6 move closer to each other, which appears to involve a major change in secondary structure at the cytoplasmic end of TM6 *(30)*.

3. 7TMR ACTIVATION INVOLVES DISRUPTION OF STABILIZING INTRAMOLECULAR INTERACTIONS

Section 2 demonstrated that the primary conformational changes during 7TMR activation were TM movements—especially a rotation and/or a tilting of TM6—and increased distance between the cytoplasmic ends of TM3 and TM6. Questions regarding the molecular events during/leading to movements of the TMs and the mechanisms underlying the conformational changes are intriguing. To provide precise molecular details, one would need X-ray crystallography data of activated forms of rhodopsin, which are not yet available. Currently available rhodopsin X-ray crystallography data are of an inactive state (refs. *3* and *4*; Fig. 1). Mutagenesis studies, combined with molecular modeling on CAMs, have been very useful to elucidate activated states of 7TMR and the molecular mechanisms of 7TMR activation *(31,32)*, which are discussed in this section.

3.1. CAMs Mimic the Active Conformations

According to the various models of 7TMR function *(18,33–35)*, receptors exist in conformational equilibriums between inactive states that are structurally constrained and unable to couple to G_a-proteins and active states that can interact productively with guanosine triphosphate (GTP)-bound G_a-subunits and can therefore activate several downstream intracellular pathways. Agonists have higher affinity to active states.

CAMs are 7TMR mutants that exhibit agonist-independent activities. CAMs have higher affinities for agonists than the wild-type receptors *(36)*, unless the mutations themselves attenuat the binding of agonists (for an example, *see* ref. *37*). Unlike those of the wild-type receptors, the affinities of CAMs for agonists were unaffected by uncoupling of the receptors from G proteins by guanine nucleotides, indicating that they are intrinsic properties of CAMs. Therefore, CAMs mimic (at least to some extents) agonist-induced activated conformations of the wild-type receptor and spontaneously adopt structural states that are able to activate G proteins *(32)*.

Agonist-independent activation of a 7TMR was first demonstrated by Kjelsberg et al. *(36)*, who showed that mutation of A6.34(293) in the α_{1B}-AR to any one of the 19 other amino acids resulted in agonist-independent activation. They proposed that there were intramolecular interactions constraining the receptor preferentially in inactive conformations in the absence of an agonist. Subsequently, CAMs of many 7TMRs have been generated, suggesting that the intramolecular constraints have been conserved during evolution to maintain receptors in inactive states. Mutations resulting in the loss of such intramolecular interactions cause receptor activation similar to agonist-induced activation. Therefore, the positions and nature of the mutations provide clues to the differences between inactive and active conformations. It has been demonstrated that CAMs of the β_2-AR *(38,39)*, α_{1B}-AR *(40)*, histamine H_2 receptor *(41)*, and µ-opioid receptor *(42,43)* are less structurally stable. Indeed, expression of some of the CAMs could be detected only after cells were grown in the presence of a ligand *(41–44)*. These results are consistent with the notion that the mutations have disrupted critical stabilizing intramolecular interactions.

3.2. R3.50 Is Constrained by Intra- and Interhelical Interactions

3.2.1. Interactions Between the D/ER3.50Y Motif in TM3 and the $X_16.30BBX_2X_36.34B$ Motif in TM6

Hydrogen bonds, ionic interactions, and Van-der-Waals contacts link the TM helices in rhodopsin structure, stabilizing the ground (inactive)-state structure *(3,4)*. The D/ERY motif at the interface of the TM3 and second intracellular loop is highly conserved in the rhodopsin family of 7TMRs (refs. *9* and *45*; Fig. 3A) and is important in receptor activation *(46,47)*. R3.50(135) in the highly conserved D/E R3.50Y motif forms a salt-bridge with the neighboring E3.49(134) and participates in two interhelical associations with E6.30(247) and T6.34(251) at the cytoplasmic end of TM6 (ref. *3*; Fig. 4A). E6.30(247) and T6.34(251) are the X_1 and X_3 residues in the conserved $X_16.30BBX_2X_36.34B$ motif (Fig. 3B), respectively, where B represents basic amino acid and X represents any other amino acid (Fig. 3B). It has been suggested that the three interactions are critical to keep rhodopsin in the inactive conformation *(3)*.

Although not as common as the R3.50–D/E3.49 interaction, the two interhelical interactions connecting TM3 and TM6, R3.50–6.30 and R3.50–6.34, were also found in some rhodopsin-like 7TMRs. Studies regarding CAMs and molecular modeling have indicated that these interactions are critical to stabilize the inactive conformations of the receptors (Fig. 4A).

3.2.2. Protonation of D/E3.49 in "D/ERY" May Be a Common Initial Event in 7TMR Activation

Experimental evidence indicates that protonation of the D/E3.49 in the highly conserved D/ERY motif is one of the initial key events in rhodopsin activation. Activation of rhodopsin by light was shown to induce uptake of two protons from the aqueous environment *(48,49)*. Parkes and Liebman *(50)* reported that the rate of the light-induced conversion between metarhodopsin I and metarhodopsin II was increased by decreasing pH from 7.7 to 6.1. Arnis et al. *(51)* found that the constitutively active E3.49(134)Q mutant displayed a loss of light-induced uptake of two protons from the

A TM3: E/DR $^{3.50}$Y motifs at its cytoplasmic end

```
              3.32                    3.49
               ▲                       ▲
ILCKIVISIDYYNMFTSIFTLCTMSV  D¹⁶⁴RY  IAV     rat μ opioid
LLCKAVLSIDYYNMFTSIFTLTMMSV  D¹⁴⁵RY  IAV     human δ opioid
VLCKIVISIDYYNMFTSIFTLTMMSV  D¹⁵⁵RY  IAV     human κ opioid
FWCEFWTSIDVLCVTASIETLCVIAV  D¹³⁰RY  FAI     β₂-AR
IFCDIWAAVDVLCCTASILSLCAISI  D¹⁴²RY  IGV     α₁ʙ-AR
TGCNLEGFFATLGGEIALWSLVVLAI  E¹³⁴RY  VVV     rhodopsin
```

B $X_1^{6.30}BBX_2X_3B$ motifs at cytoplasmic ends of TM6

```
        6.30      6.34
         ▲         ▲
KDRN  L²⁷⁵RRITR  MVL     rat μ opioid
KDRS  L²⁵⁶RRITR  MVL     human δ opioid
KDRN  L²⁶⁹RRITR  LVL     human κ opioid
TQKA  E²⁴⁷KEVTR  MVI     rhodopsin
KFSR  E²⁸⁹KKAAK  TLG     α₁ʙ-AR
RQNR  E³⁶⁹KRFTF  VLA     α₂A-AR
VALR  E³¹⁸QKALK  TLG     β₁-AR
FCLK  E²⁶⁸HKALK  TLG     β₂-AR
SISN  E³¹⁸QKACK  VLG     5-HT₂A
AINN  E³⁰⁸KKASK  VLG     5-HT₂C
MAAR  E³⁰⁹RKATK  TLG     5-HT₁B
SLVK  E³⁶⁰KKAAR  TLS     m₁ muscarinic
ATNK  D⁵⁶⁴TKIAK  KMA     lutropin/choriogonadotropin
```

Fig. 3. Amino acid sequence of TM3 (**A**) and that at the cytoplasmic ends TM6 (**B**) of the rat m-opioid receptor compared to those of several other 7TMRs. The "D/ERY" and "X₁BBX₂X₃B" motifs and variants are highlighted (B is a basic amino acid, and X is a nonbasic amino acid). D3.32 of each receptor and E3.28 of rhodopsin are in bold. The numbering indicates the amino acid numbers in the sequences of the receptors.

aqueous phase, compared with the wild-type rhodopsin. Thus, rhodopsin activation requires light-induced conformational changes that allow protonation of E3.49(134) and another residue during activation. Fahmy et al. *(52)* deduced that E3.28(113) in metarhodopsin II was protonated. The conformational changes that are likely to ensue from this change in protonation state should allow R3.50 to be repositioned to support the interaction with G proteins *(46,51)*.

Indirect evidence supports the notion that protonation is important for activation of other receptors. For the β_2-AR, lowering of pH from 8.0 to 6.5 has been demonstrated to enhance agonist-independent activity and facilitate the transition of the receptor to the activated state *(53)*. Scheer et al. *(54)* mutated D3.49(142) in the a_{1B}-AR to all possible natural amino acids and found that the level of constitutive activity correlated positively with the hydrophobicity of the residue. Because the hydrophilicity/hydrophobicity can be regulated by deprotonation/protonation of this residue, these results support the notion that protonation is an important modulator of the transition between the inactive and active states of the a_{1B}-AR. As described later, mutations of D/E3.49 residues in several 7TMRs with charge-neutralizing amino acids (which mimic the protonated state of the D/E3.49) were shown to lead to constitutive activation of the receptors. Because the D/ERY motif is almost invariably conserved in rhodopsin-like 7TMRs *(8,10)* (Fig. 3A), protonation of Asp/Glu3.49 is likely to be a common molecular event during activation of 7TMRs *(19,55)*.

3.2.3. Loss of Interactions With R3.50 Results in Constitutive Activation of 7TMRs

3.2.3.1. MUTATION OF D/E3.49 TO NEUTRAL AMINO ACIDS

The E3.49(134)Q mutant of rhodopsin was found to have enhanced constitutitive activity and to adopt a photoactivated conformation in the dark state *(56,57)*. Subsequently, mutation of D/E3.49 to a neutral amino acid has been shown to result in constitutive activation of several other 7TMRs, including μ-opioid receptor *(43)*, α_{1B}-AR *(54,58)*, β_2-AR (39), V_2 vasopressin receptor *(59)*, chemokine $CXCR_2$ receptor *(60)*, and gonadotropin-releasing hormone (GnRH) *(46)* receptor. Interestingly, the Kaposi's sarcoma-associated herpes virus 7TMR has "VRY," rather than "D/ERY," and thus does not possess and negatively charged-side chains at 3.49 and displays high levels of basal activity *(61,62)*. It is noteworthy that in the histamine H_2 receptor, of which the wild-type has high constitutive activity, mutation of D3.49(115) caused a further increase in basal signaling activity *(41)*.

In contrast, the E3.49(134)D mutation in rhodopsin *(56)* and the D3.49(164)E substitution in the μ-opioid receptor *(43)* reduced basal activity. Additionally, D3.49(142)E substitution in the α_{1B}-AR did not affect basal activity *(54)*. These results indicate that the carboxylate group of D3.49(164) is important for stabilizing the inactive state.

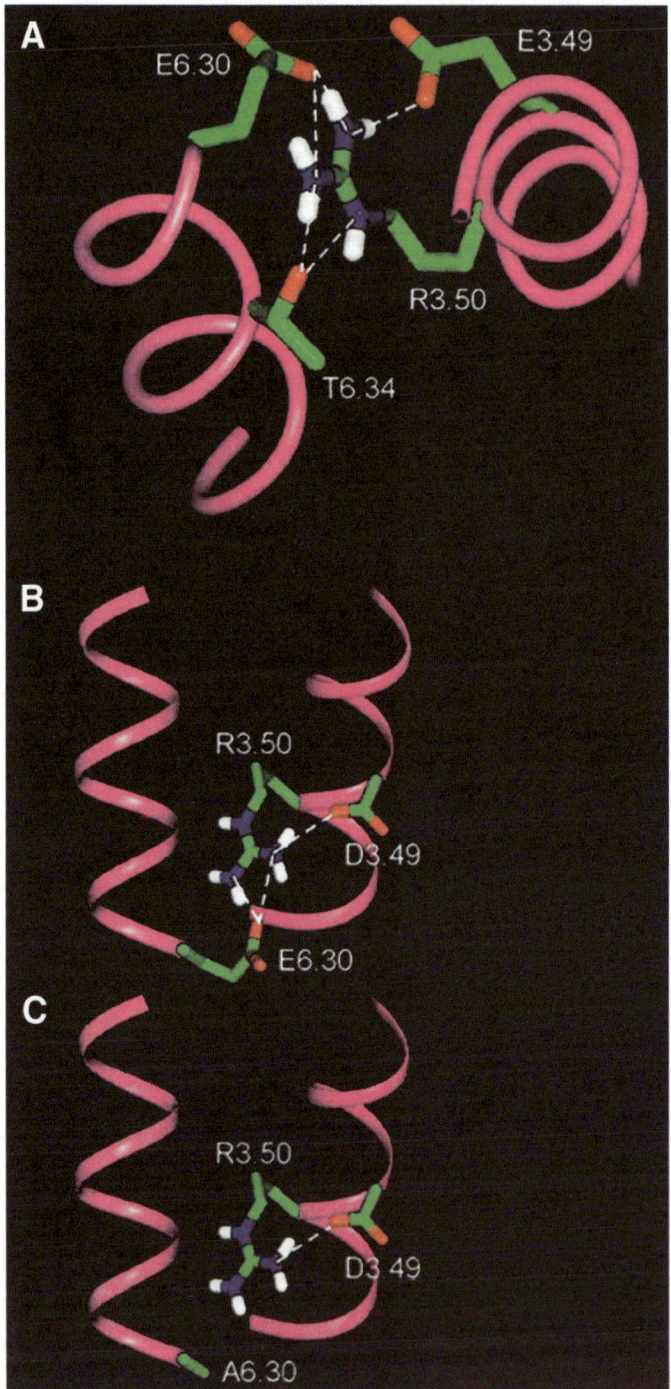

Fig. 4

Based on mutagenesis studies and molecular dynamics simulations on the GnRH receptor and β_2-AR, Ballesteros et al. *(46,63)* suggested that the conserved R3.50 of the DRY motif is constrained by an ionic interaction with the preceding D3.49 in inactive forms. The crystal structure of inactive rhodopsin suggests that the carboxylate group of E3.49(134) forms a salt-bridge with the guanidium group of R3.50(135) (ref. *3*; Fig. 4A). Therefore, it is inferred that disruption of the interaction between D/E3.49 and R3.50 would lead to activation of the receptors, mimicking protonated states of D/E3.49.

However, charge-neutralizing mutations at D/E3.49 in some 7TMRs, including those in the m1 muscarinic *(64)* and luteinizing hormone/chorionic gonadotropin *(65)* receptors, did not lead to enhanced constitutive activity. The mutation in the GnRH receptor did *(46)* or did not *(66)* lead to constitutive activation. The mutations reduced cell-surface expression of the receptors, suggesting that D/E3.49 is important for proper folding and, hence, the stability of the receptor proteins.

3.2.3.2. MUTATIONS AT 6.34(X_3) SITES

It is noteworthy that the residues at the 6.34 site within the $X_1$6.30BB$X_2X_3$6.34B motif among rhodopsin-like 7TMRs vary markedly (Fig. 3B). Therefore, a direct interaction between R3.50 and the 6.34 site may not always be present in the receptors as it is in rhodopsin. However, mutations at the 6.34(X_3) site or its variants at the junction of the third intracellular (i3) loop and TM6 have been shown to lead to constitutive activation of several 7TMRs, including A293 in α_{1B}-AR (36), T373 in α_{2A}-AR *(67)*, L322 in β_1-AR *(68)*, C322 in the 5-HT$_{2A}$ receptor *(69)*, S312 in the 5-

---

Fig. 4. *(From opposite page)* Activation of the b$_2$-AR involves disruption of an ionic lock between TM3 and TM6. The Ca traces are taken from the high resolution structure of bovine rhodopsin. Except in A, the top of each panel shows the extracellular end, and the bottom of each panel shows the intracellular end of the TMs. (A) An extracellular view of the high resolution structure of rhodopsin showing the interaction between residues at the cytoplasmic ends of TM3 and TM6. (B) A side view of the interactions between E6.30 and R3.50 as well as between R3.50 and D3.49, which are within the distance range of an ionic interaction shown with dashed lines, in a rhodopsin-based model of the b$_2$-AR. (C) After the E6.30A mutation, the ionic interaction of E6.30 and R3.50 is abolished. (Modified from Fig. 5 of Ballesteros et al., ref. *63*.)

HT$_{2C}$ receptor *(70)*, and T313 in the 5-HT$_{1B}$ receptor *(71)* (*see* Fig. 3B). In the α$_{1B}$-AR, all 19 possible amino acid substitutions at the 6.34(X$_3$) locus (A293) led to varying levels of constitutive activities, with the A6.34(293)K mutant demonstrating the highest activity *(36)*.

The effect of a mutation at the 6.34 locus can be dramatically different depending on the nature of the substitution *(44)*. For the µ-opioid receptor (Fig. 5), T6.34(279)K mutant dramatically enhanced agonist-independent activity, whereas T6.34(279)D mutation did not, although it almost abolished the G protein signaling *(44)*. The results were interpreted in the structural context of a rhodopsin-based model for the µ-opioid receptor. The interaction of T6.34(279) with R3.50(165) through a hydrogen bond in the µ-opioid receptor stabilizes the inactive conformations (Fig. 6A). The T6.34(279)K substitution disrupts this interaction because of charge repulsion and supports agonist-free activation (Fig. 6B), whereas T6.34(279)D mutation strengthens this interaction by forming an even stronger ionic bond that keeps the receptor in inactive states (ref. *44*; Fig. 6C).

However, introducing an acidic residue in the 6.34(X$_3$) locus does not always lead to inactive receptors. In contrast to the µ-opioid receptor, substitutions of the 6.34(X$_3$) locus with D or E caused agonist-independent activation of several 7TMRs, including A293D/E in the α$_{1B}$-AR *(36)*, T373E in the α$_{2A}$-AR *(67)*, L322E in the β$_1$-AR *(68)* and C322E in the 5-HT$_{2A}$ receptor *(69)*. The differences between those receptors and the µ-opioid receptor have been further studied and have been demonstrated to involve a mutation at the 6.30(X$_1$) locus as shown in Subheading 3.2.3.3.

3.2.3.3. NEUTRALIZING MUTATIONS AT ASP/GLU6.30(X$_1$) SITES

Ballesteros et al. (2001) *(63)* observed that charge-neutralizing mutations (alanine substitution) of E6.30(268)(X$_1$) alone or combined with that of Asp3.49(130) led to agonist-independent activities of the β$_2$-AR, suggesting that the ionic interactions of E6.30(X$_1$) with R3.50 in the inactive state (Fig. 4B) were disrupted by the mutations at these sites (Fig. 4C) *(63)*. Although D/E6.30 is not universally conserved among the rhodopsin-like receptors, it is nearly 100% conserved among the monoamine and glycoprotein hormone receptors and opsins *(63)*. Consistent with the observations regarding the β$_2$-AR, the E6.30(360)A mutant of the m1 muscarinic receptor displays high agonist-independent activity *(72)*. Based on computational modeling, the E6.30–R3.50 salt-bridge was proposed to occur in the 5-HT$_{2A}$ receptor *(5,73,74)* and the α$_{1B}$-AR *(75)*; this was supported by studies showing that mutations weakening this interaction led to constitutive activation. Additionally, naturally occurring D6.30(564)G mutation in the lutropin/choriogonadotropin receptor *(76,77)* and D6.30(619)G mutation in the thy-

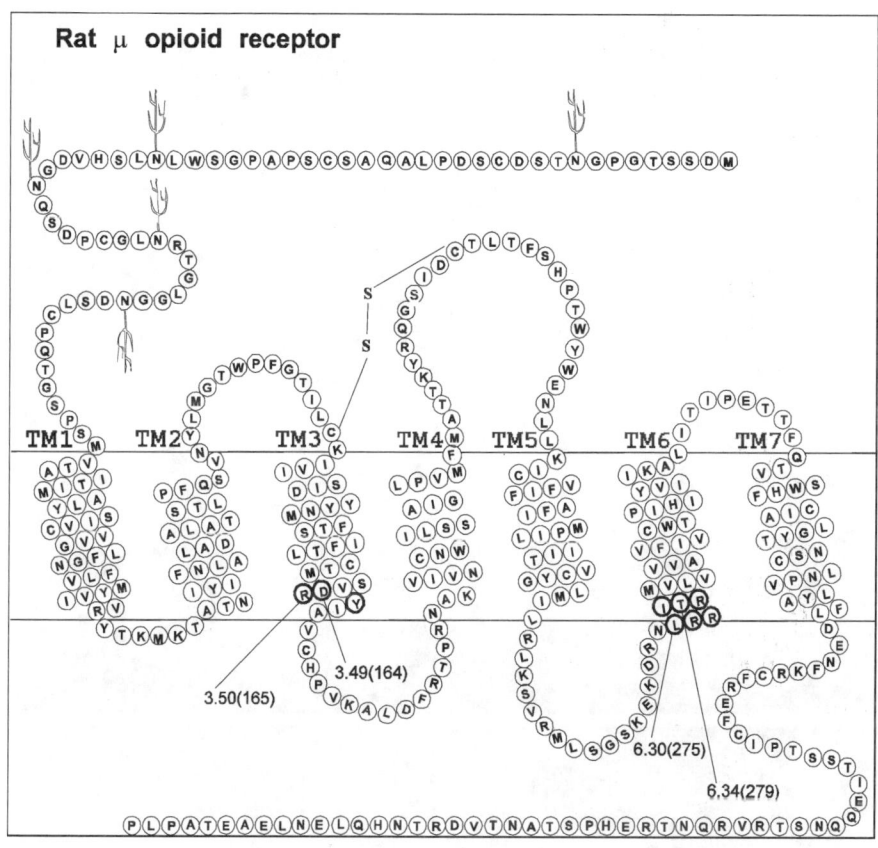

Fig. 5. Helical net representation of the rat m-opioid receptor. The D/ERY motif and $X_1BBX_2X_3B$ motif are highlighted with dark circles at the cytoplasmic ends of TM3 and TM6, respectively.

rotropin receptor *(78)* leads to constitutive receptor activation, which causes gonadotropin-independent familial male-limited precocious puberty and hyperfunctioning thyroid adenomas, respectively. The activation mechanism of the D6.30(564) mutant of the lutropin/choriogonadotropin receptor has been suggested to result from the loss of R3.50(464)–D6.30(564) ionic interaction *(79)*. The finding that substitution of D6.30(564) with Glu did not lead to constitutive activation *(77,80)* is consistent with the interpretation. Therefore, the D/E6.30(X_1)–R3.50 interaction may constitute a constraint of inactive states for many 7TMRs, which is removed in the process of receptor activation *(73–75,79)*.

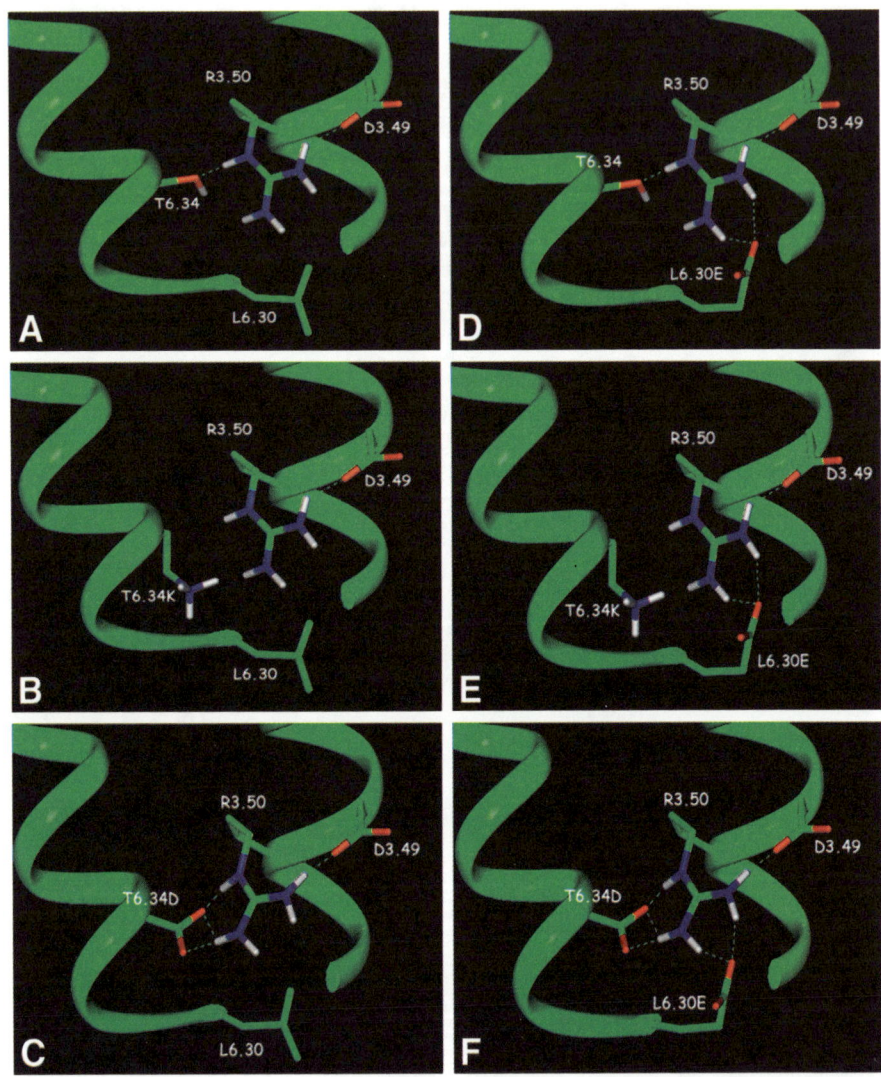

Fig. 6. Molecular three-dimensional representations of the interactions between R3.50(165) of TM3 and residues at 6.30(275) and 6.34(279) of TM6 at their cytoplasmic ends of the wild-type or mutant m-opioid receptors. (**A**) The wild-type: The T6.34(279)–R3.50(165) and D3.50(164)–R3.50(165) interactions, (shown here in a rhodopsin-based model of the m-opioid receptor) enhance the hydrogen bond network around R3.50(165), stabilizing the inactive state of the receptor. (**B**) The T6.34(279)K mutant: Because the introduced K has the same charge as that of R3.50(165), the electrostatic repulsion leads to the separation of TM3 and TM6 and enhanced constitutive activity of the receptor. (**C**) The T6.34(279)D mutant: An ionic interaction formed between R3.50(165) and D6.34(279) is stronger than the

It is noteworthy that the opioid receptors have a nonpolar residue, L6.30, at the 6.30(X_1) site (Figs. 3B and 5). Thus, the 6.30(X_1)–R3.50 ionic interaction does not exist in the opioid receptors *(81)* (Fig. 6A). The L6.30(275)E mutation inactivated the µ-opioid receptor *(81)*. Additionally, L6.30(275)E–T6.34(279)D mutants displayed no constitutive activity and could not be activated by agonists, whereas the L6.30(275)E/T6.34(279)K mutant had some constitutive activity, but much less than the T6.34(279)K CAM. Interpreted in the context of interactions with the conserved R3.50(165), when L6.30(275) is mutated to E, the favorable E6.30(275)–R3.50(165) interaction stabilizes an inactive state (as in rhodopsin) and, hence, inactivates the receptor (Fig. 6D) and reduces the activities of the T6.34(279) mutants (Fig. 6E,F). Therefore, the µ-opioid receptor is different from rhodopsin and monoamine 7TMRs, of which the wild-type receptors with the native E6.30 (Fig. 3B) can be activated and the 6.34D or 6.34K mutants display enhanced constitutive activities. Although the interaction between the cytoplasmic ends of TM3 and TM6 is conserved among receptors, there are some specific differences in the molecular mechanisms of receptor activation. It is interesting to speculate that compared with rhodopsin and receptors of monoamines, the µ-opioid receptor may have other intramolecular interactions or a specific local environment at TM6 to keep it in inactive states,

Fig. 6. *(From opposite page)* original hydrogen bond between R3.50 and T6.34 in the wild-type, keeping the receptor in inactive conformations. (D) The L6.30(275)E mutant: The L6.30(275)E mutation creates an additional E6.30–R3.50 ionic bond and reinforces the interaction between TM3 and TM6 supported by the T6.34(279)–R3.50(165) hydrogen bond to restrain the m-opioid receptor in inactive conformations. (E) The L6.30(275)E/T6.34(279)K mutant: The ionic interaction between the glutamate introduced at 6.30(275) and R3.50(165) reduces the degree of repulsion introduced by the T6.34(279)K mutation. (F) The L6.30(275)E–T6.34(279)D mutant: The ionic lock on R3.50(165) produced by the L6.30(275)E mutation is further strengthened by the introduction of an aspartate at 6.34(279), leading to enhanced stabilization of the inactive form of the receptor. (Modified from Fig. 6 of Huang et al., ref. *81*, with permission of copyright 2002 American Chemical Society.)

which compensates its loss of the 6.30(X_1)–R3.50 ionic constraint *(81)* and is likely to account for the inactivating effects of the L6.30(275)E, T6.34(279)D, and L6.30(275)E–T6.34(279)D mutations. The unique structural attributes of the μ-opioid receptor remain to be identified.

3.2.3.4. D/E6.30(X_1) IN TM6: THE SECOND PROTONATION SITE?

Rhodopsin activation is accompanied by the uptake of two protons at cytoplasmic sites *(51)*. As mentioned in Subheading 3.2.2., one site of proton uptake is E3.49(134) *(51)*, which, in the inactive state, forms an ionic bond with R3.50 in the D/ERY motif. The second protonated residue in metarhodopsin II has not been identified but was inferred to be E3.28(113) *(52)*. However, in most other 7TMRs, the corresponding 3.28 position is not an acidic residue. Lowering pH from 8.0 to 6.5 enhanced the constitutive activity of the $β_2$-AR and facilitated its transition to activated states, but mutation of D3.49(130) did not abrogate the pH sensitivity *(53)*, suggesting that there is/are other possible protonation site(s). Based on the interaction of R3.50 with Glu6.30 in rhodopsin and $β_2$-AR, Ballesteros et al. *(63)* suggested that E6.30 of the BBXXB motif may represent the second site of protonation. Thus, protonation of these two acidic residues (D/E3.49 and D/E6.30) would disrupt their ionic interactions with R3.50, thereby breaking the ionic lock holding TM3 and TM6 together in the inactive state *(63)*.

3.2.3.5. MUTATIONS AT ARG3.50 SITES

Site-directed mutagenesis studies on rhodopsin *(82)*, V_2 vasopressin *(83)*, m1 muscarinic *(84)*, GnRH *(44,66)*, CB_2 cannabinoid *(85)*, and histamine H_2 *(41)* receptors as well as $α_{1B}$-ARs (86) showed that replacement of the conserved R in the D/ERY motif by various amino acids attenuated or abolished G protein coupling, a finding consistent with Arg playing an important role in receptor activation. Interestingly, all the R3.50 mutants of the $α_{1B}$-AR *(86)* and histamine H_2 receptor *(41)* displayed increased binding affinity for agonists, which is a characteristic of CAMs *(see* Subheading 3.1.). It appears that the R3.50 mutant of the histamine H_2 receptor can adopt an active conformation but has a decreased ability to couple to or activate the G proteins *(41)*; this may result from the loss of the conserved R3.50 *(87)*.

The nature of replacing the amino acid residue, not just the loss of the conserved R3.50, significantly contributes to the functional properties of the resulting mutants. For example, mutation of R3.50(143) in the $α_{1B}$-AR resulted in mutants with a small increase in constitutive activity (H, D) and impairment (H, D) or complete loss of agonist-mediated response (E, A, N, I) *(86)*. Additionally, the same substitution in different 7TMRs differentially affected receptor function. A charge-conserving R3.50(123)K substitution in the m1 muscarinic receptor results in modest impairment of receptor function *(84)*, but the R3.50(143)K mutation in the α_{1B}-AR conferred a high degree of constitutive activity and preserved the maximal agonist-induced response *(86)*. Therefore, R3.50 of the D/ERY motif is required for efficient signal transduction for most of the rhodopsin-like 7TMRs and has been generally implicated as a central trigger of G protein activation. Ballesteros et al. found that based on molecular simulation and mutagenesis studies on the GnRH receptor, after the D3. 49(138)–R3.50(139) ionic bond was broken during activation, the unrestrained Arg was prevented from orienting itself toward the water phase by a steric clash with I3.54(143), which is necessary for receptor signaling *(46)*.

However, in the gonadotropin receptor, R3.50(464)A mutation did not disturb receptor signaling *(80)*. Additionally, R3.50(137)A mutation in the oxytocin receptor caused constitutive activation *(88)*. Thus, R3.50 may not be regarded as the indispensable molecular switch for G protein activation for all rhodopsin-like 7TMRs *(80)*.

3.2.4. Two Hypotheses: "Arginine Cage" vs "Polar Pocket"

On the basis of experimental results, molecular modeling, and computational simulations, two distinct hypotheses were proposed to define the specific role of D/E3.49 protonation and R3.50 in receptor activation. The earlier hypothesis was the "polar pocket" hypothesis, which was proposed by Scheer et al based on mutagenesis studies and simulations of the α_{1B}-AR *(54,58)*. The alternative hypothesis was the "arginine cage" hypothesis, which was proposed by Ballesteros et al. based on studies on the GnRH receptor *(46)*.

The "polar pocket" hypothesis predicted that in the inactive state of the receptor, the highly conserved R3.50 in the D/ERY motif is constrained in a pocket formed by highly conserved polar residues in TM1, -2, and -7—particularly N1.50, D2.50, and N7.49. Protonation (or charge-neutralizing mutation) of the adjacent D3.49 resulted in long-range conformational changes in the receptor molecule caused by R3.50 shifting out of the polar pocket. The highlight of this hypothesis is that in the inactive state, the specific ionic counterpart of the R3.50 is the conserved D2.50 in TM2, and this ionic interaction is broken after receptor activation.

In contrast, in the "arginine cage" hypothesis, Ballesteros et al. *(46)* proposed that in the inactive state, R3.50 interacted with the adjacent D3.49 by ionic interaction within the D/ERY motif. During receptor activation, D3.49 becomes protonated, the R3.50–D3.49 ionic bond is broken, and R3.50 forms another ionic interaction with D2.50. Thus, an ionic interaction between R3.50 and D2.50 was associated with the active state, rather than with the inactive state, as proposed in the "polar pocket" hypothesis.

The high-resolution structure of rhodopsin has revealed that R3.50(135) interacts with E3.49(134), E6.30(X_1)(247), and T6.34(X_3)(251) *(3)*, which supports the "arginine cage" hypothesis. As discussed by Ballesteros et al. *(63)*, however, in rhodopsin, the distance between R3.50(135) and D2.50(83) is approx 20 Å, which is too great a distance for an interaction between these residues in the inactive state (as proposed by the "polar pocket" hypothesis). Additionally, it is not likely that R3.50 and D2.50 interact in the activated state suggested in the "arginine cage" hypothesis, because it would require a large conformational change for the two residues to interact *(63)*.

3.3. Constraining Intramolecular Interactions Beyond the "D/ER3.50Y" Motif

In addition to the intramolecular interactions centering around R3.50 at the cytoplasmic ends of TM3 and TM6, there are other interactions important in constraining the structure of 7TMRs in inactive states. The following examples provide evidence for these interactions; disruption of these interactions leads to receptor activation. However, it is uncertain whether the results of these studies on different 7TMRs can be generalized to all rhodopsin-like 7TMRs. Some of the interactions are likely to be receptor-specific.

3.3.1. The "Aromatic Cluster" in TM6

An "aromatic cluster" formed by the aromatic residues in TM6 (F6.44, W6.48, F6.51, and F6.52) in the proximity of the highly conserved P6.50 kink are present in many 7TMRs. The structural constraints related to this aromatic cluster appear to exist in many receptors, because mutations of residues within or in the vicinity of the cluster result in alterations in receptor activity and some become constitutively active. Visiers et al. *(5)* proposed that for the 5-HT$_{2A}$ receptor, one or more of the aromatic residues are involved in agonist binding, which relieves the constraints and initiates conformational rearrangements of aromatic side-chains. Such rearrangements trigger changes along TM6 that involve P6.50 and propagate to the cytoplasmic end of TM6 and i3. The aromatic clusters are like a "toggle switch," transmitting the signal from agonist binding for receptor activation *(5)*.

F6.51(310), but not F6.52(311), of the α_{1B}-AR is involved in binding of the catechol ring. Chen et al. *(89)* postulated that activation of α_{1B}-AR involves initial disruption of the D3.32(125)–K7.36(331) ionic interaction (discussed later), followed by F6.51(310)-catechol ring-induced movement of TM6. The later interaction is essential for conformational changes from partially activated state to the fully activated state *(89)*. The F6.44(303)L mutant of the α_{1B}-AR showed markedly increased basal activity and increased potency and efficacy of agonist-stimulated signaling *(40)*. The F6.44(282) mutants of the β_2-AR display varying degrees of constitutive activity [F6.44(282)L > F6.44(282)A > F6.44(282)G] *(90)*. These results are consistent with the notion that the F6.51 residue of ARs is involved in agonist binding and the F6.44 residue is part of the toggle switch for receptor activation.

Shi et al. *(91)* reported that C6.47(285) in the β_2-AR modulates the configuration of the "aromatic cluster" and the TM6 Pro-kink through specific interactions in its different rotamer configurations; this further supports the association of the "rotamer toggle switch" with receptor activation *(91)*. These researchers proposed that in the monoamine 7TMRs, agonist binding to a cluster of aromatic residues surrounding the Pro-kink in TM6 promotes receptor activation by altering the configuration of the TM6 Pro-kink and by the subsequent movement of the cytoplasmic end of TM6 away from TM3 *(91)*.

3.3.2. Other Interhelical Interactions Between TM3 and TM6

Additional interactions between residues in TM3 and TM6 have been found to be involved in activation of 7TMRs. W6.48(256)F and Q mutants of the human B_2 bradykinin receptor display enhanced constitutive activity *(92)*. The W6.48(256)H–N3.35(113)H double-mutant exhibited higher binding affinity for Zn^{2+}, indicating their proximity. It was hypothesized that W6.48(256) interacted with N3.35(113) and stabilized the inactive state, and W6.48(256) controlled the balance between active and inactive states *(92)*. In the δ-opioid receptor, mutations of Y3.33(129)F or A in TM3 enhanced the receptor basal activity *(37)*. Based on molecular modeling, it was proposed that Y3.33(129) in TM3 forms a hydrogen bond with H6.52(278) in TM6 as part of the interhelical interactions that keep the receptor in inactive states *(37)*. This interaction may be specific for opioid receptors. It is interesting to note that for leutenizing hormone receptor, an interaction between TM3 and TM6 may be associated with active states of the receptor *(93)*. L3.43(457) is a highly conserved residue in TM3, and constitutive activation was associated with a positively charged amino acid (R, K, or H) at the L3.43(457) site. Molecular modeling showed that R, K, or H at 3.43(457) interacts electrostatically with D6.44(578) in TM6, perturbing the interhelical interactions between TM3, -6, and -7 and giving rise to the constitutive activation of the receptor *(93)*.

3.3.3. TM7: "NPXXY" Motif and Others

3.3.3.1. Y7.53-F–Y7.60 INTERACTION

Recent studies have revealed that there is an interaction between Y7.53 in the NPXXYX$_{5-6}$F/Y motif and an aromatic residue (F or Y) at the 7.60 site in H8 in the inactive states of rhodopsin *(94)* and the 5HT$_{2C}$ receptor *(95)*. This TM7–H8 interaction appears to be highly conserved in 7TMRs.

In rhodopsin, H8 is anchored to the membranes by palmitoylated residues C322 and C323, and this region is important for the interaction of light-activated rhodopsin with transducin. In rhodopsin, Ala replacement mutations within the NPXXYX$_{5-6}$F/Y domain, which eliminate an interaction between Y7.53(306) and F 7.60(313), allow the formation of metarhodopsin II. A disulfide bond linking Y7.53(306)C and F 7.60(313)C prevented metarhodopsin II formation. These observations suggest that the interaction between Y7.53(306) and F 7.60(313) is disrupted during the metarhodopsin I–metarhodopsin II transition. Therefore, NPXXY and D/ERY motifs have been proposed to provide a dual control of the activating structural changes *(94)*.

In the 5HT$_{2C}$ receptor, substitution of Y7.53(368) of the NPXXY motif with all naturally occurring amino acids resulted in three levels of constitutive activities: moderate, high, and "locked-on." Unlike the high basal activity of the other receptor mutants, the high basal activity of the locked-on Y7.53(368)N mutant was neither increased by agonists nor decreased by inverse agonists. The Y7.53(368)F mutant was uncoupled. Computational modeling suggested that Y7.53(368) interacts with the aromatic ring of Y7.60(375) in H8. These results suggest that the interaction between Y7.53 and Y7.60 connecting TM7 and H8 influences the switching of the 5HT$_{2C}$ receptor among multiple active and inactive conformations *(95)*.

3.3.3.2. 3.32–7.43 INTERACTIONS OR EQUIVALENTS

Mutation of D3.32(128) or Y7.43(308) in the δ-opioid receptor, of D3.32(125) in α$_{1B}$-AR, or of E3.28(113) (one helical turn extracellular to the 3.32 locus) in rhodopsin led to constitutive activation of the receptors *(37,96,97)*. It was hypothesized that an interaction exists that connects TM3 and TM7 between E3.28(113) and K7.43(296) in rhodopsin receptors *(97)*, D3.32(125) and K7.36(331) in the α$_{1B}$-ARs *(96)*, and D3.32(128) and Y7.43(308) in the δ-opioid receptors *(37)*, constraining the receptors in inactive states. Mutations that disrupt the interaction lead to agonist-independent activation. However, this interaction is not universal for rhodopsin-like 7TMRs. Mutation of D3.32(147) in the μ-opioid receptor or D3.32(155) in the 5-HT$_{2A}$ serotonin receptor did not result in constitutive activation *(43,98)*. For the C5a receptor, researchers have been demonstrated that

Conformational Changes in 7TMRs Activation

I3.32(116) and V7.39(286) interact with the side-chain at position 5 of the hexapeptide ligands analogous to the C-terminal eight residues of C5a. This interaction forms an activation switch, causing a change in their relative orientations, and allows TM6 to move away from the receptor core, leading to G protein activation *(99)*.

3.3.3.3. TM6–TM7 INTERACTION

In thyrotropin receptor, whereas the D6.44(633)A or N7.49(674)D mutation led to high constitutive activity, the combined N7.49(674)D–D6.44(633)N mutations resulted in a receptor that behaved like the wild-type receptor, indicating that the two residues interact and the interaction maintains the receptor in inactive states. Additionally, in the thyrotropin receptor, N7.49(674) is essential for agonist-stimulated effect; therefore, the interaction between D6.44(633) and N7.49(674) serves as an on/off switch for activation *(100)*. Although N7.49 of the NPXXY motif is highly conserved in the rhodopsin family of 7TMRs, its role may not be identical for all 7TMRs. For example, N7.49 (332)D mutation of the μ-opioid receptor eliminated detectable binding of ligands, and D2.50(114)N mutation decreased affinities for the agonists and abolished agonist-induced signaling *(101)*. Interestingly, the combined D2.50(114)N–N7.49(332)D mutations restored high-affinity binding for the antagonists and partially restored the binding affinities, potencies, and efficacies of agonists *(101)*. The results indicate that the residue N7.49(332) in the NPXXY motif of the μ-opioid receptor is adjacent in space with D2.50(114) and that the chemical functionalities are important for ligand binding and receptor activation.

3.3.3.4. TM7 MOVEMENT

Evidence from EPR spectroscopy studies in rhodopsin suggests that movements of the cytoplasmic portion of TM7 relative to TM1 and of TM2 relative to H8 may also occur in response to photoactivation *(102,103)*. The possible importance of TM7 in receptor activation is also indirectly supported by the finding that activating metal–ion binding sites can be generated between TM3 and TM7 in both the β_2-AR *(104)* and the neurokinin-1 receptor *(105)*.

3.4. Conformational Changes in CAMs Detected by Cysteine Accessibility

Methanethiosulfonate ethylammonium (MTSEA; a charged, hydrophilic, sulfhydryl-specific reagent) has been shown to react much faster with water-accessible cysteine residues than with those in the interior of proteins or facing lipids *(106,107)*. Effects of MTSEA applied extracellularly on ligand binding to receptors have been used to determine whether cysteine residues, endogenous or engineered, are exposed in the binding site crevice. MTSEA had no effect on ligand binding to the wild-type β_2-AR. In contrast, in four β_2-AR CAMs—L6.28(266)S–K6.29(267)R–H6.31(269)K–L6.34(272)A (a quadruple mutant), D3.49(130)N, F6.44(282)L, and F6.44(282)A—MTSEA significantly inhibited ligand binding, and C6.47(285) in TM6 was identified to be responsible for the inhibitory effect of MTSEA *(39,90,108)*. Furthermore, Ballestero et al. *(63)* demonstrated that among the five b_2-AR CAMs—E6.30(268)Q, E6.30(268)A, D3.49(130)N, D3.49(130N)/E6.30(268)Q, and D3.49(130)N/E6.30(268)A—the extent of MTSEA inhibition of ligand binding to these CAMs tightly correlated with the level of constitutive activity. According to their rhodopsin-based model of the β_2-AR, in the inactive state, C6.47(285) is pointing toward TM7 and, therefore, is poorly water-accessible. In contrast, in activated states, a counterclockwise rotation and/or tilting of TM6 (viewed from the intracellular side) brings C6.47(285) more toward the margin of the water-accessible ligand-binding crevice. There is a progressive increase in the accessibility of C6.47(285), with increasing activation and, hence, a corresponding increase in the rotation and/or tilting of the TM6. As described under Subheadings 2.2.1. and 2.2.2., a similar movement is predicted from analysis of light- or agonist-induced conformational changes in rhodopsin and in β_2-AR.

However, there may be a difference in the orientation of C3.44(125) in the β_2-AR in agonist-activated states and in CAMs. Using a cysteine-reactive fluorescence probe, Gether et al. *(27)* reported that agonist activation of the β_2-AR causes C3.44(125) and C6.47(285) to be exposed to a more polar environment, indicating that conformational changes occur around these two residues. However, by MTSEA reaction, C3.44(125) in the β_2-AR was not identified to become exposed in the CAMs *(39,90,108)*. Additionally, the F6.44(282)L CAM of the β_2-AR induced the movement of not only C6.47(285) but also that of C7.54(327) *(90)*, suggesting that conformational changes of TM7 are implicated in activation of the β_2-AR. These results also imply that there are multiple activated states for a given 7TMR.

3.5. Multiple Activated States of a 7TMR

A traditional view of 7TMR activation is that there is an active state stimulating all effectors and eliciting regulatory processes, such as receptor phosphorylation and internalization. However, evidence has been accumulated to suggest that more than one active conformation exists for a given 7TMR.

3.5.1. Different CAMs of the Same 7TMR Can Couple to Distinct Signal Pathways

Perez and colleagues *(109)* found that the C3.35(128)F CAM of the α_{1B}-AR exhibited a selective constitutive activation of a G_q-coupled pathway but not a $G_{i/o}$-coupled pathway. In contrast, the A6.34(293)E CAM of this receptor (*see* Subheading 3.1.) constitutively activated both pathways *(109)*. It was postulated that the α_{1B}-AR can isomerize to at least two distinct active states: one coupled to G_q and one coupled to $G_{i/o}$. Additionally, these researchers found that the C3.35(116)F CAM of the β_2-AR caused constitutive stimulation of the Na⁺/H⁺ exchanger 1, which is a G_{13}-mediated effect, but not of cyclic adenosine monophosphate production, which is a G_s-mediated event *(110)*. Although all the CAMs of the β_2-AR mentioned in Subheading 3.5. constitutively activated the G_s pathway, they displayed somewhat different movements of TMs, lending support for different activated states.

3.5.2. Some CAMs Are Not Constitutively Internalized

After agonist binding, 7TMRs not only activate G protein(s) but also initiate regulatory processes, including desensitization and internalization, that involve phosphorylation and arrestin recruitment in most receptors *(111)*. Because the structures adopted by CAMs mimic active conformations of the receptor, CAMs are likely to be substrates forGPCR kinase (GRK)-mediated phosphorylation in the absence of agonists. Constitutive phosphorylation has been demonstrated for many CAMs, including T6.34(373) of α_{2A}-AR *(67)*; L6.28(266)S–K6.29(267)R–H6.31(269)K–L6.34(272)A of β_2-AR *(112)*; and K7.43(296)G, E3.28(113)Q, A7.39(292)E, and G2.57(90)D of rhodopsin *(113)*.

However, several CAMs have been shown to not be constitutively phosphorylated or internalized. The N3.35(111)A/G CAM of the angiotensin II receptor AT_{1A} did not exhibit an increase in basal phosphorylation, nor was its phosphorylation enhanced by the agonist *(114)*. Additionally, two CAMs of the same receptor may exhibit different regulatory properties. Mhaouty-Kodja et al. *(115)* reported that although two different CAMs of the α_{1B}-AR had similar agonist-independent activities, the A6.34(293)E CAM displayed increased GRK2-mediated phosphorylation and underwent β-arrestin-mediated internalization in a constitutive manner, whereas the D3.49(142)A CAM did not. The L3.40(124)N–L3.43(127)Q CAM of the C5a receptor recruited β-arrestin and underwent constitutively internalization, whereas the F6.44(251)A CAM did not *(116)*. These observations suggest that the conformations recognized by GRKs and arrestins may be different from those that couple a 7TMR to G protein signaling.

The finding that some agonists of opioid receptors do not cause internalization or desensitization corroborates this view. Morphine and [Tyr-D-Ala-Gly-N-(Me)Phe-Gly-ol] (DAMGO) activated the μ-opioid receptor; however, unlike DAMGO, morphine did not cause phosphorylation or internalization *(117–120)*. Additionally, levorphanol or etorphine did not cause phosphorylation or internalization of the human κ-opioid receptor, although both acted as full agonists in activating G proteins *(121)*. However, levorphanol or etorphine acted as an antagonist in reducing U50,488H- or dynorphin A-induced phosphorylation and internalization *(121)*. To date, all the reported CAMs of the μ-opioid receptor are shown to, or are likely to, induce constitutive internalization of the receptor *(42,43)*; these CAMs may represent DAMGO-induced active state(s) of the receptor, whereas a CAM mimicking morphine-induced active conformation has not been found.

3.6. Information From Random Mutagenesis Studies

A large number of spontaneous and site-directed mutations resulting in constitutive activity of different 7TMRs have been identified and characterized to involve structural motifs throughout the entire receptors, well-beyond the "D/ER3.50Y," the XBBXXB, and the NPXXY motifs *(31)*. Systematic mapping of the amino acids involved in activation of a given receptor is extremely useful for complete elucidation of the molecular mechanisms of activation, without preconceived structural hypotheses. Recently, random mutagenesis studies of AT_1 *(122)*, m_3 *(123)* and m_5 *(124)* muscarinic, C5a *(125)*, and δ-opioid *(126)* receptors have shown that the locations of activating mutations are widespread, not only within the TMs but also in the second and third intracellular loops, the third extracellular loop, and even the N- and C-terminal domains.

For example, in a whole-receptor random mutagenesis study regarding the human δ-opioid receptor, Kieffer et al. *(126)* identified 30 single-mutation CAMs. The sites of mutations were distributed throughout the receptor and were mapped on a three-dimensional model of the receptor. With the exception of those located in the N- and C- terminal domains, most activating mutations were clustered spatially into four groups, and amino acid residues in each group were adjacent to each other from extracellular face to cytoplasmic face. Five CAMs (Q12L, D21G, P28L, A30D, and R41Q) were demonstrated to have mutations in the N-terminal region, suggesting that this region may be folded into the 7-TM bundle to influence receptor activation *(126)*. Four CAMs (C328R, D341N, G369S, G369D) resulted from mutations of residues in the C-terminal domain, with potential regulatory functions that may directly modify receptor interaction with G proteins *(126)*. C328 is a potential palmitoylation site. Mutations to A of C348/C353 putative palmitoylation sites in the human m-opioid receptor led to constitutive activation *(127)*. D341N mutation eliminated a negative charge at a position uniquely conserved among opioid receptors within a highly variable region. G369S introduced a potential phosphorylation site. G369D introduced a residue mimicking phosphoserine.

Group I is composed of four CAMs in the extracellular regions: K214R, V283A, I289V, and L286P. V283A, I289V, and L286P were found within a highly hydrophobic region of the third extracellular loop, and these mutations may diminish local hydrophobicity, which restrains motions of TM6 and TM7. K214R at the extracellular end of TM5 may have an indirect effect to weaken this structural constraint, favoring activation. It was postulated that the agonist may initiate the 7TMR activation by contacting the third extracellular hydrophobic core and releasing the constraint as a first step.

Group II consisted of four CAMs: T3.38(134)A, I4.47(170)T, N4.46(169)S, and W4.50(173)R. It was postulated that these residues form a microdomain surrounding W4.50, which is strictly conserved in all rhodopsin-like 7TMRs. By aromatic stacking with Y3.34(130), W4.50 may limit TM3 mobility and may represent a key switch for TM3 movements in most 7TMRs. The four mutations may perturb W4.50 orientation.

Group III included four CAMs: D3.32(128)N, S3.39(135)G, Y7.43(308)H, and W6.48(274)R. Mutation of D3.32(128) or Y7.43(308) was demonstrated to lead to constitutive activation of the δ-opioid receptor, which disrupted interactions between the two residues *(37)* and weakened interactions of TM3 and TM7 (*see* Subheading 3.3.3.). S3.39(135)G may also reduce interactions of TM3 and TM7. W6.48 is part of the aromatic cluster in TM6, acting as a "toggle switch" to transmit the signal from agonist binding for receptor activation (*5; see* Subheading 3.3.1.). In the human B_2 bradykinin receptor, it was hypothesized that W6.48(256) interacted with N3.35(113) to stabilize the inactive state (*92; see* Subheading 3.3.2.).

Group IV was comprised of five CAMs: R6.32(258)H, M6.36(262)T, Y7.53(318)H, L7.56(321)R, and E323K. These residues cluster between TM6 and TM7 at the cytoplasmic interface. An ionic bridge was proposed between R6.32(258) in the BBXXB motif and E323 in the proximal portion of C-tail. M6.36(262), Y7.53(318), and L7.56(321) may form a hydrophobic pocket. Therefore, mutations at these loci would break TM6–TM7 interactions and allow separation of TM6 and TM7 at the cytoplasmic face. Based on these results, Kieffer et al. *(126)* proposed that δ-opioid receptor activation occurs as a three-step process by an agonist. In step A (involving residues from Group I), an opioid agonist binds to the third extracellular loop and, possibly, the N-terminal domain, which perturb the third extracellular loop hydrophobic cluster and thereby destabilize TM6–TM7 interactions near the extracellular side. In step B (involving residues from Group II and III), the amphiphilic agonist enters the binding pocket, disrupting both TM3–TM6 and TM3–TM7 hydrophobic interactions from the middle of membranes to the extracellular region, provoking outward TM3–TM6–TM7 movements. In step C (mimicked by CAMs from Group IV mutants as well as from DRY, BBXXB, and NPXXY mutants), helical movements propagate downward within the receptor, disrupt cytoplasmic ionic or aromatic interactions (TM6–TM7, TM3–TM6 and TM7–H8), and possibly release the putative H8 helix. This reveals receptor intracellular determinants to interact with G proteins.

The wealth of information from such random mutagenesis studies will guide further studies to provide mechanistic explanations for each CAM, leading to better understanding regarding mechanisms of 7TMR activation.

4. CONCLUSIONS

The structural basis of rhodopsin activation is representative of receptors in the rhodopsin family of 7TMRs. The activation mechanisms of other receptors in the family are consistent with information that is available for rhodopsin involving movements of TMs. Studies regarding CAMs, supported by rhodopsin-based models, have been important in elucidating the detailed molecular events underlying the movements in TMs during receptor activation. Although receptor activation in the rhodopsin family of 7TMRs may be associated with similar motifs and conformational changes, different receptors may also employ specialized sets of molecular switches to produce these changes.

ACKNOWLEDGMENT

Preparation of this chapter was supported in part by grants from National Institutes of Health (DA04745, DA17302, and DA13429). We thank Drs. Harel Weinstein and Irache Visiers for their molecular modeling and insightful discussions in our original work. We also thank Mr. Chongguang Chen for graphic work.

REFERENCES

1. Hall RA, Premont RT, Lefkowitz RJ. Heptahelical receptor signaling: beyond the G protein paradigm. J Cell Biol 1999;145:927–932.
2. Schertler GF, Villa C, Henderson R. Projection structure of rhodopsin. Nature 1993;362:770–772.
3. Palczewski K, Kumasaka T, Hori T, et al. Crystal structure of rhodopsin: a G protein-coupled receptor. Science 2000;289:739–745.
4. Teller DC, Okada T, Behnke CA, Palczewski K, Stenkamp RE. Advances in determination of a high-resolution three-dimensional structure of rhodopsin, a model of G-protein-coupled receptors (GPCRs). Biochem 2001;40:7761–7772.
5. Visiers I, Ballesteros JA, Weinstein H. Three-dimensional representations of G protein-coupled receptor structures and mechanisms. Methods Enzymol 2002;343:329–371.
6. Ballesteros JA, Shi L, Javitch JA. Structural mimicry in G protein-coupled receptors: implications of the high-resolution structure of rhodopsin for structure–function analysis of rhodopsin-like receptors. Mol Pharmacol 2001;60:1–19.
7. Bockaert J, Pin JP. Molecular tinkering of G protein-coupled receptors: an evolutionary success. EMBO J 1999;18:1723–1729.
8. Schwartz TW, Holst B. Molecular Structure of G Protein-Coupled Receptors. In: Foreman JC, Johansen T, eds. Textbook of Receptor Pharmacology. New York: CRC Press, 2002, pp. 81–110.
9. Ballesteros JA, Weinstein H. Integrated methods for the construction of three dimensional models and computational probing of structure-function relations in G-protein coupled receptors. Methods Neurosci 1995;25:366–428.
10. Mirzadegan T, Benko G, Filipek S, Palczewski K. Sequence analyses of G-protein-coupled receptors: similarities to rhodopsin. Biochem 2003;42:2759–2767.
11. Meng EC, Bourne HR. Receptor activation: what does the rhodopsin structure tell us? Trends Pharmacol Sci 2001;22:587–593.
12. Filipek S, Stenkamp RE, Teller DC, Palczewski K. G protein-coupled receptor rhodopsin: a prospectus. Annu Rev Physiol 2003;65:851–879.
13. Sakmar TP, Menon ST, Marin EP, Awad ES. Rhodopsin: insights from recent structural studies. Annu Rev Biophys Biomol Struct 2002;31:443–484.
14. Schertler GF, Hargrave PA. Projection structure of frog rhodopsin in two crystal forms. Proc Natl Acad Sci USA 1995;92:11,578–11,582.
15. Unger VM, Hargrave PA, Baldwin JM, Schertler GF. Arrangement of rhodopsin transmembrane alpha-helices. Nature 1997;389:203–206.
16. Borhan B, Souto ML, Imai H, Shichida Y, Nakanishi K. Movement of retinal along the visual transduction path. Science 2000;288:2209–2212.
17. Filipek S, Teller DC, Palczewski K, Stenkamp R. The crystallographic model of rhodopsin and its use in studies of other G protein-coupled receptors. Annu Rev Biophys Biomol Struct 2003;32:375–397.

18. Gether U, Kobilka BK. G protein-coupled receptors. II. Mechanism of agonist activation. J Biol Chem 1998;273:17,979–17,982.
19. Gether U. Uncovering molecular mechanisms involved in activation of G protein- coupled receptors. Endocr Rev 2000;21:90–113.
20. Farrens DL, Altenbach C, Yang K, Hubbell WL, Khorana HG. Requirement of rigid-body motion of transmembrane helices for light activation of rhodopsin. Science 1996;274:768–770.
21. Cai K, Klein-Seetharaman J, Hwa J, Hubbell WL, Khorana HG. Structure and function in rhodopsin: effects of disulfide cross-links in the cytoplasmic face of rhodopsin on transducin activation and phosphorylation by rhodopsin kinase. Biochemistry 1999;38:12,893–12,898.
22. Sheikh SP, Zvyaga TA, Lichtarge O, Sakmar TP, Bourne HR. Rhodopsin activation blocked by metal–ion-binding sites linking transmembrane helices C and F. Nature 1996;383:347–350.
23. Sheikh SP, Vilardarga JP, Baranski TJ, et al. Similar structures and shared switch mechanisms of the beta2- adrenoceptor and the parathyroid hormone receptor. Zn(II) bridges between helices III and VI block activation. J Biol Chem 1999;274:17,033–17,041.
24. Hubbell WL, Cafiso DS, Altenbach C. Identifying conformational changes with site-directed spin labeling. Nat Struct Biol 2000;7:735–739.
25. Altenbach C, Yang K, Farrens DL, Farahbakhsh ZT, Khorana HG, Hubbell WL. Structural features and light-dependent changes in the cytoplasmic interhelical E-F loop region of rhodopsin: a site-directed spin-labeling study. Biochemistry 1996;35:12,470–12,478.
26. Farahbakhsh ZT, Ridge KD, Khorana HG, Hubbell WL. Mapping light-dependent structural changes in the cytoplasmic loop connecting helices C and D in rhodopsin: a site-directed spin labeling study. Biochemistry 1995;34:8812–8819.
27. Gether U, Lin S, Ghanouni P, Ballesteros JA, Weinstein H, Kobilka BK. Agonists induce conformational changes in transmembrane domains III and VI of the beta2 adrenoceptor. EMBO J 1997;16:6737–6747.
28. Gether U, Lin S, Kobilka BK. Fluorescent labeling of purified beta 2 adrenergic receptor. Evidence for ligand-specific conformational changes. J Biol Chem 1995;270:28,268–28,275.
29. Kobilka BK, Gether U. Use of fluorescence spectroscopy to study conformational changes in the beta 2-adrenoceptor. Methods Enzymol 2002;343:170–182.
30. Ward SD, Hamdan FF, Bloodworth LM, Wess J. Conformational changes that occur during m3 muscarinic acetylcholine receptor activation probed by the use of an *in situ* disulfide cross-linking strategy. J Biol Chem 2002;277:2247–2257.
31. Pauwels PJ, Wurch T. Review: amino acid domains involved in constitutive activation of G-protein-coupled receptors. Mol Neurobiol 1998;17:109–135.
32. Parnot C, Miserey-Lenkei S, Bardin S, Corvol P, Clauser E. Lessons from constitutively active mutants of G protein-coupled receptors. Trends Endocrinol Metab 2002;13:336–343.

33. De Lean A, Stadel JM, Lefkowitz RJ. A ternary complex model explains the agonist-specific binding properties of the adenylate cyclase-coupled beta-adrenergic receptor. J Biol Chem 1980;255:7108–7117.
34. Samama P, Cotecchia S, Costa T, Lefkowitz RJ. A mutation-induced activated state of the beta 2-adrenergic receptor. Extending the ternary complex model. J Biol Chem 1993;268:4625–4636.
35. Fahmy K, Sakmar TP, Siebert F. Structural determinants of active state conformation of rhodopsin: molecular biophysics approaches. Methods Enzymol 2000;315:178–196.
36. Kjelsberg MA, Cotecchia S, Ostrowski J, Caron MG, Lefkowitz RJ. Constitutive activation of the alpha 1B-adrenergic receptor by all amino acid substitutions at a single site. Evidence for a region which constrains receptor activation. J Biol Chem 1992;267:1430–1433.
37. Befort K, Zilliox C, Filliol D, Yue S, Kieffer BL. Constitutive activation of the delta opioid receptor by mutations in transmembrane domains III and VII. J Biol Chem 1999;274:18,574–18,581 [published erratum appears in J Biol Chem 1999;274(39):28,058].
38. Gether U, Ballesteros JA, Seifert R, Sanders-Bush E, Weinstein H, Kobilka BK. Structural instability of a constitutively active G protein-coupled receptor. Agonist-independent activation due to conformational flexibility. J Biol Chem 1997;272:2587–2590.
39. Rasmussen SG, Jensen AD, Liapakis G, Ghanouni P, Javitch JA, Gether U. Mutation of a highly conserved aspartic acid in the beta2 adrenergic receptor: constitutive activation, structural instability, and conformational rearrangement of transmembrane segment 6. Mol Pharmacol 1999;56:175–184.
40. Chen S, Lin F, Xu M, Graham RM. Phe(303) in TMVI of the alpha(1B)-adrenergic receptor is a key residue coupling TM helical movements to G-protein activation. Biochem 2002;41:588–596.
41. Alewijnse AE, Timmerman H, Jacobs EH, et al. The effect of mutations in the DRY motif on the constitutive activity and structural instability of the histamine H-2 receptor. Mol Pharmacol 2000;57:890–898.
42. Li J, Chen C, Huang P, Liu-Chen L-Y. Inverse agonist up-regulates the constitutively active D3.49(164)Q mutant of the rat μ opioid receptor by stabilizing the structure and blocking constitutive internalization and down-regulation. Mol Pharmacol 2001;60:1064–1075.
43. Li J, Huang P, Chen C, de Riel JK, Weinstein H, Liu-Chen L-Y. Constitutive activation of the mu opioid receptor by mutation of D3.49(164), but not D3.32(147): D3.49(164) is critical for stabilization of the inactive form of the receptor and for its expression. Biochemistry 2001;40:12,039–12,050.
44. Huang P, Li J, Chen C, Visiers I, Weinstein H, Liu-Chen L-Y. Functional role of a conserved motif in TM6 of the rat mu opioid receptor: constitutively active and inactive receptors result from substitutions of Thr6.34(279) with Lys and Asp. Biochem 2001;40:13,501–13,509.
45. Probst WC, Snyder LA, Schuster DI, Brosius J, Sealfon SC. Sequence alignment of the G-protein coupled receptor superfamily. DNA Cell Biol 1992;11:1–20.

46. Ballesteros J, Kitanovic S, Guarnieri F, et al. Functional microdomains in G-protein-coupled receptors. The conserved arginine-cage motif in the gonadotropin-releasing hormone receptor. J Biol Chem 1998;273:10,445–10,453.
47. Oliveira L, Paiva AC, Sander C, Vriend G. A common step for signal transduction in G protein-coupled receptors. Trends Pharmacol Sci 1994;15:170–172.
48. Radding CM, Wald G. The stability of rhodopsin and opsin; effects of pH and aging. J Gen Physiol 1956;39:923–933.
49. Radding CM, Wald G. Acid-base properties of rhodopsin and opsin. J Gen Physiol 1956;39:909–922.
50. Parkes JH, Liebman PA. Temperature and pH dependence of the metarhodopsin I-metarhodopsin II kinetics and equilibria in bovine rod disk membrane suspensions. Biochem 1984;23:5054–5061.
51. Arnis S, Fahmy K, Hofmann KP, Sakmar TP. A conserved carboxylic acid group mediates light-dependent proton uptake and signaling by rhodopsin. J Biol Chem 1994;269:23,879–23,881.
52. Fahmy K, Jager F, Beck M, Zvyaga TA, Sakmar TP, Siebert F. Protonation states of membrane-embedded carboxylic acid groups in rhodopsin and metarhodopsin II: a Fourier-transform infrared spectroscopy study of site-directed mutants. Proc Natl Acad Sci USA 1993;90:10,206–10,210.
53. Ghanouni P, Schambye H, Seifert R, et al. The effect of pH on beta(2) adrenoceptor function. Evidence for protonation-dependent activation. J Biol Chem 2000;275:3121–3127.
54. Scheer A, Fanelli F, Costa T, De Benedetti PG, Cotecchia S. The activation process of the alpha1B-adrenergic receptor: potential role of protonation and hydrophobicity of a highly conserved aspartate. Proc Natl Acad Sci USA 1997;94:808–813.
55. Gether U, Asmar F, Meinild AK, Rasmussen SG. Structural basis for activation of G-protein-coupled receptors. Pharmacol Toxicol 2002;91:304–312.
56. Cohen GB, Yang T, Robinson PR, Oprian DD. Constitutive activation of opsin: influence of charge at position 134 and size at position 296. Biochem 1993;32:6111–6115.
57. Kim JM, Altenbach C, Thurmond RL, Khorana HG, Hubbell WL. Structure and function in rhodopsin: rhodopsin mutants with a neutral amino acid at E134 have a partially activated conformation in the dark state. Proc Natl Acad Sci USA 1997;94:14,273–14,278.
58. Scheer A, Fanelli F, Costa T, De Benedetti PG, Cotecchia S. Constitutively active mutants of the alpha 1B-adrenergic receptor: role of highly conserved polar amino acids in receptor activation. EMBO J 1996;15:3566–3578.
59. Morin D, Cotte N, Balestre MN, et al. The D136A mutation of the V2 vasopressin receptor induces a constitutive activity which permits discrimination between antagonists with partial agonist and inverse agonist activities. FEBS Lett 1998;441:470–475.
60. Burger M, Burger JA, Hoch RC, Oades Z, Takamori H, Schraufstatter IU. Point mutation causing constitutive signaling of CXCR2 leads to transform-

ing activity similar to Kaposi's sarcoma herpesvirus-G protein-coupled receptor. J Immunol 1999;163:2017–2022.
61. Ho HH, Du D, Gershengorn MC. The N terminus of Kaposi's sarcoma-associated herpesvirus G protein-coupled receptor is necessary for high affinity chemokine binding but not for constitutive activity. J Biol Chem 1999;274:31,327–31,332.
62. Ho HH, Ganeshalingam N, Rosenhouse-Dantsker A, Osman R, Gershengorn MC. Charged residues at the intracellular boundary of transmembrane helices 2 and 3 independently affect constitutive activity of Kaposi's sarcoma-associated herpesvirus G protein-coupled receptor. J Biol Chem 2001;276:1376–1382.
63. Ballesteros JA, Jensen AD, Liapakis G, et al. Activation of the beta 2-adrenergic receptor involves disruption of an ionic lock between the cytoplasmic ends of transmembrane segments 3 and 6. J Biol Chem 2001;276:29,171–29,177.
64. Lu ZL, Curtis CA, Jones PG, Pavia J, Hulme EC. The role of the aspartate-arginine-tyrosine triad in the m1 muscarinic receptor: mutations of aspartate 122 and tyrosine 124 decrease receptor expression but do not abolish signaling. Mol Pharmacol 1997;51:234–241.
65. Wang Z, Wang H, Ascoli M. Mutation of a highly conserved acidic residue present in the second intracellular loop of G-protein-coupled receptors does not impair hormone binding or signal transduction of the luteinizing hormone/chorionic gonadotropin receptor. Mol Endocrinol 1993;7:85–93.
66. Arora KK, Cheng Z, Catt KJ. Mutations of the conserved DRS motif in the second intracellular loop of the gonadotropin-releasing hormone receptor affect expression, activation, and internalization. Mol Endocrinol 1997;11:1203–1212.
67. Ren Q., Kurose H, Lefkowitz RJ, Cotecchia S. Constitutively active mutants of the alpha 2-adrenergic receptor. J Biol Chem 1993;268:16,483–16,487 [published erratum appears in J Biol Chem 1994;269(2):1566].
68. Lattion A, Abuin L, Nenniger-Tosato M, Cotecchia S. Constitutively active mutants of the beta1-adrenergic receptor. FEBS Lett 1999;457:302–306.
69. Egan CT, Herrick-Davis K, Teitler M. Creation of a constitutively activated state of the 5-hydroxytryptamine2A receptor by site-directed mutagenesis: inverse agonist activity of antipsychotic drugs. J Pharmacol Exp Ther 1998;286:85–90.
70. Herrick-Davis K, Egan C, Teitler M. Activating mutations of the serotonin 5-HT2C receptor. J Neurochem 1997;69:1138–1144.
71. Pauwels PJ, Gouble A, Wurch T. Activation of constitutive 5-hydroxytryptamine(1B) receptor by a series of mutations in the BBXXB motif: positioning of the third intracellular loop distal junction and its G(o)alpha protein interactions. Biochem J 1999;343(pt 2):435–442.
72. Hogger P, Shockley MS, Lameh J, Sadee W. Activating and inactivating mutations in N- and C-terminal i3 loop junctions of muscarinic acetylcholine Hm1 receptors. J Biol Chem 1995;270:7405–7410.

73. Visiers I, Ebersole BJ, Dracheva S, Ballesteros JA, Sealfon SC, Weinstein H. Structural motifs as functional microdomains in G protein-coupled receptors: energetic considerations in the mechanism of activation of the serotonin 5HT$_{2A}$ receptor by disruption of the ionic lock of the arginine cage. Int J Quant Chem 2002;88:65–75.
74. Shapiro DA, Kristiansen K, Weiner DM, Kroeze WK, Roth BL. Evidence for a model of agonist-induced activation of 5-hydroxytryptamine 2A serotonin receptors that involves the disruption of a strong ionic interaction between helices 3 and 6. J Biol Chem 2002;277:11,441–11,449.
75. Greasley PJ, Fanelli F, Rossier O, Abuin L, Cotecchia S. Mutagenesis and modelling of the alpha(1b)-adrenergic receptor highlight the role of the helix 3/helix 6 interface in receptor activation. Mol Pharmacol 2002;61:1025–1032.
76. Laue L, Chan WY, Hsueh AJ, et al. Genetic heterogeneity of constitutively activating mutations of the human luteinizing hormone receptor in familial male-limited precocious puberty. Proc Natl Acad Sci USA 1995;92:1906–1910.
77. Kosugi S, Mori T, Shenker A. An anionic residue at position 564 is important for maintaining the inactive conformation of the human lutropin/choriogonadotropin receptor. Mol Pharmacol 1998;53:894–901.
78. Parma J, Duprez L, Van Sande J, et al. Somatic mutations in the thyrotropin receptor gene cause hyperfunctioning thyroid adenomas [see comments]. Nature 1993;365:649–651.
79. Ascoli M, Fanelli F, Segaloff DL. The lutropin/choriogonadotropin receptor, a 2002 perspective. Endocr Rev 2002;23:141–174.
80. Schulz A, Schoneberg T, Paschke R, Schultz G, Gudermann T. Role of the third intracellular loop for the activation of gonadotropin receptors. Mol Endocrinol 1999;13:181–190.
81. Huang P, Visiers I, Weinstein H, Liu-Chen LY. The local environment at the cytoplasmic end of TM6 of the mu opioid receptor differs from those of rhodopsin and monoamine receptors: introduction of an ionic lock between the cytoplasmic ends of helices 3 and 6 by a L6.30(275)E mutation inactivates the mu opioid receptor and reduces the constitutive activity of its T6.34(279)K mutant. Biochemistry 2002;41:11,972–11,980.
82. Franke RR, Sakmar TP, Graham RM, Khorana HG. Structure and function in rhodopsin. Studies of the interaction between the rhodopsin cytoplasmic domain and transducin. J Biol Chem 1992;267:14,767–14,774.
83. Rosenthal W, Antaramian A, Gilbert S, Birnbaumer M. Nephrogenic diabetes insipidus. A V2 vasopressin receptor unable to stimulate adenylyl cyclase. J Biol Chem 1993;268:13,030–13,033.
84. Jones PG, Curtis CA, Hulme EC. The function of a highly-conserved arginine residue in activation of the muscarinic m1 receptor. Eur J Pharmacol 1995;288:251–257.
85. Rhee MH, Nevo I, Levy R, Vogel Z. Role of the highly conserved Asp-Arg-Tyr motif in signal transduction of the CB2 cannabinoid receptor. FEBS Lett 2000;466:300–304.

86. Scheer A, Costa T, Fanelli F, et al. Mutational analysis of the highly conserved arginine within the Glu/Asp-Arg-Tyr motif of the alpha(1b)-adrenergic receptor: effects on receptor isomerization and activation. Mol Pharmacol 2000;57:219–231.
87. Acharya S, Karnik SS. Modulation of GDP release from transducin by the conserved Glu134-Arg135 sequence in rhodopsin. J Biol Chem 1996;271:25,406–25,411.
88. Fanelli F, Barbier P, Zanchetta D, De Benedetti PG, Chini B. Activation mechanism of human oxytocin receptor: a combined study of experimental and computer-simulated mutagenesis. Mol Pharmacol 1999;56:214–225.
89. Chen S, Xu M, Lin F, Lee D, Riek P, Graham RM. Phe310 in transmembrane VI of the alpha1B-adrenergic receptor is a key switch residue involved in activation and catecholamine ring aromatic bonding. J Biol Chem 1999;274:16,320–16,330.
90. Chen S, Lin F, Xu M, Riek RP, Novotny J, Graham RM. Mutation of a single TMVI residue, Phe(282), in the beta(2)-adrenergic receptor results in structurally distinct activated receptor conformations. Biochemistry 2002; 41: 6045–6053.
91. Shi L, Liapakis G, Xu R, Guarnieri F, Ballesteros JA, Javitch JA. Beta2 adrenergic receptor activation. Modulation of the proline kink in transmembrane 6 by a rotamer toggle switch. J Biol Chem 2002;277:40,989–40,996.
92. Marie J, Richard E, Pruneau D, et al. Control of conformational equilibria in the human B2 bradykinin receptor. Modeling of nonpeptidic ligand action and comparison to the rhodopsin structure. J Biol Chem 2001;276:41,100–41,111.
93. Shinozaki H, Fanelli F, Liu X, Jaquette J, Nakamura K, Segaloff DL. Pleiotropic effects of substitutions of a highly conserved leucine in transmembrane helix III of the human lutropin/choriogonadotropin receptor with respect to constitutive activation and hormone responsiveness. Mol Endocrinol 2001;15:972–984.
94. Fritze O, Filipek S, Kuksa V, Palczewski K, Hofmann KP, Ernst OP. Role of the conserved NPxxY(x)5,6F motif in the rhodopsin ground state and during activation. Proc Natl Acad Sci USA 2003;100:2290–2295.
95. Prioleau C, Visiers I, Ebersole BJ, Weinstein H, Sealfon SC. Conserved helix 7 tyrosine acts as a multistate conformational switch in the 5HT2C receptor. Identification of a novel "locked-on" phenotype and double revertant mutations. J Biol Chem 2002;277:36,577–36,584.
96. Porter JE, Hwa J, Perez DM. Activation of the alpha1b-adrenergic receptor is initiated by disruption of an interhelical salt bridge constraint. J Biol Chem 1996;271:28,318–28,323.
97. Robinson PR, Cohen GB, Zhukovsky EA, Oprian DD. Constitutively active mutants of rhodopsin. Neuron 1992;9:719–725.
98. Kristiansen K, Kroeze WK, Willins DL, et al. A highly conserved aspartic acid (Asp-155) anchors the terminal amine moiety of tryptamines and is involved in membrane targeting of the 5- HT(2A) serotonin receptor but does

not participate in activation via a "salt-bridge disruption" mechanism. J Pharmacol Exp Ther 2000;293:735–746.
99. Gerber BO, Meng EC, Dotsch V, Baranski TJ, Bourne HR. An activation switch in the ligand binding pocket of the C5a receptor. J Biol Chem 2001;276:3394–3400.
100. Govaerts C, Lefort A, Costagliola S, et al. A conserved Asn in transmembrane helix 7 is an on/off switch in the activation of the thyrotropin receptor. J Biol Chem 2001;276:22,991–22,999.
101. Xu W, Ozdener F, Li J-G, et al. Functional role of the spatial proximity of Asp114(2.50) in TMH 2 and Asn332(7.49) in TMH 7 of the mu opioid receptor. FEBS Lett 1999;447:318–324.
102. Altenbach C, Cai K, Klein-Seetharaman J, Khorana HG, Hubbell WL. Structure and function in rhodopsin: mapping light-dependent changes in distance between residue 65 in helix TM1 and residues in the sequence 306–319 at the cytoplasmic end of helix TM7 and in helix H8. Biochemistry 2001;40:15,483–15,492.
103. Altenbach C, Oh KJ, Trabanino RJ, Hideg K, Hubbell WL. Estimation of inter-residue distances in spin labeled proteins at physiological temperatures: experimental strategies and practical limitations. Biochem 2001;40:15,471–15,482.
104. Elling CE, Thirstrup K, Holst B, Schwartz TW. Conversion of agonist site to metal-ion chelator site in the beta(2)-adrenergic receptor. Proc Natl Acad Sci USA 1999;96:12,322–12,327.
105. Holst B, Elling CE, Schwartz TW. Partial agonism through a zinc–Ion switch constructed between transmembrane domains III and VII in the tachykinin NK(1) receptor. Mol Pharmacol 2000;58:263–270.
106. Roberts DD, Lewis SD, Ballou DP, Olson ST, Shafer JA. Reactivity of small thiolate anions and cysteine-25 in papain toward methyl methanethiosulfonate. Biochemistry 1986;25:5595–5601.
107. Karlin A, Akabas MH. Substituted-cysteine accessibility method. Methods Enzymol 1998;293:123–145.
108. Javitch JA, Fu D, Liapakis G, Chen J. Constitutive activation of the beta2 adrenergic receptor alters the orientation of its sixth membrane-spanning segment. J Biol Chem 1997;272:18,546–18,549.
109. Perez DM, Hwa J, Gaivin R, Mathur M, Brown F, Graham RM. Constitutive activation of a single effector pathway: evidence for multiple activation states of a G protein-coupled receptor. Mol Pharmacol 1996;49:112–122.
110. Zuscik MJ, Porter JE, Gaivin R, Perez DM. Identification of a conserved switch residue responsible for selective constitutive activation of the beta2-adrenergic receptor. J Biol Chem 1998;273:3401–3407.
111. Ferguson SS. Evolving concepts in G protein-coupled receptor endocytosis: the role in receptor desensitization and signaling. Pharmacol Rev 2001;53:1–24.
112. Pei G, Samama P, Lohse M, et al. A constitutively active mutant beta 2-adrenergic receptor is constitutively desensitized and phosphorylated. Proc Natl Acad Sci USA 1994;91:2699–2702.

113. Rim J, Oprian DD. Constitutive activation of opsin: interaction of mutants with rhodopsin kinase and arrestin. Biochemistry 1995;34:11,938–11,945.
114. Thomas WG, Qian H, Chang CS, Karnik S. Agonist-induced phosphorylation of the angiotensin II (AT(1A)) receptor requires generation of a conformation that is distinct from the inositol phosphate-signaling state. J Biol Chem 2000;275:2893–2900.
115. Mhaouty-Kodja S, Barak LS, Scheer A, et al. Constitutively active alpha-1b adrenergic receptor mutants display different phosphorylation and internalization features. Mol Pharmacol 1999;55:339–347.
116. Whistler JL, Gerber BO, Meng EC, Baranski TJ, von Zastrow M, Bourne HR. Constitutive activation and endocytosis of the complement factor 5a receptor: evidence for multiple activated conformations of a G protein-coupled receptor. Traffic 2002;3:866–877.
117. Arden JR, Segredo V, Wang Z, Lameh J, Sadee W. Phosphorylation and agonist-specific intracellular trafficking of an epitope-tagged mu-opioid receptor expressed in HEK 293 cells. J Neurochem 1995;65:1636–1645.
118. Keith DE, Murray SR, Zaki PA, et al. Morphine activates opioid receptors without causing their rapid internalization. J Biol Chem 1996;271:19,021–19,024.
119. Sternini C, Spann M, Anton B, et al. Agonist-selective endocytosis of mu opioid receptor by neurons in vivo. Proc Natl Acad Sci USA 1996;93:9241–9246.
120. Zhang J, Ferguson SS, Barak LS, et al. Role for G protein-coupled receptor kinase in agonist-specific regulation of mu-opioid receptor responsiveness. Proc Natl Acad Sci USA 1998;95:7157–7162.
121. Li JG, Zhang F, Jin XL, Liu-Chen LY. Differential regulation of the human kappa opioid receptor by agonists: etorphine and levorphanol reduced dynorphin A- and U50,488H-induced internalization and phosphorylation. J Pharmacol Exp Ther 2003;305:531 540.
122. Parnot C, Bardin S, Miserey-Lenkei S, Guedin D, Corvol P, Clauser E. Systematic identification of mutations that constitutively activate the angiotensin II type 1A receptor by screening a randomly mutated cDNA library with an original pharmacological bioassay. Proc Natl Acad Sci USA 2000;97:7615–7620.
123. Schmidt C, Li B, Bloodworth L, Erlenbach I, Zeng FY, Wess J. Random mutagenesis of the m3 muscarinic acetylcholine receptor expressed in yeast. Identification of point mutations that "silence" a constitutively active mutant M3 receptor and greatly impair receptor/G protein coupling. J Biol Chem 2003;278:30,248–30,260.
124. Spalding TA, Burstein ES. Constitutively active muscarinic receptors. Life Sci 2001;68:2511–2516.
125. Baranski TJ, Herzmark P, Lichtarge O, et al. C5a receptor activation. Genetic identification of critical residues in four transmembrane helices. J Biol Chem 1999;274:15,757–15,765.
126. Decaillot FM, Befort K, Filliol D, Yue S, Walker P, Kieffer BL. Opioid receptor random mutagenesis reveals a mechanism for G protein-coupled receptor activation. Nat Struct Biol 2003;10:629–636.

127. Brillet K, Kieffer BL, Massotte D. Enhanced spontaneous activity of the mu opioid receptor by cysteine mutations: characterization of a tool for inverse agonist screening. BMC Pharmacol 2003;3:14.

3
GPCR Folding and Maturation

The Effect of Pharmacological Chaperones

Ulla E. Petäjä-Repo and Michel Bouvier

1. INTRODUCTION

Folding and maturation of G protein-coupled receptors (GPCRs) are complex events, which take place before these integral membrane proteins are transported from their site of synthesis in the endoplasmic reticulum (ER) to their site of action at the plasma membrane. Problems in these events (e. g., as a result of minor mutations) often result in the inability of the newly synthesized receptors to function properly because of mislocalization in the cell. Therefore, ways to circumvent difficulties in GPCR folding, maturation, and trafficking could be of primary importance for alleviating diseases related to GPCR mislocalization. This chapter first provides an overview on the biogenesis of GPCRs and then describes the recently discovered concept of pharmacological chaperones. These membrane-permeable compounds have been found to enhance processing and maturation of several wild-type and mutant GPCRs as well as of some other proteins that are normally retained in the ER.

2. BIOGENESIS OF GPCRS

The GPCRs do not share extensive sequence identity but are believed to have a similar three-dimensional structure with seven hydrophobic transmembrane (TM) domains connected by three extracellular and three intracellular loops *(1,2)*. Similarly to other membrane proteins, the nascent GPCRs are cotranslationally targeted and translocated to the ER membrane. Only about 10% of GPCRs contain a cleavable signal sequence (stop–transfer sequence), whereas the majority rely on the first or the second TM domain,

From: *Contemporary Clinical Neuroscience: The G Protein-Coupled Receptors Handbook*
Edited by: L. A. Devi © Humana Press Inc., Totowa, NJ

which is believed to function as a targeting sequence (reverse signal anchor) *(3)*. In the ER membrane, the newly synthesized receptor molecules acquire their native three-dimensional conformations and undergo extensive processing, such as signal sequence cleavage, *N*-glycosylation, and disulfide bond formation. Final processing, such as further *N*-glycan modifications, *O*-glycosylation, palmitoylation, phosphorylation, and sulfation take place as the receptors are transported through the Golgi or have reached their final destination at the cell surface. Thus, folding and processing of GPCRs are complex events, and mechanisms that control their biogenesis have a critical role in governing cellular responsiveness to an array of extracellular signals.

2.1. Folding and Posttranslational Modifications

Folding of GPCRs in the ER is likely to follow the two-stage model for polytopic membrane proteins proposed by Popot and Engelman *(4)*. According to this model, the individual TM α-helical domains fold independently upon insertion into the ER membrane and assemble without significant rearrangement of their secondary structure. This hypothesis is supported by the finding that both folding and assembly of GPCRs do not appear to require covalent linkage of the polypeptide chain, because individually expressed TM domains can form functional receptors *(5–7)*. Factors that direct the assembly of the individual TM domains probably include packing of the α-helices and formation of intramolecular interactions, which involve residues in different TM domains. Additionally, the extra- and intracellular loops connecting the helices are likely to be important—especially the conserved cysteine residues in the first and second extracellular loops, which apparently form a constraining disulfide bond in more than 90% of GPCRs *(8)*. Direct evidence for the existence of such a bond has been obtained for the muscarinic m1 acetylcholine receptor *(9)*, rhodopsin *(10,11)*, and substance P receptor *(12)*. Increasing evidence also suggests that dimerization or higher oligomeric arrangements between GPCR molecules or with other accessory proteins may contribute to the quaternary structure of the receptors *(13–16)*.

The functional activity of GPCRs is critically dependent on the ability of the protein to change its conformation upon activation *(8,17)*. Therefore, it can be hypothesized that the apparent need to maintain conformational flexibility may lead to inherent instability of the assembled TM bundle. This notion is supported by the observation that several constitutively active mutant GPCRs are conformationally unstable *(18–20)*. The vital need to maintain conformational flexibility may also result in high susceptibility to misfolding. This is exemplified by the numerous naturally occurring mutations among GPCRs that result in intracellular trapping of the misfolded protein *(21–29)*. Not only truncations and short in-frame deletions and inversions, but also single amino acid changes, can lead to ER retention of the affected receptor, leading to human diseases such as retinitis pigmentosa, nephrogenic diabetes insipidus, and hypogonadotropic hypogonadism (reviewed in ref. *30*). For example, to date, 178 different V2 vasopressin receptor-inactivating mutations causing nephrogenic diabetes insipidus have been identified (http://www.medicine.mcgill.ca/nephros/), most of which appear to be defective in intracellular transport *(26)*. In addition to GPCRs, many other membrane-bound and secretory proteins may contain mutations that produce minor changes in the primary structure and that result in aberrant intracellular retention of the mutant proteins. Examples include minor mutations in the cystic fibrosis TM conductance regulator (CFTR) and α_1-antitrypsin that have been shown as the underlying causes for cystic fibrosis and some forms of emphysema, respectively *(31,32)*.

Folding difficulties do not necessarily follow only from deviations from the wild-type structure; they may be characteristic for the wild-type GPCRs as well, as suggested by our studies regarding the human δ-opioid receptor *(33,34)*. We found that in stably transfected human embronic kidney (HEK)293 cells, conversion of the receptor precursor to the mature cell-surface form was slow ($t_{1/2}$: approx 120 min), and only about 40% of the precursors were exported out of the ER. This did not result from overloading of the capacity of the cells to process the newly synthesized receptors, because there was no correlation between receptor expression level and maturation efficiency. Therefore, it is quite reasonable to speculate that the low efficiency of maturation of the δ-opioid receptor is an intrinsic property of the protein molecule itself and is related to folding difficulties. Increasing evidence supports the notion that such inefficient maturation is not restricted to the δ-opioid receptor but may represent a general feature shared by many GPCRs. For example, gonadotropin receptors (the luteinizing hormone and follicle-stimulating hormone receptors) have been shown to mature inefficiently in heterologous expression systems *(35–37)*, and a large pool of precursor forms of these receptors has been detected in natural tissues *(35,36,38,39)*, suggesting that processing inefficiency may characterize in vivo systems as well. It has also been shown that naturally occurring GPCR variants may differ in maturation efficiency; in pulse–chase studies, Fishburn et al. *(40)* observed that about 20% of the long isoform of the D2 dopamine receptor was in an immature form after 3 h of chase, whereas the short isoform was fully processed to the mature form during that time. Inefficient folding of polytopic membrane proteins is not limited to GPCRs, because only about 20% of the wild-type CFTR was found to be processed to the mature form *(41)*.

Nascent glycoproteins are cotranslationally modified by the addition of *N*-linked glycans. This applies also to GPCRs, because they contain several putative sites for N-linked glycosylation (Asn–X–Thr/Ser, where X is any amino acid except Pro) in their extracellular N-terminal domain or in the extracellular loops. Only a few isolated exceptions have been reported that are devoid of any consensus sites for N-linked glycosylation, such as the α_{2B}-adrenergic receptor *(42)*. The oligosaccharyl transferase transfers the glycan precursor $Glc_3Man_9GlcNAc_2$ from a lipid carrier to the nascent protein. Processing of the precursor is initiated immediately by removal of the terminal glucoses *(43,44)*, final processing of the N-linked glycans occurs during transport of the protein through the Golgi. In contrast to N-glycosylation, O-glycosylation and palmitoylation are late posttranslational modifications. Thus far, only a few GPCRs have been shown to be modified by O-glycosylation on Thr or Ser residues *(33,45,46)*, but addition of palmitate to cysteines on the proximal end of the C-terminal domain of GPCRs is a common modification (reviewed in ref. *47*). Both O-glycosylation and palmitoylation appear to occur after the newly synthesized receptors have been exported from the ER (*[33]*, U. Petäjä-Repo, Mireille Hogue, and M. Bouvier, manuscript in preparation).

The N-linked glycans of GPCRs have been studied extensively, but their functional role has remained elusive. It appears that at the cell surface, their role may vary depending on the receptor and the cell type. Nevertheless, it has become increasingly apparent that addition of the N-linked glycans to the nascent GPCRs may be important for folding of these proteins, because several receptors have been demonstrated to display decreased cell-surface expression upon mutation of their putative N-glycosylation sites *(48–52)*. This most likely results from the important role of the N-linked glycans in the ER quality control that oversees the folding of glycoproteins (*see* Subheading 2.2.). Addition of palmitate may also be important for the transport of newly synthesized receptors to the plasma membrane *(53–56)*, but the mechanisms behind this observation have not been investigated.

2.2. ER Quality Control

Similarly to all other proteins traversing the secretory pathway, folding and maturation of GPCRs are monitored by the ER quality control, which allows only correctly folded and assembled proteins to leave this cellular compartment and progress to their final destinations *(57–59)*. The primary quality control is based on common structural features that distinguish native protein conformations from non-native conformations and relies on ubiquitous molecular chaperones and folding factors, such as BiP, calnexin, caltericulin, protein disulfide isomerase, and ERp57. Important features for recognition include exposure of hydrophobic regions, unpaired cysteine residues, and tendency to aggregate *(58)*.

The mechanisms that distinguish native GPCR conformations from nonnative ones and assist in folding are undoubtedly crucial for the biogenesis and cell-surface expression of GPCRs, but they have remained largely uncharacterized. Only three molecular chaperones, calnexin, calreticulin, and BiP (which are known to be involved in the primary ER quality control), have been shown to interact with GPCRs. The lectin calnexin was found in a complex with the V2 vasopressin, glycoprotein hormone, and olfactory receptors *(60–63)*, and the thyroid stimulating hormone receptor was also discovered to interact with calreticulin and BiP *(62)*. Importantly, calnexin was shown to be involved in the ER retention of mutant V2 vasopressin receptors *(61)*, indicating that the calnexin/calreticulin cycle may be intimately involved in GPCR folding. The calnexin/calreticulin cycle is one of the most thoroughly characterized primary quality control systems that are responsible for assessing the folding of glycoproteins *(44,57–59)*. The monoglucosylated form of the N-linked oligosaccharides confers the capacity to bind to calnexin and/or calreticulin to the glycoproteins, and a cycle of binding to and release from these chaperones is determined by the sequential action of two enzymes: the UDP-glucose-glycoprotein glucosyltransferase and glucosidase II, which add and remove the terminal glucose from the N-glycan, respectively. The former enzyme can assess the folding status of the substrate and adds the glucose to the N-glycan only if the substrate protein has not folded correctly, thereby promoting rebinding of calnexin/calreticulin. Both calnexin and calreticulin are known to form a complex with the ERp57, thus coupling folding assistance to disulfide bond formation.

Most of the ER molecular chaperones and folding factors are lumenal proteins and, therefore, possess a direct access only to soluble, newly synthesized proteins in the ER lumen. This raises the important issue regarding how folding of the GPCR TM and cytosolic domains is monitored, because a large portion of GPCRs is buried within the membrane bilayer or lies in the cytoplasm. Molecular chaperones that could directly oversee folding and assembly of TM domains of GPCRs are completely unknown. On the other hand, folding of GPCR cytosolic domains might be assisted by cytosolic chaperones such as the heat-shock protein Hsp70, as has been shown for another polytopic membrane protein, CFTR *(64)*. The recent observation that rhodopsin can interact with cytosolic Hsp40 proteins HSJ1a and HSJ1b and with Hsp70 is consistent with this hypothesis *(65)*.

Unlike the primary quality control that relies on common structural features, the secondary ER quality control involves protein-, cell- and tissue-specific factors that regulate folding and transport of individual proteins or protein families *(57)*. Some secondary ER quality control factors involved in GPCR intracellular transport have been identified (reviewed in ref. *66*). These include the DRiP78 *(67)*, cis–trans prolyl isomerase homologs NinaA *(68)* and RanBP2 *(69)*, ORD-4 *(70)*, M10 and M1 families of major histocompatibility complex class 1b molecules *(71)*, and receptor activity-modifying proteins (RAMPs) *(72,73)*, which have been demonstrated to be crucial for the expression of the D1 dopamine receptor, some opsins, olfactory receptors, V2R pheromone receptors, and the calcitonin receptor-like receptor, respectively. The last four protein groups function not only in the ER but may also have a role as escort proteins during transit through the Golgi. In the case of RAMPs, the interaction with the receptors is maintained at the plasma membrane governing the functional activity of the receptors *(74)*. Interestingly, in the case of the γ-aminobutyric acid (GABA) type B1 receptor, heterodimerization with another GPCR, the GABA type B2 receptor, was shown to be required for both proper trafficking and function *(75)*. This requirement for heterodimerization may also apply to some taste receptors *(76–78)*.

2.3. ER-Associated Degradation

If folding of the nascent proteins fails or subunits of multimeric proteins are unable to assemble correctly, the ER quality control recognizes them as aberrant and targets them for degradation. This ER-associated degradation (ERAD) involves polyubiquitination and dislocation through the Sec61 translocon to the cytosol, where the misfolded and assembled proteins are degraded by the 26S proteasomes *(79–81)*. In the case of glycoproteins, the ER α1,2-mannosidase I has been proposed to act as a timer for the exit of substrate glycoproteins from the calnexin/calreticulin cycle by producing a $Man_8GlcNAc_2$ form of the oligosaccharide *(82,83)*. This N-linked glycan structure mediates binding to the Man_8-specific ER lectin EDEM, and this interaction apparently plays a role in the delivery of the substrate glycoproteins to the degradation machinery *(84–86)*.

The human δ-opioid receptor was the first GPCR that was demonstrated to be polyubiquitinated and targeted for degradation by the proteasomes *(34)*. Subsequently, several other ER-retained wild-type or mutant GPCRs have been shown to be degraded by this pathway *(63,87–90)*. The mechanism by which ER quality control recognizes these proteins as aberrant and targets them for degradation is currently unknown. Upon proteasomal blockade, the δ-opioid and thyrotropin-releasing hormone receptors were found to accumulate in the cytosol in a deglycosylated soluble form *(34,90)*. In contrast, misfolded opsin mutants accumulate in large insoluble cytosolic aggregates called aggresomes *(87,88)*, and murine olfactory receptors (ml7 and mOREG) appear to form aggregates in the ER and are at least partially degraded by autophagy *(63)*. The reasons for these apparent differences are presently unknown. However, it is possible that the propensity of the mutant opsins and olfactory receptors to aggregate may relate to the fact that the entire pool of these receptors is targeted for degradation, whereas the majority of newly synthesized δ-opioid and thyrotropin-releasing hormone receptors can exit the ER and are transported to the cell surface. In any case, it is apparent that accumulation of misfolded proteins in the ER is detrimental to the cell, and the quality control mechanisms that dispose misfolded proteins need to function very efficiently. Therefore, conformationally unstable proteins, such as GPCRs, may be targeted for degradation prematurely—even in the case of wild-type proteins. This notion is consistent with the findings that the amount of mature δ-opioid receptors, rhodopsin, and luteinizing hormone receptors appear to increase following blockade of the proteasomal degradation pathway (refs. *34* and *87*; E. Maritta Pietilä, Jussi T. Tuusa, Pirjo M. Apaja, Jyrki T. Aatsinki, Hannu J. Rajaniemi, and U. Petäjä-Repo, manuscript in preparation).

3. CHEMICAL AND PHARMACOLOGICAL CHAPERONES

A large fraction of cytosolic proteins apparently fail to fold correctly and are degraded *(91)*, and accumulating evidence suggests that this may also be the case for proteins that traverse the secretory pathway, including GPCRs. Therefore, it could be envisaged that promoting the release of ER-retained proteins might be sufficient to enhance processing and cell-surface targeting of these proteins and even recover function in some loss-of-function diseases. This is based on the fact that the ER-retention relies on conformational, rather than functional, criteria *(58)*, possibly preventing many potentially functional proteins from reaching their correct cellular location.

3.1. Chemical Chaperones

The initial attempts to enhance folding and maturation of ER-retained proteins relied on low-molecular-weight compounds such as glycerol, which were known to increase the stability of native proteins and assist in refolding of unfolded proteins *(92,93)*. These compounds, called chemical chaperones, were first used to correct the mutant phenotype in cells expressing the ΔF^{508} form of the CFTR *(94,95)*. This mutation, which is responsible for the majority of cases of cystic fibrosis, leads to ER retention and degradation of the affected protein so that it is unable to reach the apical plasma membrane of epithelial cells *(41)*. Glycerol, deuterated water, dimethylsulfoxide, or trimethylamine *N*-oxide enhanced posttranslational maturation of the mutant channel, leading to an increased cyclic adenosine monophosphate (cAMP)-dependent chloride transport *(94,95)*. Since these pioneering studies were performed, chemical chaperones, many of which have the property of increasing cellular osmolar activity, have been discovered to partially correct mislocalization of several other proteins, such as the mutant aquaporin-2 (which is associated with nephrogenic diabetes insipidus *[96]*) and the mutant α_1-antitrypsin Z (which is associated with emphysema and liver disease *[97]*).

Several mechanisms have been proposed for the chemical chaperone action. For example, it has been envisioned that these compounds might stabilize incorrectly folded proteins, reduce aggregation, or prevent nonproductive interactions with other proteins *(98)*. Although the precise mechanism of action is unknown, it has been proposed that osmolytes (amino acids such as proline, sugars such as trehalose, and polyols such as glycerol) tend to raise the free energy of the misfolded species through their unfavorable interactions with the protein backbone, thus favoring the native completely folded state of the protein *(99)*. In any case, the chemical chaperones are nonspecific and are active only in high concentrations, which hinders their use in in vivo settings.

3.2. Pharmacological Chaperones

An exciting and novel twist on the concept of chemical chaperones was introduced by Loo and Clarke *(100)*, who studied the effects of more specific compounds on P-glycoprotein transporter mutants encoded by the multidrug resistance 1 gene (Table 1). The authors found that various substrates and modulators of the transporter enhanced folding of the mutants and increased the amount of fully processed mature protein at the cell surface. Subsequently, Fan et al. *(101)* applied the same paradigm for mutant forms of the lysosomal enzyme α-galactosidase A that are associated with Fabry disease. In this study, a competitive inhibitor, 1-deoxygalactonojirimycin, was found to facilitate transport of the $R^{301}Q$ mutant to its correct destination within lysosomes and to significantly increase galactosidase activity in lymphoblasts carrying the $R^{301}Q$ or $Q^{279}E$ mutations. Similarly, the mutant $N^{470}D$ HERG potassium channel associated with congenital long QT syndrome was rescued to the plasma membrane following treatment with the channel blockers E-4031, astemizole, and cisapride *(105)*. These observations paved the way to more specific approaches to try to alleviate problems in protein mislocalization and misfolding. In contrast to chemical chaperones, the compounds that have the ability to bind specifically to the affected protein have a much higher potential to be useful in in vivo settings and are less likely to result in toxic effects. Therefore, we defined these specific compounds as pharmacological chaperones *(114,123)*. The concept of pharmacological chaperones was first applied to GPCRs in an attempt to enhance cell-surface expression of mutant V2 vasopressin receptors that are retained in the ER *(114)* and are responsible for the development of X-linked nephrogenic diabetes insipidus. The cell-permeable receptor-specific antagonists SR121463A and VPA-985 converted the precursor form of the $\Delta L^{62}AR^{64}$ V2 vasopressin receptor (a deletion mutant lacking three amino acids in the first cytoplasmic loop) into a fully glycosylated mature protein that was now targeted to the cell surface in African green monkey fibro blasts COS and HEK293 cells, as determined by pulse–chase analysis, cell-surface immunofluorescence microscopy, and flow cytometry. Once these receptors were at their correct cellular location, they were functional and conferred arginine vasopressin-stimulated cAMP responses. Importantly, this effect could not be mediated by nor competed with an antagonist that was membrane impermeable, indicating that SR121463A and VPA-985 were mediating their effects intracellularly. To date, a total of 11 of 18 intracellularly retained V2 vasopressin receptor mutants have demonstrated cell-surface targeting and agonist-stimulated signaling following SR121463A treatment. In a subsequent study, two other V2 vasopressin receptor mutants, $L^{292}P$ and ΔV^{278}, were found to be responsive to SR121463B, which rescued their surface expression in both COS cells and polarized Madin–Darby canine kidney II cells *(115)*. In the latter cell line, the mutants $L^{292}P$ and $R^{337}X$ were appropriately delivered to the basolateral surface following the antagonist treatment.

Because the receptor-specific ligands were able to rescue ER-retained V2 vasopressin receptor mutants to the cell surface, we tested whether the same strategy could enhance processing and cell-surface targeting of the wild-type δ-opioid receptor, which has an inherently low maturation efficiency *(116)*. Addition of the lipophilic opioid antagonist naltrexone to the culture medium in pulse–chase experiments significantly increased the turnover rate of receptor precursors in stably transfected HEK293 cells, leading to a twofold increase in the processing of this species to the mature form. The naltrexone-mediated enhancement in receptor maturation was dose-dependent, and the EC_{50} was similar to the estimated K_i for this ligand, indicating requirement for receptor occupancy. Additionally, the antagonist was able to enhance receptor maturation, even when protein transport to the cell surface was blocked by brefeldin A and the membrane-impermeable peptidic opioid agonist Leu-enkephalin (at saturating concentration) was not able to block its effects, confirming the intracellular site of action of the antagonist.

Table 1
Rescue of Misfolded or Incompletely Folded Proteins by Pharmacological Chaperones

Protein	Disease	Mutant rescued	Reagent	Reference
P-glycoprotein	—	$G^{54}V$, $G^{251}V$, $G^{268}V$, $G^{300}V$, $G^{427}C$, $S^{434}C$, ΔY^{490}, $E^{707}A$, $A^{718}L$, $W^{803}A$, $A^{841}L$, $G^{854}V$	Capsaicin, cyclosporine A, vinblastine, verapamil (substrates and modulators)	100
Lysosomal α-galactosidase A	Fabry disease	$R^{301}Q$, $Q^{279}E$	1-deoxy-galactonojirimycin and its derivatives (competitive inhibitors)	101,102
Lysosomal β-galactosidase	β-galactosidosis (G_{M1}-gangliosidosis and Morquio B disease)	Wild-type, $R^{201}C$, $R^{201}H$, $R^{457}Q$, $W^{273}L$, $Y^{83}F$	N-octyl-4-epi-β-valienamine (competitive inhibitor)	103
Lysosomal β-glucosidase	Gaucher disease	Wild-type, $N^{370}S$	N-(n-nonyl)deoxynojirimycin	104
HERG potassium channel	Long QT syndrome	$T^{65}P$, $N^{470}D$, $G^{601}S$	Alkyltriethylammonium derivatives, astemizole, cisapride, E-4031, fexofenadine, quinidine, terfenadine (channel blockers)	105–108
Sulfonylurea receptor	Hyperinsulinemic hypoglycemia of infancy	Wild-type, $A^{116}P$, $V^{187}D$	glibenclamide, tolbutamide (inhibitors)	109
Tyrosinase	—	Wild-type	DOPA, tyrosine (substrates)	110
Anti-phenyl-phosphocholine antibody	—	—	Phenylphosphocholine, (hapten ligand)	111

(continued)

Protein	Disease	Mutations	Ligands	Refs
p53	cancer	$I^{95}T, R^{175}H, R^{273}H$	CDB3 (peptide)	112,113
V2 vasopressin receptor	nephrogenic diabetes insipidus	$\Delta L^{62}AR^{64}, L^{59}P, L^{83}Q, Y^{128}S, S^{167}L, \Delta V^{278}, L^{292}P, A^{294}P, P^{322}H, R^{337}\overline{X}$	SR121463A, SR121463B, VPA-985 (antagonists)	114,115
δ-opioid receptor	—	Wild-type, $D^{95}A$		116
receptor			naloxone, naltrexone, naltriben, naltrindole, TICP (antagonists); buprenorphine, bremazocine, nalbuphine, SNC-80, TAN-67, tonazocine (agonists)	
μ-opioid receptor	—	$\Delta R^{258}LSKV^{262}, \Delta K^{344}FCTR^{348}$	Diprenorphine, naloxone, naltrindole (antagonists); buprenorphine, etorphine, levorphanol, L-methadone, morphine, nalorphine, oxymorphine (agonists)	117
Gonadotropin-releasing hormone receptor	Hypogonadotropic hypogonadism	Wilc-type, $N^{10}K, T^{32}I, E^{90}K, Q^{106}R, A^{129}D, R^{139}H, C^{200}Y, R^{262}Q, L^{266}R, C^{279}Y, Y^{284}C$ and several artifical mutants	IN31b, IN3, IN30 (indoles); Q08, Q76, Q89 (quinolones); A-7662.0, A-177775.0, A-64755.0, A-222509.0 (erythromycin macrolides) (antagonists)	28,118,119
Rhodopsin	Autosomal dominant retinitis, pigmentosa	$T^{17}M, P^{23}H$	9-cis-retinal, 11-cis-retinal, 11-cis-7 ring retinal (inverse agonists)	88,120–122

Importantly, the chaperone action was found to be pharmacologically selective because unrelated GPCR ligands, such as the β-adrenergic antagonist propranolol, were unable to mimic the effects mediated by the opioid ligand. The effects also could not be mimicked by SR121463A, a compound that promotes functional rescue of ER-retained V2 vasopressin receptor mutants *(114)*. These data strongly support the hypothesis that the receptor-specific ligands act as pharmacological chaperones by binding to newly synthesized receptors in the ER. Interestingly, the signaling efficacy of the ligands does not seem to be an important aspect of their pharmacological chaperone action, because both antagonists and agonists were found to have comparable activity. This suggests that stabilization of several distinct receptor conformations can facilitate adequate folding and/or export of the receptor protein from the ER.

The subsequent observations on another opioid receptor subtype, the μ-opioid receptor, are consistent with the findings obtained for the δ-opioid receptor. Chaipatikul et al. *(117)* showed that intracellularly retained mutants were transported to the cell surface after incubating transiently transfected HEK293 cells with μ-opioid receptor-specific ligands. This was demonstrated by ligand binding, flow cytometry, and immunofluorescence. Importantly, as was observed for the δ-opioid receptor *(116)*, only membrane-permeable ligands were found to be effective, and both agonists and antagonists were able to rescue the intracellularly trapped receptors.

Subsequently, the pharmacological chaperone paradigm has been extended to other GPCRs. Conn and colleagues demonstrated that it is possible to use peptidomimetics indoles, quinololes, and erythromycin macrolides to enhance cell-surface expression of gonadotropin-releasing hormone receptor mutants that cause hypogonadotropic hypogonadism *(28,118,119,124)*. For 11 of 14 mutant receptors, addition of the ligand at the time of transfection conferred both ligand binding and effector coupling to COS cells. Ligands with a high affinity for the wild-type receptor were observed to be the most efficient. In an analogous approach, ER-retained rhodopsin mutants related to retinitis pigmentosa were rescued to the cell surface using retinal-based ligands *(88,120–122)*.

The concept of pharmacological chaperones has also been extended to several other misfolded proteins that cause human diseases. In addition to receptor ligands, several small pharmacologically selective molecules have been used to rescue misfolded ion channel, enzyme, and transcriptional factor mutants (Table 1). The paradigm has also been applied to enhance assembly and secretion of immunoglobulins using the cognate hapten ligand as a pharmacological chaperone *(111)*.

3.3. The Mechanism of Action of Pharmacological Chaperones

The mechanisms of the pharmacological chaperone action on GPCRs are not fully understood; however, it can be hypothesized that the most likely mechanism by which receptor-specific ligands mediate their effect is stabilization of the newly synthesized receptors in the native or intermediate state of their folding pathway. Binding of a ligand may alter the thermodynamic equilibrium in favor of the correctly folded protein. This would then enhance the possibility of the protein to escape the stringent ER quality control, ultimately leading to an increase in the steady-state level of functional receptors at the cell surface. This hypothesis is consistent with the observation that there is a clear correlation between the magnitude of ligand-mediated rescue and binding affinity of ligands that mediate the effects *(116,119)*. A similar correlation has been observed for enzyme inhibitors and blockers that rescue mutant forms of the α-galactosidase A and HERG potassium channel, respectively *(102,106)*. The link between the binding affinity and the chaperone effect can be rationalized according to either an induced fit or a kinetic selection model. In the induced fit paradigm, the higher binding energy provided by the high-affinity interaction contributes to promote the native protein conformation that is compatible with ER export. In the selection model, the longer average binding time ($>k_{on}/k_{off}$) of the higher affinity ligands allows more time for the ligand-bound stabilized conformation to reach its native form and leave the ER. Obviously, these are not mutually exclusive models, and both mechanisms could contribute to the pharmacological chaperone action of the GPCR ligands.

The precise molecular mechanisms by which ligands may enhance conformational stabilization of the newly synthesized GPCRs are still unknown. Nevertheless, it can be hypothesized that these compounds may promote more stable packing of the TM α-helices of the receptor, analogously to other small molecular ligands that have been shown to induce changes in protein thermostability and flexibility *(125–127)*. Although it has not been directly shown that the ligands are able to bind to receptor precursors in the ER, the indirect evidence is compelling. Therefore, it is reasonable to speculate that binding of a ligand to the newly synthesized receptor might stabilize the labile protein by inducing additional conformational constraints within the α-helical bundle. Interestingly, all the membrane-permeable opioid ligands that were tested were observed to enhance opioid receptor cell-surface expression *(116,117)*, suggesting that both the inactive receptor conformation (stabilized by antagonists or inverse agonists) and the active one(s) (stabilized by agonists) are recognized as export competent forms by the ER quality control.

4. CONCLUSIONS

As the molecular mechanisms of genetically inherited diseases are uncovered, it becomes apparent that errors in folding and localization are the underlying causes for an increasing number of diseases and disorders, including those relating to GPCRs. Therefore, assessing the mechanisms that are responsible for the errors in folding and localization and development of novel approaches to alleviate these problems have become increasingly important. Thus, the observation that specific cell-permeable compounds can stabilize protein conformations(s) that are compatible with their export from the ER might represent a generally applicable rescue strategy for many mutant proteins that are responsible for human diseases.

Potential clinical implications of pharmacological chaperones have already been evoked in several cases of inherited diseases, and the therapeutic potential of pharmacological chaperones has provided a rationale for designing novel drugs to treat these disorders. For example, pharmacological chaperones are proposed as potential drugs for the treatment of lysosomal storage disorders such as Fabry disease and Gaucher syndrome, which are caused by ER-retained lysosomal enzymes *(128,129)*. Similar scenarios could be envisioned for treating patients that suffer from GPCR-related disorders. For example, patients that suffer from nephrogenic diabetes insipidus, hypogonadotropic hypogonadism, or retinitis pigmentosa could benefit from treatment with pharmacological chaperones. In the case of GPCRs, many pharmacological chaperones that could be used clinically are likely to be among existing drugs that are already used. However, further studies are required to determine whether the pharmacological chaperone activity is an intrinsic property of GPCR ligands in general. In any case, the interesting results reviewed in this chapter provide a compelling argument for further research on this strategy. Screening campaigns based on the ability to rescue cell-surface targeting and function of otherwise ER-retained mutant proteins are likely to uncover new pharmacological chaperones with various potential therapeutic applications.

Importantly, it should also be remembered that the clinical implications of pharmacological chaperones might not be limited to conditions resulting from mutated genes. This was suggested by the observations that receptor-specific ligands can promote maturation and cell-surface expression of the wild-type δ-opioid and gonadotropin-releasing hormone receptors *(116,130)*. If future studies reveal that this concept applies to a larger number of wild-type GPCRs, then the pharmacological chaperone activity of the corresponding ligands may prove a useful parameter in the design of better therapeutic drugs and may help to explain and predict the functional consequences that follow chronic treatment with these agents in vivo. Indeed, compounds that have pharmacological chaperone activity could regulate the expression level of their target receptors and modulate their responsiveness in ways that are independent from their direct signaling efficacy. In summary, pharmacological chaperone activity represents a recently uncovered property that permits specific facilitation of cell-surface expression of membrane proteins and may have therapeutic applications in many clinical conditions.

REFERENCES

1. Palczewski K, Kumasaka T, Hori T, et al. Crystal structure of rhodopsin: a G protein-coupled receptor. Science 2000;289:739–745.
2. Trabanino RJ, Hall SE, Vaidehi N, Floriano WB, Kam VWT, Goddard WA 3rd. First principles predictions of the structure and function of G-protein-coupled receptors: validation for bovine rhodopsin. Biophys J 2004;86:1904–1921.
3. Wallin E, von Heijne G. Properties of N-terminal tails in G-protein coupled receptors: a statistical study. Protein Eng 1995;8:693–698.
4. Popot JL, Engelman DM. Helical membrane protein folding, stability, and evolution. Annu Rev Biochem 2000;69:881–922.
5. Kobilka BK, Kobilka TS, Daniel K, Regan JW, Caron MG, Lefkowitz RJ. Chimeric α2-,β2-adrenergic receptors: delineation of domains involved in effector coupling and ligand binding specificity. Science 1988;240:1310–1316.
6. Ridge KD, Lee SSJ, Yao LL. In vivo assembly of rhodopsin from expressed polypeptide fragments. Proc Natl Acad Sci USA 1995;92:3204–3208.
7. Schöneberg T, Liu J, Wess J. Plasma membrane localization and functional rescue of truncated forms of a G protein-coupled receptor. J Biol Chem 1995;270:18,000–18,006.
8. Karnik SS, Gogonea C, Patil S, Saad Y, Takezako T. Activation of G-protein-coupled receptors: a common molecular mechanism. Trends Endocrinol Metab 2003;14:431–437.
9. Kurtenbach E, Curtis CA, Pedder EK, Aitken A, Harris ACM, Hulme EC. Muscarinic acetylcholine receptors. Peptide sequencing identifies residues involved in antagonist binding and disulfide bond formation. J Biol Chem 1990;265:13,702–13,708.
10. Karnik SS, Khorana HG. Assembly of functional rhodopsin requires a disulfide bond between cysteine residues 110 and 187. J Biol Chem 1990;265:17,520–17,524.
11. Hwa J, Klein-Seetharaman J, Khorana HG. Structure and function in rhodopsin: mass spectrometric identification of the abnormal intradiscal disulfide bond in misfolded retinitis pigmentosa mutants. Proc Natl Acad Sci USA 2001;98:4872–4876.
12. Boyd ND, Kage R, Dumas JJ, Krause JE, Leeman SE. The peptide binding site of the substance P (NK-1) receptor localized by a photoreactive analogue

of substance P: presence of a disulfide bond. Proc Natl Acad Sci USA 1996;93:433–437.
13. Rios CD, Jordan BA, Gomes I, Devi LA. G-protein-coupled receptor dimerization: modulation of receptor function. Pharmacol Ther 2001;92:71–87.
14. Angers S, Salahpour A, Bouvier M. Dimerization: an emerging concept for G protein-coupled receptor ontogeny and function. Annu Rev Pharmacol Toxicol 2002;42:409–435.
15. George SR, O'Dowd BF, Lee SP. G-protein-coupled receptor oligomerization and its potential for drug discovery. Nat Rev Drug Discov 2002;1:808–820.
16. Milligan G, Ramsay D, Pascal G, Carrillo JJ. GPCR dimerisation. Life Sci 2003;74:181–188.
17. Gether U. Uncovering molecular mechanisms involved in activation of G protein-coupled receptors. Endocr Rev 2000;21:90–113.
18. Gether U, Ballesteros JA, Seifert R, Sanders-Bush E, Weinstein H, Kobilka BK. Structural instability of a constitutively active G protein-coupled receptor. Agonist-independent activation due to conformational flexibility. J Biol Chem 1997;272:2587–2590.
19. Alewijnse AE, Timmerman H, Jacobs EH, et al. The effect of mutations in the DRY motif on the constitutive activity and structural instability of the histamine H_2 receptor. Mol Pharmacol 2000;57:890–898.
20. Wilson MH, Highfield HA, Limbird LE. The role of a conserved inter-transmembrane domain interface in regulating α_{2a}-adrenergic receptor conformational stability and cell-surface turnover. Mol Pharmacol 2001;59:929–938.
21. Sung CH, Schneider BG, Agarwal N, Papermaster DS, Nathans J. Functional heterogeneity of mutant rhodopsins responsible for autosomal dominant retinitis pigmentosa. Proc Natl Acad Sci USA 1991;88:8840–8844.
22. Bai M, Quinn S, Trivedi S, et al. Expression and characterization of inactivating and activating mutations in the human $Ca^{2+}o$-sensing receptor. J Biol Chem 1996;271:19,537–19,545.
23. Tanaka H, Moroi K, Iwai J, et al. Novel mutations of the endothelin B receptor gene in patients with Hirschsprung's disease and their characterization. J Biol Chem 1998;273:11,378–11,383.
24. d'Addio M, Pizzigoni A, Bassi MT, et al. Defective intracellular transport and processing of OA1 is a major cause of ocular albinism type 1. Hum Mol Genet 2000;9:3011–3018.
25. Themmen APN, Huhtaniemi IT. Mutations of gonadotropins and gonadotropin receptors: elucidating the physiology and pathophysiology of pituitary-gonadal function. Endocr Rev 2000;21:551–583.
26. Morello JP, Bichet DG. Nephrogenic diabetes insipidus. Annu Rev Physiol 2001;63:607–630.
27. Wonerow P, Neumann S, Gudermann T, Paschke R. Thyrotropin receptor mutations as a tool to understand thyrotropin receptor action. J Mol Med 2001;79:707–721.
28. Leaños-Miranda A, Janovick JA, Conn PM. Receptor-misrouting: an unexpectedly prevalent and rescuable etiology in GnRHR-mediated

hypogonadotropic hypogonadism. J Clin Endocrinol Metab 2002;87:4825–4828.
29. Lubrano-Berthelier C, Durand E, Dubern B, et al. Intracellular retention is a common characteristic of childhood obesity-associated MC4R mutations. Hum Mol Genet 2003;12:145–153.
30. Spiegel AM, Weinstein LS. Inherited diseases involving G proteins and G protein-coupled receptors. Annu Rev Med 2004;55:27–39.
31. Aridor M, Hannan LA. Traffic jam: a compendium of human diseases that affect intracellular transport processes. Traffic 2000;1:836–851.
32. Aridor M, Hannan LA. Traffic jams II: an update of diseases of intracellular transport. Traffic 2002;3:781–790.
33. Petäjä-Repo UE, Hogue M, Laperrière A, Walker P, Bouvier M. Export from the endoplasmic reticulum represents the limiting step in the maturation and cell surface expression of the human δ opioid receptor. J Biol Chem 2000;275:13,727–13,736.
34. Petäjä-Repo UE, Hogue M, Laperrière A, Bhalla S, Walker P, Bouvier M. Newly synthesized human δ opioid receptors retained in the endoplasmic reticulum are retrotranslocated to the cytosol, deglycosylated, ubiquitinated, and degraded by the proteasome. J Biol Chem 2001;276:4416–4423.
35. Hipkin RW, Sánchez-Yagüe J, Ascoli M. Identification and characterization of a luteinizing hormone/chorionic gonadotropin (LH/CG) receptor precursor in a human kidney cell line stably transfected with the rat luteal LH/CG receptor complementary DNA. Mol Endocrinol 1992;6:2210–2218.
36. Vannier B, Loosfelt H, Meduri G, Pichon C, Milgrom E. Anti-human FSH receptor monoclonal antibodies: immunochemical and immunocytochemical characterization of the receptor. Biochemistry 1996;35:1358–1366.
37. Beau I, Misrahi M, Gross B, et al. Basolateral localization and transcytosis of gonadotropin and thyrotropin receptors expressed in Madin-Darby canine kidney cells. J Biol Chem 1997;272:5241–5248.
38. VuHai-LuuThi MT, Misrahi M, Houllier A, Jolivet A, Milgrom E. Variant forms of the pig lutropin/choriogonadotropin receptor. Biochemistry 1992;31:8377–8383.
39. Apaja PM, Harju KT, Aatsinki JT, Petäjä-Repo UE, Rajaniemi HJ. Identification and structural characterization of the neuronal luteinizing hormone receptor associated with sensory systems. J Biol Chem 2004;279:1899–1906.
40. Fishburn CS, Elazar Z, Fuchs S. Differential glycosylation and intracellular trafficking for the long and short isoforms of the D_2 dopamine receptor. J Biol Chem 1995;270:29,819–29,824.
41. Kopito RR. Biosynthesis and degradation of CFTR. Physiol Rev 1999;79:S167–S173.
42. Lomasney JW, Lorenz W, Allen LF, et al. Expansion of the α_2-adrenergic receptor family: cloning and characterization of a human α_2-adrenergic receptor subtype, the gene for which is located on chromosome 2. Proc Natl Acad Sci USA 1990;87:5094–5098.

43. Parodi AJ. Protein glucosylation and its role in protein folding. Annu Rev Biochem 2000;69:69–93.
44. Helenius A, Aebi M. Intracellular functions of N-linked glycans. Science 2001;291:2364–2369.
45. Sadeghi H, Birnbaumer M. O-Glycosylation of the V2 vasopressin receptor. Glycobiology 1999;9:731–737.
46. Nakagawa M, Miyamoto T, Kusakabe R, et al. O-Glycosylation of G-protein-coupled receptor, octopus rhodopsin. Direct analysis by FAB mass spectrometry. FEBS Lett 2001;496:19–24.
47. Qanbar R, Bouvier M. Role of palmitoylation/depalmitoylation reactions in G-protein-coupled receptor function. Pharmacol Ther 2003;97:1–33.
48. Rands E, Candelore MR, Cheung AH, Hill WS, Strader CD, Dixon RA. Mutational analysis of β-adrenergic receptor glycosylation. J Biol Chem 1990;265:10,759–10,764.
49. Kaushal S, Ridge KD, Khorana HG. Structure and function in rhodopsin: the role of asparagine-linked glycosylation. Proc Natl Acad Sci USA 1994;91:4024–4028.
50. Couvineau A, Fabre C, Gaudin P, Maoret JJ, Laburthe M. Mutagenesis of N-glycosylation sites in the human vasoactive intestinal peptide 1 receptor. Evidence that asparagine 58 or 69 is crucial for correct delivery of the receptor to plasma membrane. Biochemistry 1996;35:1745–1752.
51. Lanctôt PM, Leclerc PC, Escher E, Leduc R, Guillemette G. Role of N-glycosylation in the expression and functional properties of human AT_1 receptor. Biochemistry 1999;38:8621–8627.
52. Böer U, Neuschäfer-Rube F, Möller U, Püschel GP. Requirement of N-glycosylation of the prostaglandin E_2 receptor EP3b for correct sorting to the plasma membrane but not for correct folding. Biochem J 2000;350:839–847.
53. Zhu H, Wang H, Ascoli M. The lutropin/choriogonadotropin receptor is palmitoylated at intracellular cysteine residues. Mol Endocrinol 1995;9:141–150.
54. Tanaka K, Nagayama Y, Nishihara E, Namba H, Yamashita S, Niwa M. Palmitoylation of human thyrotropin receptor: slower intracellular trafficking of the palmitoylation-defective mutant. Endocrinology 1998;139:803–806.
55. Blanpain C, Wittamer V, Vanderwinden JM, et al. Palmitoylation of CCR5 is critical for receptor trafficking and efficient activation of intracellular signaling pathways. J Biol Chem 2001;276:23,795–23,804.
56. Percherancier Y, Planchenault T, Valenzuela-Fernandez A, Virelizier JL, Arenzana-Seisdedos F, Bachelerie F. Palmitoylation-dependent control of degradation, life span, and membrane expression of the CCR5 receptor. J Biol Chem 2001;276:31,936–31,944.
57. Ellgaard L, Molinari M, Helenius A. Setting the standards: quality control in the secretory pathway. Science 1999;286:1882–1888.
58. Ellgaard L, Helenius A. Quality control in the endoplasmic reticulum. Nat Rev Mol Cell Biol 2003;4:181–191.

59. Trombetta ES, Parodi AJ. Quality control and protein folding in the secretory pathway. Annu Rev Cell Dev Biol 2003;19:649–676.
60. Rozell TG, Davis DP, Chai Y, Segaloff DL. Association of gonadotropin receptor precursors with the protein folding chaperone calnexin. Endocrinology 1998;139:1588–1593.
61. Morello JP, Salahpour A, Petäjä-Repo UE, et al. Association of calnexin with wild type and mutant AVPR2 that causes nephrogenic diabetes insipidus. Biochemistry 2001;40:6766–6775.
62. Siffroi-Fernandez S, Giraud A, Lanet J, Franc JL. Association of the thyrotropin receptor with calnexin, calreticulin and BiP. Effects on the maturation of the receptor. Eur J Biochem 2002;269:4930–4937.
63. Lu M, Echeverri F, Moyer BD. Endoplasmic reticulum retention, degradation, and aggregation of olfactory G-protein coupled receptors. Traffic 2003;4:416–433.
64. Meacham GC, Lu Z, King S, Sorscher E, Tousson A, Cyr DM. The Hdj-2/Hsc70 chaperone pair facilitates early steps in CFTR biogenesis. EMBO J 1999;18:1492–1505.
65. Chapple JP, Cheetham ME. The chaperone environment at the cytoplasmic face of the endoplasmic reticulum can modulate rhodopsin processing and inclusion formation. J Biol Chem 2003;278:19,087–19,094.
66. Tan CM, Brady AE, Nickols HH, Wang Q, Limbird LE. Membrane trafficking of G protein-coupled receptors. Annu Rev Pharmacol Toxicol 2004;44:559–609.
67. Bermak JC, Li M, Bullock C, Zhou QY. Regulation of transport of the dopamine D1 receptor by a new membrane-associated ER protein. Nat Cell Biol 2001;3:492–498.
68. Colley NJ, Baker EK, Stamnes MA, Zuker CS. The cyclophilin homolog ninaA is required in the secretory pathway. Cell 1991;67:255–263.
69. Ferreira PA, Nakayama TA, Pak WL, Travis GH. Cyclophilin-related protein RanBP2 acts as chaperone for red/green opsin. Nature 1996;383:637–640.
70. Dwyer ND, Troemel ER, Sengupta P, Bargmann CI. Odorant receptor localization to olfactory cilia is mediated by ODR-4, a novel membrane-associated protein. Cell 1998;93:455–466.
71. Loconto J, Papes F, Chang E, et al. Functional expression of murine V2R pheromone receptors involves selective association with the M10 and M1 families of MHC class Ib molecules. Cell 2003;112:607–618.
72. McLatchie LM, Fraser NJ, Main MJ, et al. RAMPs regulate the transport and ligand specificity of the calcitonin-receptor-like receptor. Nature 1998;393:333–339.
73. Fraser NJ, Wise A, Brown J, McLatchie LM, Main MJ, Foord SM. The amino terminus of receptor activity modifying proteins is a critical determinant of glycosylation state and ligand binding of calcitonin receptor-like receptor. Mol Pharmacol 1999;55:1054–1059.
74. Morfis M, Christopoulos A, Sexton PM. RAMPs: 5 years on, where to now? Trends Pharmacol Sci 2003;24:596–601.

75. White JH, Wise A, Main MJ, et al. Heterodimerization is required for the formation of a functional $GABA_B$ receptor. Nature 1998;396:679–682.
76. Nelson G, Hoon MA, Chandrashekar J, Zhang Y, Ryba NJP, Zuker CS. Mammalian sweet taste receptors. Cell 2001;106:381–390.
77. Nelson G, Chandrashekar J, Hoon MA, et al. An amino-acid taste receptor. Nature 2002;416:199–202.
78. Li X, Staszewski L, Xu H, Durick K, Zoller M, Adler E. Human receptors for sweet and umami taste. Proc Natl Acad Sci USA 2002;99:4692–4696.
79. Kopito RR. ER quality control: the cytoplasmic connection. Cell 1997;88:427–430.
80. Bonifacino JS, Weissman AM. Ubiquitin and the control of protein fate in the secretory and endocytic pathways. Annu Rev Cell Dev Biol 1998;14:19–57.
81. Jarosch E, Lenk U, Sommer T. Endoplasmic reticulum-associated protein degradation. Int Rev Cytol 2003;223:39–81.
82. Yang M, Omura S, Bonifacino JS, Weissman AM. Novel aspects of degradation of T cell receptor subunits from the endoplasmic reticulum (ER) in T cells: importance of oligosaccharide processing, ubiquitination, and proteasome-dependent removal from ER membranes. J Exp Med 1998;187:835–846.
83. Liu Y, Choudhury P, Cabral CM, Sifers RN. Oligosaccharide modification in the early secretory pathway directs the selection of a misfolded glycoprotein for degradation by the proteasome. J Biol Chem 1999;274:5861–5867.
84. Hosokawa N, Wada I, Hasegawa K, et al. A novel ER a-mannosidase-like protein accelerates ER-associated degradation. EMBO Rep 2001;2:415–422.
85. Molinari M, Calanca V, Galli C, Lucca P, Paganetti P. Role of EDEM in the release of misfolded glycoproteins from the calnexin cycle. Science 2003;299:1397–1400.
86. Oda Y, Hosokawa N, Wada I, Nagata K. EDEM as an acceptor of terminally misfolded glycoproteins released from calnexin. Science 2003;299:1394–1397.
87. Illing ME, Rajan RS, Bence NF, Kopito RR. A rhodopsin mutant linked to autosomal dominant retinitis pigmentosa is prone to aggregate and interacts with the ubiquitin proteasome system. J Biol Chem 2002;277:34,150–34,160.
88. Saliba RS, Munro PMG, Luthert PJ, Cheetham ME. The cellular fate of mutant rhodopsin: quality control, degradation and aggresome formation. J Cell Sci 2002;115:2907–2918.
89. Andersson H, D'Antona AM, Kendall DA, Von Heijne G, Chin CN. Membrane assembly of the cannabinoid receptor 1: impact of a long N-terminal tail. Mol Pharmacol 2003;64:570–577.
90. Cook LB, Zhu CC, Hinkle PM. Thyrotropin-releasing hormone receptor processing: role of ubiquitination and proteasomal degradation. Mol Endocrinol 2003;17:1777–1791.
91. Schubert U, Antón LC, Gibbs J, Norbury CC, Yewdell JW, Bennink JR. Rapid degradation of a large fraction of newly synthesized proteins by proteasomes. Nature 2000;404:770–774.

92. Gekko K, Timasheff SN. Mechanism of protein stabilization by glycerol: preferential hydration in glycerol–water mixtures. Biochemistry 1981;20:4667–4676.
93. Sawano H, Koumoto Y, Ohta K, Sasaki Y, Segawa S, Tachibana H. Efficient in vitro folding of the three-disulfide derivatives of hen lysozyme in the presence of glycerol. FEBS Lett 1992;303:11–14.
94. Brown CR, Hong-Brown LQ, Biwersi J, Verkman AS, Welch WJ. Chemical chaperones correct the mutant phenotype of the ΔF508 cystic fibrosis transmembrane conductance regulator protein. Cell Stress Chaperones 1996;1:117–125.
95. Sato S, Ward CL, Krouse ME, Wine JJ, Kopito RR. Glycerol reverses the misfolding phenotype of the most common cystic fibrosis mutation. J Biol Chem 1996;271:635–638.
96. Tamarappoo BK, Verkman AS. Defective aquaporin-2 trafficking in nephrogenic diabetes insipidus and correction by chemical chaperones. J Clin Invest 1998; 101:2257–2267.
97. Burrows JAJ, Willis LK, Perlmutter DH. Chemical chaperones mediate increased secretion of mutant a1- antitrypsin (α1-AT) Z: a potential pharmacological strategy for prevention of liver injury and emphysema in α1-AT deficiency. Proc Natl Acad Sci USA 2000;97:1796–1801.
98. Perlmutter DH. Chemical chaperones: a pharmacological strategy for disorders of protein folding and trafficking. Pediatr Res 2002;52:832–836.
99. Bolen DW, Baskakov IV. The osmophobic effect: natural selection of a thermodynamic force in protein folding. J Mol Biol 2001;310:955–963.
100. Loo TW, Clarke DM. Correction of defective protein kinesis of human P-glycoprotein mutants by substrates and modulators. J Biol Chem 1997;272:709–712.
101. Fan JQ, Ishii S, Asano N, Suzuki Y. Accelerated transport and maturation of lysosomal α-galactosidase A in Fabry lymphoblasts by an enzyme inhibitor. Nat Med 1999;5:112–115.
102. Asano N, Ishii S, Kizu H, et al. In vitro inhibition and intracellular enhancement of lysosomal α-galactosidase A activity in Fabry lymphoblasts by 1-deoxygalactonojirimycin and its derivatives. Eur J Biochem 2000;267:4179–4186.
103. Matsuda J, Suzuki O, Oshima A, et al. Chemical chaperone therapy for brain pathology in G_{M1}-gangliosidosis. Proc Natl Acad Sci USA 2003;100:15,912–15,917.
104. Sawkar AR, Cheng WC, Beutler E, Wong CH, Balch WE, Kelly JW. Chemical chaperones increase the cellular activity of N370S β-glucosidase: a therapeutic strategy for Gaucher disease. Proc Natl Acad Sci USA 2002;99:15,428–15,433.
105. Zhou Z, Gong Q, January CT. Correction of defective protein trafficking of a mutant HERG potassium channel in human long QT syndrome. Pharmacological and temperature effects. J Biol Chem 1999;274:31,123–31,126.
106. Ficker E, Obejero-Paz CA, Zhao S, Brown AM. The binding site for channel blockers that rescue misprocessed human long QT syndrome type 2 *ether-a-gogo*-related gene (HERG) mutations. J Biol Chem 2002;277:4989–4998.

107. Paulussen A, Raes A, Matthijs G, Snyders DJ, Cohen N, Aerssens J. A novel mutation (T65P) in the PAS domain of the human potassium channel HERG results in the long QT syndrome by trafficking deficiency. J Biol Chem 2002;277:48,610–48,616.
108. Rajamani S, Anderson CL, Anson BD, January CT. Pharmacological rescue of human K^+ channel long-QT2 mutations: human ether-a-go-go-related gene rescue without block. Circulation 2002;105:2830–2835.
109. Yan F, Lin CW, Weisiger E, Cartier EA, Taschenberger G, Shyng SL. Sulfonylureas correct trafficking defects of ATP-sensitive potassium channels caused by mutations in the sulfonylurea receptor. J Biol Chem 2004;279:11,096–11,105.
110. Halaban R, Cheng E, Svedine S, Aron R, Hebert DN. Proper folding and endoplasmic reticulum to Golgi transport of tyrosinase are induced by its substrates, DOPA and tyrosine. J Biol Chem 2001;276:11,933–11,938.
111. Wiens GD, O'Hare T, Rittenberg MB. Recovering antibody secretion using a hapten ligand as a chemical chaperone. J Biol Chem 2001;276:40,933–40,939.
112. Friedler A, Hansson LO, Veprintsev DB, et al. A peptide that binds and stabilizes p53 core domain: chaperone strategy for rescue of oncogenic mutants. Proc Natl Acad Sci USA 2002;99:937–942.
113. Issaeva N, Friedler A, Bozko P, Wiman KG, Fersht AR, Selivanova G. Rescue of mutants of the tumor suppressor p53 in cancer cells by a designed peptide. Proc Natl Acad Sci USA 2003;100:13,303–13,307.
114. Morello JP, Salahpour A, Laperrière A, et al. Pharmacological chaperones rescue cell-surface expression and function of misfolded V2 vasopressin receptor mutants. J Clin Invest 2000;105:887–895.
115. Tan CM, Nickols HH, Limbird LE. Appropriate polarization following pharmacological rescue of V2 vasopressin receptors encoded by X-linked nephrogenic diabetes insipidus alleles involves a conformation of the receptor that also attains mature glycosylation. J Biol Chem 2003;278:35,678–35,686.
116. Petäjä-Repo UE, Hogue M, Bhalla S, Laperrière A, Morello JP, Bouvier M. Ligands act as pharmacological chaperones and increase the efficiency of δ opioid receptor maturation. EMBO J 2002;21:1628–1637.
117. Chaipatikul V, Erickson-Herbrandson LJ, Loh HH, Law PY. Rescuing the traffic-deficient mutants of rat μ-opioid receptors with hydrophobic ligands. Mol Pharmacol 2003;64:32–41.
118. Janovick JA, Maya-Nunez G, Conn PM. Rescue of hypogonadotropic hypogonadism-causing and manufactured GnRH receptor mutants by a specific protein-folding template: misrouted proteins as a novel disease etiology and therapeutic target. J Clin Endocrinol Metab 2002;87:3255–3262.
119. Janovick JA, Goulet M, Bush E, Greer J, Wettlaufer DG, Conn PM. Structure–activity relations of successful pharmacologic chaperones for rescue of naturally occurring and manufactured mutants of the gonadotropin-releasing hormone receptor. J Pharmacol Exp Ther 2003;305:608–614.
120. Li T, Sandberg MA, Pawlyk BS, et al. Effect of vitamin A supplementation on rhodopsin mutants threonine-17 → methionine and proline-347 → serine

in transgenic mice and in cell cultures. Proc Natl Acad Sci USA 1998;95:11,933–11,938.
121. Noorwez SM, Kuksa V, Imanishi Y, et al. Pharmacological chaperone-mediated *in vivo* folding and stabilization of the P23H-opsin mutant associated with autosomal dominant retinitis pigmentosa. J Biol Chem 2003;278:14,442–14,450.
122. Noorwez SM, Malhotra R, McDowell JH, Smith KA, Krebs MP, Kaushal S. Retinoids assist the cellular folding of the autosomal dominant retinitis pigmentosa opsin mutant P23H. J Biol Chem 2004;279:16,278–16,284.
123. Morello JP, Petäjä-Repo UE, Bichet DG, Bouvier M. Pharmacological chaperones: a new twist on receptor folding. Trends Pharmacol Sci 2000;21:466–469.
124. Conn PM, Leaños-Miranda A, Janovick JA. Protein origami: therapeutic rescue of misfolded gene products. Mol Intervent 2002;2:308–316.
125. Kahn TW, Sturtevant JM, Engelman DM. Thermodynamic measurements of the contributions of helix-connecting loops and of retinal to the stability of bacteriorhodopsin. Biochemistry 1992;31:8829–8839.
126. Villaverde J, Cladera J, Padrós E, Rigaud JL, Duñach M. Effect of nucleotides on the thermal stability and on the deuteration kinetics of the thermophilic F_0F_1 ATP synthase. Eur J Biochem 1997;244:441–448.
127. Celej MS, Montich GG, Fidelio GD. Protein stability induced by ligand binding correlates with changes in protein flexibility. Protein Sci 2003;12:1496–1506.
128. Desnick RJ, Schuchman EH. Enzyme replacement and enhancement therapies: lessons from lysosomal disorders. Nat Rev Genet 2002;3:954–966.
129. Fan JQ. A contradictory treatment for lysosomal storage disorders: inhibitors enhance mutant enzyme activity. Trends Pharmacol Sci 2003;24:355–360.
130. Leaños-Miranda A, Ulloa-Aguirre A, Ji TH, Janovick JA, Conn PM. Dominant-negative action of disease-causing gonadotropin-releasing hormone receptor (GnRHR) mutants: a trait that potentially coevolved with decreased plasma membrane expression of GnRHR in humans. J Clin Endocrinol Metab 2003;88:3360–3367.

4
Regulated Membrane Trafficking and Proteolysis of GPCRs

James N. Hislop and Mark von Zastrow

1. INTRODUCTION

Multiple mechanisms contribute to the physiological regulation of G protein-coupled receptors (GPCRs) present in the plasma membrane, the main site where ligand-induced signaling events are initiated. Early studies delineated the existence of distinct functional processes of receptor regulation in natively expressing cells and tissues *(1,2)*. More recent studies have led to an explosion of new information regarding cellular and molecular mechanisms of receptor regulation.

1.1. Functional Uncoupling of GPCRs From Heterotrimeric G Proteins Mediated by Receptor Phosphorylation

Extensive studies of certain GPCRs, such as rhodopsin (a light-activated GPCR) and the β_2-adrenergic receptor (β_2-AR; a ligand-activated GPCR), established a highly conserved mechanism that regulates the functional activity of many GPCRs *(3)*. This mechanism involves the phosphorylation of receptors by a specific family of G protein-coupled receptor kinases (GRKs), followed by the interaction of phosphorylated receptors with cytoplasmic accessory proteins called arrestins. Arrestin-bound receptors are unable to couple to heterotrimeric G proteins, disrupting the pathway of GPCR-mediated signal transduction at the earliest stage.

Biochemical studies of signal transduction in isolated rod outer segment preparations identified the protein rhodopsin kinase (or GRK1), which inhibited the ability of light-activated rhodopsin to stimulate its cognate heterotrimeric G protein (transducin). Light-activated rhodopsin is a good

From: *Contemporary Clinical Neuroscience: The G Protein-Coupled Receptors Handbook*
Edited by: L. A. Devi © Humana Press Inc., Totowa, NJ

substrate for phosphorylation by rhodopsin kinase, whereas rhodopsin that has not been activated by light is a poor substrate *(4)*. Phosphorylation rhodopsin only partially inhibited the activation of transducin. A second protein, visual arrestin, which is present in cytoplasmic fractions of rod cells, was able to bind to the phosphorylated rhodopsin and completely inhibit ("arrest") activation of transducin *(5)*.

Studies using functional reconstitution of β_2-AR-mediated activation of adenylyl cyclase have provided strong evidence for a role of phosphorylation in mediating rapid desensitization of a ligand-activated GPCR *(6)*. Biochemical purification of the cytoplasmic activity responsible identified the protein β-AR kinase (βARK, or GRK2), which preferentially phosphorylates agonist-occupied receptors and has similar properties to rhodopsin kinase *(7)*. Biochemical reconstitution studies indicated that increasingly purified fractions of βARK exhibited reduced ability to attenuate β_2-AR-mediated signal transduction in reconstituted membrane preparations. Further analysis of this effect led to the identification of a distinct protein component that was lost in increasingly purified fractions and that increased functional desensitization when re-added to highly purified fractions of βARK *(7,8)*. This protein cofactor was similar to visual arrestin and, therefore, was named "nonvisual" arrestin or β-arrestin. Complementary DNA cloning has identified a family of arrestins involved in regulating the function of phosphorylated GPCRs *(9)*.

Agonists not only regulate phosphorylation of GPCRs by GRKs but also regulate the affinity with which phosphorylated receptors bind to arrestins *(10)*. This dual-control mechanism assures that only those receptors actually activated by agonist are desensitized. In this way, other receptors that are not activated, including co-expressed GPCRs that recognize other ligands and are potentially desensitized by the same mechanism, are not affected. Indeed, GRK-mediated phosphorylation and subsequent binding of arrestins is generally considered to be a paradigm for homologous desensitization, a form of desensitization that is specific only to the specified activated GPCR and is not influenced by activation of other receptors in the same cell *(3)*.

1.2. Agonist-Induced Endocytosis of GPCRs

Pharmacological studies of the process of sequestration led to the hypothesis that certain GPCRs are removed from the plasma membrane within minutes after agonist-induced activation *(11,12)*. Biochemical and immunochemical methods have verified this finding in both cultured cells and certain native tissues *(13–15)*. Rapid endocytosis of the β_2-AR is mediated by an agonist-dependent lateral redistribution into clathrin-coated pits *(16)*. Coated pits then pinch off from the plasma membrane (Fig. 1) to form endocytic vesicles, a process that is dependent on the cytoplasmic protein dynamin *(17–20)*. Subsequent studies have demonstrated that regulated endocytosis of several other GPCRs is also mediated by a dynamin-dependent mechanism, suggesting a conserved role of clathrin-coated pits in mediating endocytosis of many GPCRs.

Clathrin-coated pits play a general role in mediating rapid endocytosis of a large number of cell-surface components other than signaling receptors, many of which are endocytosed constitutively (i. e., in a ligand-independent manner). This has raised the question: How is GPCR endocytosis regulated by ligands? GRKs and arrestins (in addition to their previously established role in mediating functional uncoupling of receptors from heterotrimeric G proteins) play an important role in regulating endocytosis of certain GPCRs. Particularly, β-arrestins can promote the concentration of phosphorylated receptors in coated pits by binding simultaneously to the receptors and to the clathrin-containing lattice structure via distinct protein interaction domains, thus functioning as "adapters" that link specific GPCRs to endocytic membranes *(21,22)*. This is true for many, but not all, GPCRs. There are also examples of GPCRs that either do not endocytose rapidly or endocytose by a different mechanism *(23–25)*. Although this diversity of GPCR membrane trafficking is not yet fully understood at the mechanistic level, it may have important implications for the physiological regulation of distinct GPCRs.

2. FUNCTIONAL CONSEQUENCES OF GPCR ENDOCYTOSIS

2.1. Role in Rapid Desensitization of GPCRs

In many cases, endocytosis is not believed to play a primary role in mediating rapid desensitization of GPCRs, although the precise role of endocytosis in this process may depend on receptor expression level. Endocytosis of μ-opioid peptide (MOP) receptors does not contribute significantly to functional desensitization in cells expressing relatively high levels of receptor protein but does appear to cause desensitization in cells expressing lower levels of receptor *(26)*. Studies of the $β_2$-AR emphasize that GRK/arrestin-dependent uncoupling of receptor from G protein occurs in the plasma membrane before endocytosis begins, and desensitization of the $β_2$-AR is not prevented by blockade of receptor endocytosis *(27)*.

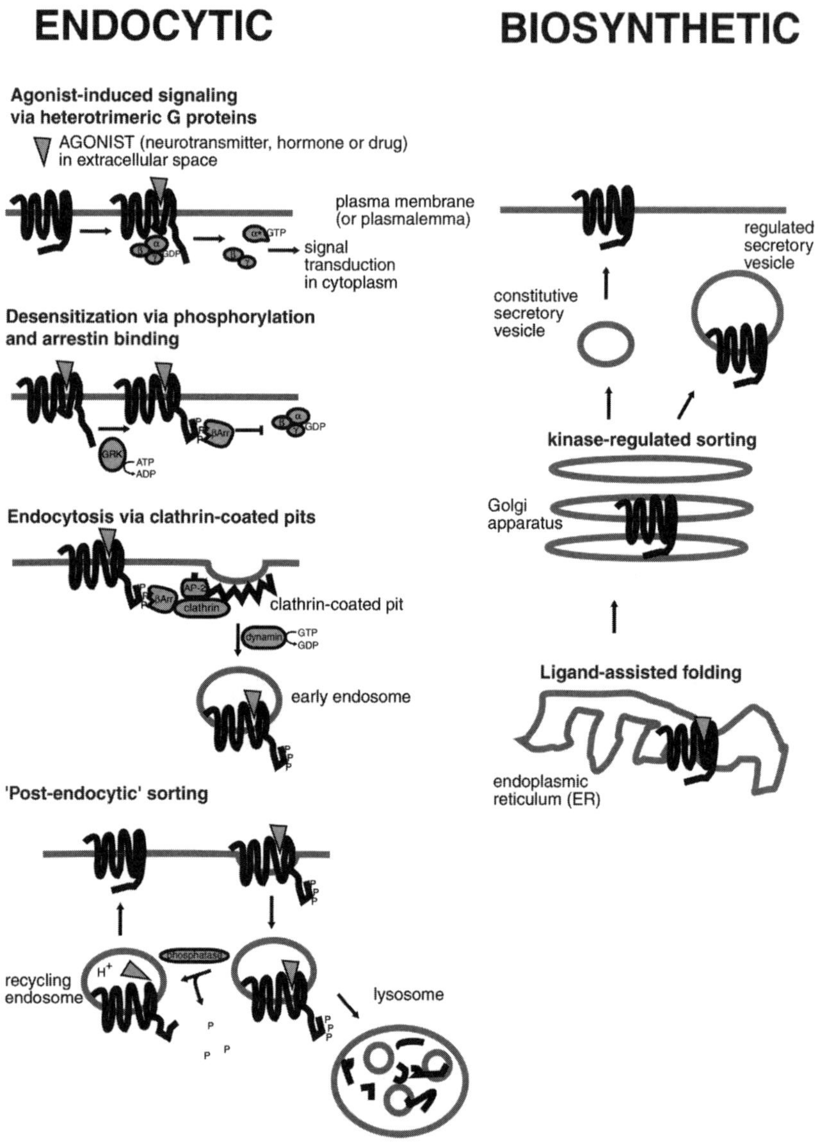

Fig. 1. Examples of events controlling the membrane trafficking of GPCRs. In the endocytic pathway, agonist-activated GPCRs can be endocytosed via clathrin-coated pits, followed by sorting to lysosomes or recycling to the plasma membrane. In the biosynthetic pathway, GPCRs synthesized in the endoplasmic reticulum can undergo ligand-assisted folding followed by kinase regulated sorting to either constitutive or regulated secretory vesicles.

2.2. Role in Resensitization of GPCR Signaling

In contrast to its limited role in mediating rapid desensitization, endocytosis of certain GPCRs is believed to play a major role in mediating the distinct process of receptor resensitization *(28,29)*. The reason for this major role is believed to be that endocytosis brings receptors in close proximity to an endosome-associated phosphatase, which mediates dephosphorylation of receptors previously phosphorylated (hence, "desensitized") at the cell surface. Dephosphorylated receptors are then recycled back to the plasma membrane in a "resensitized" state, which is fully functional to mediate subsequent rounds of signal transduction upon re-exposure to agonist *(3,27)*. A similar role of endocytic trafficking in promoting functional desensitization has been described for other GPCRs, such as the MOP receptor *(30)*, although this may not be the case for all GPCRs *(3)*.

2.3. Role in Mediating Proteolytic Downregulation of GPCRs

Endocytosis is also believed to play an important role in mediating downregulation of many GPCRs by promoting proteolysis of receptors. Although downregulation of GPCRs can occur via multiple mechanisms *(31)*, one mechanism involves endocytosis of receptors followed by membrane trafficking to lysosomes. GPCRs can be targeted to lysosomes after initial endocytosis by clathrin-coated pits or alternate mechanism(s) of endocytosis *(32,33)*. Certain GPCRs efficiently recycle to the plasma membrane following endocytosis, whereas other GPCRs are sorted preferentially to lysosomes *(34,35)*. The sorting decision between tha plasma membrane and lysosomes is important because it can determine whether agonist-induced endocytosis promotes the distinct functional consequences of receptor resensitization or downregulation, respectively.

3. MEMBRANE TRAFFICKING OF GPCRS AFTER ENDOCYTOSIS

3.1. Mechanisms of Recycling

Many integral membrane proteins are believed to recycle to the plasma membrane after endocytosis by bulk membrane flow, without requiring specific sorting information of the membrane protein itself *(36,37)*. This does not appear to be true for some GPCRs, such as the β_2-AR and the MOP receptor, in which specific sequences present in the C-terminal cytoplasmic domain have been identified that are required for efficient recycling of endocytosed receptors to the plasma membrane *(38–40)*. The "recycling signal" present in the C-terminal cytoplasmic domain of the β_2-AR binds both to a family of PDZ domain-containing proteins capable of interacting with the cortical actin cytoskeleton and to the N-ethyl-maleimide-sensitive fusion factor (NSF); both types of protein interaction have been proposed to function in controlling endocytic trafficking of receptors *(38,39)*. The recycling signal defined in the MOP receptor tail does not bind detectably to either the PDZ domains or NSF and, presumably, functions via distinct cytoplasmic protein interaction(s) *(40)*. It is not yet known how widespread the phenomenon of "signal-mediated" recycling is in the GPCR superfamily, although there is evidence for cytoplasmic sequences in several other GPCRs that either enhance or inhibit recycling of internalized receptors. Certain GPCRs remain persistently associated with arrestins after endocytosis, which has been proposed to inhibit receptor recycling *(41)* or may mediate other regulatory effects *(42)*.

3.2. Mechanisms of GPCR Sorting to Lysosomes

Certain GPCRs, including the human β_2-AR and the CXCR4 chemokine receptor, can be sorted to lysosomes after prolonged stimulation by a mechanism that requires covalent attachment of ubiquitin to the receptor protein *(43–45)*. Ubiquitin is a small (76-residue) polypeptide attached to the ϵ-amino group of lysine residue(s) present on the cytoplasmic surface of receptors. The ubiquitin isopeptide functions as a covalent "tag," promoting the sorting of many integral membrane proteins to lysosomes via ubiquitin-mediated binding to a specialized protein complex associated with the endosome membrane *(46)*. Ubiquitination of the δ-opioid peptide (DOP) receptor is not required for efficient trafficking to lysosomes in a human cell culture model *(47)*. A cytoplasmic protein has been identified that can bind to the DOP receptor and modulate its trafficking to lysosomes, and binding of this protein to receptors does not appear to require receptor ubiquitination *(48)*. Together, these observations suggest the existence of additional machinery mediating lysosomal trafficking of GPCRs in mammalian cells. The details of this additional machinery and relationships to the conserved ubiquitination-dependent sorting mechanism remain to be elucidated.

4. REGULATION OF GPCR TRAFFIC IN THE BIOSYNTHETIC PATHWAY

4.1. Regulation of GPCR Folding and Export From the Endoplasmic Reticulum

Much of what is known about GPCR membrane traffic involves receptor transport through the endocytic pathway (Fig.1). However, the number of

GPCRs available for ligand binding at the cell surface can also be regulated by receptor trafficking through the biosynthetic pathway (Fig.1). Studies of vasopressin and opioid receptors expressed in heterologous cell models have revealed that a substantial fraction of newly synthesized receptors are improperly folded and proteolyzed by a mechanism involving proteosomes (rather than lysosomes) before exiting the endoplasmic reticulum *(49)*. Certain membrane-permeant (nonpeptide) ligands are capable of increasing the surface delivery of newly synthesized receptors, apparently by promoting the productive folding of receptors in the endoplasmic reticulum. This role of opiate drugs as "pharmacological chaperones" may help to explain the reason that certain opioid antagonists and partial agonist drugs produce upregulation of opioid receptors in tissue *(50)*. This is discussed in detail in Chapter 4.

4.2. Regulation of GPCR Traffic After Exit From the Golgi Apparatus

Another mechanism controlling the delivery of recently synthesized GPCRs to the plasma membrane appears to involve the regulation of receptor trafficking at a later stage in the biosynthetic pathway, after receptors exit the endoplasmic reticulum and are delivered to the Golgi apparatus. Tyrosine kinase (Trk)-mediated signaling (via Trk-family neurotrophin receptors) is sufficient to retain DOP expressed in cultured neurosecretory cells in an intracellular membrane pool, localized adjacent to the *trans*-Golgi network, which does not constitutively traffic to the plasma membrane but is able to mediate rapid receptor insertion to the plasma membrane in response to depolarization. This mechanism selectively regulates biosynthetic membrane trafficking of DOP receptors, whereas MOP receptors appear to constitutively traffic to the plasma membrane *(51)*. It is suggested that this mechanism may mediate activity-dependent changes in the responsiveness of neurons to DOP receptor agonists and may also contribute to specific changes in DOP receptor surface localization observed in opiate-dependent animals *(52)*.

5. CONCLUSION

Considerable progress has been made in defining GPCR membrane trafficking itineraries in the endocytic and biosynthetic pathways. There has also been progress in elucidating molecular mechanisms that mediate specific GPCR trafficking events; however, many questions remain. Functional consequences of specific GPCR endocytic trafficking events have been identified most compellingly for the processes of resensitization and downregulation of receptors. It is likely that endocytic trafficking of GPCRs has other important functional effects that remain to be fully elucidated. It also appears that certain GPCRs undergo regulated membrane trafficking in the biosynthetic pathway, but the physiological functions of these events are currently unknown. A limitation of our current understanding is that it is derived largely from studies of model cell systems expressing recombinant receptors. An important future challenge is to define membrane trafficking mechanisms that regulate GPCRs in native cells and to elucidate physiological consequences of these trafficking events in vivo. It is conceivable that progress in this area will identify novel targets for therapeutic intervention in disease states associated with dysregulation of GPCR signaling.

REFERENCES

1. Clark RB. Receptor desensitization. Adv Cyclic Nuc Prot Phos Res 1986;20:151–209.
2. Perkins JP, Hausdorff WP, Lefkowitz RJ. Mechanisms of Ligand-Induced Desensitization of β-Adrenergic Receptors. In: Perkins JP, ed. The Beta-Adrenergic Receptor. Clifton, NJ: Humana Press, 1991, pp. 73–124.
3. Lefkowitz RJ, Pitcher J, Krueger K, Daaka Y. Mechanisms of β-adrenergic receptor desensitization and resensitization. Adv Pharmacol 1998;42:416–420.
4. McDowell JH, Kuhn H. Light-induced phosphorylation of rhodopsin in cattle photoreceptor membranes: substrate activation and inactivation. Biochemistry 1977;16:4054–4060.
5. Bennett N, Sitaramayya A. Inactivation of photoexcited rhodopsin in retinal rods: the roles of rhodopsin kinase and 48-kDa protein (arrestin). Biochemistry 1988;27:1710–1715.
6. Sibley DR, Strasser RH, Caron MG, Lefkowitz RJ. Homologous desensitization of adenylate cyclase is associated with phosphorylation of the β-adrenergic receptor. J Biol Chem 1985;260:3883–3886.
7. Benovic JL, Strasser RH, Caron MG, Lefkowitz RJ. β-adrenergic receptor kinase: identification of a novel protein kinase that phosphorylates the agonist-occupied form of the receptor. Proc Natl Acad Sci USA 1986;83:2797–2801.
8. Benovic JL, Kuhn H, Weyand I, Codina J, Caron MG, Lefkowitz RJ. Functional desensitization of the isolated β-adrenergic receptor by the beta-adrenergic receptor kinase: potential role of an analog of the retinal protein arrestin (48-kDa protein). Proc Natl Acad Sci USA 1987;84:8879–8882.
9. Carman CV, Benovic JL. G-protein-coupled receptors: turn-ons and turn-offs. Curr Opin Neurobiol 1998;8:335–344.
10. Gurevich VV, Benovic JL. Mechanism of phosphorylation-recognition by visual arrestin and the transition of arrestin into a high affinity binding state. Mol Pharmacol 1997;51:161–169.
11. Staehelin M, Simons P. Rapid and reversible disappearance of beta-adrenergic cell surface receptors. EMBO J 1982;1:187–190.

12. Toews ML, Perkins JP. Agonist-induced changes in beta-adrenergic receptors on intact cells. J Biol Chem 1984;259:2227–2235.
13. von Zastrow M, Kobilka BK. Ligand-regulated internalization and recycling of human beta 2-adrenergic receptors between the plasma membrane and endosomes containing transferrin receptors. J Biol Chem 1992;267:3530–3538.
14. Kurz JB, Perkins JP. Isoproterenol-initiated β-adrenergic receptor diacytosis in cultured cells. Mol Pharmacol 1992;41:375–381.
15. Keith DE, Anton B, Murray SR, et al. μ-Opioid receptor internalization: opiate drugs have differential effects on a conserved endocytic mechanism in vitro and in the mammalian brain. Mol Pharmacol 1998;53:377–384.
16. von Zastrow M, Kobilka BK. Antagonist-dependent and -independent steps in the mechanism of adrenergic receptor internalization. J Biol Chem 1994;269:1,8448–18,452.
17. van der Bliek AM, Redelmeier TE, Damke H, Tisdale EJ, Meyerowitz EM, Schmid SL. Mutations in human dynamin block an intermediate stage in coated vesicle formation. J Cell Biol 1993;122:553–563.
18. Herskovits JS, Burgess CC, Obar RA, Vallee RB. Effects of mutant rat dynamin on endocytosis. J Cell Biol 1993;122:565–578.
19. Zhang J, Ferguson S, Barak LS, Menard L, Caron MG. Dynamin and β-arrestin reveal distinct mechanisms for G protein-coupled receptor internalization. J Biol Chem 1996;271:18,302–18,305.
20. Cao TC, Mays RW, von Zastrow M. Regulated endocytosis of G protein-coupled receptors by a biochemically and functionally distinct subpopulation of clathrin-coated pits. J Biol Chem 1998;273:24,592–24,602.
21. Goodman OB, Jr., Krupnick JG, Santini F, et al. Beta-arrestin acts as a clathrin adaptor in endocytosis of the β2- adrenergic receptor. Nature 1996;383:447–450.
22. Laporte SA, Oakley RH, Holt JA, Barak LS, Caron MG. The interaction of β-arrestin with the AP-2 adaptor is required for the clustering of β2-adrenergic receptor into clathrin-coated pits. J Biol Chem 2000;275:23,120–23,126.
23. von Zastrow M, Link R, Daunt D, Barsh G, Kobilka B. Subtype-specific differences in the intracellular sorting of G protein-coupled receptors. J Biol Chem 1993;268:763–766.
24. Roettger BF, Rentsch RU, Pinon D, et al. Dual pathways of internalization of the cholecystokinin receptor. J Cell Biol 1995;128:1029–1041.
25. Lee KB, Pals RR, Benovic JL, Hosey MM. Arrestin-independent internalization of the m1, m3, and m4 subtypes of muscarinic cholinergic receptors. J Biol Chem 1998;273:12967–12972.
26. Pak Y, Kouvelas A, Scheideler MA, Rasmussen J, O'Dowd BF, George SR. Agonist-induced functional desensitization of the mu-opioid receptor is mediated by loss of membrane receptors rather than uncoupling from G protein. Mol Pharmacol 1996;50:1214–1222.
27. Pippig S, Andexinger S, Lohse MJ. Sequestration and recycling of β2-adrenergic receptors permit receptor resensitization. Mol Pharmacol 1995;47:666–676.

28. Pippig S, Andexinger S, Daniel K, et al. Overexpression of beta-arrestin and β-adrenergic receptor kinase augment desensitization of β2-adrenergic receptors. J Biol Chem 1993;268:3201–3208.
29. Yu SS, Lefkowitz RJ, Hausdorff WP. β-adrenergic receptor sequestration. A potential mechanism of receptor resensitization. J Biol Chem 1993;268:337–341.
30. Koch T, Schulz S, Schröder H, Wolf R, Raulf E, Hollt V. Carboxyl-terminal splicing of the rat μ opioid receptor modulates agonist-mediated internalization and receptor resensitization. J Biol Chem 1998;273:13,652–13,657.
31. Tsao P, von Zastrow M. Downregulation of G protein-coupled receptors. Curr Opin Neurobiol 2000;10:365–369.
32. Koenig JA, Edwardson JM. Endocytosis and recycling of G protein-coupled receptors. Trends Pharmacol Sci 1997;18:276–287.
33. Tsao PI, von Zastrow M. Diversity and specificity in the regulated endocytic membrane trafficking of G-protein-coupled receptors. Pharmacol Ther 2001;89:139–147.
34. Gagnon AW, Kallal L, Benovic JL. Role of clathrin-mediated endocytosis in agonist-induced down-regulation of the β2-adrenergic receptor. J Biol Chem 1998;273:6976–6981.
35. Tsao PI, von Zastrow M. Type-specific sorting of G protein-coupled receptors after endocytosis. J Biol Chem 2000;275:11,130–11,140.
36. Dunn KW, McGraw TE, Maxfield FR. Iterative fractionation of recycling receptors from lysosomally destined ligands in an early sorting endosome. J Cell Biol 1989;109:3303–3314.
37. Gruenberg J. The endocytic pathway: a mosaic of domains. Nat Rev Mol Cell Biol 2001;2:721–730.
38. Cao TT, Deacon HW, Reczek D, Bretscher A, von Zastrow M. A kinase-regulated PDZ-domain interaction controls endocytic sorting of the β2-adrenergic receptor. Nature 1999;401:286–290.
39. Cong M, Perry SJ, Hu LA, Hanson PI, Claing A, Lefkowitz RJ. Binding of the β2 adrenergic receptor to N-ethylmaleimide-sensitive factor regulates receptor recycling. J Biol Chem 2001;276:45,145–45,152.
40. Tanowitz M, von Zastrow M. A novel endocytic recycling signal that distinguishes the membrane trafficking of naturally occurring opioid receptors. J Biol Chem 2003;278:45,978–45,986.
41. Oakley RH, Laporte SA, Holt JA, Barak LS, Caron MG. Association of β-arrestin with G protein-coupled receptors during clathrin-mediated endocytosis dictates the profile of receptor resensitization. J Biol Chem 1999;274:32,248–32,257.
42. Klein U, Muller C, Chu P, Birnbaumer M, von Zastrow M. Heterologous inhibition of G protein-coupled receptor endocytosis mediated by receptor-specific trafficking of β-arrestins. J Biol Chem 2001;276:17,442–17,447.
43. Hicke L. Protein regulation by monoubiquitin. Nat Rev Mol Cell Biol 2001;2:195–201.

44. Marchese A, Benovic JL. Agonist-promoted ubiquitination of the G protein-coupled receptor CXCR4 mediates lysosomal sorting. J Biol Chem 2001;276:45,509–45,512.
45. Shenoy SK, McDonald PH, Kohout TA, Lefkowitz RJ. Regulation of receptor fate by ubiquitination of activated β2- adrenergic receptor and β-arrestin. Science 2001;294:1307–1313.
46. Katzmann DJ, Babst M, Emr SD. Ubiquitin-dependent sorting into the multivesicular body pathway requires the function of a conserved endosomal protein sorting complex, ESCRT-I. Cell 2001;106:145–155.
47. Tanowitz M, von Zastrow M. Ubiquitination-independent trafficking of G protein-coupled receptors to lysosomes. J Biol Chem 2002;277:50,219–50,222.
48. Whistler JL, Enquist J, Marley A, et al. Modulation of post-endocytic sorting of G protein-coupled receptors. Science 2002;297:615–620.
49. Petaja-Repo UE, Hogue M, Laperriere A, Bhalla S, Walker P, Bouvier M. Newly synthesized human δ opioid receptors retained in the endoplasmic reticulum are retrotranslocated to the cytosol, deglycosylated, ubiquitinated, and degraded by the proteasome. J Biol Chem 2000;276:4416–4423.
50. Petaja-Repo UE, Hogue M, Laperriere A, Walker P, Bouvier M. Export from the endoplasmic reticulum represents the limiting step in the maturation and cell surface expression of the human δ opioid receptor. J Biol Chem 2000;275:13,727–13,736.
51. Kim KA, von Zastrow M. Neurotrophin-regulated sorting of opioid receptors in the biosynthetic pathway of neurosecretory cells. J Neurosci 2003;23:2075–2085.
52. Cahill CM, Morinville A, Lee MC, Vincent JP, Collier B, Beaudet A. Prolonged morphine treatment targets δ opioid receptors to neuronal plasma membranes and enhances δ-mediated antinociception. J Neurosci 2001;21:7598–7607.

II
GPCR ACTIVITY AND ITS REGULATORS

5
Heterotrimeric G Proteins and Their Effector Pathways

Tracy Nguyen Hwangpo, and Ravi Iyengar

1. INTRODUCTION

Almost 30 yr have passed since the discovery of the heptahelical transmembrane (TM) receptors and their connection to heterotrimeric G proteins and sequential signal flow to intracellular effectors *(1–3)*. Many hormones, sensory stimuli, and neurotransmitters use this signaling system to convert chemical or physical information from the G protein-coupled receptor (GPCR) through a transducer (G protein) to an effector into an intracellular language that the cell can comprehend and to which it can respond. In the liver, epinephrine signals via the β-adrenergic receptor (AR) through $G\alpha_s$ to adenylyl cyclase to increase cyclic adenosine monophosphate (cAMP) production such that it leads to stimulation of glycogen breakdown and inhibition of glycogen synthesis, resulting in glucose production. In the eye, light stimulates the GPCR rhodopsin, which activates the G protein transducin, to stimulate the activity of cyclic guanosine monophosphate (cGMP) phosphodiesterase (PDE). This results in decreased cGMP levels and changes in the activity of the cyclic nucleotide-gated Na^{2+} channels, thereby converting photons into electrical impulses and transmitting information to the visual cortex. These cascades of events allow for processing of the initial signal, including amplification.

2. G PROTEIN-COUPLED RECEPTORS

The heptahelical TM receptor, also known as the GPCR, comprises the largest group of TM receptor proteins involved in signal transduction. They couple to heterotrimeric G proteins and induce a conformational change in the G protein upon ligand binding and receptor activation. All GPCRs share a common molecular architecture consisting of seven-TM (7TM) helices that are connected by three extracellular and three intracellular loops. Although GPCRs share similar overall secondary structures, they vary widely in amino acid identity and in the manner by which the different receptors are activated by ligands.

To date, over 1200 members of the GPCR family have been identified in the human genome, and about 40 to 60% have orthologs in other species *(4)*. Agonists or potential ligands have been assigned to approx 190 GPCRs, and more than 900 are olfactory GPCRs *(4)*. The remaining GPCRs are called "orphan" receptors because their ligands have yet to be identified.

GPCRs are grouped into five major families based on sequence identity, of which the first three comprise mammalian GPCRs. Family A contains receptors related to rhodospin receptors and ARs that bind ligands as diverse as biogenic amines (such as histamine and serotonin), peptides (opioid and somatostatin), hormone proteins (follicle-stimulating hormone), olfactory molecules, lipids, phospholipids (cannabinoids), and viral proteins. They are the largest family of GPCRs and are subdivided according to structural similarities. They are characterized by the presence of several highly conserved amino acids and contain a disulfide bridge that connects the first and second extracellular loops *(5)*.

Family B, the second largest family of GPCRs, contains receptors for peptide hormones that are similar to secretin. They include receptors for secretin, glucagon, calcitonin, gastric inhibitory peptide, and vasoactive intestinal peptide (VIP). They possess about 60 members and are characterized by the presence of a large N-terminus, which contains several cysteines that presumably form a network of disulfide bridges *(4,5)*. These receptors usually are coupled to more than one G protein, with the $G\alpha_s$-adenylyl cyclase pathway predominating *(6)*.

Family C contains receptors related to the metabotropic receptors. It contains over 2 dozen members and includes the metabotropic glutamate receptors, the calcium-sensing receptors, and γ-aminobutyric acid $(GABA)_B$ receptors. Family C is defined as a group of receptors comprising at least three different subfamilies that share 20% or greater amino acid sequence over their 7TM regions *(4)*. This family contains a long amino terminus and carboxyl tail, of which the N-terminus is the ligand-binding region *(5)*.

The last two GPCR families are relatively small. Family D contains receptors related to the fungal pheromone receptors; in July 2004, there were about 24 members in this class (from the GPCR database [GPCRDB]) *(7,8)*. This family is divided into STE2- and STE3-like receptor subfamilies. Family E, the smallest family, contains receptors related to the slime mold cAMP receptors; in July 2004, there were about five members (according to the GPCRDB) *(7,8)*. Within each of these structural classes of receptors, there are several functional classes that couple to the various heterotrimeric G proteins.

3. HETEROTRIMERIC G PROTEINS

Heterotrimeric G proteins are composed of an α-subunit, a β-subunit, and a γ-subunit. The β-subunit and γ-subunit are considered to be a single functional complex, because they do not dissociate in nondenaturing conditions. The α-subunit can bind and hydrolyze guanosine triphosphate (GTP). They are called heterotrimeric because of the three different subunits and G proteins because they display highly selective binding for guanine nucleotides. In the basal state, the α-subunit, which is bound to guanosine diphosphate (GDP), associates with the βγ complex. Upon ligand binding to the GPCR, the latter undergoes a conformational change such that it promotes the exchange of GDP for GTP on the α-subunit. In the GTP-bound state, the α-subunit dissociates from the βγ complex, and both the α-subunit and the βγ complex can interact with and regulate downstream effectors to evoke physiological responses. However, the hydrolysis of GTP on the α-subunit results in its re-association with the βγ complex, leading to the dissipation of the intracellular response. There are several other levels of regulation for this system, such as regulators of G protein signaling (RGS), activators of G proteins signaling (AGS), and G protein receptor kinases (GRKs), which have been described in detail in many recent reviews *(9–11)*. Briefly, RGS proteins increase the intrinsic GTPase activity of the α-subunit; AGS proteins activate G proteins independent of GPCR-mediated signaling, and GRKs phosphorylate key residues on the GPCR, leading to desensitization and/or endocytosis of the receptor.

4. DIFFERENT CLASSES OF HETEROTRIMERIC G PROTEINS AND THEIR COUPLING PROPERTIES

The different classes of G proteins are defined by their sequence identity as well as their downstream effector coupling specificity. There are over 20 Gα-subunits known to date, and they are divided into four major Gα subfamilies. Because the coupling of the receptor to the different classes of G proteins specifies the signaling pathways that are activated, GPCRs are functionally classified according to their coupling specificity, such as G_s- or G_q-coupled receptors. Several recent reviews have detailed the expression and biological significance of G protein pathways *(12–14)*.

4.1. Gα$_s$ Family

The Gα$_s$ family includes Gα$_s$, Gα$_{sXL}$, and Gα$_{olf}$. Members of this family of Gα-subunits interact with and stimulate the activity of adenylyl cyclase to increase cAMP levels. Additionally, the Gα-subunits are sensitive to the cholera toxin made by *Vibrio cholera*. The A1 subunit of cholera toxin is an intracellular adenosine diphosphate (ADP)-ribosyl transferase that catalyzes the covalent addition of the ADP-ribose moiety from NAD$^+$ to Gα$_s$. ADP-ribosylated Gα$_s$·GTP can activate adenylyl cyclase but cannot hydrolyze the GTP to GDP; therefore, Gα$_s$ remains constitutively active. There are four splice variants of Gα$_s$, and they are all known to stimulate adenylyl cyclase activity *(15)*.

The "extra large" G protein (Gα$_{sXL}$), which is expressed in neuroendocrine cells, is a plasma membrane-associated protein; it consists of a novel 37-kDa XL domain followed by a 41-kDA α$_s$-domain encoded by exons 2 to 13 of the Gα$_s$ gene *(16,17)*. Similarly to Gα$_s$, Gα$_{sXL}$ can bind GTP and stimulate adenylyl cyclase activity, but it remains to be determined whether Gα$_{sXL}$ can interact with GPCRs that are known to interact with Gα$_s$ in vivo *(17,18)*.

The final member, Gα$_{olf}$, which was initially discovered in the neuroepthilium and striatum and considered to be the G protein of the olfactory system, shows 88% homology to Gα$_s$. It has been shown to stimulate adenylyl cyclase, can be constitutively activated by cholera toxin, and can interact with the β-AR *(19,20)*. Gα$_{olf}$ is involved in odor-evoked signal transduction because Gα$_{olf}$ knockout mice are anosmic *(21)*. It is believed that upon odorant binding, the odorant receptor activates Gα$_{olf}$, which consequently activates adenylyl cyclase III (AC3). AC3 then raises cAMP levels, causing cyclic nucleotide-gated channels to open, which leads to an influx of cations and, eventually, the formation of an action potential that signals to the brain *(22)*. Gα$_{olf}$ has also been shown to be present in peripheral tissues; more recently, it has been implicated in oncogenic transformation of digestive and urogenital epithelial cells *(23)*.

There are three major pathways by which G$_s$-coupled receptors take effect. All of these involve adenylyl cyclases and cAMP. Most of the effects of cAMP occur through the activation of protein kinase A (PKA), which can phosphorylate and regulate diverse substrates such as transcription factors, metabolic enzymes, and channels. In recent years, it has been demonstrated that cAMP can bind directly and regulate the activity of Epac (the exchange factor for the small GTPase Rap), thereby stimulating Rap activity. In sensory organs such as the nasal neuroepithelium, cAMP also binds and regulates the activity of cyclic nucleotide-gated channels. The various effector pathways for Gα$_s$ are summarized in Fig. 1.

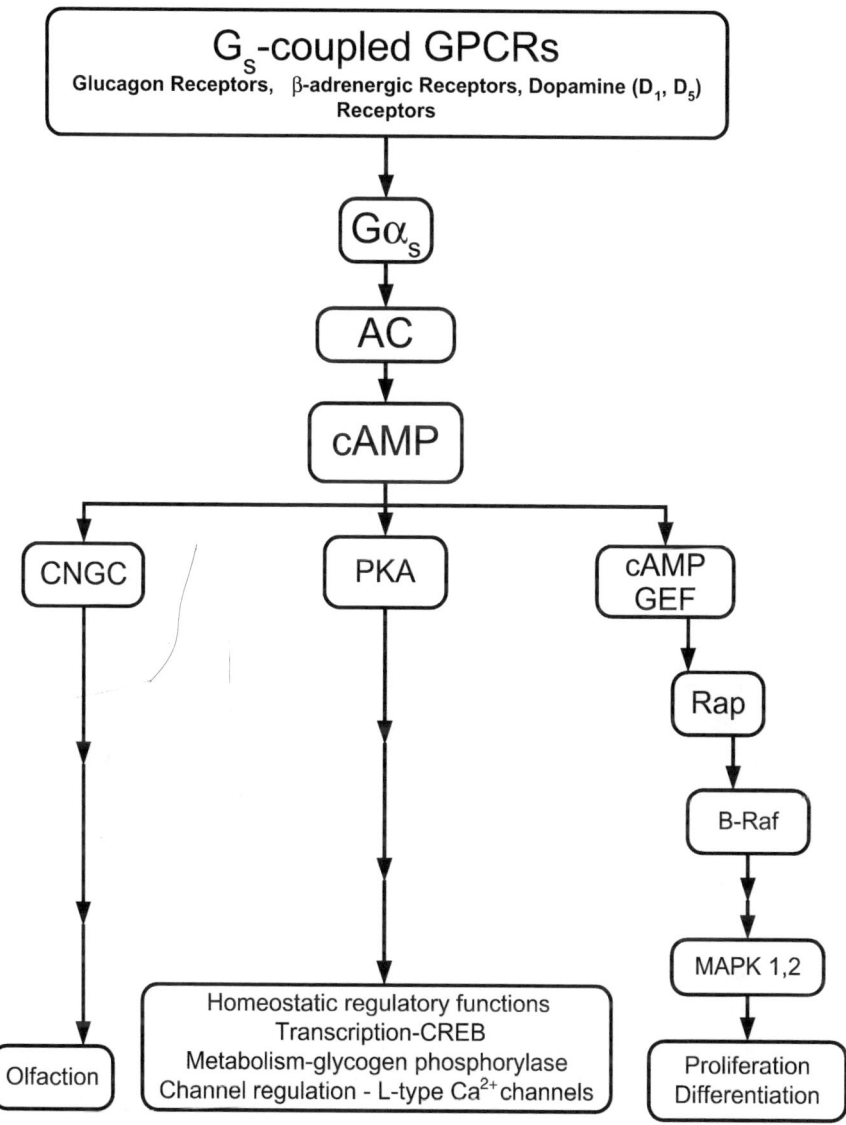

Fig. 1. The Gα$_s$ signaling pathway. This schematic diagram demonstrates how the Gα$_s$–AC pathway regulates multiple physiological processes, including olfaction, homeostatic regulatory functions, proliferation, and differentiation. AC, adenylyl cyclase; PKA, protein kinase A; CNGC, cyclic nucleotide-gated channel; GEF, guanine exchange factor; Rap1, a small GTPase; MAPK, mitogen-activated protein kinase; B-Raf, MAP Kinase Kinase Kinase for MAP Kinase Kinase 1,2.

4.2. $G\alpha_{i/o}$ Family

The $G\alpha_{i/o}$ family includes $G\alpha_{i1}$, $G\alpha_{i2}$, $G\alpha_{i3}$, $G\alpha_o$, $G\alpha_z$, $G\alpha_{gust}$, $G\alpha_{t-r}$, and $G\alpha_{t-c}$. This family, with the exception of $G\alpha_z$, is pertussis toxin-sensitive. This toxin, which is also an ADP-ribosyl-transferase, catalyzes the addition of ADP-ribose to the α-subunit of $G\alpha_i$. This irreversible modification occurs on a cysteine, which is at the fourth position from the C-terminus. This region is also involved in coupling to receptors; hence, ADP-ribosylated $G\alpha_{i/o}$-subunits do not interact with receptors. Consequently, pertussis toxin inhibits receptor activation of the $G\alpha_{i/o}$ pathway.

$G\alpha_{i1}$, $G\alpha_{i2}$, and $G\alpha_{i3}$ are products of different *Gnai* (name for genes encoding $G\alpha_i$ members) genes, but all mediate inhibition of various adenylyl cyclases *(24,25)*. They are partly functionally redundant because $G\alpha_{i1}$- and $G\alpha_{i3}$-deficient mice do not shown gross phenotypic changes *(13)*. However, $G\alpha_{i2}$-deficient mice show defects in B- and T-cell signaling *(26,27)*.

The most abundant G protein in the mammalian brain is $G\alpha_o$. Three isoforms of $G\alpha_o$ exist; two of which are generated by alternative splicing, and the third is generated by posttranslational modification *(28–30)*. This G protein is enriched in growth cones and has been implicated in neurite outgrowth *(31,32)*. Recent studies using yeast two-hybrid screens and complementary DNA (cDNA) expression cloning have yielded some candidate proteins that are direct effectors of $G\alpha_o$. They include the GTPase-activating protein (GAP) for the small G protein Rap (RapGAP), the GAP from $G\alpha_z$ ($G\alpha_z$–GAP), RGS-17, and the G protein-regulated inducer of neurite outgrowth (GRIN) *(33,34)*.

An excellent review on $G\alpha_z$ signaling was recently published *(35)*. Similarly to $G\alpha_i$, $G\alpha_z$ can inhibit adenylyl cyclase and stimulate K^+ channels *(36,37)*. However, $G\alpha_z$ lacks a consensus site for ADP-ribosylation by pertussis toxin and is thus unaffected by it *(38,39)*. Additionally, $G\alpha_z$ hydrolyzes GTP at a much slower rate compared to other Gα-subunits, so it is not surprising that it interacts with several RGS proteins *(35,40)*. $G\alpha_z$ can be phosphorylated by protein kinase C (PKC) and p21-activated kinase (PAK)1, leading to a decrease in $G\alpha_z$'s affinity for the Gβγ complex and thereby maintaining the G protein in an active state for a longer period of time *(41,42)*.

The last three members of the $G\alpha_i$ family are involved in sensory systems. Gα-gustducin ($G\alpha_{gust}$), expressed mainly in taste cells, is responsible for transducing the bitter and sweet taste qualities to the brain because $G\alpha_{gust}$ knockout mice show deficiency in bitter and sweet taste recognition *(43)*. Activation of $G\alpha_{gust}$ leads to a rise in intracellular Ca^{2+} followed by neurotransmitter release *(44)*. A recent report by Zhang et al. *(45)* identified TRPM5 (a taste receptor cell-specific ion channel) and PLC-β2 as necessary for bitter, sweet, and umami signal transduction. The report demonstrated that TRPM5 and PLC-β2 knockout mice could not recognize bitter and sweet tastes as well as L-amino acids *(45)*.

Rod transducin (Gα_{t-r}) and cone transducin (Gα_{t-c}) are the G proteins involved in visual transduction. The first type is important in dim illumination, and the latter is involved in color and sharp vision. In this system, a photon of light causes the isomerization of a visual pigment in rhodopsin, the GPCR involved in phototransduction. This conformation change stimulates the exchange of GTP for GDP in transducin. Transducin then activates cGMP PDE, also known as PDE8, by binding to the regulatory γ-subunit of PDE8. PDE8 activation results in lower cytoplasmic cGMP levels, which leads to closure of cGMP-gated cation channels and membrane hyperpolarization. The hyperpolarization leads to a decrease in the release of the neurotransmitter glutamate at the photoreceptor terminal *(46–48)*. Therefore, a sensory stimulus is translated into an electrical signal that can be communicated to connecting neurons.

The various pathways regulated by G$\alpha_{i/o}$ are shown in Fig. 2. Because many of the biological effects of the G$\alpha_{i/o}$ pathway are mediated through G$\beta\gamma$-subunits, these pathways are also shown in Fig. 2.

4.3. G$\alpha_{q/11}$ Family

The G$\alpha_{q/11}$ family includes Gα_q, Gα_{11}, Gα_{14}, Gα_{15}, and Gα_{16}. This family of G proteins is coupled to the activation of PLC-β *(49–51)*. Activation of PLC-β leads to the hydrolysis of phosphatidylinositol 4,5-bisphosphate (PIP$_2$) and the production of inositol triphosphate (IP$_3$) and diacylglycerol (DAG).

Gα_q and Gα_{11} (which are 88% identical in amino acid sequence) are widely distributed in mammalian tissues and can activate PLC-β1, -β3, and -β2 in decreasing affinity as well as PLC-β4 *(51)*. Gα_{14} (which is 81% identical to Gα_q) is found in the spleen, lung, kidney, and testis *(52)*. The human Gα_{16} protein and the mouse Gα_{15} homolog (which share 57% homology to Gα_q) are expressed in hematopoietic cells *(52,53)*. Although members of the G$\alpha_{q/11}$ family are indistinguishable regarding the PLC-β isozymes they activate, some GPCRs can discriminate among them. For example, the macrophage chemotactic protein-1 receptor B (MCP-1Rb) can couple to Gα_{14} and Gα_{16} but not Gα_q or Gα_{11}, whereas the C-C chemokine receptor-1 (CKR-1) can couple to Gα_{14} but not Gα_{16} *(54)*. Signaling through this path-

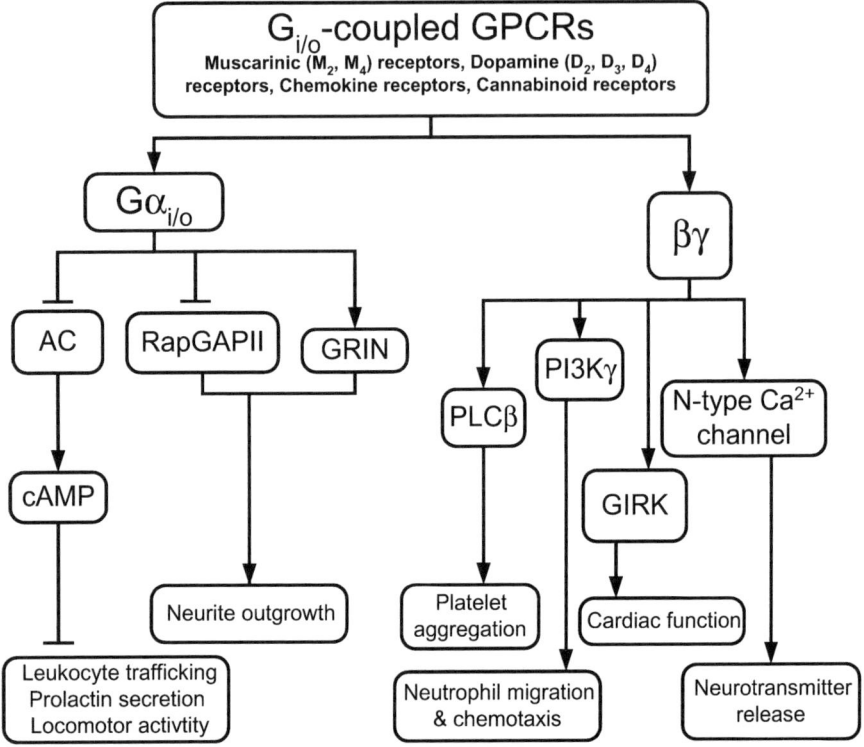

Fig. 2. The Gα$_{i/o}$ and Gbg signaling pathway. This schematic diagram demonstrates how the Gα$_{i/o}$ pathway regulates multiple physiological processes such as leukocyte trafficking, locomotor activity, and neurite outgrowth. In contrast, Gβγ regulates other signaling pathways to affect physiological processes such as platelet aggregation, neutrophil chemotaxis, and neuro-transmitter release. AC, adenylyl cyclase; GAP, a GTPase-activating protein; GRIN, G protein regulator of neurite outgrowth; PLC-β, phospholipase-b; PI3Kγ, phosphoinositide-3 kinase; GIRK, G protein-gated inwardly rectifying potassium channel.

way leads to responses that are mediated through Ca^{2+} and PKC and is illustrated in Fig. 3.

4.4. Gα$_{12/13}$ Family

The Gα$_{12}$ family includes Gα$_{12}$ and Gα$_{13}$ and defines the fourth class of G proteins. This family of G protein shares less than 45% sequence identity to other α-subunits and 67% homology to each other. Both proteins are ubiquitously expressed, although Gα$_{13}$ is especially abundant in human platelets *(55,56)*.

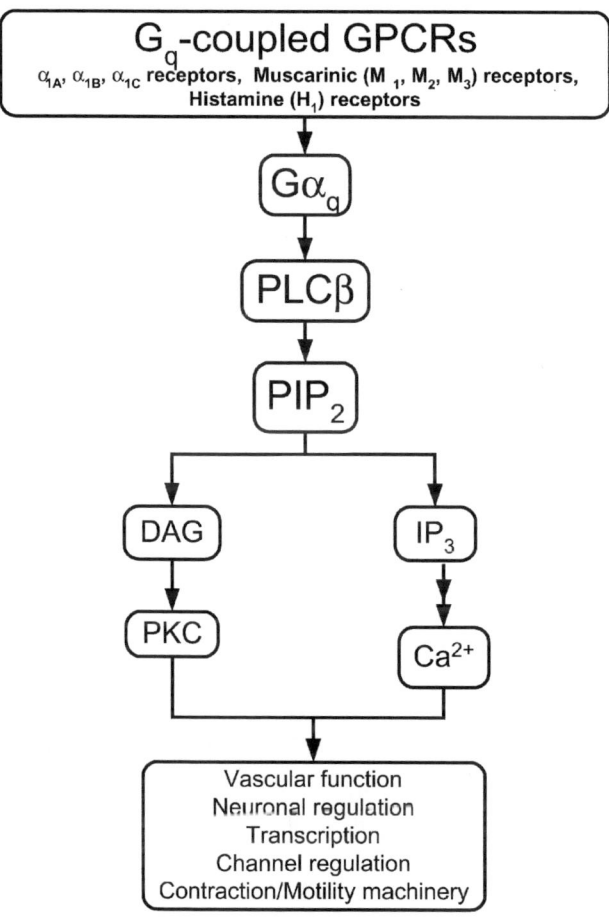

Fig. 3. The Gα$_q$ signaling pathway. This schematic diagram demonstrates how the PLC-β pathway regulates multiple physiological processes including channel regulation, contraction, and neuronal regulation. PLC-β, phospholipase-β; PIP$_2$, phosphatidylinositol 4,5-bisphosphate; DAG, diacylglycerol; PKC, protein kinase C; Ca^{2+}, calcium ions.

Several interacting proteins to either Gα$_{12}$ or Gα$_{13}$ have been identified by yeast two-hybrid screening, including HSP90, AKAP, nonreceptor tyrosine kinases, rasGAP, Rho guanine exchange factor (RhoGEF), radixin, protein phosphatase type 5 (pp5), and cadherin (57). The diversity of these interacting proteins suggests that the Gα$_{12}$ family is involved in multiple signaling pathways.

However, most of the studies conducted on Gα$_{12}$ and Gα$_{13}$ have used the activated form of these proteins because many GPCRs that couple to this family can also activate $G\alpha_q$ family members. From these studies, numerous proteins have been observed to be downstream of $G\alpha_{12}$ and $G\alpha_{13}$ signaling. For example, c-Jun N-terminal kinase (JNK), the Na^+/H^+ exchanger (NHE), and phospholipase D (PLD) have been shown to be activated in the presence of these mutant proteins *(58–62)*. In particular, overexpression of activated $G\alpha_{12}$ leads to increased JNK activity through various small G proteins (Ras, Rac, Cdc42, and Rho) and the Src family tyrosine kinases, because dominant-negatives or specific inhibitors of these proteins inhibit JNK activation *(58,59,63)*. In the case of the NHE, signaling via $G\alpha_{12}$ or $G\alpha_{13}$ shows distinct differences. For example, $G\alpha_{13}$ can activate all three isoforms of NHE, whereas $G\alpha_{12}$ activates NHE2 and NHE3 isoforms but will inhibit NHE1 *(61)*. In the case of PLD, the activation of the rat PLD1 via $G\alpha_{13}$ Q226L involves the Rho-dependent pathway because the dominant-negative N19RhoA and a Rho inhibitor can block this activation *(62)*. These pathways and their biological functions are summarized in Fig. 4.

Interestingly, the constitutively active $G\alpha_{12}$ and $G\alpha_{13}$ forms are oncogenic. Overexpression of these proteins leads to the transformation of cultured fibroblasts *(64)*. The growth-promoting activity and oncogenicity of the activated forms is dependent on Rho signaling because their capability to form foci is inhibited when the Rho-sensitive *Clostridium botulinum* toxin C3 exoenzyme is present *(65)*. The link between $G\alpha_{12/13}$ and Rho signaling has occurred through RhoGEFs such as p115-RhoGEF, PDZ-RhoGEF, and leukemia-associated RhoGEF (LARG). $G\alpha_{12/13}$ can physically interact with these RhoGEFs through their RGS domain and may stimulate their guanine nucleotide exchange activity toward Rho *(66)*. Nevertheless, $G\alpha_{12}$ and $G\alpha_{13}$ display distinct signaling pathways toward Rho. Activated $G\alpha_{13}$ can stimulate the GEF activity of p115-RhoGEF, whereas $G\alpha_{12}$ cannot. The difference is also shown in $G\alpha_{13}$ knockout mice. The $G\alpha_{13}$-deficient mice show embryonic lethality resulting from defects in the vascular system, and $G\alpha_{12}$ cannot rescue the function of $G\alpha_{13}$ in these mice *(67)*. Recent experiments have shown that $G\alpha_{12}$ can stimulate the GEF activity of LARG only when this protein is phosphorylated by Tec, a nonreceptor tyrosine kinase *(68)*. Therefore, $G\alpha_{12}$ may be routed to Rho signaling via tyrosine kinases such as Tec.

4.5. Gβγ Complex

There are over five Gβ-isoforms, of which two are splice variants. They include $G\beta_1$, $G\beta_2$, $G\beta_3$, $G\beta_{3S}$, $G\beta_4$, $G\beta_5$, and $G\beta_{5L}$ *(69–71)*. The least similar of the Gβ-subunits is $G\beta_5$. $G\beta_5$ is expressed mainly in the central nervous system, whereas $G\beta_{5L}$ is expressed only in the retina *(70)*. In contrast, there are more than 13 different Gγ-subunits. The Gγ-subunits can be modified by a number of processes, such as isoprenylation, methylation, farnesylation, and geranylgeranylation. Several good reviews have been published regarding the structure and function of the Gβγ complex *(12,69–71)*.

The Gβ- and Gγ-subunits are considered to form one functional unit. They exist in several specific combinations and various techniques involving purified proteins, functional expression, yeast two-hybrid systems, and ribozyme approaches have been used to determine the various combinations *(69,72–74)*. Some of the residues that determine the specificity of these interactions have been localized. For example, the specificity of the assembly of the $Gβ_1γ_1$ complex and $Gβ_2γ_1$ complex appears to be determined by a stretch of 14 residues lying within the middle of the Gγ-subunit *(69)*. Additionally, the residues in Gβ that contact the region of Gγ are clustered on blade 5 and a small section of the N-terminal region of Gβ *(69)*.

The Gβγ complex can affect downstream effectors just as well and as much as the Gα-subunit. In 1987, the first effector of the Gβγ complex, the G protein gated inwardly rectifying potassium channel (GIRK), was discovered *(75)*. Since then, a number of downstream effectors of Gβγ have been identified. For example, adenylyl cyclase, phospholipases Cβ1 through Cβ3, and PI3K are directly affected by the Gβγ-subunit *(12,69,71)*.

5. DIRECT DOWNSTREAM EFFECTORS OF THE G PROTEIN PATHWAY

5.1. Adenylyl Cyclases

Adenylyl cyclase (AC), a membrane protein, is important in promoting the conversion of adenosine triphosphate (ATP) to cAMP. Nine membrane-bound isoforms of AC and two spliced variants of AC8 have been cloned and characterized in mammals, and all are expressed in the brain *(76)*. cAMP is an important second messenger in the cell. It stimulates the activation of PKA by binding to the regulatory subunit of PKA and removing the inhibition by that subunit.

PKA can then phosphorylate several substrates important in glucose metabolism and other vital physiological processes. cAMP can also interact with many other proteins to promote protein–protein interactions that are independent of PKA signaling. For example, cAMP can directly bind a RasGEF called Epac and stimulate its GEF activity toward the small G protein Rap1 *(77,78)*.

Several G proteins modulate AC activity directly. $Gα_s$, $Gα_i$, and Gβγ can interact physically with various ACs to affect their activity. In particular, $Gα_s$ can activate all nine isoforms of ACs to various degrees, whereas the $Gα_i$ family can inhibit some ACs, especially AC5 through AC6 *(25,76)*.

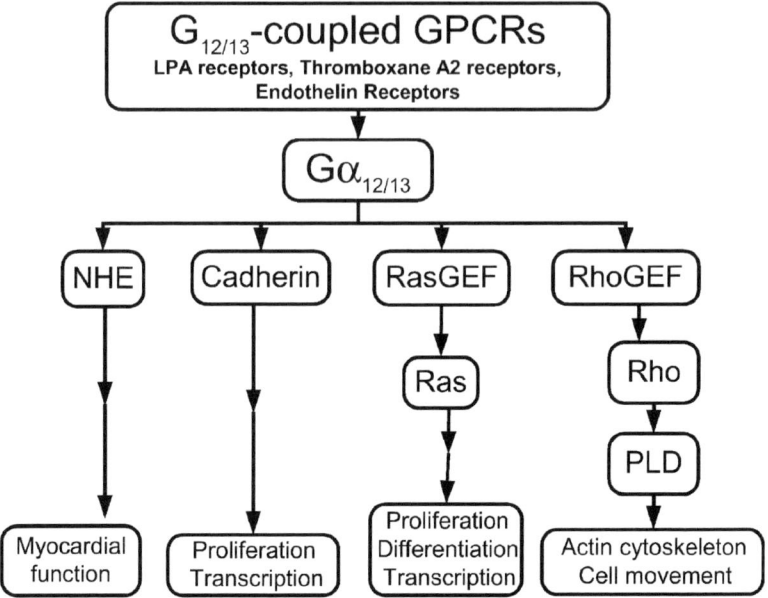

Fig. 4. The $G\alpha_{12/13}$ signaling pathways. This schematic diagram demonstrates how the $G\alpha_{12/13}$ pathway regulates multiple cellular pathways to affect physiological processes such as transcription, proliferation, and cell movement. LPA, lysophosphatidic acid ; NHE, sodium-hydrogen exchanger; GEF, guanine exchange factor; Ras, a small GTPase; Rho, a small GTPase; PLD, phospholipase D.

The βγ-subunit has also been shown to be able to activate or inhibit some AC isoforms, depending on the βγ combination or on the presence of activated $G\alpha_s$. For example, $G\beta_1\gamma_2$ can inhibit AC1 and AC3, but some βγ complexes can directly activate AC2, AC4, and AC7 in the presence of activated $G\alpha_s$ *(12,76,79)*.

AC has also been shown to be important in numerous processes, such as long-term potentiation (LTP) and long-term memory, cell differentiation, development, and drug dependence *(76,79)*. In some of these processes, the AC isoform involved is either activated or inhibited by calcium. For example, AC1 activity is stimulated by Ca^{2+}/camodulin, and this AC isoform is important in LTP and long-term memory *(76)*. In relation to GPCRs and $G\alpha_s$ activation, AC in which $G\alpha_s$ is coupled to β_1-ARs functions in increasing cardiac rate and force of contraction; $G\alpha_s$ coupled to β_2-ARs functions in smooth muscle relaxation; $G\alpha_s$ that is coupled to β_3-ARs functions in lipolysis of white adipose tissue *(80)*.

5.2. Phospholipase C

Various G proteins can activate phospholipases. Phospholipases such as phospholipase C (PLC) can cleave the polar head group of inositol phospholipids. All members of the Gα_q family can activate PLC-β1 to β4 isoforms; Gα_{12} can activate PLC-ε; Gβγ can activate all four PLC-β isoforms as well as PLA$_2$ *(12)*. In particular, PLC-β can hydrolyze the phosphorylated lipid PIP$_2$ to generate two intracellular products: inositol 1,4,5-trisphosphate (a universal calcium-mobilizing second messenger) and DAG (an activator of PKC). These products are important in raising intracellular Ca^{2+} levels in the cells and in activating Ca^{2+}-sensitive proteins such as calmodulin, which regulates other proteins within the cell.

PLC-β is composed of the N-terminal pleckstrin homology (PH) domain, EF-hand domain, catalytic X and Y domains, and regulatory C2 domain *(81)*. The PH domains of PLC-β2 and PLC-β3 bind the heterotrimeric G protein subunit Gβγ. The C2 domains of PLC-β1 and PLC-β2 bind the GTP-bound Gα_q. Comparison of the ligand-binding affinities of different PLC-β isozymes shows that each isozyme is regulated differently by both subunits of Gα_q. PLC-β2 and PLC-β3 are more sensitive to the βγ-subunit than PLC-β1 and PLC-β4, whereas the affinities of Gα_q for PLC-β1 and PLC-β3 are higher than that for PLC-β2 *(81)*.

Another phospholipase family member that can be modulated by G proteins is PLC-ε. PLC-ε can function as a phospholipase as well as a RasGEF *(82)*. It can interact with both large and small G proteins. To do so, PLC-ε contains conserved catalytic and regulatory domains common to other eukaryotic PLCs, but it also contains two Ras-associating domains and a RasGEF motif. This isoform can hydrolyze PIP$_2$, and this activity is selectively stimulated by a constitutively active form of Gα_{12} as well as various Gβγ-dimers *(82,83)*. Additionally, PLC-ε's lipase activity can be inhibited by pertussis toxin, suggesting that Gα_i/Gα_o signaling may be involved *(84)*. Furthermore, PLC-ε can promote formation of Ras–GTP through its RasGEF domain. However, it has been suggested that PLC-ε is a Ras effector because the Ras-associating domain of PLC-ε can bind to the H-Ras in a GTP-dependent manner that correlates with stimulation of PLC-ε's lipase activity of PLC-ε *(85,86)*.

5.3. Ion Channels

Several ion channels are directly affected by G protein activation. GIRK, voltage-gated Ca^{2+} channels (VDCCs), cardiac and epithelial chloride (Cl$^-$) channels, and cardiac and epithelial sodium channels (Na$^+$) are affected by G proteins. Such interactions have been deduced on the basis of several criteria. The channel activity must be conditionally and reversibly modified by the activation of a relevant GPCR. The use of a nonhydrolyzable GTP analog should allow for channel modulation, even in the absence of receptor stimulation. The addition of a purified and active G protein subunit should be sufficient to trigger changes in channel activity. Finally, physical association between the G protein and the channel subunits must be demonstrated within the confines of an intact cell.

The most well-characterized channel that is directly affected by G protein activation is the GIRK, or Kir3, family. Direct activation of GIRKs by G proteins is involved in the rapid inhibition of membrane excitability resulting in the slowing of heart rate by the vagus nerve or the autoinhibitory release of dopamine by midbrain neurons. Therefore, these GIRKs couple GPCR signaling to membrane excitability. There are four isoforms of Kir3s that are affected by G proteins: Kir3.1 (GIRK1), Kir3.2 (GIRK2), Kir3.3 (GIRK3), and Kir3.4 (GIRK4). They are all expressed in the brain. GIRK4 is also expressed in the heart. The Kir3s were the first effectors demonstrated to be directly activated by the Gβγ-subunit *(75)*. The current view is that the interaction of the Gβγ-subunits with the GIRK channel involves multiple binding domains that synergistically control channel gating. Several studies have been conducted using fusion proteins, mutagenesis, and peptides to identify the sites that are important for GIRK activity *(87)*. Gβγ directly binds to both the carboxy- and amino- cytoplasmic segments of the channel protein. It is also believed that Gβγ binding stabilizes channel–PIP_2 interactions that open the channel gate *(88)*.

VDCCs are another type of ion channels that are directly affected by G protein signaling. These channels, located near vesicle docking sites, are important because they are responsible for the influx of Ca^{2+} into the presynaptic neuron, which allows for the release of neurotransmitters from the synaptic nerve terminals. Calcium ions act in concert with distinct components of the presynaptic machinery to facilitate fusion of synaptic vesicles within the plasma membrane. Modulating the entry of Ca^{2+} into the nerve terminal thus represents a major means by which neurotransmitter secretion can be controlled. There are four major families of VDCCs: N-type, L-type, P/Q-type, and T-type. The N-type Ca^{2+} (Ca2.2v) and L-type Ca^{2+} (Cav3.1) channels are regulated by G proteins. $G\alpha_{i1/i2/z}$ and Gβγ have been shown to inhibit the N-type Ca^{2+} channels, whereas $G\alpha_s$ and Gβγ can stimulate the L-type Ca^{2+} channel *(89–93)*. Of these, the Gβγ-complex has been shown to directly interact with and modulate N-type Ca^{2+} channel activity *(89)*. In fact, the Gβγ-complex can bind several contact sites on the α_{1B}-subunit of the N-type Ca^{2+} channel. The interaction between Gβγ- and α_{1B}-subunits has been mapped to the N-terminal, C-terminal, and I–II loop of the $\alpha_{1\beta}$-subunit of the N-type Ca^{2+} channel *(89)*.

5.5. Regulators of Small GTPases

Several small G protein regulators are direct effectors of G proteins. Ras-GRF (also known as $CDC25^{Mm}$), Raf-1, and Shc are believed to be modulated by the Gβγ-complex, whereas Gap^m and RhoGEFs are affected by the Gα-subunit *(68,94–100)*. Of these, Raf-1 is believed to directly interact with the Gβγ-complex, whereas RapGAP and RhoGEFs are known to directly interact with the Gα-subunit.

The serine/threonine (Ser/Thr) protein kinase Raf-1, directly downstream of Ras, has been reported to interact with the $Gβ_2$-subunit *in vitro* and *in vivo (96)*. In competition assays, only β-AR kinase can inhibit $Gβ_2$ binding to Raf-1 but can not inhibit Ras or 14-3-3. The significance of this interaction has not been determined, although overexpression of $Gβ_1γ_2$ in HEK293 cells can stimulate the phosphorylation of mitogen-activated protein kinase (MAPK) and enhance the MAP/ERK kinase (MEK) kinase activity of c-Raf *(101)*. These studies were performed in cell lines in which these proteins were overexpressed; therefore, the relevance of this interaction and the kinase assays need to be further investigated.

RasGRF has GEF activity toward the small G proteins Ras and Rac *(95,102,103)*. RasGRF can be activated via two different pathways. The Gβγ-complex has been implicated in one of these pathways. Carbachol-treated fibroblasts transfected with the muscarinic receptor type 1 ($Gα_q$-coupled) or type 2 ($Gα_i$-coupled) increased the phosphorylation state of RasGRF as well as its GEF activity toward Ras. This increase can also be observed in COS-1 cells transfected with $Gβ_1γ_2$ complex, suggesting that it is the Gβγ-complex that is significant in activating RasGRF. Additionally, carbachol treatment of neonatal rat brain explants increased RasGRF's GEF activity and phosphorylation state *(94)*. Furthermore, RasGRF immunoprecipitated from HEK293 cells overexpressing Gβγ showed enhanced GEF activity toward the Rac1 protein *(95)*.

The $Gα_{12}$ family can also modulate the activity of the small G proteins Rho and Ras via their direct interaction with RhoGEF and RasGAP *(99,100)*. $Gα_{13}$ can physically interact with three RhoGEFs (p115PhoGEF, PDZORhoGEF, and LARG) and stimulate their guanine exchange activity *(99,104,105)*. In contrast, $Gα_{12}$ has been shown to physically interact with $Gap1^m$, a RasGAP, via the PH–BM domain of RasGAP in vitro and in vivo. This interaction can stimulate the GAP activity of $Gap1^m$ toward Ras *(100)*.

Both $Gα_i$ and $Gα_o$ can modulate the small G protein Rap via binding to RapGAP directly *(33,106)*. In particular, $Gα_i$ binds the Rap1GAPII isoform directly in its N-terminus. The activated form of $G\alpha_i$ binds Rap1GAPII more efficiently than wild-type $G\alpha_i$, and stimulation of G_i-coupled receptors recruits RapIGAPII from the cytosol to the membrane, which then leads to a decrease in the levels of RapGTP. Consequently, the decrease in Rap1 leads to the activation of the MAPK, because there is less of Rap1 to inhibit Ras activity *(106)*. In contrast, wild-type $G\alpha_o$ can bind RapIGAPII directly and inhibit RapIGapII from acting on Rap1 *(33)*. Consequently, the increased levels of Rap1 can stimulate MAPK1/2 activity. The mechanism by which this occurs can be attributed to the promotion of the ubiquination and protease degradation of Rap1GAP upon binding to $G\alpha_o$ *(107,108)*.

Other small G proteins affected by G protein signaling include Rac and Cdc42. The activity of Rac can be modulated by $G\alpha_q$ signaling because $G\alpha_q$-deficient platelets cannot activate Rac upon stimulation of thromboxane A2 receptor using the agonist U46619 *(109)*. Constitutively active $G\alpha_q$ can stimulate Cdc42 activity toward insulin signaling to GLUT4 translocation in adipoctyes *(110)*.

6. NEURONAL REGULATION: HETEROTRIMERIC G PROTEIN PATHWAYS AND NEURONAL PLASTICITY

Neurite outgrowth is an important process by which neurons achieve neuronal connectivity during brain development. The mechanism involved in this process requires coordination of signals coming from outside and inside the cell. The growth cone is a critical structure in the neuron that is important for such function. The nerve growth cone is the motile structure at the tip of elongating axons and dendrites and is believed to be responsible for recognizing pathways and targets and transducing such information into directed movement *(111)*. Growing axons are guided to appropriate targets by responses of their motile growth cones to environmental cues. Many signals participate in the regulation of neuronal outgrowth, and heterotrimeric G protein pathways play an important role.

The major proteins that are present in growth cone membranes include tubulin, actin, GAP43, and $G\alpha_o$ and its $G\beta$ counterpart *(31,112,113)*. Of these, GAP43 and $G\alpha_o$ (the noncytoskeletal proteins) have been implicated in neurite outgrowth. GAP43 is a neuron-specific protein whose expression is closely related to axonal growth and can regulate $G\alpha_o$ activity *(31,114)*. However, in GAP43-deficient mice, GAP43 is not essential for axonal outgrowth or growth cone formation but is required at certain decision points, suggesting a model in which GAP43 is important for amplifying pathfinding signals from the growth cone *(115)*.

Similarly, $G\alpha_o$ is expressed predominantly in the brain, accounts for 1% of membrane proteins, and is highly enriched in growth cone membranes *(28,29,31)*. Therefore, many investigators have studied this G protein to determine its significance in neurite outgrowth. They have shown that the collapse of growth cones is mediated by GPCRs because pertussis toxin inhibits growth cone collapse *(116)*. Additionally, the expression of constitutively active $G\alpha_o$ results in an increase in the number of neurites per cell *(32)*. However, little else is known about how $G\alpha_o$ can affect neurite outgrowth.

The downstream effectors of $G\alpha_o$ signaling for neurite outgrowth are under investigation. GRIN is one membrane protein that has been found to directly interact preferentially with activated $G\alpha_o$ in vitro and in vivo *(33,34)*. Its expression pattern is similar to that of $G\alpha_o$, because it is expressed largely in the brain and is enriched in growth cone membranes *(34)*. Overexpression of GRIN in the presence of activated $G\alpha_o$ results in long neurites and many hairlike processes *(34)*. GRIN1 is the mouse protein, whereas GRIN2 is the human ortholog. These studies have prompted our laboratory to study the downstream effectors of GRIN. Using the yeast two-hybrid system with GRIN2 as a bait, we found many interacting candidates; of these candidates, Sprouty2 was discovered as a possible partner (unpublished data, 2003).

Sprouty was originally identified by genetic analysis as a regulator of tracheal branching in *Drosophila* and was implicated as an inhibitor of the MAPK pathway *(117,118)*. Four mammalian genes (*mSpry 1–4*) encoding protein homologs of dSpry have been identified *(118)*. All sprouty proteins share a highly conserved cysteine-rich domain at the C-terminus *(118)*. The different mechanisms by which Sprouty proteins inhibit signaling pathways may rely on the differences in their amino terminals, which are weakly homologous and assumed to interact with different effectors *(118)*. In fact, it was a region of the N-terminal of Sprouty2 that was isolated from the yeast two-hybrid system. In vitro and in vivo experiments confirmed the interaction between Sprouty2 and GRIN2 (unpublished data, 2004).

To further study the significance of these interactions, the cannabinoid receptors (CB_1/CB_2) endogenously expressed in Neuro2A cells are being used as a prototypical endogenous system for study of $G\alpha_o$, GRIN, and Sprouty2. In initial experiments, Win515,2-2mesylate, a CB1/CB2 receptor agonist, was used to stimulate $G\alpha_o$ signaling followed by basic fibroblast growth factor (bFGF) stimulation to examine the MAPK pathway. Preliminary data have revealed that prestimulation of $G\alpha_i/G\alpha_o$ signaling suppressed MAPK activation via bFGF stimulation, compared to controls. We hypothesize that activation of $G\alpha_o$ and, in turn, GRIN potentiates Sprouty2 inhibition of the MAPK pathway. The inhibition of MAPK may be one method that neurons can use to guide axonal growth. Further experiments are being conducted to delineate the importance of GRIN and Sprouty in neurite outgrowth. Dominant-negatives and small interfering RNA (siRNA) of both proteins will be used to determine their significance in neurite outgrowth upon $G\alpha_o$ and FGF signaling.

Another interacting protein of $G\alpha_o$ is RapIGAPII *(108)*. In vitro and in vivo experiments confirm the interaction and show that RapIGAPII preferentially binds the inactivated $G\alpha_o$ *(33)*. RapIGAPII is the GAP of the small protein Rap1. By sequestering RapIGAPII, there is an increase in Rap activity *(33)*. Follow-up studies showed that $G\alpha_o$ signaling can promote the ubiquitination and proteasomal degradation of RapIGAPII, because overexpression of $G\alpha_o$ reduced the protein stability of RapIGAPII, whereas the presence of proteasomal inhibitors such as lactacystin increased protein levels *(108)*. Another group showed that RapIGAP can be ubiquinated and degraded by the proteasome *(107)*.

Agonist stimulation of CB_1/CB_2 receptors endogenously expressed in Neuro2A cells promoted neurite outgrowth, whereas the presence of a dominant-negative of Rap1 and siRNA of Rap1 inhibited neurite outgrowth, supporting a role for Rap1 in neurite extension. Additionally, the expression of RapIGapII blocked $G\alpha_o$-induced neurite outgrowth, and lactacystin potentiated this inhibition *(108)*. These results suggest that by regulating the proteasomal degradation of Rap1GAPII, the CB1 receptor activates Rap to induce neurite outgrowth.

7. CONCLUSIONS

It has been known for some time that heterotrimeric G protein pathways play pivotal roles in almost every cell type. They are involved in many acute, important physiological processes such as glucose regulation, phototransduction, and cardiac contractility. The approx 20 $G\alpha$-subunits, the 5 $G\beta$-subunits, and the 13 $G\gamma$-subunits can transduce sensory, chemical, and peptide signals into a language that cells can comprehend and to which they can react. Recent research, as described earlier, indicates that the G protein pathways also play major roles in regulating long-term processes. In addition to the structural plasticity described in this chapter, the G protein pathways—especially the cAMP signaling pathway—play a major role in learning and memory processes *(79,119)* and in addictive behavior *(120)*. Therefore, the functioning of and interactions between heterotrimeric G protein pathways as well as with other pathways are major determinants of neuronal functions across timescales.

ACKNOWLEDGMENTS

Research in the Iyengar laboratory regarding G proteins is supported by GM 54508 and DK38761. T. Nguyen Hwangpo is supported by the Mount Sinai MSTP Training Program and the Cardiovascular Sciences (HL-07824).

REFERENCES

1. Ross EM, Gilman AG. Biochemical properties of hormone-sensitive adenylate cyclase. Annu Rev Biochem 1980;49:533–564.
2. Rodbell M. The role of hormone receptors and GTP-regulatory proteins in membrane transduction. Nature 1980;284:17–22.
3. Gilman AG. G proteins: transducers of receptor-generated signals. Annu Rev Biochem 1987;56:615–649.
4. Schoneberg T. GPCR Superfamily and Its Structural Characterization. In: Pangalos MaD, CH, ed. Understanding G Protein-Coupled Receptors and Their Role in the CNS. 1st ed. New York: Oxford University Press, 2002, pp. 3–27.
5. George SR, O'Dowd BF, Lee SP. G-protein-coupled receptor oligomerization and its potential for drug discovery. Nat Rev Drug Discov 2002;1:808–820.
6. Ulrich CD 2nd, Holtmann M, Miller LJ. Secretin and vasoactive intestinal peptide receptors: members of a unique family of G protein-coupled receptors. Gastroenterology 1998;114:382–897.
7. Horn F, Weare J, Beukers MW, et al. GPCRDB: an information system for G protein-coupled receptors. Nucleic Acids Res 1998;26:275–279.
8. Horn F, Bettler E, Oliveira L, Campagne F, Cohen FE, Vriend G. GPCRDB information system for G protein-coupled receptors. Nucleic Acids Res 2003;31:294–297.
9. Hollinger S, Hepler JR. Cellular regulation of RGS proteins: modulators and integrators of G protein signaling. Pharmacol Rev 2002;54:527–559.
10. Cismowski MJ, Takesono A, Bernard ML, Duzic E, Lanier SM. Receptor-independent activators of heterotrimeric G-proteins. Life Sci 2001;68:2301–2308.
11. Ferguson SS. Evolving concepts in G protein-coupled receptor endocytosis: the role in receptor desensitization and signaling. Pharmacol Rev 2001;53:1–24.
12. Ho MaW, YH. G Protein Structure Diversity. In: Pangalos MaD, CH, ed. Understanding G Protein-Coupled Receptors and Their Role in the CNS. 1st ed. New York: Oxford University Press,2002, pp. 63–86.
13. Offermanns S. G-proteins as transducers in transmembrane signalling. Prog Biophys Mol Biol 2003;83:101–130.
14. Cabrera-Vera TM, Vanhauwe J, Thomas TO, et al. Insights into G protein structure, function, and regulation. Endocr Rev 2003;24:765–781.
15. Weinstein LS, Chen M, Liu J. Gs(alpha) mutations and imprinting defects in human disease. Ann NY Acad Sci 2002;968:173–197.
16. Kehlenbach RH, Matthey J, Huttner WB. XL alpha s is a new type of G protein. Nature 1994;372:804–809.

17. Klemke M, Pasolli HA, Kehlenbach RH, Offermanns S, Schultz G, Huttner WB. Characterization of the extra-large G protein alpha-subunit XLalphas. II. Signal transduction properties. J Biol Chem 2000;275:33,633–33,640.
18. Bastepe M, Gunes Y, Perez-Villamil B, Hunzelman J, Weinstein LS, Juppner H. Receptor-mediated adenylyl cyclase activation through XLalpha(s), the extra-large variant of the stimulatory G protein alpha-subunit. Mol Endocrinol 2002;16:1912–1919.
19. Jones DT, Reed RR. Golf: an olfactory neuron specific-G protein involved in odorant signal transduction. Science 1989;244:790–795.
20. Jones DT, Masters SB, Bourne HR, Reed RR. Biochemical characterization of three stimulatory GTP-binding proteins. The large and small forms of Gs and the olfactory-specific G-protein, Golf. J Biol Chem 1990;265:2671–2676.
21. Belluscio L, Gold GH, Nemes A, Axel R. Mice deficient in G(olf) are anosmic. Neuron 1998;20:69–81.
22. Ebrahimi FA, Chess A. Olfactory G proteins: simple and complex signal transduction. Curr Biol 1998;8:R431–R433.
23. Regnauld K, Nguyen QD, Vakaet L, et al. G-protein alpha(olf) subunit promotes cellular invasion, survival, and neuroendocrine differentiation in digestive and urogenital epithelial cells. Oncogene 2002;21:4020–4031.
24. Itoh H, Toyama R, Kozasa T, Tsukamoto T, Matsuoka M, Kaziro Y. Presence of three distinct molecular species of Gi protein alpha subunit. Structure of rat cDNAs and human genomic DNAs. J Biol Chem 1988;263:6656–6664.
25. Sunahara RK, Dessauer CW, Gilman AG. Complexity and diversity of mammalian adenylyl cyclases. Annu Rev Pharmacol Toxicol 1996;36:461–480.
26. Dalwadi H, Wei B, Schrage M, Su TT, Rawlings DJ, Braun J. B cell developmental requirement for the G alpha i2 gene. J Immunol 2003;170:1707–1715.
27. Huang TT, Zong Y, Dalwadi H, et al. TCR-mediated hyper-responsiveness of autoimmune Galphai2(-/-) mice is an intrinsic naive CD4(+) T cell disorder selective for the Galphai2 subunit. Int Immunol 2003;15:1359–1367.
28. Strathmann M, Wilkie TM, Simon MI. Alternative splicing produces transcripts encoding two forms of the alpha subunit of GTP-binding protein Go. Proc Natl Acad Sci USA 1990;87:6477–6481.
29. Hsu WH, Rudolph U, Sanford J, et al. Molecular cloning of a novel splice variant of the alpha subunit of the mammalian Go protein. J Biol Chem 1990;265:11,220–11,226.
30. Exner T, Jensen ON, Mann M, Kleuss C, Nurnberg B. Posttranslational modification of Galphao1 generates Galphao3, an abundant G protein in brain. Proc Natl Acad Sci USA 1999;96:1327–1332.
31. Strittmatter SM, Valenzuela D, Kennedy TE, Neer EJ, Fishman MC. G0 is a major growth cone protein subject to regulation by GAP-43. Nature 1990;344:836–841.
32. Strittmatter SM, Fishman MC, Zhu XP. Activated mutants of the alpha subunit of G(o) promote an increased number of neurites per cell. J Neurosci 1994;14:2327–2338.

33. Jordan JD, Carey KD, Stork PJ, Iyengar R. Modulation of rap activity by direct interaction of Galpha(o) with Rap1 GTPase-activating protein. J Biol Chem 1999;274:21,507–21,510.
34. Chen LT, Gilman AG, Kozasa T. A candidate target for G protein action in brain. J Biol Chem 1999;274:26,931–26,938.
35. Ho MK, Wong YH. G(z) signaling: emerging divergence from G(i) signaling. Oncogene 2001;20:1615–1625.
36. Wong YH, Conklin BR, Bourne HR. Gz-mediated hormonal inhibition of cyclic AMP accumulation. Science 1992;255:339–342.
37. Jeong SW, Ikeda SR. G protein alpha subunit G alpha z couples neurotransmitter receptors to ion channels in sympathetic neurons. Neuron 1998;21:1201–1212.
38. Fong HK, Yoshimoto KK, Eversole-Cire P, Simon MI. Identification of a GTP-binding protein alpha subunit that lacks an apparent ADP-ribosylation site for pertussis toxin. Proc Natl Acad Sci USA 1988;85:3066–3070.
39. Matsuoka M, Itoh H, Kozasa T, Kaziro Y. Sequence analysis of cDNA and genomic DNA for a putative pertussis toxin-insensitive guanine nucleotide-binding regulatory protein alpha subunit. Proc Natl Acad Sci USA 1988;85:5384–5388.
40. Casey PJ, Fong HK, Simon MI, Gilman AG. Gz, a guanine nucleotide-binding protein with unique biochemical properties. J Biol Chem 1990;265:2383–2390.
41. Fields TA, Casey PJ. Phosphorylation of Gz alpha by protein kinase C blocks interaction with the beta gamma complex. J Biol Chem 1995;270:23,119–23,125.
42. Wang J, Frost JA, Cobb MH, Ross EM. Reciprocal signaling between heterotrimeric G proteins and the p21-stimulated protein kinase. J Biol Chem 1999;274:31,641–31,647.
43. Wong GT, Gannon KS, Margolskee RF. Transduction of bitter and sweet taste by gustducin. Nature 1996;381:796–800.
44. Margolskee RF. Molecular mechanisms of bitter and sweet taste transduction. J Biol Chem 2002;277:1–4.
45. Zhang Y, Hoon MA, Chandrashekar J, et al. Coding of sweet, bitter, and umami tastes: different receptor cells sharing similar signaling pathways. Cell 2003;112:293–301.
46. Burns ME, Baylor DA. Activation, deactivation, and adaptation in vertebrate photoreceptor cells. Annu Rev Neurosci 2001;24:779–805.
47. Fain GL, Matthews HR, Cornwall MC, Koutalos Y. Adaptation in vertebrate photoreceptors. Physiol Rev 2001;81:117–151.
48. Arshavsky VY, Lamb TD, Pugh EN Jr. G proteins and phototransduction. Annu Rev Physiol 2002;64:153–187.
49. Exton JH. Regulation of phosphoinositide phospholipases by hormones, neurotransmitters, and other agonists linked to G proteins. Annu Rev Pharmacol Toxicol 1996;36:481–509.
50. Rebecchi MJ, Pentyala SN. Structure, function, and control of phosphoinositide-specific phospholipase C. Physiol Rev 2000;80:1291–1335.

51. Rhee SG. Regulation of phosphoinositide-specific phospholipase C. Annu Rev Biochem 2001;70:281–312.
52. Wilkie TM, Scherle PA, Strathmann MP, Slepak VZ, Simon MI. Characterization of G-protein alpha subunits in the Gq class: expression in murine tissues and in stromal and hematopoietic cell lines. Proc Natl Acad Sci USA 1991;88:10,049–10,053.
53. Amatruda TT 3rd, Steele DA, Slepak VZ, Simon MI. G alpha 16, a G protein alpha subunit specifically expressed in hematopoietic cells. Proc Natl Acad Sci USA 1991; 88:5587–5591.
54. Kuang Y, Wu Y, Jiang H, Wu D. Selective G protein coupling by C-C chemokine receptors. J Biol Chem 1996;271:3975–3978.
55. Strathmann MP, Simon MI. G alpha 12 and G alpha 13 subunits define a fourth class of G protein alpha subunits. Proc Natl Acad Sci USA 1991;88:5582–5586.
56. Milligan G, Mullaney I, Mitchell FM. Immunological identification of the alpha subunit of G13, a novel guanine nucleotide binding protein. FEBS Lett 1992;297:186–188.
57. Kurose H. Galpha12 and Galpha13 as key regulatory mediator in signal transduction. Life Sci 2003;74:155–161.
58. Collins LR, Minden A, Karin M, Brown JH. Galpha12 stimulates c-Jun NH2-terminal kinase through the small G proteins Ras and Rac. J Biol Chem 1996;271:17,349–17,353.
59. Voyno-Yasenetskaya TA, Faure MP, Ahn NG, Bourne HR. Galpha12 and Galpha13 regulate extracellular signal-regulated kinase and c-Jun kinase pathways by different mechanisms in COS-7 cells. J Biol Chem 1996;271:21,081–21,087.
60. Hooley R, Yu CY, Symons M, Barber DL. G alpha 13 stimulates Na+-H+ exchange through distinct Cdc42-dependent and RhoA-dependent pathways. J Biol Chem 1996;271:6152–6158.
61. Lin X, Voyno-Yasenetskaya TA, Hooley R, Lin CY, Orlowski J, Barber DL. Galpha12 differentially regulates Na+-H+ exchanger isoforms. J Biol Chem 1996;271:22,604–22,610.
62. Plonk SG, Park SK, Exton JH. The alpha-subunit of the heterotrimeric G protein G13 activates a phospholipase D isozyme by a pathway requiring Rho family GTPases. J Biol Chem 1998;273:4823–4826.
63. Nagao M, Kaziro Y, Itoh H. The Src family tyrosine kinase is involved in Rho-dependent activation of c-Jun N-terminal kinase by Galpha12. Oncogene 1999;18:4425–4434.
64. Dhanasekaran N, Dermott JM. Signaling by the G12 class of G proteins. Cell Signal 1996;8:235–245.
65. Fromm C, Coso OA, Montaner S, Xu N, Gutkind JS. The small GTP-binding protein Rho links G protein-coupled receptors and Galpha12 to the serum response element and to cellular transformation. Proc Natl Acad Sci USA 1997;94:10,098–10,103.
66. Fukuhara S, Chikumi H, Gutkind JS. RGS-containing RhoGEFs: the missing link between transforming G proteins and Rho? Oncogene 2001;20:1661–1668.

67. Offermanns S. In vivo functions of heterotrimeric G-proteins: studies in Galpha-deficient mice. Oncogene 2001;20:1635–1642.
68. Suzuki N, Nakamura S, Mano H, Kozasa T. Galpha 12 activates Rho GTPase through tyrosine-phosphorylated leukemia-associated RhoGEF. Proc Natl Acad Sci USA 2003;100:733–738.
69. Clapham DE, Neer EJ. G protein beta gamma subunits. Annu Rev Pharmacol Toxicol 1997;37:167–203.
70. Gautam N, Downes GB, Yan K, Kisselev O. The G-protein betagamma complex. Cell Signal 1998;10:447–455.
71. Schwindinger WF, Robishaw JD. Heterotrimeric G-protein betagamma-dimers in growth and differentiation. Oncogene 2001;20:1653–1660.
72. Yan K, Kalyanaraman V, Gautam N. Differential ability to form the G protein betagamma complex among members of the beta and gamma subunit families. J Biol Chem 1996;271:7141–7146.
73. Wang Q, Mullah BK, Robishaw JD. Ribozyme approach identifies a functional association between the G protein beta1gamma7 subunits in the beta-adrenergic receptor signaling pathway. J Biol Chem 1999;274:17,365–17,371.
74. Asano T, Morishita R, Ueda H, Kato K. Selective association of G protein beta(4) with gamma(5) and gamma(12) subunits in bovine tissues. J Biol Chem 1999;274:21,425–21,429.
75. Logothetis DE, Kurachi Y, Galper J, Neer EJ, Clapham DE. The beta gamma subunits of GTP-binding proteins activate the muscarinic K+ channel in heart. Nature 1987;325:321–326.
76. Hanoune J, Defer N. Regulation and role of adenylyl cyclase isoforms. Annu Rev Pharmacol Toxicol 2001;41:145–174.
77. de Rooij J, Zwartkruis FJ, Verheijen MH, et al. Epac is a Rap1 guanine-nucleotide-exchange factor directly activated by cyclic AMP. Nature 1998;396:474–477.
78. Kawasaki H, Springett GM, Mochizuki N, et al. A family of cAMP-binding proteins that directly activate Rap1. Science 1998;282:2275–2279.
79. Defer N, Best-Belpomme M, Hanoune J. Tissue specificity and physiological relevance of various isoforms of adenylyl cyclase. Am J Physiol Renal Physiol 2000;279:F400–F416.
80. Hieble JaRJ, RR. Adrenergic Receptors. In: Pangalos MN and Davies CH, ed. Understanding G Protein-Coupled Receptors and Their Role in the CNS. 1st ed. New York: Oxford University Press, 2002, pp. 205–220.
81. Fukami K. Structure, regulation, and function of phospholipase C isozymes. J Biochem (Tokyo) 2002;131:293–299.
82. Lopez I, Mak EC, Ding J, Hamm HE, Lomasney JW. A novel bifunctional phospholipase c that is regulated by Galpha 12 and stimulates the Ras/mitogen-activated protein kinase pathway. J Biol Chem 2001;276:2758–2765.
83. Wing MR, Houston D, Kelley GG, Der CJ, Siderovski DP, Harden TK. Activation of phospholipase C-epsilon by heterotrimeric G protein betagamma-subunits. J Biol Chem 2001;276:48,257–48,261.
84. Kelley GG, Reks SE, Smrcka AV. Hormonal regulation of phospholipase Cepsilon through distinct and overlapping pathways involving G12 and Ras family G proteins. Biochem J 2004;378:129–139.

85. Song C, Hu CD, Masago M, et al. Regulation of a novel human phospholipase C, PLCepsilon, through membrane targeting by Ras. J Biol Chem 2001;276:2752–2757.
86. Kelley GG, Reks SE, Ondrako JM, Smrcka AV. Phospholipase C(epsilon): a novel Ras effector. EMBO J 2001;20:743–754.
87. Sadja R, Alagem N, Reuveny E. Gating of GIRK channels: details of an intricate, membrane-delimited signaling complex. Neuron 2003;39:9–12.
88. Mirshahi T, Jin T, Logothetis DE. G beta gamma and KACh: old story, new insights. Sci STKE 2003;2003(194):E32.
89. Kaneko S, Akaike A, Satoh M. Receptor-mediated modulation of voltage-dependent Ca2+ channels via heterotrimeric G-proteins in neurons. Jpn J Pharmacol 1999;81:324–331.
90. Ikeda SR. Voltage-dependent modulation of N-type calcium channels by G-protein beta gamma subunits. Nature 1996;380:255–258.
91. Herlitze S, Garcia DE, Mackie K, Hille B, Scheuer T, Catterall WA. Modulation of Ca2+ channels by G-protein beta gamma subunits. Nature 1996;380:258–262.
92. Zhong J, Hume JR, Keef KD. beta-Adrenergic receptor stimulation of L-type Ca2+ channels in rabbit portal vein myocytes involves both alphas and betagamma G protein subunits. J Physiol 2001;531:105–115.
93. Viard P, Macrez N, Mironneau C, Mironneau J. Involvement of both G protein alphas and beta gamma subunits in beta-adrenergic stimulation of vascular L-type Ca(2+) channels. Br J Pharmacol 2001;132:669–676.
94. Mattingly RR, Macara IG. Phosphorylation-dependent activation of the Ras-GRF/CDC25Mm exchange factor by muscarinic receptors and G-protein beta gamma subunits. Nature 1996;382:268–272.
95. Kiyono M, Satoh T, Kaziro Y. G protein beta gamma subunit-dependent Rac-guanine nucleotide exchange activity of Ras-GRF1/CDC25(Mm). Proc Natl Acad Sci USA 1999;96:4826–4831.
96. Pumiglia KM, LeVine H, Haske T, Habib T, Jove R, Decker SJ. A direct interaction between G-protein beta gamma subunits and the Raf-1 protein kinase. J Biol Chem 1995;270:14,251–14,254.
97. Luttrell LM, Hawes BE, van Biesen T, Luttrell DK, Lansing TJ, Lefkowitz RJ. Role of c-Src tyrosine kinase in G protein-coupled receptor- and Gbetagamma subunit-mediated activation of mitogen-activated protein kinases. J Biol Chem 1996;271:19,443–19,450.
98. Kozasa T, Jiang X, Hart MJ, et al. p115 RhoGEF, a GTPase activating protein for Galpha12 and Galpha13. Science 1998;280:2109–2111.
99. Hart MJ, Jiang X, Kozasa T, et al. Direct stimulation of the guanine nucleotide exchange activity of p115 RhoGEF by Galpha13. Science 1998;280:2112–2114.
100. Jiang Y, Ma W, Wan Y, Kozasa T, Hattori S, Huang XY. The G protein G alpha12 stimulates Bruton's tyrosine kinase and a rasGAP through a conserved PH/BM domain. Nature 1998;395:808–813.

101. Ito A, Satoh T, Kaziro Y, Itoh H. G protein beta gamma subunit activates Ras, Raf, and MAP kinase in HEK 293 cells. FEBS Lett 1995;368:183–187.
102. Martegani E, Vanoni M, Zippel R, et al. Cloning by functional complementation of a mouse cDNA encoding a homologue of CDC25, a *Saccharomyces cerevisiae* RAS activator. EMBO J 1992;11:2151–2157.
103. Shou C, Farnsworth CL, Neel BG, Feig LA. Molecular cloning of cDNAs encoding a guanine-nucleotide-releasing factor for Ras p21. Nature 1992;358:351–354.
104. Fukuhara S, Murga C, Zohar M, Igishi T, Gutkind JS. A novel PDZ domain containing guanine nucleotide exchange factor links heterotrimeric G proteins to Rho. J Biol Chem 1999;274:5868–5879.
105. Fukuhara S, Chikumi H, Gutkind JS. Leukemia-associated Rho guanine nucleotide exchange factor (LARG) links heterotrimeric G proteins of the G(12) family to Rho. FEBS Lett 2000;485:183–188.
106. Mochizuki N, Ohba Y, Kiyokawa E, et al. Activation of the ERK/MAPK pathway by an isoform of rap1GAP associated with G alpha(i). Nature 1999;400:891–894.
107. Tsygankova OM, Feshchenko E, Klein PS, Meinkoth JL. TSH/cAMP and GSK3beta elicit opposing effects on Rap1GAP stability. J Biol Chem 2004;279:5501–5507.
108. Jordan JD, He CJ, Eungdamrong NJ, et al. Cannabinoid receptor induced neurite outgrowth is mediated by Rap1 activation through Galphao/i-triggered proteasomal degradation of Rap1GAPII. J Biol Chem, 2005; in press.
109. Gratacap MP, Payrastre B, Nieswandt B, Offermanns S. Differential regulation of Rho and Rac through heterotrimeric G-proteins and cyclic nucleotides. J Biol Chem 2001;276:47,906–47,913. [Epub Sep 17, 2001.]
110. Usui I, Imamura T, Huang J, Satoh H, Olefsky JM. Cdc42 is a Rho GTPase family member that can mediate insulin signaling to glucose transport in 3T3-L1 adipocytes. J Biol Chem 2003;278:13,765–13,774.
111. Strittmatter SM, Fishman MC. The neuronal growth cone as a specialized transduction system. Bioessays 1991;13:127–134.
112. Simkowitz P, Ellis L, Pfenninger KH. Membrane proteins of the nerve growth cone and their developmental regulation. J Neurosci 1989;9:1004–1017.
113. Edmonds BT, Moomaw CR, Hsu JT, Slaughter C, Ellis L. The p38 and p34 polypeptides of growth cone particle membranes are the alpha- and beta-subunits of G proteins. Brain Res Dev Brain Res 1990;56:131–136.
114. Strittmatter SM, Vartanian T, Fishman MC. GAP-43 as a plasticity protein in neuronal form and repair. J Neurobiol 1992;23:507–520.
115. Strittmatter SM, Fankhauser C, Huang PL, Mashimo H, Fishman MC. Neuronal pathfinding is abnormal in mice lacking the neuronal growth cone protein GAP-43. Cell 1995;80:445–452.
116. Igarashi M, Strittmatter SM, Vartanian T, Fishman MC. Mediation by G proteins of signals that cause collapse of growth cones. Science 1993;259:77–79.
117. Christofori G. Split personalities: the agonistic antagonist Sprouty. Nat Cell Biol 2003;5:377–379.

118. Dikic I, Giordano S. Negative receptor signalling. Curr Opin Cell Biol 2003;15:128–135.
119. Silva AJ, Kogan JH, Frankland PW, Kida S. CREB and memory. Annu Rev Neurosci 1998;21:127–148.
120. Nestler EJ. Molecular basis of long-term plasticity underlying addiction. Nat Rev Neurosci 2001;2:119–128.

6
RGS Proteins

Orchestration of Multiple Signaling Pathways

Ryan W. Richman and María A. Diversé-Pierluissi

1. INTRODUCTION

For several years, the model for the transduction of G protein-mediated signals consisted of three components: a heptahelical receptor, a heterotrimeric G protein, and an effector *(1)*. The heptahelical receptor, which spans the membrane seven times, is coupled to a G protein complex consisting of an α-subunit in the guanosine-5'-diphosphate-bound form (α-GDP) and a βγ-dimer. Upon agonist binding to the receptor, a conformational change occurs in the G protein α-subunit, which leads to the release of the GDP and the binding to guanosine-5'-triphosphate (GTP). The α-GTP has lower affinity toward the βγ-dimer, releasing it from the G protein heterotrimer complex (Fig. 1) *(1)*. Both α-GTP and βγ-dimer are known to regulate a wide range of effectors *(2)*.

All subunits of Gα possess intrinsic GTPase activity, but the rate of GTP hydrolysis from the α-subunit alone is too low to account for the duration of G protein signaling observed in many physiological processes, such as visual transduction or ion channel modulation. The discovery of the regulators of G protein signaling (RGS) has helped to explain the difference in timing of the G protein-mediated responses. G protein-coupled receptor (GPCR) kinases (GRKs) and RGS proteins are involved in the termination or desensitization of G protein-mediated responses (Fig. 2). GRKs work at the receptor level by phosphorylating receptors in their active, ligand-bound form, uncoupling receptors from G proteins *(3)*. RGS proteins act at the G protein level by accelerating the rate of GTP hydrolysis (Fig. 1) *(4)*.

From: *Contemporary Clinical Neuroscience: The G Protein-Coupled Receptors Handbook*
Edited by: L. A. Devi © Humana Press Inc., Totowa, NJ

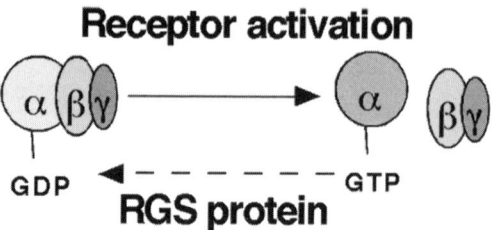

Fig. 1. Exchange of nucleotides from the Gα subunit. Upon receptor activation, the Gα subunit goes from the GDP-bound form to the GTP-bound form. Gα-GTP has lower affinity towards Gβγ subunits, causing dissociation of the complex. Termination of the response takes place when the G protein returns to the basal state as a consequence of GTP hydrolysis. RGS proteins accelerate the rate of hydrolysis.

The initial observations that suggested that RGS proteins have a role in signal termination were performed in yeast *(5)* and nematode worms *(6)*. The *sst2* gene was discovered in a genetic screen of mutants showing an arrest in the G1 to S-phase transition cycle—an indication of a termination failure of the mating pheromone response *(7)*. Loss of the *Caenorhabditis elegans egl-10* gene by mutation resulted in a decrease in egg-laying, which is a behavior inhibited by serotonin *(6)*. These genes from yeast and nematodes shared a 120-amino acid region also found in GOS8 (an immediate early gene in T-cell activation) *(8)*. Subsequent studies by Kehrl and colleagues *(9)* and the laboratories of Peralta and Casey showed that mammalian members of the RGS protein family impaired mitogen-activated protein kinase (MAPK) activity and could bind to the α-subunit of the G protein *(10)*. A yeast two-hybrid screen for proteins that interact with $G\alpha_{i3}$ resulted in the characterization of another RGS protein, GAIP (Gα-interacting protein) *(11)*.

The mechanism of action of RGS proteins became better understood when it was reported that GAIP and RGS4, two members of the RGS family, stabilize the GTP-to-GDP transition state of the G protein α-subunit *(12)*. RGS proteins bind to the switch regions of Gα-subunits *(13)*. To date, more than 30 members of the RGS protein family have been characterized. Based on structural similarities in their RGS box, the RGS proteins have been divided into subfamilies *(14,15)* (Fig. 3). For example, RGS4 has an Asn residue in position 128, which is conserved in subfamilies B, C, and D. This residue is believed to be involved in the stabilization of the transition state of Gα-subunits *(13)*. Other subfamilies have a Glu (subfamily F), Gln (subfamily E), or Ser (subfamily A) residues in the equivalent position.

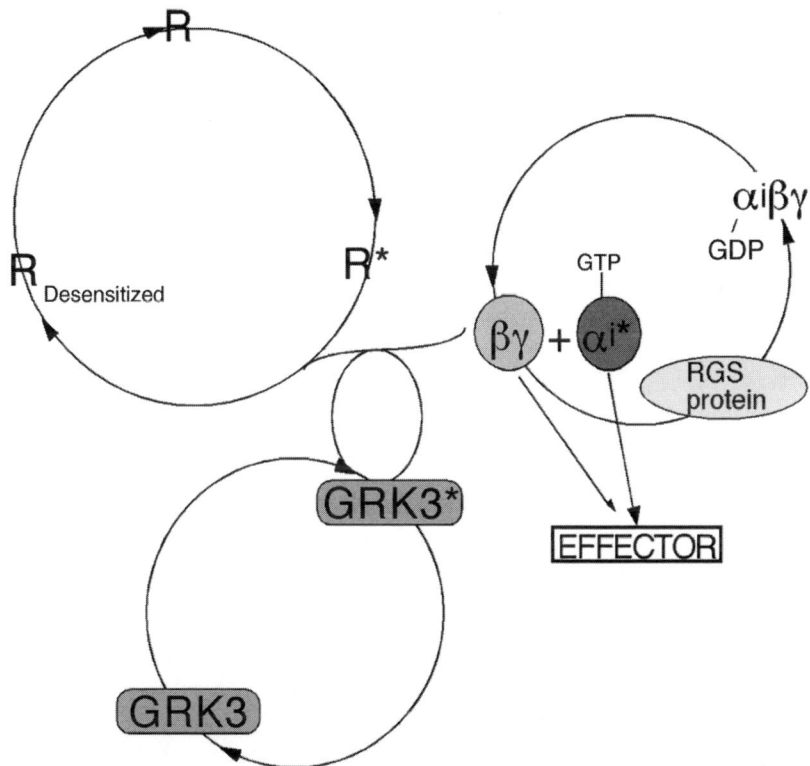

Fig. 2. Cycle of onset and termination of the G protein-mediated response. Gα-GTP and Gβγ subunits dissociate as a consequence of receptor activation and interact with a wide range of effectors. Gβγ subunits can bind to GRK3, a G protein-coupled receptor kinase, and phosphorylate the receptor, turning it into a desensitized state. RGS proteins can bind to the G α subunit and accelerate the rate of GTP hydrolysis.

1.1. Expression of RGS Proteins

Multiple RGS messenger RNAs (mRNAs) have been found in a wide range of tissues. Some RGS proteins, such as GAIP *(16)*, RGS2 *(17)*, RGS 3 *(18)*, RGS5 *(18)*, RGS16 *(19)*, axin *(20)*, p115Rho-guanine nucleotide exchange factor (GEF) *(21)*, and PSD-95, Dlg, and ZO-1 proteins (PDZ)-Rho GEF *(22)*, are ubiquitous in their expression pattern. Some RGS proteins show limited expression to certain tissues, suggesting they might serve specialized roles. For example, RET-RGS1 is only expressed in retina *(23)*, whereas RGS1 is expressed in lymphocytes *(24)*. Early studies showed that

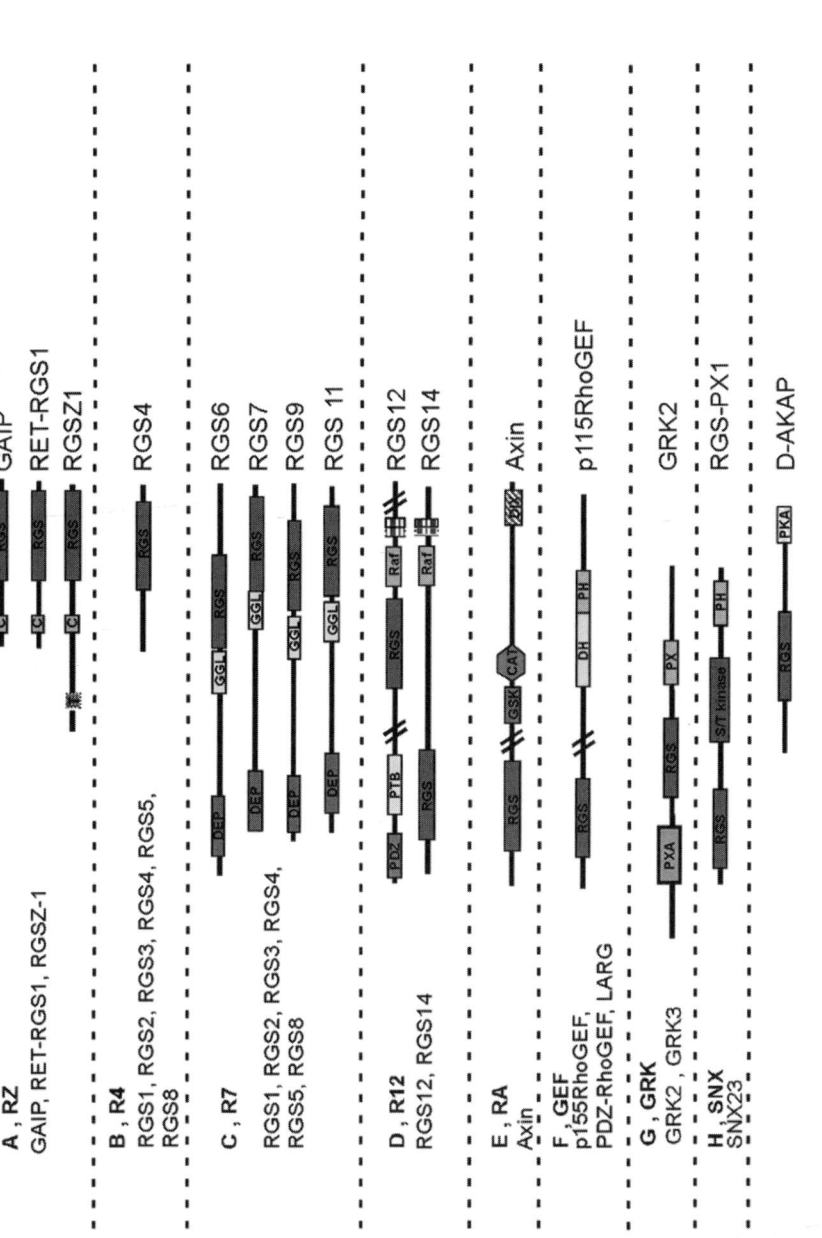

Fig. 3

neuronal tissues could express multiple RGS proteins. Embryonic chick dorsal root ganglion (DRG) neurons express 10 RGS protein transcripts *(25)*. Koelle and Horvitz *(6)* detected nine RGS family members in rat brain complementary DNA using polymerase chain reaction primers based on the nucleotide sequence of RGS proteins from the B subfamily.

The diversity of the RGS protein family is increased by the existence of alternatively spliced forms of RGS proteins. Four alternatively spliced forms of RGS12 are characterized by the presence or absence of the N-terminus PDZ domain and the C terminus PDZ-interacting motif *(26)*. The *RGS9* gene has two splice variants: (a) RGS9-1, a short variant which is expressed only in retina *(27)*, and (b) RGS9-2, which is 191 amino acids longer and is expressed in striatum *(27)*.

Some RGS proteins can be regulated at the level of transcription. The mRNA levels of RGS2 in hippocampus, cortex, and striatum are regulated by synaptic activity *(28)*. Drugs of abuse, such as cocaine and amphetamines, can regulate the levels of RGS protein mRNA. RGS1 levels are upregulated in B cells upon stimulation with phorbol esters *(29)*.

1.2. Subcellular Localization of RGS Proteins

The subcellular localization of RGS proteins remains poorly understood. The generation of antibodies against the different members of this protein family to detect the localization of endogenous RGS proteins will provide useful information. The RGS proteins (whose subcellular distribution has been studied) seem to have cytosolic and membrane-associated pools. The membrane-associated pool of GAIP is phosphorylated on Ser-24 and is found on clathrin-coated vesicles close to the plasma membrane and the *trans*-Golgi network *(30)*. It is believed that RET-RGS1 is membrane bound

---

Fig. 3. Subfamilies of RGS proteins. The different subfamilies and their members are listed. A schematic representation of the domain architecture for representative members of the different classes is included. RGS box represents the RGS domain common to all members of the family. Abbreviations: C represents the cysteine string region; CAT, β-catenin; DEP, dishevelled/EGL-10/plecstrin homology domain; DH, dbl homology domain, DIX, disheveled homology domain; GGL, G protein γ-like domain; GSK, glycogen synthase3b binding domain; PDZ, PSD95/ Disk Large/ Zona occludens domain; PKA, protein kinase A-anchoring domain; PTB, phosphotyrosine binding domain; PX, phosphatidylinositol-binding domain; PXA, PX-associate domains; Raf, Raf B homology domain; SNX, sorting nexin; S/ T, kinase serine-threonine kinase; T, transmembrane domain. Boxes in RGS12 and RGS14 represent goloco motifs.

because it has a cysteine-string motif and a putative transmembrane domain *(31)*. Overexpression of the constitutive active form of $G\alpha_{i2}$ results in increased association of RGS4 with the plasma membrane *(32)*.

Lipid modification of RGS proteins provides another potential mechanism of regulation of subcellular targeting and activity of these proteins. Palmitoylation of an N-terminal cysteine in RGS16 protein promotes its lipid raft targeting and allows palmitoylation of a poorly accessible cysteine residue in the RGS box (residue 98) *(33)*. Some RGS proteins contain myristoylation sites, but it is not known whether this lipid modification plays a role on their membrane association.

Another potential mechanism for targeting RGS proteins to specific subcellular compartments is their association with non-G protein partners. For example, the PDZ domain of RGS3 binds to the C-terminus of ephrin B (ephB) *(34)*. The ephB receptor, a receptor of tyrosine kinase, is involved in axonal guidance and other developmental processes. The ligand for this receptor, eph B, is a single transmembrane-spanning protein. In migration assays, ephrin B has been shown to inhibit the effects of the chemokine receptor CXCR4, a GPCR *(34)*. This inhibition is mediated by RGS3. This signaling model, proposed by Lu et al. *(34)*, is important in understanding the localization of granule neurons in the cerebellum *(34)*. Granule neurons are retained in the pia by chemo-attraction mediated by the CXCR4 receptor. At postnatal day 3, ephrin B is upregulated, RGS3 is recruited, and the RGS domain of RGS3 inhibits CXCR4 signaling (Fig. 4). Granular cells are then free to migrate through the cerebellum.

Many of the RGS proteins are localized in the nucleus. In the case of RGS10, phosphorylation by cyclic adenosine monophosphate-dependent kinase causes its translocation to the nucleus, making RGS10 unavailable to limit G protein-mediated signaling at the plasma membrane *(35)*. Splicing variants of RGS12 with a short C-terminal region are localized in the nucleus with a punctate foci distribution *(36)*. The functional implications of this distribution are not known.

2. RGS PROTEINS AS MULTIFUNCTIONAL MOLECULES

All RGS proteins possess the 120-amino acid RGS domain, but they can vary in length from 217 (GAIP and RGSZ1) to 1387 amino acids (RGS12). In addition to the RGS domain, RGS proteins have many different domains, suggesting selective regulation or multifunctional activity (Fig. 3). The multiplicity of RGS proteins suggests they might exhibit selectivity in the pathways on which they exert their actions. The following sections discuss different roles of RGS proteins.

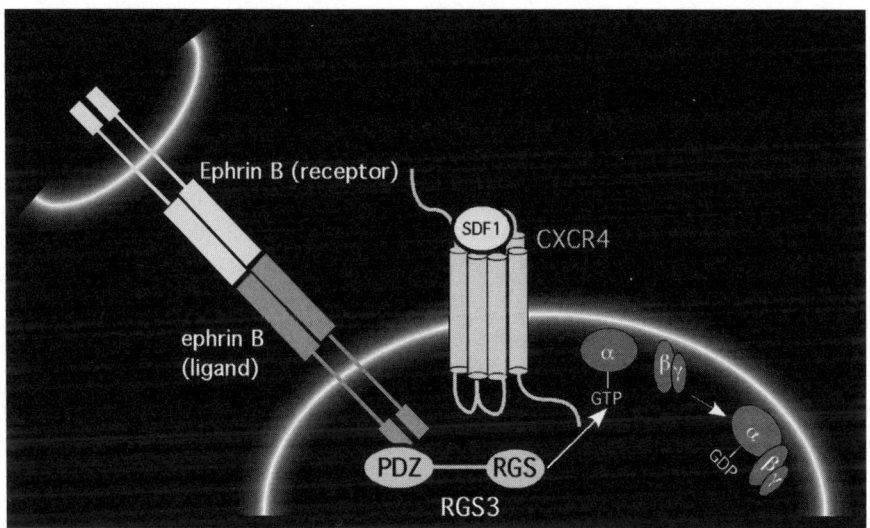

Fig. 4. Inhibition of chemoattraction by ephrin B. PDZ-RGS3 is recruited by binding to the PDZ interacting domain of ephrin B and the RGS domain of RGS3 inhibits CXCR4 signaling by turning off the G protein.

2.1. RGS Proteins as Partners for G Protein β-Subunits

A subfamily of RGS proteins, C or R7 *(14,15)* (which include RGS6, -7, -9 and –11), binds to the G protein β_5-subunit. These RGS proteins contain a homologous domain, with Gγ-subunits termed G protein γ-like domains *(37)* Numerous studies have demonstrated the physiological importance of this subfamily of RGS proteins in the kinetics of light response *(38–41)*. RGS7 has been shown to form a complex with Gβ_5 in retina *(42)*. Gβ_5 knockout mice exhibit lower levels of RGS9 mRNA *(43)*. The RGS9–Gβ_5 complex has been shown to stimulate the GTPase activity of transducin in its bound state to the effector, cyclic guanosine monophosphate (cGMP) phosphodiesterase.

A large percentage of RGS9–Gβ_5 complexes are tethered to the plasma membrane by an anchoring protein, R9AP *(44)*. R9AP has only been detected in photoreceptors *(44)*. The anchored complex shows a fourfold increase in GTPase activity *(45)*. The N-terminus of RGS9, which contains a disheveled/EGL-10/pleckstrin domain, is believed to be important for this interaction *(46)*. Mice lacking RGS9 or lacking an RGS9–R9AP interaction show a delay in the recovery from the light response *(47)*. The physiological relevance of RGS9 in the visual system was underscored by a recent study in which it was demonstrated that patients exhibiting a condition known as bradyopsia had recessive mutations in the *RGS9* or *R9AP* genes. These patients experienced difficulty in adapting to sudden changes in light levels and in seeing low-contrast moving objects *(48)*.

Members of the C or R7 subfamily are also expressed in other parts of the central nervous system. R7 proteins can increase tolerance at the level of μ-opioid receptors. Knockdown of RGS9 and -11 by antisense oligonucleotides increased the duration of morphine-induced analgesia *(49)*. The RGS11-deficient mice showed reduced analgesic response to the δ-opioid receptor agonist [D-Ala (2)] deltorphin II, whereas the mice deficient in RGS6 and -9 showed alterations in the time-course of the effects of this agonist *(50)*.

2.2. RGS Proteins Bridge Signaling Between Heterotrimeric and Monomeric G Proteins

A direct link between heterotrimeric and monomeric G protein signaling was discovered when it was discovered that the Rho exchange factor p115RhoGEF has an N-terminus RGS domain that exhibits GTPase-activating protein (GAP) activity on $G\alpha_{12}$ and $G\alpha_{13}$. Activation of $G\alpha_{13}$ results in translocation of p115RhoGEF to the plasma membrane *(51)*. A K-to-L mutation in residue 677 in the pleckstrin homology domain of this protein was sufficient to abolish Rho-mediated gene transcription but did not alter translocation to the membrane. This link helped to explain the molecular mechanisms by which activation of a receptor coupled to a heterotrimeric G protein can result in cytoskeletal rearrangement and changes in cell morphology *(52)*. These members of the F (or GEF) family include p115RhoGEF, PDZ-RhoGEF, and leukemia-associated RhoGEF (LARG) *(53)*

$G\alpha_{12}$ and $G\alpha_{13}$ are known to activate growth-promoting responses and activation of c-fos through regulation of its serum response element (SRE). Activation of muscarinic receptors expressed in HeLa cells results in activation of PYK2 (a tyrosine kinase member of the Src-related family) and downstream activation of SRE-mediated transcription. These responses are blocked by a kinase-deficient form of PYK2. The GTPase-deficient form of $G\alpha_{13}$ results in a potentiation of the response. This potentiation was blocked by co-expression of the RGS box of p115RhoGEF *(54)*.

Similar links to tyrosine kinase pathways have been demonstrated in studies of PDZ-RhoGEF and LARG. Activation of thrombin receptors result in the activation of the focal adhesion kinase, which in turn phosphorylates PDZ-RhoGEF and LARG *(55)*. The tyrosine phosphorylation of these RGS proteins results in a sustained activation of Rho and creates a positive feedback loop for the activation of Rho by GPCRs *(55)*.

2.3. RGS Proteins As Elements in G Protein and Tyrosine Kinase Pathways

Modulation of calcium channels by GPCRs is a transient phenomenon; neurons become unresponsive upon prolonged exposure to neurotransmitters. Despite the common requirement of GRK activation for the onset of desensitization of transmitter-induced inhibition of calcium current in chick DRG neurons *(56)*, G_i- and G_o-mediated inhibition desensitize at different rates *(25)*. Activation of γ-aminobutyric acid $(GABA)_B$ receptors in chick DRG neurons inhibits the $Ca_v2.2$ calcium channel in a voltage-independent manner through activation of a tyrosine kinase of the Src-related family *(57)*. This inhibition desensitizes within 100 seconds *(58)*. The voltage-independent inhibition requires activation of a tyrosine kinase that phosphorylates the α1-subunit of the channel and thereby recruits the binding of RGS12 *(58)*. Introduction of a recombinant protein containing the sequence of the phosphotyrosine binding domain from RGS12 slows the desensitization rate of GABA-induced voltage-independent inhibition of $Ca_v2.2$ calcium channels, whereas the PDZ domain is not affected. RGS12 coprecipitates with the tyrosine-phosphorylated calcium channel. This RGS12–calcium channel association is decreased by pretreatment with genistein, a tyrosine kinase inhibitor *(58)*.

Another example of the role of RGS proteins in providing a link between GPCRs and receptor tyrosine kinases comes from studies of nerve growth factor (NGF) signaling in PC12 cells and the role of GAIP in bridging these signaling pathways *(59)*. NGF binding to trkA receptors produces signals that are important for neuronal survival, axonal guidance, and differentiation *(59)*. GIPC (a PDZ domain containing protein that binds to GAIP) can also bind to and form a complex with the trkA receptor. Immunofluorescence experiments have shown that in retrograde transport vesicles, GIPC colocalizes with the tyrosine-phosphorylated form of the trkA receptor *(59)*. Overexpression of GIPC in PC12 cells inhibits NGF-induced increases in phosphorylation of MAPKs. Interestingly, no change has been observed on the phosphorylation of other signaling molecules, such as phospholipase C-γ1, Shc, or Akt.

2.4. Modulation of Ion Channels in Neurons by RGS Proteins

One of the first indications that RGS proteins could be more than just GAPs for heterotrimeric G proteins came from studies of the modulation of G protein inward-rectifying potassium channels *(60)*. These channels are opened by direct binding of G protein βγ-subunits *(61)*. RGS proteins increase the speed of the deactivation of these channels, as is expected from a GAP; the α-GDP-subunit will bind to the βγ-subunit, which is then no longer available to bind to the channel.

In addition to the effects on the kinetics of termination of the response, RGS4 and -8 accelerated the rate of current activation *(60)*. One potential explanation for these results is that accelerating the GTPase cycle can lead to a more sustained response by increasing the number of G proteins that can be activated.

The role of endogenous RGS proteins in the modulation of voltage-dependent calcium channels was studied by Ikeda et al. *(62)*, who used an approach that mutated a glycine to serine in the switch region so that the Gα-subunit became insensitive to the GAP activity of the RGS proteins. Additionally, they introduced a mutation that rendered the α-subunit of $G_{i/o}$ proteins insensitive to pertussis toxin. In this case, both activation and deactivation kinetics of the ion channels were slowed. A similar approach used adenoviral delivery of RGS- and pertussis toxin-insensitive Gα-subunits to hippocampal neurons and yielded similar results *(63)*.

Experiments in which antibodies have been introduced into the cell bodies of neurons by microinjection have shown that RGS4 and GAIP have differential effects on the coupling of $α_2$-adrenergic receptors to the inhibition of $Ca_v2.2$ calcium channels in chick DRG neurons *(25)*. Removal of the N- and C-terminus domains of GAIP abolished this selectivity.

RGS proteins might be regulated by calcium influx. Studies in which RGS3 was overexpressed in chick DRG neurons showed that deletion of the EF-hand of RGS3 abolished desensitization of transmitter-mediated inhibition of $Ca_v2.2$ calcium channels *(64)*. The RGS3-mediated effects are blocked by a calmodulin antagonist.

3. CONCLUSIONS

RGS proteins have been demonstrated to play a role in the cellular transition between excited and inhibited states in a wide range of physiological processes, such as lymphocyte chemotaxis, ion channel modulation, cytoskeletal rearrangement, visual transduction, adaptation to drugs of abuse, and membrane trafficking. The mechanisms by which RGS proteins are regulated have yet to be completely elucidated. It has become clear that RGS proteins are more than just GAPs; they serve as bridges between heterotrimeric G proteins and a multiplicity of pathways.

ACKNOWLEDGMENTS

This work was supported by NS 37443 (National Institues of Health) and a Hirschl Trust Fund Career Development Award to M. Diversé-Pierluissi.

REFERENCES

1. Hamm HE. The many faces of G protein signaling. J Biol Chem 1998;273:669–672.
2. Clapham DE, Neer EJ. G protein beta gamma subunits. Annu Rev Pharmacol Toxicol 1997;37:167–203.
3. Premont RT, Inglese J, Lefkowitz RJ. Protein kinases that phosphorylate activated G protein-coupled receptors. Fascb J 1995;9:175–182
4. Ross EM, Wilkie TM. GTPase-activating proteins for heterotrimeric G proteins: regulators of G protein signaling (RGS) and RGS-like proteins. Annu Rev Biochem 2000;69:795–827.
5. Dohlman HG, Song J, Apanovitch DM, DiBello PR, Gillen KM. Regulation of G protein signalling in yeast. Semin Cell Dev Biol 1998;9:135–141.
6. Koelle MR, Horvitz HR. EGL-10 regulates G protein signaling in the C. elegans nervous system and shares a conserved domain with many mammalian proteins. Cell 1996;84:115–125.
7. Dohlman HG, Song J, Ma D, Courchesne WE, Thorner J. Sst2, a negative regulator of pheromone signaling in the yeast Saccharomyces cerevisiae: expression, localization, and genetic interaction and physical association with Gpa1 (the G-protein alpha subunit). Mol Cell Biol 1996;16:5194–5209.
8. Siderovski DP, Heximer SP, Forsdyke DR. A human gene encoding a putative basic helix-loop-helix phosphoprotein whose mRNA increases rapidly in cycloheximide-treated blood mononuclear cells. DNA Cell Biol 1994;13:125–147.
9. Druey KM, Blumer KJ, Kang VH, Kehrl JH. Inhibition of G-protein-mediated MAP kinase activation by a new mammalian gene family. Nature 1996;379:742–746.
10. Hunt TW, Fields TA, Casey PJ, Peralta EG. RGS10 is a selective activator of G alpha i GTPase activity. Nature 1996;383:175–177.
11. De Vries L, Mousli M, Wurmser A, Farquhar MG. GAIP, a protein that specifically interacts with the trimeric G protein G alpha i3, is a member of a protein family with a highly conserved core domain. Proc Natl Acad Sci USA 1995;92:11,916–11920.
12. Berman DM, Wilkie TM, Gilman AG. GAIP and RGS4 are GTPase-activating proteins for the Gi subfamily of G protein alpha subunits. Cell 1996;86:445–452.
13. Tesmer JJ, Berman DM, Gilman AG, Sprang SR. Structure of RGS4 bound to AlF4—activated G (i alpha1): stabilization of the transition state for GTP hydrolysis. Cell 1997;89:251–261.
14. De Vries L, Zheng B, Fischer T, Elenko E, Farquhar MG. The regulator of G protein signaling family. Annu Rev Pharmacol Toxicol 2000;40:235–271.
15. Sierra DA, Gilbert DJ, Householder D, et al. Evolution of the regulators of G-protein signaling multigene family in mouse and human. Genomics 2002;79:177–185.
16. Grafstein-Dunn E, Young KH, Cockett MI, Khawaja XZ. Regional distribution of regulators of G-protein signaling (RGS) 1, 2, 13, 14, 16, and GAIP messenger ribonucleic acids by in situ hybridization in rat brain. Brain Res Mol Brain Res 2001;88:113–123.

17. Gold SJ, Han MH, Herman AE, et al. Regulation of RGS proteins by chronic morphine in rat locus coeruleus. Eur J Neurosci 2003;17:971–980.
18. Cho H, Kozasa T, Bondjers C, Betsholtz C, Kehrl JH. Pericyte-specific expression of Rgs5: implications for PDGF and EDG receptor signaling during vascular maturation. FASEB J 2003;17:440–442.
19. Chen C, Zheng B, Han J, Lin SC. Characterization of a novel mammalian RGS protein that binds to Galpha proteins and inhibits pheromone signaling in yeast. J Biol Chem 1997;272:8679–8685.
20. Zeng L, Fagotto F, Zhang T, et al. The mouse Fused locus encodes Axin, an inhibitor of the Wnt signaling pathway that regulates embryonic axis formation. Cell 1997;90:181–192.
21. Hart MJ, Sharma S, elMasry N, et al. Identification of a novel guanine nucleotide exchange factor for the Rho GTPase. J Biol Chem 1996;271:25,452–25,458.
22. Fukuhara S, Murga C, Zohar M, Igishi T, Gutkind JS. A novel PDZ domain containing guanine nucleotide exchange factor links heterotrimeric G proteins to Rho. J Biol Chem 1999;274:5868–5879.
23. Faurobert E, Hurley JB. The core domain of a new retina specific RGS protein stimulates the GTPase activity of transducin in vitro. Proc Natl Acad Sci USA 1997;94:2945–2950.
24. Druey KM, Blumer KJ, Kang VH, Kehrl JH. Inhibition of G-protein-mediated MAP kinase activation by a new mammalian gene family. Nature 1996; 379: 742–746.
25. Diverse-Pierluissi MA, Fischer T, Jordan JD, et al. Regulators of G protein signaling proteins as determinants of the rate of desensitization of presynaptic calcium channels. J Biol Chem 1999; 274: 14,490–14,494.
26. Snow BE, Hall RA, Krumins AM, et al. GTPase activating specificity of RGS12 and binding specificity of an alternatively spliced PDZ (PSD-95/Dlg/ZO-1) domain. J Biol Chem 1998;273:17,749–17455.
27. Zhang K, Howes KA, He W, et al. Structure, alternative splicing, and expression of the human RGS9 gene. Gene 1999;240:23–34.
28. Ingi T, Krumins AM, Chidiac P, et al. Dynamic regulation of RGS2 suggests a novel mechanism in G-protein signaling and neuronal plasticity. J Neurosci 1998;18:7178–7188.
29. Heximer SP, Cristillo AD, Forsdyke DR. Comparison of mRNA expression of two regulators of G-protein signaling, RGS1/BL34/1R20 and RGS2/G0S8, in cultured human blood mononuclear cells. DNA Cell Biol 1997;16:589–598.
30. Fischer T, Elenko E, Wan L, Thomas G, Farquhar MG. Membrane-associated GAIP is a phosphoprotein and can be phosphorylated by clathrin-coated vesicles. Proc Natl Acad Sci USA 2000;97:4040–4045.
31. Faurobert E, Hurley JB. The core domain of a new retina specific RGS protein stimulates the GTPase activity of transducin in vitro. Proc Natl Acad Sci USA 1997;94:2945–2950.
32. Druey KM, Sullivan BM, Brown D, Fischer ER, Watson N, Blumer KJ, Gerfen CR, Scheschonka A, Kehrl JH. Expression of GTPase-deficient Gialpha2

results in translocation of cytoplasmic RGS4 to the plasma membrane. J Biol Chem 1998;273:18,405–18,410.
33. Hiol A, Davey PC, Osterhout JL, Waheed AA, Fischer ER, Chen CK, Milligan G, Druey KM, Jones TL. Palmitoylation regulates regulators of G-protein signaling (RGS) 16 function. I. Mutation of amino-terminal cysteine residues on RGS16 prevents its targeting to lipid rafts and palmitoylation of an internal cysteine residue. J Biol Chem 2003;278:19,301–19,308.
34. Lu Q, Sun EE, Klein RS, Flanagan JG. Ephrin-B reverse signaling is mediated by a novel PDZ-RGS protein and selectively inhibits G protein-coupled chemoattraction. Cell 2001;105:69–79.
35. Burgon PG, Lee WL, Nixon AB, Peralta EG, Casey PJ. Phosphorylation and nuclear translocation of a regulator of G protein signaling (RGS10). J Biol Chem 2001;276:32,828–32,834. Epub 2001 Jul 06.
36. Chatterjee TK, Fisher RA. RGS12TS-S localizes at nuclear matrix-associated subnuclear structures and represses transcription: structural requirements for subnuclear targeting and transcriptional repression. Mol Cell Biol 2002; 22:4334–4345.
37. Snow BE, Krumins AM, Brothers GM, et al. A G protein gamma subunit-like domain shared between RGS11 and other RGS proteins specifies binding to Gbeta5 subunits. Proc Natl Acad Sci USA 1998;95:13,307–13,312.
38. Cowan CW, He W, Wensel TG. RGS proteins: lessons from the RGS9 subfamily. Prog Nucleic Acid Res Mol Biol 2001;65:341–359.
39. Cabrera JL, de Freitas F, Satpaev DK, Slepak VZ. Identification of the Gbeta5-RGS7 protein complex in the retina. Biochem Biophys Res Commun 1998;249:898–902.
40. Chen CK, Burns ME, He W, Wensel TG, Baylor DA, Simon MI. Slowed recovery of rod photoresponse in mice lacking the GTPase accelerating protein RGS9-1. Nature 2000;403:557–560.
41. He W, Cowan CW, Wensel TG. RGS9, a GTPase accelerator for phototransduction. Neuron 1998;20:95–102.
42. Zhang JH, Simonds WF. Copurification of brain G-protein beta5 with RGS6 and RGS7. J Neurosci 2000;20:RC59.
43. Krispel CM, Chen CK, Simon MI, Burns ME. Prolonged photoresponses and defective adaptation in rods of Gbeta5-/- mice. J Neurosci 2003;23:6965–6971.
44. Hu G, Wensel TG. R9AP, a membrane anchor for the photoreceptor GTPase accelerating protein, RGS9-1. Proc Natl Acad Sci USA 2002;99:9755–9760.
45. Hu G, Zhang Z, Wensel TG. Activation of RGS9-1GTPase acceleration by its membrane anchor, R9AP. J Biol Chem 2003;278:14,550–14,554.
46. Martemyanov KA, Lishko PV, Calero N, et al. The DEP domain determines subcellular targeting of the GTPase activating protein RGS9 in vivo. J Neurosci 2003;23:10,175–10,181.
47. Keresztes G, Martemyanov KA, Krispel CM, et al. Absence of the RGS9.Gbeta5 GTPase-activating complex in photoreceptors of the R9AP knockout mouse. J Biol Chem 2004;279:1581–1584.
48. Nishiguchi KM, Sandberg MA, Kooijman AC, et al. Defects in RGS9 or its anchor protein R9AP in patients with slow photoreceptor deactivation. Nature 2004;427:75–78.

49. Garson J, Rodriguez-Diaz M, Lopez-Fando A, Sanchez-Blazquez P. RGS9 proteins facilitate acute tolerance to mu-opioid effects. Eur J Neurosci 2001;13:801–811.
50. Garzon J, Lopez-Fando A, Sanchez-Blazquez P. The R7 subfamily of RGS proteins assists tachyphylaxis and acute tolerance at mu-opioid receptors. Neuropsychopharmacology 2003;28:1983–1990.
51. Bhattacharyya R, Wedegaertner PB. Galpha 13 requires palmitoylation for plasma membrane localization, Rho-dependent signaling, and promotion of p115-RhoGEF membrane binding. J Biol Chem 2000;275:14,992–14,999.
52. Wells CD, Liu MY, Jackson M, et al. Mechanisms for reversible regulation between G13 and Rho exchange factors. J Biol Chem 2002;277:1174–1181.
53. Booden MA, Siderovski DP, Der CJ. Leukemia-associated Rho guanine nucleotide exchange factor promotes G alpha q-coupled activation of RhoA. Mol Cell Biol 2002;22:4053–4061.
54. Shi CS, Sinnarajah S, Cho H, Kozasa T, Kehrl JH. G13alpha-mediated PYK2 activation. PYK2 is a mediator of G13alpha -induced serum response element-dependent transcription. J Biol Chem 2000;275:24,470–24,476.
55. Chikumi H, Fukuhara S, Gutkind JS. Regulation of G protein-linked guanine nucleotide exchange factors for Rho, PDZ-RhoGEF, and LARG by tyrosine phosphorylation: evidence of a role for focal adhesion kinase. J Biol Chem 2002;277:12,463–12,473.
56. Diverse-Pierluissi M, Inglese J, Stoffel RH, Lefkowitz RJ, Dunlap K. G protein-coupled receptor kinase mediates desensitization of norepinephrine-induced Ca2+ channel inhibition. Neuron 1996;16:579–585.
57. Diverse-Pierluissi M, Remmers AE, Neubig RR, Dunlap K. Novel form of crosstalk between G protein and tyrosine kinase pathways. Proc Natl Acad Sci USA 1997;94:5417–5421.
58. Schiff ML, Siderovski DP, Jordan JD, et al. Tyrosine-kinase-dependent recruitment of RGS12 to the N-type calcium channel. Nature 2000;408:723–727.
59. Lou X, Yano H, Lee F, Chao MV, Farquhar MG. GIPC and GAIP form a complex with TrkA: a putative link between G protein and receptor tyrosine kinase pathways. Mol Biol Cell 2001;12:615–627.
60. Chuang HH, Yu M, Jan YN, Jan LY. Evidence that the nucleotide exchange and hydrolysis cycle of G proteins causes acute desensitization of G-protein gated inward rectifier K+ channels. Proc Natl Acad Sci USA 1998;95:11,727–11,732.
61. Logothetis DE, Kurachi Y, Galper J, Neer EJ, Clapham DE. The beta gamma subunits of GTP-binding proteins activate the muscarinic K+ channel in heart. Nature 1987 325:321–326.
62. Jeong SW, Ikeda SR. Endogenous regulator of G-protein signaling proteins modify N-type calcium channel modulation in rat sympathetic neurons. J Neurosci 2000;20:4489–4496.
63. Chen H, Lambert NA. Endogenous regulators of G protein signaling proteins regulate presynaptic inhibition at rat hippocampal synapses. Proc Natl Acad Sci USA 2000;97:12,810–12,815.
64. Tosetti P, Pathak N, Jacob MH, Dunlap K. RGS3 mediates a calcium-dependent termination of G protein signaling in sensory neurons. Proc Natl Acad Sci USA 2003;100:7337–7342.

7
G Protein-Coupled Receptor Kinases

Lan Ma, Jingxia Gao, and Xiaoqing Chen

1. INTRODUCTION

As the largest family of membrane receptors, G protein-coupled receptors (GPCRs) transduce a large number of extracellular signals from hormones, neurotransmitters, chemokines, and other environmental stimuli to the interior of cells and play fundamental roles in the regulation of cellular functions. One of the most important features of GPCR-mediated signal transduction is that the activation of the receptor also triggers a feedback regulatory mechanism to attenuate GPCR-mediated signal transduction (homologous desensitization) in the cell. Homologous receptor desensitization is a common mechanism employed by the cell to prevent potential harmful effects that result from persistent activation of the signaling pathways. The initial event of GPCR desensitization occurs within seconds to minutes after agonist binding and is induced by receptor phosphorylation that is catalyzed by GPCR kinases (GRKs) *(1)*.

2. FAMILY MEMBERS AND STRUCTURAL FEATURES

Studies regarding the mechanisms involved in the homologous desensitization of rhodopsin and the β2-adrenergic receptor led to the discovery of rhodopsin kinase (GRK1) and β2-adrenergic receptor kinase (βARK or GRK2) and their cloning from bovine tissue *(1)*. Thus far, at least seven members (GRK1–7) of the GRK family have been cloned *(1–3)*. Among them, GRK1, -7, and -4 are expressed exclusively in retina and testis, whereas GRK2, -3, -5, and -6 are expressed in a wide range of tissues such as heart, brain, lung, and placenta *(1,3)*.

From: *Contemporary Clinical Neuroscience: The G Protein-Coupled Receptors Handbook*
Edited by: L. A. Devi © Humana Press Inc., Totowa, NJ

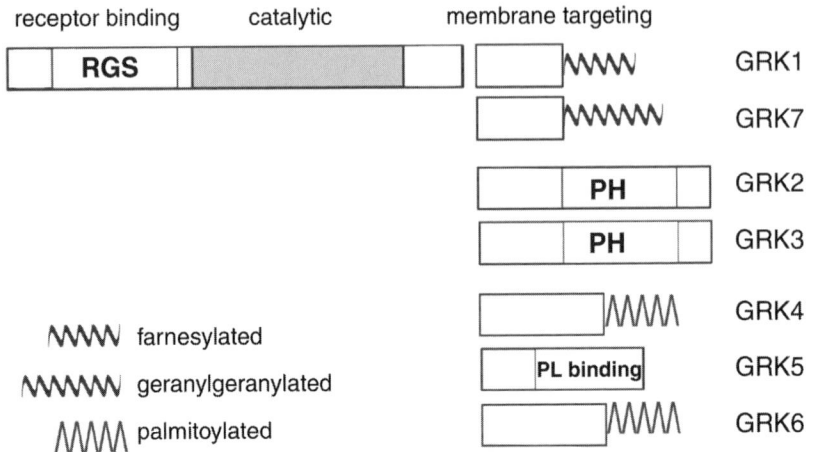

Fig. 1. Schematic presentation of the structural organization of seven GRK subtypes. The N-terminal domain of GRK 1–7 contains a conserved RGS domain and is involved in receptor binding. The central domain of GRK 1–7 is responsible for the catalytic activity of the kinase. The C-terminal domain of GRK is involved in membrane targeting. GRK1 is farnesylated and GRK7 is geranylgeranylated at their C-termini. The C-terminal domain of GRK2 and -3 contains a PH domain that is required for Gβγ binding. GRK4 and -6 are palmitoylated at their C-termini. The C-terminal domain of GRK5 contains basic amino acid residues for phospholipid binding.

GRKs consist of a single, 60- to 80-kDa polypeptide chain that contains the following three domains (Fig. 1): (a) an N-terminal domain (approximately 185 residues), which is involved in receptor binding and contains a regulator of G protein signaling (RGS)-like domain; (b) a central domain, which exerts kinase catalytic function; and (c) a less conserved C-terminal domain, which contains structures responsible for plasma membrane targeting of the kinase. Caveolin and the α-subunit of Gq interact with GRKs through the RGS domain at the N-terminal region of GRKs. Gβγ-subunits and acidic lipids bind to the pleckstrin homology (PH) domain, and clathrin binds to the clathrin binding motif at the C-terminal domain of the GRK *(4)*. GRKs are divided into rhodopsin kinase (GRK1 and -7), βARK (GRK2 and -3), and GRK4 (GRK4, -5, and -6) subfamilies based on their structural and functional similarities, *(2,5,6)*.

3. ACTIVITY AND ROLES IN REGULATION OF RECEPTOR SIGNALING

GRKs are serine/threonine kinases. They are activated upon agonist stimulation and catalyze phosphorylation of agonist-occupied GPCRs. GRK-mediated receptor phosphorylation inhibits GPCR coupling to G proteins and recruits signal molecules such as β-arrestins to the phosphorylated receptor, resulting in desensitization of receptor-mediated signal transduction as well as receptor internalization (1,7). Overexpression of GRKs in transfected cells or transgenic animals enhances agonist-induced receptor phosphorylation and desensitization, whereas inhibition of GRK activity or expression in cells or transgenic animals attenuates receptor desensitization. Mutation of GRK phosphorylation sites reduces agonist-stimulated receptor phosphorylation and desensitization (1,8). Studies have revealed a critical role for GRK-catalyzed receptor phosphorylation in the regulation of GPCR signals (Fig. 2).

GRKs have been shown to be involved in phosphorylation-independent desensitization of certain GPCRs (4). GRKs mediate phosphorylation of Gq-coupled metabotropic glutamate receptor 1a (mGluR1a) and contribute to the desensitization of mGluR1a (9,10). However, Dhami et al. (11) recently demonstrated that overexpression of GRK2 inhibits signaling mediated by a C-terminal-truncated form of mGluR1a that is incapable of being phosphorylated. Additionally, the expression of the GRK2 N-terminal domain or the catalytically inactive GRK mutant GRK2-K220R is sufficient to inhibit agonist-stimulated mGluR1a signaling. These observations suggest that GRK2-mediated regulation of mGluR1a signaling is not dependent on the kinase activity of GRK (phosphorylation-independent). It is likely that binding of the RGS domain of GRK to the activated Gαq or other signal molecules contributes to the phosphorylation-independent regulation of GPCR signal transduction, because the RGS domain at the N-terminus of GRK2 can bind to Gαq and inhibit its activity. The binding of GRKs to many other proteins, such as the GRK-interacting protein GIT1 (12), hints that GRKs may also function as scaffolds in the regulation of signals mediated by certain GPCRs, just like their counterpart arrestins.

4. SUBSTRATE AND GRK SPECIFICITY

GRKs preferentially phosphorylate GPCR in an active (agonist-occupied) state at serine and threonine residues localized within either the third intra-

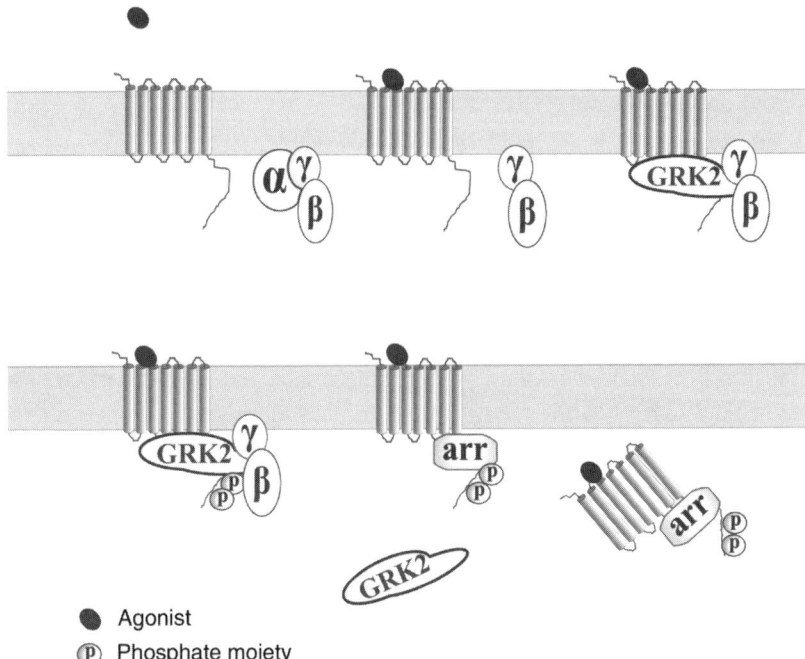

Fig. 2. Schematic presentation of the role of GRK2-catalyzed receptor phosphorylation in agonist-stimulated GPCR desensitization. GPCR activation induces GRK2 binding to free Gβγ on the membrane. This is followed by binding of the GRK2-Gβγ complex to the activated receptor to form a receptor-GRK2-Gβγ ternary complex, resulting in phosphorylation of the activated receptor and receptor–G protein uncoupling. The phosphorylation of GPCR by GRKs recruits β-arrestins to the receptor complex and promotes receptor internalization. The formation of the receptor-GRK2-Gβγ ternary complex is important for stabilization of the GRK2 membrane localization and catalytic function.

cellular loop or the C-terminal domains. GRK phosphorylation sites have been identified for only a few GPCR substrates, and no clear consensus substrate sequence has been found. Studies with synthetic peptide substrates and purified GRKs suggest that the kinases in the GRK2 family preferentially phosphorylate peptides containing acidic residues that flank the target serines or threonines *(1)*. Studies with the M2 muscarinic receptor and α_1AR have shown that acidic amino acid residues are important in the agonist-dependent phosphorylation and desensitization of these receptors *(13,14)*.

Data obtained from opioid receptors have demonstrated that the negatively charged acidic residues flanking GRK phosphorylation sites are critically involved in the interaction of the receptor with GRKs of both the βARK and GRK4 subfamilies in vivo and, therefore, are required for GRK-mediated receptor phosphorylation and desensitization *(15,16)*.

GRKs mediate the phosphorylation and desensitization of a variety of agonist activated GPCRs, including opioid receptors, dopamine receptors, substance P receptors, chemokine receptors, and so on *(17)*. However, the phosphorylation of non-GPCR substrates (such as receptor tyrosine kinases) by GRK2 has also been observed *(18)*. Additionally, activation of β2-AR induces GRK2-mediated phosphorylation of nonreceptor substrates such as tubulin, synucleins, phosducin, and ribosomal protein-2 *(19)*.

More than 600 GPCR genes have been identified in the human genome and more than 1000 of 19,000 open reading frames in the genome of *Caenorhabditis elegans* encode GPCRs *(20,21)*. To date, however, only seven members of the GRK family have been identified. The mechanism by which this limited number of GRKs regulates the huge number of receptors is not well-understood. Studies on different GRK subtypes expressed in heterologous systems have demonstrated both the functional redundancy and specificity of GRKs. In recent years, insightful information regarding GRK specificity has been gained from GRK transgenic and knockout mice *(8)*. Studies of animals with altered *GRK* gene have revealed that GRK1 phosphorylates rhodopsin and regulates light response of retinal cells; GRK2 is responsible for regulation of β-AR signaling in heart and vasculature; GRK3 controls the functions of odorant and muscarinic receptors; GRK4 is able to regulate D1 dopamine receptor; GRK5 targets the β-AR and muscarinic receptors; and GRK6 is responsible for phosphorylation of CXCR4 and regulation of D2 dopamine receptors *(8,22)*. Thus far, these in vivo studies have demonstrated that defined roles exist for each of the GRK subtypes and have indicated that the in vivo functions of GPCRs may be regulated by a particular GRK subtype.

5. REGULATION OF LOCALIZATION AND ACTIVITY

The predominant physiological significance of GPCR desensitization mediated by GRKs is to protect cells from overstimulation in the persistent presence of agonists and, therefore, regulate signaling. This requires tight regulation of the activation, cellular localization, and gene expression of GRKs. Subheadings 5.1.–5.3. describe the various mechanisms involved in regulating GRK activity.

5.1. Regulation by G Proteins and Phospholipids

Interaction between GRKs and the activated GPCR on the plasma membrane is required for GRK-catalyzed receptor phosphorylation. Membrane targeting of GRK1, -4, -6, and -7 is through posttranslational modification of the GRK C-terminus. Membrane association of GRKs occurs via farnesylation (GRK1), palmitoylation (GRK4, GRK6), or geranylgeranylation (GRK7). GRK5 associates with the cell membrane through interactions between their negatively charged C-terminal domain and phospholipids such as phosphatydylinositol-4,5-bisphosphate (PIP2), and these interactions enhance its activity *(1,6)*.

Studies have revealed that the members of the β-AR kinase subfamily (GRK2 and -3) are present primarily in the cytosol of unstimulated cells and do not undergo posttranslational lipid modification. GRK2 and -3 translocate to the plasma membrane upon agonist stimulation, and both possess PH domains through which they interact with PIP2 in the plasma membrane. The membrane translocation of GRK2 and -3 also requires the activation of G proteins, because, to be recruited to the membrane, GRK2 and -3 must bind free Gβγ-subunits that are anchored to the membrane *(1,8)*. Furthermore, the interaction of GRK2 and -3 with the Gβγ-subunits stimulates the phosphorylation of GPCRs *(1)*. A recent study indicated that the agonist-dependent association of GRK2 and opioid receptors requires Gβγ-subunits *(16)*, and coimmunoprecipitation studies suggested that the formation of receptor–GRK–Gβγ complex is required for GRK2 membrane translocation and catalytic activity (Fig. 2).

GRK2 and -3 have been shown to interact with the activated form of the Gαq subunit and to inhibit Gαq-mediated phospholipase C activity *(6)*. This effect on phospholipase C is mediated via the RGS domain and occurs independently of the catalytic activity of GRK2 and -3.

5.2. Regulation by Other Kinases

5.2.1. Regulation by Extracellular Signal-Regulated Protein Kinase-1/2

Most of the GRKs in the cytoplasm are in a basally phosphorylated and inactive form. In the cytosol, GRK2 is primarily phosphorylated at serine 670, which is a putative phosphorylation site for extracellular signal-regulated protein kinase (ERK)1/2 *(8)*. ERK activation stimulates GRK2 phosphorylation and reduces both its kinase activity and binding to Gβγ-subunits *(8)*. Activation of β2-AR stimulates the association of GRK2 with ERK1 and this is enhanced in the presence of both agonist-occupied receptor and Gβγ-subunits *(8)*. These data suggest that GRK2 activation is negatively

G Protein-Coupled Receptor Kinases 155

regulated by the mitogen-activated protein kinase pathway and phosphorylation of GRK2 by ERK1/2 may work as a switch to turn off GRKs and keep them in an inactive state in the cytoplasm.

5.2.2. Regulation by c-Src

GRK2 can also be phosphorylated at tyrosine residues in the RGS domain upon activation of β2-AR; this depends on the ability of β-arrestin to bind to and recruit c-Src to the receptor. Phosphorylation of GRK by c-Src increases the activity of GRK2 toward receptor substrates, enhances the interaction of GRK2 with Gαq, and potentiates receptor desensitization. However, tyrosine phosphorylation of GRK2 promotes degradation of GRK in the proteasome *(19)*.

5.2.3. Regulation by Kinase A and Protein C

Second messenger-dependent kinases such as protein kinase A (PKA) and protein kinase C (PKC) have been shown to regulate GRK activity and membrane targeting. Both GRK2 and -5 are substrates of PKC. Phosphorylation of GRK2 by PKC stimulates GRK-mediated receptor phosphorylation most likely through enhancement of its membrane translocation, whereas phosphorylation of GRK5 by PKC inhibits GRK-mediated receptor phosphorylation and its binding to the membrane *(19)*. The PKC phosphorylation site on GRK2 is located at the N-terminus of GRK2, whereas phosphorylation of GRK5 by PKC occurs at two sites within the C-terminal region of GRK5 *(19)*. Activation of a Gs-coupled receptor that binds A-kinase anchoring protein 79 (AKAP79) induces phosphorylation of GRK2 at serine 685 by PKA, which increases the affinity of GRK2 for Gβγ-subunits and stimulates its translocation to the membrane *(19)*.

5.3. Regulation by Calcium-Binding Proteins

Calcium ions, as a universal second messenger, play important roles in neuronal signaling and signal transduction processes that occur in non-neuronal cells. Increase in the cytosolic calcium ion concentration activates calcium sensor proteins, of which recoverin and calmodulin are well-studied GRK interacting proteins. Recoverin is predominantly present in photoreceptor cells. It binds to GRK1 and inhibits its activity in the dark when [Ca^{2+}] is high. However, upon light stimulation (which lowers [Ca^{2+}]), recoverin dissociates from GRK1 and allows calmodulin to phosphorylate rhodopsin *(8)*.

Calmodulin is a ubiquitously expressed universal calcium sensor protein and its binding to both the N- and C-terminal domains of GRKs inhibits the activity of GRKs 2–6; however, this inhibition occurs at different potencies. GRK5 is very sensitive to the calcium-bound calmodulin (IC_{50}, 40–50 n*M*), whereas GRK2 has a much lower sensitivity (IC_{50}, 2 μM) *(19)*. The different affinities of various GRKs for calcium sensor proteins may provide the basis for the specificity of signal transduction modulation by GRKs that are activated by different receptors.

6. REGULATION OF PROTEIN LEVELS

6.1. Stability

Regulation of GRK enzymatic activity provides a mechanism to exert rapid control on GPCR signaling. Regulation of GRK expression adjusts cellular GRK protein levels and may produce a long-lasting effect on receptor-mediated signal transduction. Cellular levels of GRKs are under stringent control. GRK2 undergoes ubiquitination and is rapidly degraded in the proteasome (with a half-life of less than 2 h), and agonist stimulation promotes the degradation of GRK2 *(19)*. The mechanisms underlying GRK2 degradation involve β-arrestin binding and subsequent phosphorylation of GRK2 by c-Src recruited by β-arrestins *(19)*.

6.2. Expression

Transgenic studies have indicated that the regulation of GPCR signaling is sensitive to the level of GRKs expressed in vivo. Alteration in GPCR signaling and changes in GRK expression and activity are correlated. Overexpression of GRK2 or -5 in cardiac tissue impairs β-AR agonist-stimulated contractile functions *(8)*. The abnormal activity and expression of GRK2, -5, or -6 have been observed under many pathological conditions, including heart failure, cystic fibrosis, hypertension, and rheumatoid arthritis *(6,8,19)*. Chronic activation of GPCRs induces changes in GRK expression. For example, chronic infusion of β-AR agonist results in increased levels of GRK2 messenger RNA (mRNA) and protein, β-AR desensitization, and myocardial hypertrophy. Additionally, chronic administration of morphine resulting in morphine tolerance causes changes in the levels of GRK2 protein and GRK2 and -5 mRNA levels *(6,8,23–25)*. The mechanisms underlying regulation of GRK gene expression are not well-understood. One study showed that mitogenic stimulation of T cells causes an increase in GRK mRNA levels, which could be partially mimicked by PKC activators *(19)*. This study also showed that transcriptional regulation of the promoter of *GRK2* gene is cell type-specific and is regulated by phorbol esters as well as activation of Gαq or α1-AR pathways in aortic smooth muscle cells *(19)*.

REFERENCES

1. Pitcher JA, Freedman NJ, Lefkowitz RJ. G protein-coupled receptor kinases. Annu Rev Biochem 1998; 67:653–692.
2. Hisatomi O, Matsuda S, Satoh T, Kotaka S, Imanishi Y, Tokunaga F. A novel subtype of G-protein-coupled receptor kinase, GRK7, in teleost cone photoreceptors. FEBS Lett 1998; 424:159–164.
3. Weiss ER, Raman D, Shirakawa S, et al. The cloning of GRK7, a candidate cone opsin kinase, from cone- and rod-dominant mammalian retinas. Mol Vis 1998; 4:27.
4. Pao CS, Benovic JL. Phosphorylation-independent desensitization of G protein-coupled receptors? Sci STKE 2002;153:pe42.
5. Premont RT, Inglese J, Lefkowitz RJ. Protein kinases that phosphorylate activated G protein-coupled receptors. FASEB J 1995; 9:175–182.
6. Penn RB, Pronin AN, Benovic JL. Regulation of G protein-coupled receptor kinases. Trends Cardiovasc Med 2000; 10:81–89.
7. Ferguson SS. Evolving concepts in G protein-coupled receptor endocytosis: the role in receptor desensitization and signaling. Pharmacol Rev 2001; 53:1–24.
8. Kohout TA, Lefkowitz RJ. Regulation of G protein-coupled receptor kinases and arrestins during receptor desensitization. Mol Pharmacol 2003; 63:9–18.
9. Dale LB, Bhattacharya M, Anborgh PH, et al. G protein-coupled receptor kinase-mediated desensitization of metabotropic glutamate receptor 1A protects against cell death. J Biol Chem 2000; 275:38,213–38,220.
10. Sallese M, Salvatore L, D'Urbano E, et al. The G-protein-coupled receptor kinase GRK4 mediates homologous desensitization of metabotropic glutamate receptor 1. FASEB J 2000; 14:2569–2580.
11. Dhami GK, Anborgh PH, Dale LB, Sterne-marr R, Ferguson SS. Phosphorylation-independent regulation of metabotropic glutamate receptor signaling by G protein-coupled receptor kinase 2. J Biol Chem 2002; 277:25,266–25,272.
12. Premont RT, Claing A, Vitale N, et al. A G protein-coupled receptor kinase-associated ADP ribosylation factor GTPase-activating protein. Proc Natl Acad Sci 1998; 95:14,082–14,087.
13. Lee KB, Ptasienski JA, Bunemann M, Hosey MM. Acidic amino acids flanking phosphorylation sites in the M2 muscarinic receptor regulate receptor phosphorylation, internalization, and interaction with arrestins. J Biol Chem 2000; 275:35,767–35,777.
14. Small KM, Brown KM, Forbes SL, Liggett SB. Polymorphic deletion of three intracellular acidic residues of the alpha 2B-adrenergic receptor decreases G protein-coupled receptor kinase-mediated phosphorylation and desensitization. J Biol Chem 2001; 276:4917–4922.
15. Guo J, Wu YL, Zhang WB, et al. Identification of G protein-coupled receptor kinase 2 phosphorylation sites responsible for agonist-stimulated δ-opioid receptor phosphorylation. Mol Pharmacol 2000; 58:1050–1056.

16. Li JL, Xiang B, Su WJ, ZhangXQ, Huang YL, Ma L. Agonist-induced formation of opioid receptor-G protein-coupled receptor kinase (GRK)-Gβγ complex on membrane is required for GRK2 function in vivo. J Biol Chem 2003; 278:30,219–30,226.
17. Carman CV, Benovic JL. G-protein-coupled receptors: turn-ons and turn-offs. Curr Opin Neurobiol 1998; 8335–344.
18. Freedman NJ, Kim LK, Murray JP, et al. Phosphorylation of the platelet-derived growth factor receptor-beta and epidermal growth factor receptor by G protein-coupled receptor kinase-2. Mechanisms for selectivity of desensitization. J Biol Chem 2002; 277:48,261–48,269.
19. Penela P, Ribas C, Mayor F Jr. Mechanisms of regulation of the expression and function of G protein-coupled receptor kinases. Cell Signal 2003; 15:973–981.
20. Venter JC, Adams MD, Myers EW, et al. The sequence of the human genome. Science 2001; 291:1304–1351.
21. Bargmann CI. Neurobiology of the *Caenorhabditis elegans* genome. Science 1998; 282:2028–2033.
22. Gainetdinov RR, Bohn LM, Sotnikova TD, et al. Dopaminergic supersensitivity in G protein-coupled receptor kinase 6-deficient mice. Neuron 2003; 38:291–303.
23. Ozaita A, Escriba PV, Ventayol P, Murga C, Mayor F Jr, Garcia-Sevilla JA. Regulation of G protein-coupled receptor kinase 2 in brains of opiate-treated rats and human opiate addicts. J Neurochem 1998; 70:1249–1257.
24. Terwilliger RZ, Ortiz J, Guitart X, Nestler EJ. Chronic morphine administration increases beta-adrenergic receptor kinase (beta ARK) levels in the rat locus coeruleus. J Neurochem 1994; 63:1983–1986.
25. Fan XL, Zhang JS, Zhang XQ, Yue W, Ma L. Acute and chronic treatments and morphine withdrawal differentially regulate *GRK2* and *GRK5* gene expression in rat brain. Neuropharmacology 2002; 43:809—816.

8
Regulators of GPCR Activity

The Arrestins

Louis M. Luttrell

1. INTRODUCTION

A single photon of light, triggering the physicochemical isomerization of an 11-*cis*-retinal to its all-*trans* form, initiates the process of vision. The retinal moiety serves as a tethered ligand for the photoreceptor rhodopsin, a G protein-coupled receptor (GPCR); within milliseconds, the conformational change in rhodopsin that occurs as a result of the retinal isomerization catalyzes guanosine 5'-triphosphate (GTP) for guanosine 5'-diphosphate (GDP) exchange on the heterotrimeric G protein transducin, transducin-dependent activation of a cyclic guanosine monophosphate (cGMP) phosphodiesterase (PDE), and the closure of cGMP-gated ion channels. The resulting hyperpolarization of the rod outer segment membrane inhibits release of the neurotransmitter glutamate from the photoreceptor terminal, and the light-induced stimulus is transmitted through the neural network of the retina to the central nervous system (CNS) for processing.

In contrast to the rapidity of photoreceptor activation, reversal of the light-induced conformational change in 11-*cis*-retinal is a slow process. Thus, the ability to retain light sensitivity beyond an initial round of photobleaching absolutely depends on the existence of a mechanism for the rapid termination of rhodopsin–transducin coupling. In the retina, this is accomplished through a two-step process. First, the bleached rhodopsin is phosphorylated by rhodopsin kinase, a membrane-associated serine/threonine kinase that specifically targets the activated conformation of the receptor. Phosphorylated rhodopsin then binds tightly to an abundant 48-kDa cytosolic protein, originally termed S antigen. Binding of S antigen, now called visual arrestin

From: *Contemporary Clinical Neuroscience: The G Protein-Coupled Receptors Handbook*
Edited by: L. A. Devi © Humana Press Inc., Totowa, NJ

or arrestin 1, sterically interdicts rhodopsin–transducin coupling to prevent persistent signaling in the absence of an ongoing stimulus *(1,2)*.

This basic paradigm in which an agonist-induced change in GPCR conformation initiates two antagonistic processes (G protein activation and receptor desensitization) is not confined to the retina. Nearly all GPCRs studied to date undergo agonist-dependent phosphorylation by GPCR kinases (GRKs) and bind to members of the arrestin family *(3,4)*. It is now recognized that the arrestins are central players not only in desensitization but also in GPCR internalization, intracellular trafficking, and signaling. They function not only as "arresting proteins" but also as adapters targeting GPCRs for removal from the cell surface via clathrin-coated pits and as scaffolding proteins recruiting signaling proteins to ligand-bound GPCRs (Fig. 1). They may even function as "G protein-independent" signal transducers and play roles in the trafficking and signaling of membrane receptors other than GPCRs. This chapter reviews what is currently known about the structure of these versatile proteins and the diverse roles they play in the termination and transmission of receptor-mediated signals.

2. THE ARRESTIN FAMILY OF PROTEINS

2.1. Visual and Nonvisual Arrestins

Although simple eukaryotes such as yeast possess GPCRs and heterotrimeric G proteins, they do not contain arrestins. However, arrestins are expressed in many invertebrates, including *Caenorhabditis elegans*, *Drosophila*, and squid *(5–8)*. Vertebrates (including amphibians, birds, and mammals) possess two types of arrestin: (a) the visual arrestins, which are expressed almost exclusively in the retina and pineal gland, and (b) the nonvisual arrestins (β-arrestins), which are ubiquitously expressed. Genomic analysis of the early ascidian chordate *Ciona intestinalis* revealed a single arrestin gene possessing features of both visual and nonvisual arrestins, suggesting that the two types may have arisen from gene duplication of an ancestral arrestin early in vertebrate evolution *(9)*.

To date, four functional members of the vertebrate arrestin gene family have been cloned *(3,4)*. The two arrestins expressed in the retina, visual arrestin (S antigen or arrestin 1 *[10,11]*) and cone arrestin (X-arrestin or C-arrestin *[12,13]*), primarily exist to regulate photoreceptor function. The nonvisual arrestins, β-arrestin 1 (arrestin 2 *[14]*) and β-arrestin 2 (arrestin 3*[15]*), regulate the activity of most of the other 600-plus GPCRs in the genome. Partial complementary DNA clones of two additional arrestins, D-arrestin and E-arrestin, have been reported *(13)*, but it remains unclear if functional D- and E-arrestin proteins are expressed.

Alternative splice variants exist for each of the arrestins *(4)*. Visual arrestin, a 404-amino acid protein, has two splice variants—one in which the last 34 amino acids are truncated and replaced by an alanine residue and another that lacks exon 13, which encodes residues 338 to 345. The truncated form of visual arrestin localizes specifically to the rod outer segment and is a more potent inhibitor of rhodopsin signaling than the longer form. β-arrestin 1, a 418-amino acid protein, has a splice variant consisting of the insertion of 8 residues between amino acids 333 and 334; β-arrestin 2, a 410-amino acid protein, has a splice variant consisting of the insertion of 11 residues between amino acids 361 and 362. To date, no functional differences have been ascribed to the two β-arrestin splice variants.

All four arrestins bind specifically to light-activated or agonist-occupied GPCRs that have been phosphorylated by GRKs and block the receptor–G protein interaction. In addition to their role in GPCR desensitization, the β-arrestins have other functions that are not shared with the visual arrestins. The β-arrestin C-terminal tail (distal to residue 374) contains binding motifs for clathrin *(16,17)* and the $β_2$-adaptin subunit of the AP-2 complex *(18–20)*, which allow β-arrestins to act as adapter proteins and target GPCRs to clathrin-coated pits for endocytosis. These are the primary interactions that distinguish the two arrestin subfamilies and make β-arrestin binding integral to the interrelated processes of GPCR endocytosis, intracellular trafficking, resensitization, and downregulation.

2.2. Structure–Function Relationships

High-resolution crystal structures are currently available for visual arrestin *(21,22)* and β-arrestin 1 *(23)*. They reveal proteins of high structural homology. The arrestins contain two major domains: an N domain (residues 8–180 of visual arrestin) and a C domain (residues 188–362), each of which is composed of a seven-stranded β-sandwich (Fig. 2). A polar core, which lies embedded between the N- and C-domains in the basal state, links the two major domains. This region, called the phosphate sensor domain, forms the fulcrum of the arrestin molecule and plays a key role in the conformational changes that occur when an arrestin encounters a phosphorylated GPCR. Residues from the free N- and C-terminal tails of the protein also contribute to the polar core of unbound arrestin.

Extensive mutagenesis studies performed using visual arrestin have divided the protein into three functional and two regulatory domains that coincide closely to the structural elements defined crystallographically *(24)*. The functional domains include a receptor activation recognition domain in the globular N-domain (residues 24–180), a secondary receptor-binding re-

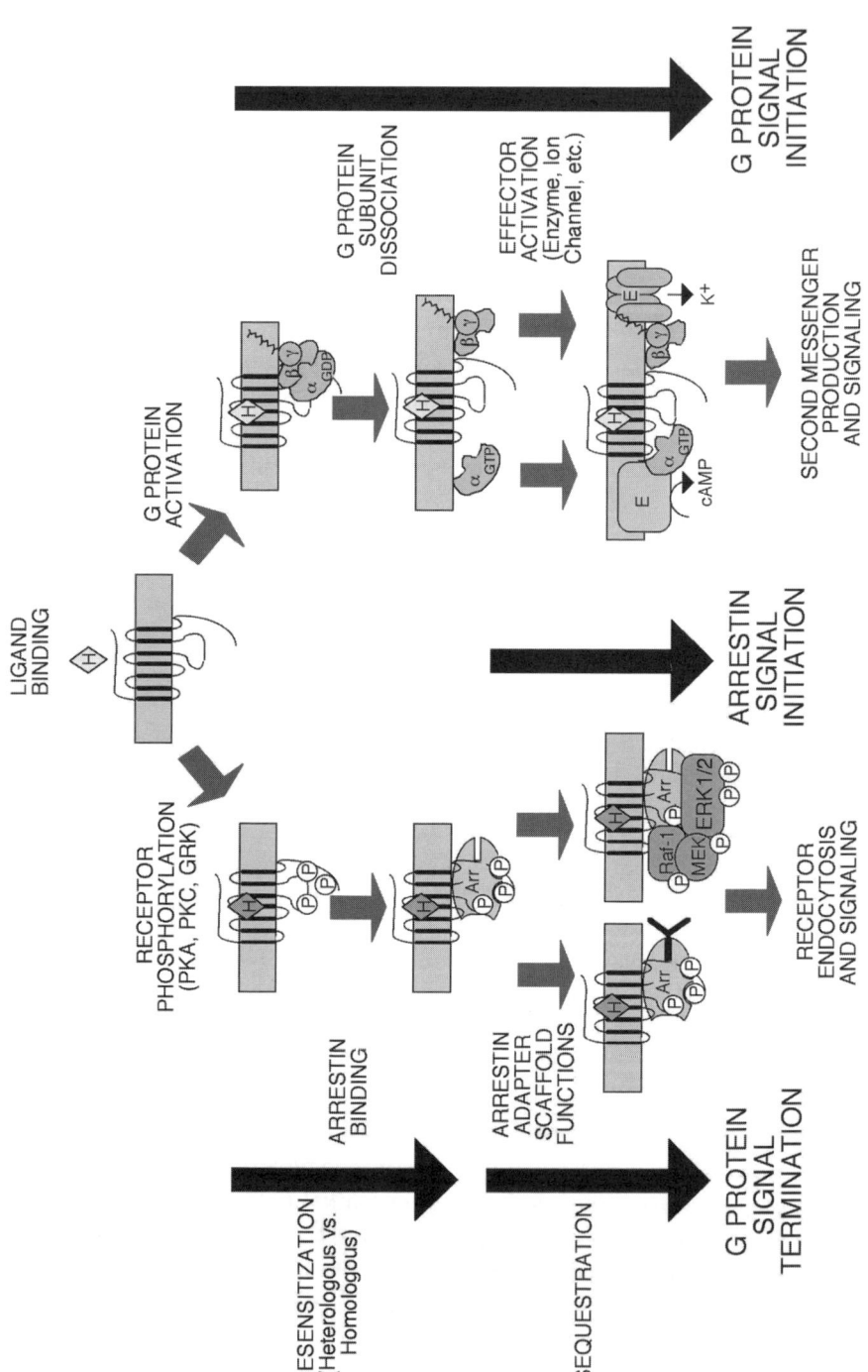

Fig 1.

gion in the C-domain (residues 180–330), and the phosphate sensor domain (residues 163–182). Data obtained from visual arrestin–β-arrestin 1 chimeras indicate that the specificity of receptor recognition resides within residues 49 to 90 and 237 to 268 of visual arrestin and the homologous regions of β-arrestin *(25)*. Interactions among the polar core, the C-terminal tail, and the phosphate sensor region maintain arrestin in the inactive state. Upon binding to an activated receptor, the phosphorylated receptor tail is believed to invade the polar core, displacing the arrestin C-terminus and allowing for conformational changes that promote tight binding to the receptor *(26)*. Consistent with this model, mutations in arrestin or β-arrestin that disrupt intramolecular interactions within the polar core produce arrestins that bind with high affinity to agonist-occupied GPCRs without requiring receptor phosphorylation *(27,28)*.

As noted, it is the C-terminus of the β-arrestins distal to the globular C-domain that confers properties lacking in visual arrestin. This region of the protein contains two well-defined motifs that link β-arrestin-bound GPCRs to the clathrin-dependent endocytic machinery. A LIEF/L sequence located between residues 374 and 377 of β-arrestin 2 binds to a region located between amino acids 89 and 100 of the N-terminal domain of the clathrin heavy chain *(16,17)*. Additionally, β-arrestins bind directly to the β_2-adaptin subunit of the heterotetrameric AP-2 adaptor complex through an RxR sequence, located at residues 394 to 396 of β-arrestin 2 *(18–20)*. The AP-2 complex links many receptors to the clathrin endocytic machinery by binding to clathrin, dynamin, and EPS-15 and is involved in the initiation of clathrin-coated pit formation *(29)*. Both interactions appear to be important for efficient β-arrestin-mediated GPCR internalization *(19,30)*.

Additional interactions modulate the endocytic function of β-arrestins. β-arrestins bind to phosphoinositides—particularly InsP6 *(31)*. The

---

Fig.1. The alternative fates of ligand-occupied GPCRs. Upon binding agonist (H), GPCRs undergo conformational changes that initiate heterotrimeric G protein activation and signaling through G protein-regulated effectors (E). Alternatively, receptor phosphorylation by second-messenger-dependent protein kinases (PKA, PKC) produces heterologous desensitization, whereas phosphorylation by GRKs promotes arrestin (Arr) binding and homologous desensitization. Once bound to the receptor, β-arrestins engage the endocytic machinery and target desensitized GPCRs for clathrin-mediated sequestration. By acting as adapters or scaffolds, β-arrestins also recruit signaling proteins, such as components of the ERK1/2 MAPK cascade (Raf-1, MEK, ERK1/2) to the receptor to initiate a "second wave" of GPCR signaling through protein complex assembly on the desensitized GPCR.

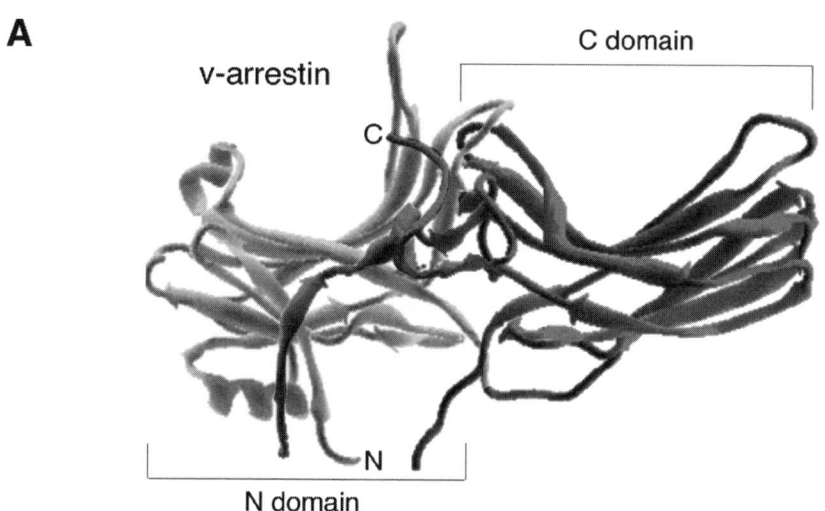

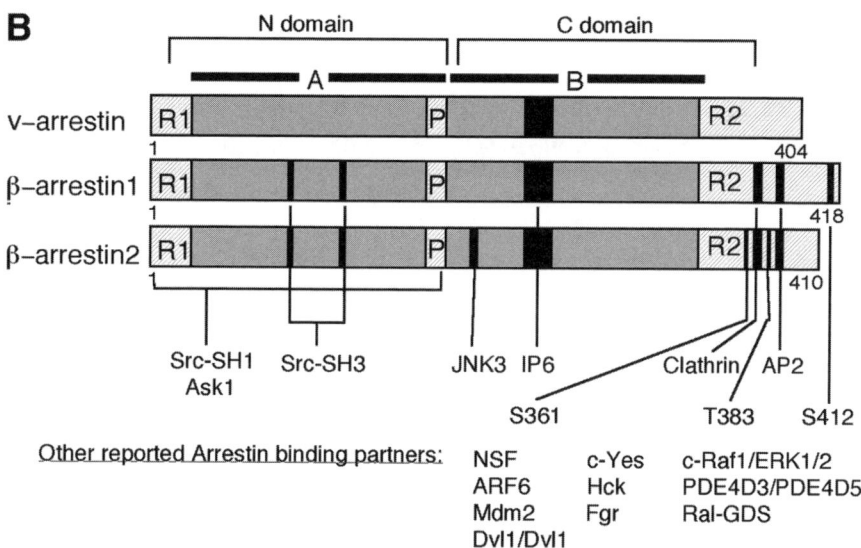

Fig. 2. Arrestin structure and binding partners. (**A**) Ribbon diagram of the crystal structure of visual arrestin. The globular N- and C-domains that surround the polar core of the protein are indicated. In the inactive state the N-terminus (N) and C-terminus (C) contribute to the polar core. (**B**) Line diagram comparing visual arrestin with β-arrestin 1 and β-arrestin 2. The location of the crystallographically defined N- and C-domains are shown in relation to the functionally defined amino terminal (A) domain responsible for recognition of activated GPCRs, the carboxy terminal (B) domain responsible for secondary receptor recognition, the phosphate sensor

phosphoinositide-binding region of β-arrestin 2 resides within residues 233 to 251. Mutation of basic residues within this region produces a protein that translocates to the membrane but that fails to target $β_2$-adrenergic receptors (ARs) to clathrin-coated pits, thereby inhibiting endocytosis of the receptor. The N-ethylmaleimide-sensitive fusion (NSF) protein also binds to β-arrestin 1 in vitro and in vivo *(32)*. NSF is an adenosine triphosphate (ATP)ase involved in intracellular transport. Overexpression of NSF enhances $β_2$-AR endocytosis in HEK293 cells, which suggests that the interaction between β-arrestin and NSF is important for receptor endocytosis. β-arrestins also form complexes with the small GTP-binding protein ARF6 and the ARF guanine nucleotide exchange factor ARNO *(33)*. Expression of constitutively activated or dominant inhibitory mutants of ARF6 inhibits GPCR internalization, whereas overexpression of ARNO enhances GPCR internalization, suggesting that these interactions also contribute to β-arrestin-mediated endocytosis.

3. ARRESTINS IN GPCR DESENSITIZATION

3.1. Heterologous vs Homologous Desensitization

The waning of GPCR signals in the continuous presence of agonist is accomplished by a coordinated series of events that are typically considered as three distinct processes: receptor desensitization, sequestration, and downregulation (Fig. 3). Desensitization, which begins within seconds of agonist exposure, is initiated by phosphorylation of the receptor. Second-messenger-dependent protein kinases, including cyclic adenosine monophosphate (cAMP)-dependent protein kinase A (PKA) and protein kinase C (PKC), phosphorylate serine and threonine residues within the cytoplasmic loops and C-terminal tail domains of many GPCRs. Phosphorylation of these sites is sufficient to impair receptor–G protein coupling efficiency in the absence of β-arrestin. For example, phosphorylation of the $β_2$-AR in vitro by PKA markedly impairs receptor-stimulated GTPase activity *(34)*, whereas removal of the PKA phosphorylation sites delays the onset of desensitization in intact cells *(35)*. Agonist occupancy of the target GPCR is

---

Fig. 2. *(From opposite page)* domain (P), and the amino (R1) and carboxy (R2) terminal regulatory domains. The locations of binding sites for Src SH1 and SH3 domains, Ask1, JNK3, inositol 6-phosphate (IP6), clathrin and AP2, as well as the major regulatory phosphorylation sites of β-arrestin 1 and β-arrestin 2 are also indicated. Other putative arrestin binding proteins whose interacting domains have not been mapped are listed.

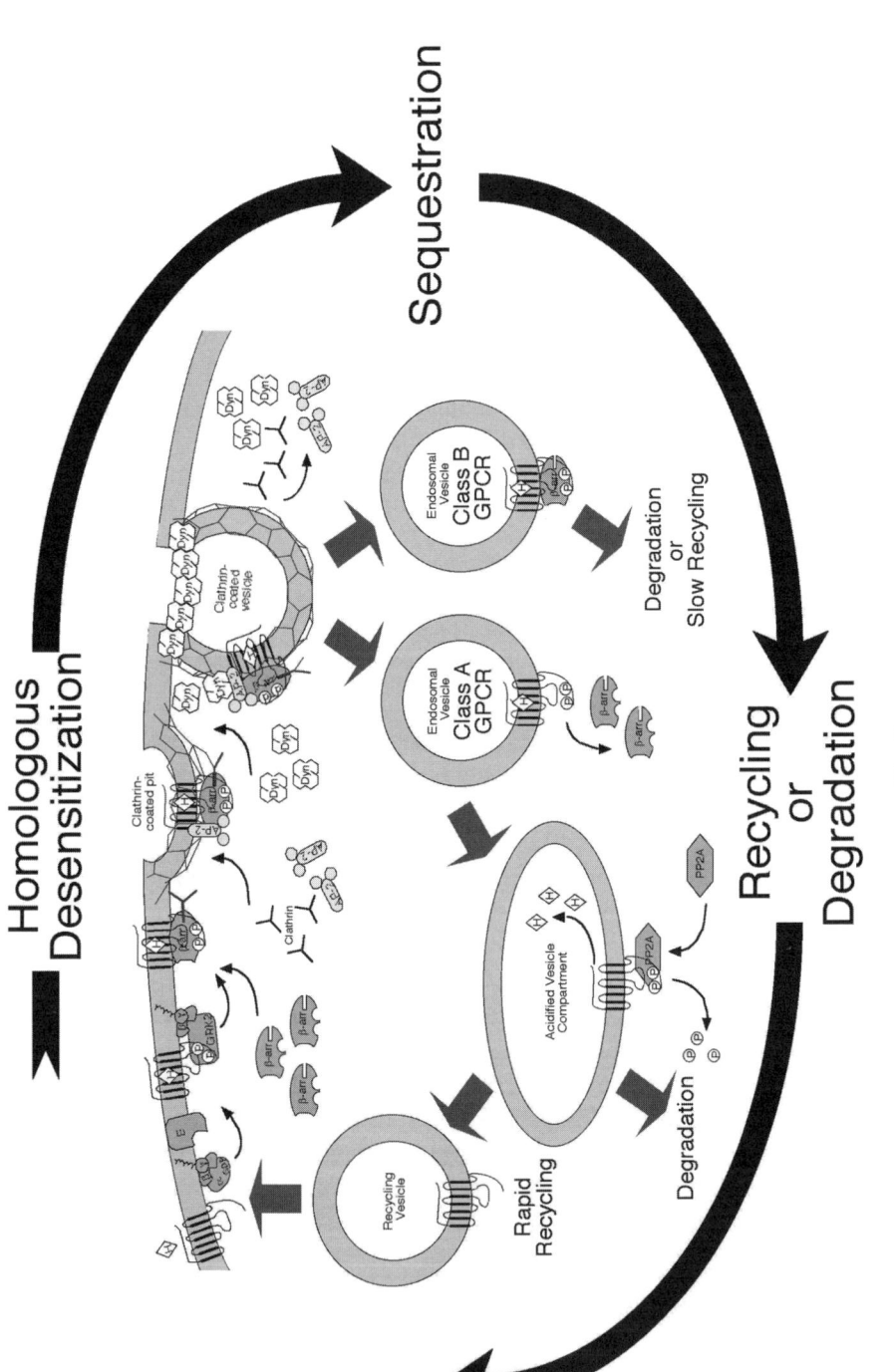

Fig 3

not required for this process; thus, receptors that have not bound ligand, including receptors for other ligands, can be desensitized by the activation of second-messenger-dependent protein kinases. This lack of requirement for receptor occupancy has led to the use of the term heterologous desensitization to describe the process *(36)*. In some cases, such as the β_2-ARs *(37,38)* and murine prostacyclin receptors *(39)*, PKA phosphorylation also alters the G protein-coupling specificity of the receptor to favor coupling to the adenylyl cyclase inhibitory G protein Gi over the stimulatory G protein Gs, causing the PKA-phosphorylated receptor to "reverse direction" regarding cAMP production *(40)*.

In contrast to heterologous desensitization, homologous desensitization is specific for agonist-occupied receptors and involves phosphorylation of the receptor by GRKs as well as subsequent binding of β-arrestin. GRK phosphorylation alone has little effect on receptor-G protein coupling in vitro. In fact, the initial attempts to characterize β-AR kinase 1 (GRK2) were confounded by the finding that as the purity and specific activity of the kinase for agonist-occupied β_2-ARs increased, its ability to inactivate receptor-Gs coupling decreased. The finding that GRK2-mediated β_2-AR desensitization could be restored by the addition of visual arrestin led to the hypothesis that additional arrestin-like proteins existed outside of the retina *(41)*. The removal of an essential cofactor during GRK2 purification (subsequently identified as β-arrestin 1) accounted for the loss of GRK2-dependent desensitization by the highly purified kinase preparations. We now

---

Fig. 3. Role of β-arrestins in the desensitization, sequestration, and intracellular trafficking of GPCRs. Homologous desensitization of GPCRs results from the binding of β-arrestins (β-arr) to agonist (H)-occupied receptors following phosphorylation of the receptor by GRKs. β-arrestin binding precludes further coupling between the receptor and heterotrimeric G proteins, leading to termination of signaling by G protein effectors (E). Receptor-bound β-arrestins direct GPCR sequestration by linking the receptor to components of the clathrin endocytic machinery including clathrin and β_2-adaptin (AP-2). Receptor sequestration reflects the dynamin (Dyn)-dependent endocytosis of GPCRs via clathrin-coated pits. Once internalized, GPCRs exhibit two distinct patterns of β-arrestin interaction. "Class A" GPCRs rapidly dissociate from β-arrestin and are trafficked to an acidified endosomal compartment, wherein the ligand is dissociated and the receptor is dephosphorylated by a GPCR-specific protein phosphatase PP2-A isoform. These receptors are rapidly recycled to the plasma membrane. "Class B" receptors form stable receptor-β-arrestin complexes. These receptors accumulate in endocytic vesicles and are either targeted for degradation or slowly recycled to the membrane via as yet poorly defined routes.

recognize that the principal function of the GRK in GPCR desensitization is to increase receptor affinity for β-arrestin. In vitro phosphorylation of the $β_2$-AR by GRK2 increases receptor affinity for-β-arrestin 1 by 10- to 30-fold *(42)*. Binding of arrestin to receptor domains involved in G protein coupling, rather than GRK phosphorylation *per se*, leads to homologous desensitization.

3.2. The GRKs

There are seven known GRKs. Of these, rhodopsin kinase (GRK1) and GRK7 (a candidate cone opsin kinase *[43]*) are retinal kinases involved in photoreceptor regulation, whereas GRK2 to GRK6 are more widely expressed. Membrane targeting of all the GRKs is critical to their function and is conferred by distinctive C-terminal tail domains *(44)*. GRK1 and GRK7 possess a C-terminal CAAX motif. Light-induced translocation of GRK1 from the cytosol to the plasma membrane is facilitated by posttranslational farnesylation of this site. The β-AR kinases (GRK2 and GRK3) have C-terminal Gβγ-subunit-binding and pleckstrin-homology domains and translocate to the membrane because of interactions between these domains and free Gβγ-subunits and inositol phospholipids. Palmitoylation of GRK4 and GRK6 on C-terminal cysteine residues leads to constitutive membrane localization. Targeting of GRK5 to the membrane is believed to involve the electrostatic interaction of a highly basic 46-residue C-terminal domain with membrane phospholipids.

Similarly to second-messenger-dependent protein kinases, GRKs phosphorylate GPCRs on serine and threonine residues in their third intracellular loop and C-terminal domains. The significant difference is that GRKs preferentially phosphorylate receptors that are in the agonist-occupied conformation. The subsequent recruitment of β-arrestins only to activated receptors accounts for the specificity of homologous desensitization.

3.3. β-Arrestins and Homologous Desensitization In Vivo

Studies performed using β-arrestin knockout mice have provided evidence of the in vivo role of β-arrestin-mediated desensitization in modulating GPCR function. Because simultaneous knockout of both β-arrestin 1 and 2 results in embryonic lethality, assays of the physiological function of β-arrestins have been confined to single knockout lines. Homozygous β-arrestin 1 knockout animals appear phenotypically normal and exhibit normal resting cardiac parameters, such as heart rate, blood pressure, and left ventricular ejection fraction. However, acute challenge with β-adrenergic agonists provokes an exaggerated increase in heart rate and ventricular ejection fraction in the knockouts, suggesting that β-arrestin 1 is important for cardiac β-AR desensitization *(45)*. Similarly, homozygous β-arrestin 2 knockout mice are phenotypically normal but exhibit a dramatic potentiation and prolongation of the analgesic effect of morphine, which is consistent with impaired μ-opioid receptor desensitization in the CNS *(46)*. In these animals, the loss of opioid receptor desensitization correlates with an inability to develop tolerance to the antinociceptive effects of morphine but does not prevent the development of opioid dependence *(47)*.

4. β-ARRESTINS IN GPCR INTERNALIZATION AND TRAFFICKING

4.1. β-Arrestins and Receptor Sequestration

Internalization of GPCRs (also termed receptor sequestration or endocytosis) occurs more slowly than desensitization, taking place over a period of several minutes after agonist exposure. Most, but not all, GPCRs undergo sequestration, and for many, the process can be blocked by expressing a dominant inhibitory mutant of dynamin, a large GTPase necessary for the fission of clathrin-coated vesicles from the plasma membrane *(48)*. It is now clear that GRK-mediated GPCR phosphorylation and binding of β-arrestin to the receptor facilitates the clathrin-dependent endocytosis of many GPCRs, including the $β_2$-ARs, angiotensin II type 1a (AT_{1A}), m2–m5 muscarinic cholinergic, endothelin A, D2 dopamine, follitropin, monocyte chemoattractant protein-1, and the CCR-5 and CXCR4 chemokine receptors (ref. *4*; Fig. 3). Once β-arrestin translocates from the cytosol to bind a GRK-phosphorylated GPCR, the LIEF/L and RxR motifs in the C-terminal regulatory domain engage clathrin and $β_2$-adaptin, respectively, leading to the clustering of receptors in clathrin-coated pits. The clathrin-coated pits that mediate GPCR endocytosis appear to comprise a distinct subpopulation from those that mediate the constitutive endocytosis of transferrin receptors *(49)*.

The physiological relevance of β-arrestin-dependent GPCR endocytosis in vivo is illustrated by the behavior of a naturally occurring R137H mutation of the V2 vasopressin receptor that is associated with familial nephrogenic diabetes insipidus. Familial nephrogenic diabetes insipidus results from a loss of vasopressin responsiveness in the renal tubule and is most commonly caused by mutations in the V2 vasopressin receptor that lead to absent expression or misfolding of the receptor. The R137H mutation lies within the highly conserved DRY motif in the second intracellular domain of the receptor and produces a receptor that is constitutively phosphorylated and bound to β-arrestin *(50)*. As a result, the R137H V2 receptor is continuously desensitized and sequestered in arrestin-associated intracellular vesicles, even in the absence of agonist. Experimental introduction of secondary mutations that disrupt the receptor-arrestin interaction re-establishes plasma membrane localization of the receptor and its ability to respond to agonist, indicating that the clinical loss of function phenotype derives from constitutive β-arrestin-mediated sequestration.

4.2. Variations on the Theme

Although the preceding model is true for the majority of GPCRs studied to date, the process of GPCR sequestration can be a heterogenous process, and the extent of β-arrestin involvement appears to vary significantly with the receptor, agonist, and cell type. This probably reflects variation in endogeous patterns of GRK and β-arrestin expression, the specific effects of agonist and partial agonist drugs on receptor conformation, and the availability of alternative pathways for GPCR endocytosis.

Several examples illustrate that β-arrestin-dependent GPCR desensitization and sequestration are not inextricably linked. The thrombin receptor protease-activated receptor (PAR)-1 undergoes agonist-dependent phosphorylation and binds to β-arrestin 1. In murine embryo fibroblasts (MEFs) lacking expression of either β-arrestin 1 or 2, PAR-1 receptor desensitization is markedly impaired, which is consistent with the loss of homologous receptor desensitization, but clathrin-dependent receptor endocytosis proceeds normally. Interestingly, a C-terminal phosphorylation site mutant of PAR-1 fails to internalize in either the β-arrestin replete or β-arrestin null background, suggesting that PAR-1 receptors use a phosphorylation-dependent, but β-arrestin-independent, mechanism for endocytosis *(51)*. Similar results have been obtained using the *N*-formyl peptide receptor *(52)*, and the somatostatin (SST) receptor type 2A *(53)*. The human cytomegalovirus GPCR US28, a homolog of the human chemokine receptor family, provides another example. US28 is a constitutively active GPCR and, therefore, is constitutively phosphorylated and bound to β-arrestin. Mutation of Ser 323, the critical C-terminal residue for β-arrestin binding, enhances G_q-coupling and inositol phosphate accumulation, confirming the link among constitutive activity, phosphorylation, and β-arrestin-dependent desensitization *(54)*. Nonetheless, US28 internalization (which is clathrin-mediated) is similar in both control and β-arrestin null fibroblasts *(55)*.

Other GPCRs that undergo β-arrestin-mediated desensitization appear to internalize via clathrin-independent mechanisms. Internalization of AT_{1A} and m2 muscarinic acetylcholine receptors is insensitive to expression of dominant inhibitory mutants of either β-arrestin or dynamin *(48,56)*. $β_1$-AR mutants lacking GRK phosphorylation sites exhibit impaired β-arrestin recruitment and clathrin-mediated endocytosis but still internalize through an alternative pathway that involves PKA phosphorylation that can be blocked by pharmacological inhibition of caveoli *(57)*. Although the physiological relevance of these β-arrestin and/or clathrin-independent mechanisms of GPCR endocytosis is uncertain, their existence clearly points to the potential for independent regulation of receptor desensitization and sequestration.

4.3. Functional Specialization of β-Arrestin Isoforms

Early work with overexpression systems and purified proteins in vitro revealed little regarding functional differences between β-arrestin 1 and β-arrestin 2. However, the use of chimeric β-arrestin–green fluorescent protein (GFP) fusion proteins that permit visualization of β-arrestin binding and receptor trafficking in live cells has revealed significant differences in the interaction of GPCRs with the two β-arrestin isoforms *(58)*. Using this approach, it has been demonstrated that GPCRs exhibit distinctive patterns of β-arrestin interaction, with most receptors falling into one of two distinct classes *(59)*. Class A receptors include the $β_2$- and $α_{1B}$-ARs, μ-opioid receptors, endothelin A receptors, and D1A dopamine receptors. These receptors bind to β-arrestin 2 with higher affinity than they do with β-arrestin 1 and do not bind to visual arrestin. Additionally, their interaction with β-arrestin is transient. β-arrestin is recruited to the receptor at the plasma membrane and translocates with it to clathrin-coated pits. Upon internalization of the receptor, the receptor-β-arrestin complex dissociates, so that the β-arrestin recycles to the plasma membrane, and the receptor proceeds into an endosomal pool *(60)*. Class B receptors, represented by the AT_{1A}, neurotensin 1, V2vasopressin, thyrotropin-releasing hormone, and neurokinin (NK)-1 receptors, bind β-arrestin 1 and β-arrestin 2 with equal affinity and interact with visual arrestin. These receptors form stable complexes with β-arrestin, so that the receptor-β-arrestin complex internalizes as a unit that is targeted to endosomes. The structural features of the receptor that dictate the stability of the receptor-β-arrestin complex reside within specific clusters of serine and threonine residues in the C-terminal tail of the receptor *(61)*. The C-terminus of β-arrestin also determines the stability of the interaction, because a β-arrestin mutant truncated at residue 383 binds to the $β_2$-AR (a class A GPCR) with high affinity and traffics with it into endosomes.

Studies done using MEFs that lack either or both β-arrestins further support the hypothesis that β-arrestin 1 and β-arrestin 2 exhibit functional specialization *(62)*. Knockout of either β-arrestin 1 or β-arrestin 2 is sufficient to impair desensitization of both the $β_2$-ARs and AT_{1A} receptors. Desensitization is further reduced in double-knockout β-arrestin 1/2 null cells, suggesting that the two isoforms are equally effective at inducing desensitization. In contrast, $β_2$-AR sequestration is inhibited only in β-arrestin 2 knockout and double-knockout MEFs, not in β-arrestin 1 knockouts. Reconstitution of β-arrestin expression in double-knockout MEFs revealed that β-arrestin 2 is 100-fold more potent than β-arrestin 1 in supporting $β_2$-AR endocytosis. AT_{1A} receptor sequestration is minimally affected by the absence of either β-arrestin 1 or β-arrestin 2 and is markedly impaired only in the double-knockout MEFs. Consistent with data obtained using the β-arrestin–GFP chimeras, these results suggest that $β_2$-AR endocytosis primarily involves β-arrestin 2, whereas either β-arrestin can support AT_{1A} receptor sequestration.

4.4. Posttranslational Modifications Affecting β-Arrestin Function

β-arrestin function is regulated by posttranslational modification, notably phosphorylation and ubiquitination. Cytoplasmic β-arrestin 1 is almost stoichiometrically phosphorylated on S412, which lies within the C-terminal regulatory domain *(63)*. Upon translocation to the membrane, β-arrestin 1 undergoes rapid C-terminal dephosphorylation. An S412D mutant of β-arrestin 1 that mimics the phosphorylated state binds to agonist-occupied $β_2$-ARs and supports receptor desensitization but does not interact well with clathrin and, therefore, inhibits receptor sequestration. The corresponding S412A mutation associates constitutively with clathrin and is capable of supporting GPCR sequestration. Thus, dephosphorylation of S412 appears to regulate the ability of the receptor–β-arrestin complex to engage the endocytic machinery. β-Arrestin 2, which lacks an S412 equivalent, undergoes a similar pattern of regulation by phosphorylation-dephosphorylation. In this case, the phosphorylated residues are S361 and T383. As with β-arrestin 1, dephosphorylation occurs upon receptor binding, and phosphomimetic mutations at these sites impair clathrin binding and GPCR endocytosis without affecting β-arrestin 2-mediated receptor desensitization *(64)*.

Regulated ubiquitination of β-arrestin 2 has also been shown to play an important role in GPCR endocytosis. Both β-arrestin 2 and $β_2$-ARs are rapidly and transiently ubiquitinated in response to agonist *(65)*. β-arrestin 2 ubiquitination is catalyzed by the E3 ubiquitin ligase Mdm2, which binds directly to the β-arrestin. Ubiquitination of the receptor is catalyzed by an as yet unidentified ubiquitin ligase but still requires the presence of β-arrestin. Ubiquitination of β-arrestin apparently is required for $β_2$-AR internalization, whereas ubiquitination of the receptor is involved in receptor degradation but not internalization. The V2 vasopressin receptor also undergoes β-arrestin-dependent ubiquitination. Unlike the β_2-AR, the V2 receptor-associated β-arrestin 2 remains stably ubiquitinated as it trafficks with the receptor into the endosomal pool. Similarly to the β_2-AR, ubiquitination of the V2 receptor itself is not required for endocytosis but does accelerate receptor degradation *(66)*. Interestingly, the time-course of β-arrestin 2 ubiquitination when associated with class A or class B GPCRs correlates with stability of the receptor–β-arrestin interaction. In cells expressing a β-arrestin 2–ubiquitin chimera, the "stably ubiquitinated" β-arrestin 2 remains associated with both β_2-ARs and V2 vasopressin receptors as they internalize. Thus, regarding its trafficking pattern, irreversible ubiquitination of β-arrestin 2 converts the class A β_2-AR into a class B receptor. These data suggest that de-ubiquitination of β-arrestin 2 is a prerequisite for dissociation of the receptor–β-arrestin 2 complex *(67)*.

5. β-ARRESTINS IN GPCR DOWNREGULATION, RESENSITIZATION, AND RECYCLING

Downregulation of GPCRs, the persistent loss of cell-surface receptors that occurs over a period of hours to days, is the least understood of the processes controlling GPCR responsiveness. Control of cell-surface receptor density occurs partially at the transcriptional level, but removal of agonist-occupied receptors from the cell surface and their sorting for either degradation or recycling to the membrane is also important, at least in the early stages of downregulation (Fig. 3).

For a fast-recycling GPCR, such as the β_2-AR, it is clear that β-arrestin-dependent endocytosis plays a key role in the resensitization process *(68)*. In COS-7 cells, overexpression of β-arrestins enhances the rate of β_2-AR endocytosis and resensitization *(69)*. Conversely, downregulation of β_2-ARs does not occur in β-arrestin 1/2 double null MEFs *(62)*.

Resensitization of a sequestered GPCR requires that β-arrestin is dissociated, the receptor is dephosphorylated, and bound ligand is removed. Shortly after stimulation, phosphorylated β_2-ARs appear in an endosomal vesicle fraction that is enriched in GPCR-specific protein phosphatase (PP2A) activity *(70)*. Dephosphorylation of the receptor occurs in an acidified vesicle compartment, because treatment of cells with ammonium chloride (which neutralizes the acidity of endosomal vesicles) blocks association of the receptor with the phosphatase and prevents receptor dephosphorylation *(71)*. Dephosphorylation of the GPCR may be the rate limiting in determining the rate of GPCR recycling. Overexpression of phosphorylation-state insensitive mutants of β-arrestin 1 reduces the extent of GRK phosphorylation of β_2-ARs, presumably by binding agonist-occupied receptors prior to GRK phosphorylation, and markedly accelerates the process of recycling *(72)*.

The stability of the receptor-β-arrestin interaction also dictates the fate of the internalized receptor. Although the class A $β_2$-AR is rapidly dephosphorylated and recycled to the plasma membrane, the class B V2 vasopressin receptor recycles slowly. Switching the C-terminal tails of these two receptors, which converts the $β_2$-AR into a class B receptor and the V2 vasopressin receptor into a class A receptor, reverses the pattern of dephosphorylation and recycling *(73)*. Therefore, the formation of a transient receptor–β-arrestin complex favors rapid dephosphoryation and a return to the plasma membrane, whereas the formation of a stable receptor–β-arrestin complex retards resensitization and may favor targeting of the receptor for degradation. The *N*-formyl peptide receptor, which binds β-arrestin but does not require it for internalization, fails to recycle when expressed in β-arrestin 1/2 null MEFs, instead becoming trapped in a perinuclear vesicle compartment *(52)*. Thus, the protein–protein interactions that determine β-arrestin's role in receptor trafficking may be distinct from those that mediate endocytosis.

Other protein–protein interactions involving the GPCRs also contribute specificity to the process of endocytic sorting and determination of the ultimate fate of an internalized GPCR. Transplantation of a PDZ-binding motif, DSLL, from the C-terminus of the $β_2$-AR onto the δ-opioid receptor causes the latter to reroute from a degradative pathway into the rapid recycling pathway *(74)*. Binding between the DSLL motif and the PDZ-domain containing protein, NHERF-1/EBP50, has been implicated in directing the $β_2$-AR into the recycling pathway *(75)*. Conversely, a candidate GPCR-associated sorting protein (GASP) has been identified that binds to the C-terminus of the δ-opioid receptor and appears to preferentially target the receptor to lysosomes for proteolytic degradation *(76)*.

6. ARRESTINS AS SIGNAL TRANSDUCERS

6.1. GPCR-Arrestin As an Alternative Ternary Complex

The basic paradigm of GPCR signaling predicts that activated receptors function catalytically to stimulate guanine nucleotide exchange on heterotrimeric G protein $G_α$-subunits and to promote $G_α$- and $G_{βγ}$-subunit dissociation. In contrast, arrestin-bound receptors exist in relatively stable protein complexes of defined stoichiometry. In this state, β-arrestins act as adapter proteins that physically link the receptor to the clathrin-mediated endocytic machinery and promote receptor endocytosis. Although sterically precluded from further G protein coupling, evolving literature suggests that these arrestin-bound receptors are still involved in signal transduction.

Rather than functioning enzymatically to catalyze G protein activation, the desensitized GPCR and receptor-associated proteins (such as the arrestins) appear to serve as scaffolds that recruit signaling molecules into complexes with ligand-bound receptors. Although this model, in which the receptor essentially functions as a ligand-activated docking protein, is thoroughly established for other types of membrane receptors, it has only recently been invoked to explain aspects of GPCR signaling *(77)*.

It is quite clear that GPCRs exist in a conformationally distinct state when bound to arrestins. Classical GPCR pharmacology has demonstrated that in the absence of GTP, agonist, receptor, and heterotrimeric G proteins can exist as a preformed "ternary complex" that exhibits higher agonist affinity than the GPCR alone. More recent studies of GRK-phosphorylated, β-arrestin-bound GPCRs have demonstrated that arrestin binding also induces conformational changes in the receptor that affect agonist binding affinity, leading to the suggestion that the agonist–receptor–arrestin complex represents an "alternative ternary complex" *(78)*. For example, m2 muscarinic acetylcholine and *N*-formyl peptide receptors each bind agonist, but not antagonist, with increased affinity in the β-arrestin-bound state *(78,79)*. Similarly, a chimeric NK-1 receptor–β-arrestin 1 fusion protein appears to exist in a stable, high-affinity agonist-binding form *(80)*. Interestingly, the receptor conformations that favor β-arrestin interaction appear to be different from those that promote productive G protein coupling. For example, certain synthetic angiotensin II analogs that act as antagonists of phospholipase C signaling are nonetheless capable of inducing β-arrestin recruitment, receptor sequestration, and mitogen-activated protein kinase (MAPK) activation *(81)*. Similarly, point mutations within the conserved DRY motif in the second transmembrane domain of the AT_{1A} receptor *(82)* can uncouple the receptor from heterotrimeric G proteins but preserve agonist-induced β-arrestin binding and internalization. Collectively, such observations suggest that distinct points of contact between agonist and GPCR select or induce distinct receptor conformations that favor particular receptor interactions.

6.2. Arrestins As Scaffolds for GPCR Signaling

Data from yeast two-hybrid screens using β-arrestins, from biochemical characterization of receptor–arrestin complexes, from study of GPCR signaling in β-arrestin 1/2 null MEFS, and from siRNA-mediated silencing of β-arrestin expression have identified several potential arrestin-binding partners that may play a role in GPCR signaling beyond their dampening effects on receptor—G protein coupling. The following sections summarize some of the arrestin interactions that may permit the "alternative ternary complex" to function as a signaling entity (refs. *83* and *84*; Table 1).

Table 1
Arrestin-Interacting Signal Transducers and Their Possible Functions

Arrestin-binding partner	Arrestins bound	Possible functions	References
Src family tyrosine kinases			
c-Src	β-arrestin 1, visual arrestin	Phosphorylation of dynamin I, GRK2 Regulation of GPCR desensitization and endocytosis ERK1/2 activation NFkB activation	85,86,88,90–93,98
c-Yes	β-arrestin 1	Phosphorylation of $G_{aq}/11$ GLUT4 translocation	94
Hck, Fgr	β-arrestin 1	Neutrophil degranulation	87
MAPKs			
Ask1/JNK3	β-arrestin 2	Activation and cytosolic retention of JNK3	105–108
c-Raf1/ERK1/2	β-arrestin 1/2	Activation and cytosolic retention of ERK1/2 Phosphorylation of β-arrestin 1 on S412 Regulation of GRK2 turnover	86,109–113,122,130
Ask1/p38 MAPK	β-arrestin 2	Actin cytoskeletal reorganization and chemotaxis Activation of p38 MAPK Chemokine-mediated chemotaxis	54,130
Mdm-2	β-arrestin 2	Regulation of GPCR endocytosis Stabilization of the GPCR-arrestin complex p53 regulation and apoptosis	66,67,131
PDE4D3/PDE4D5	β-arrestin 1/2	Local inhibition of PKA activity Modulation of $β_2$-adrenergic receptor G_i coupling	132,133
Ral-GDS	β-arrestin 2	Actin cytoskeletal reorganization	134
Dvl1/Dvl2	β-arrestin 1	Regulation of Frizzled endocytosis Activation of LEF transcription	151,152

6.2.1. Arrestins and Src Family Tyrosine Kinases

The initial evidence suggesting that β-arrestins function as signal transducers came from the observation that β-arrestin 1 and β-arrestin 2 can bind directly to Src family kinases and recruit them to agonist-occupied GPCRs. In HEK293 cells, stimulation of $β_2$-ARs triggers the colocalization of the receptor with both endogenous β-arrestins and Src kinases in clathrin-coated pits *(85)*. This colocalization reflects the assembly of a protein complex containing activated Src, β-arrestin, and the receptor. Similar results have been obtained in KNRK cells, in which β-arrestins are involved in recruiting Src to the NK-1 receptor *(86)*; in neutrophils, in which β-arrestins recruit the Src family kinases Hck and Fgr to the CXC chemokine receptor CXCR1 *(87)*; and in the rod outer segment, in which bleached rhodopsin, visual arrestin, and Src assemble to form a multimeric complex *(88)*.

Comparison of the functionally defined arrestin "docking" domains with the crystal structures of visual arrestin and β-arrestin 1 suggests that arrestin could simultaneously accommodate binding to a phosphorylated GPCR, Src, phosphoinositides, and clathrin/β-adaptin *(89)*. Src binding to arrestins is partially mediated by an interaction between the Src homology (SH)3 domain of the kinase and proline-rich PXXP motifs in the β-arrestin N-domain. β-arrestin 1 has three such motifs, spanning residues 88 to 91, 121 to 124, and 175 to 178, whereas visual arrestin has only a single motif of this type. All three of the β-arrestin 1 PXXP motifs reside on the solvent-exposed surface of the molecule, where they would be available for binding, and a P91G/P121E mutant of β-arrestin 1 is impaired in c-Src binding *(85)*, suggesting that at least two of the motifs contribute to Src binding. However, a second major site of interaction appears to involve the N-terminal portion of the catalytic (SH1) domain of Src and additional epitopes located within the N-terminal domain of β-arrestin 1 *(90)*. Interestingly, complexes containing bleached rhodopsin and arrestin associate specifically with the immobilized Src SH2 domain in vitro (88), suggesting that multiple points of contact exist.

GPCR stimulation results in the Src-mediated phosphorylation of several proteins directly involved in the modulation of GPCR signaling, including dynamin, GRK2, and $G_{αq/11}$-subunits. Some evidence suggests that β-arrestin–Src complexes mediate these events. Activation of $β_2$-ARs results in rapid Src-dependent tyrosine phosphorylation of dynamin on Y597, which stimulates dynamin self-assembly and increases its GTPase activity *(91,92)*. Mutation of the Src phosphorylation sites produces a dominant inhibitory form of dynamin that impairs GPCR endocytosis. Expression of an inactive Src SH1 domain that binds selectively to β-arrestin inhibits $β_2$-AR-stimulated tyrosine phosphorylation of dynamin and receptor internalization *(90)*, suggesting that the β-arrestin–Src interaction may regulate tyrosine phosphorylation of dynamin, thereby enhancing GPCR endocytosis.

GRK2 is another potential substrate for β-arrestin-bound Src *(93)*. $β_2$-AR or CXCR4 stimulation in HEK293, Jurkat, or C6 cells stimulates Src-mediated GRK2 phosphorylation, followed by rapid ubiquitination and proteosomal degradation of the GRK. Expression of the β-arrestin 1 P91G/P121E mutant inhibits GRK2 phosphorylation and degradation, again suggesting that targeting of Src may occur through its binding to β-arrestin. In this case, the role of Src appears to be to downregulate of GRK activity, which could provide a feedback mechanism for regulating GRK levels and GPCR signaling.

Controlling of the rate of $G_{αq/11}$-mediated vesicular trafficking to the plasma membrane is another potential role *(94)*. Stimulation of endothelin ETA receptors in 3T3-L1 adipocytes leads to Src kinase activation, Src-dependent tyrosine phosphorylation of $G_{αq/11}$-subunits, translocation of insulin-sensitive glucose transporter (GLUT4)-containing vesicles to the plasma membrane, and an increase in GLUT4-mediated glucose uptake. Treatment with Src inhibitors or microinjection of antibodies against either the Src family kinase c-Yes or β-arrestin 1 blocks endothelin 1-stimulated glucose uptake. Furthermore, β-arrestin 1 can be demonstrated to recruit Src into a molecular complex with the ETA receptor in 3T3-L1 cells. Therefore, stimulation of GLUT4 translocation by GPCRs appears to involve β-arrestin-mediated Src activation and, possibly, Src-mediated tyrosine phosphorylation of heterotrimeric G protein subunits.

Further evidence for a role of β-arrestin–Src complexes in the trafficking of exocytic vesicles comes from granulocytic neutrophils. In these, activation of CXCR1 by interleukin (IL)-8 stimulates the rapid formation of complexes containing endogenous β-arrestin and the Src family kinases Hck or Fgr *(87)*. The formation of β-arrestin–Hck complexes leads to Hck activation and trafficking of the complexes to granule-rich regions. Granulocytes expressing β-arrestin 1 P91G/P121E fail to activate tyrosine kinases and demonstrate reduced chemoattractant-stimulated granule release after IL-8 stimulation.

Ras-dependent activation of the extracellular signal-regulated kinase (ERK)1/2 MAPK cascade by many GPCRs requires Src kinase activity *(95–97)*. In some cases, the interaction between β-arrestin and Src appears to play a role in the process. In HEK293 cells, overexpression of β-arrestin 1 mutants that exhibit either impaired Src binding or that are unable to target receptors to clathrin-coated pits blocks $β_2$-AR-mediated activation of ERK1/

2 *(85)*. In KNRK cells, activation of NK-1 receptors by substance P leads to the assembly of a scaffolding complex containing the internalized receptor, β-arrestin, Src, and ERK1/2. Expression of either a dominant-negative β-arrestin 1 mutant or a truncated NK-1 receptor that fails to bind β-arrestin blocks complex formation and inhibits both substance P-stimulated endocytosis of the receptor and activation of ERK1/2 *(86)*.

Activation of nuclear factor-κB (NF-κB) is another downstream signaling event that may involve β-arrestin-dependent Src activation. D2 dopamine receptors expressed in HeLa cells activate NF-κB through a pathway involving pertussis toxin-sensitive G proteins, $G_{\beta\gamma}$-subunits, and Src family tyrosine kinases. The signal is independent of phospholipase C and phosphatidylinositol-3-kinase activity but is enhanced by overexpression of β-arrestin 1, suggesting that β-arrestin recruitment, rather than activation of traditional heterotrimeric G protein effectors, may be the initiating event in the process *(98)*.

6.2.2. Arrestins and MAPKs

Most GPCRs can stimulate the activation of MAPKs. The mechanisms underlying these signals are highly diverse and vary both with receptor and cell type. Furthermore, the mechanism underlying MAPK activation has a significant impact on the duration and subcellular distribution of MAPK activity and, hence, its function. A growing body of evidence suggests that under certain circumstances, β-arrestins can act as positive regulators of MAPK activity and, in so doing, can promote the formation of functionally discrete MAPK pools within the cell *(95–97)*.

The MAPKs are a family of evolutionarily conserved serine/threonine kinases that are involved in the transduction of externally derived signals regulating cell growth, division, differentiation, and apoptosis. Mammalian cells contain at least three major classes of MAPK: ERKs, c-Jun N-terminal kinases (JNKs) (also known as stress-activated protein kinase [SAPK]) and p38/HOG1 MAPKs *(99,100)*. MAPK activity in cells is regulated by a series of parallel kinase cascades comprised of three kinases that successively phosphorylate and activate the downstream component. In each cascade, the most proximal element, a MAP kinase kinase kinase (MAPKKK), phosphorylates a MAP kinase kinases (MAPKK), which, in turn, carries out the phosphorylation and activation of the MAPK. Once activated, MAPKs phosphorylate various membrane, cytoplasmic, nuclear, and cytoskeletal substrates. Activated MAPKs can translocate to the nucleus, where they phosphorylate and activate nuclear transcription factors involved in DNA synthesis and cell division.

In many cases, the activation of a MAPK cascade is controlled by binding of the component kinases to a scaffolding protein *(100,101)*. These scaffolds serve at least three functions in cells: to increase the efficiency of signaling between successive kinases in the phosphorylation cascade, to ensure signaling fidelity by dampening crosstalk between parallel MAPK cascades, and to target MAPKs to specific subcellular locations. The prototypic MAPK scaffold is the *Saccharomyces cervisiae* protein Ste5p *(102)*, which controls activation of the yeast pheromone mating pathway, a GPCR signaling cascade. Although no structural homologs of Ste5p exist in mammalian cells, several potential functional homologs have been identified that can bind to two or more components of a MAPK module. For example, the JNK-interacting protein (JIP) family of proteins act as scaffolds for regulation of the JNK–SAPK pathway *(103,104)*. Current data suggest that β-arrestins can serve as GPCR-regulated scaffolds for some of the MAPK cascades through direct interaction with MAPKKKs and MAPKs.

6.2.2.1. JNK–STRESS-ACTIVATED PROTEIN KINASE

In whole-brain lysates and yeast two-hybrid assays, β-arrestin 2 binds to the neuronal JNK–SAPK isoform JNK3 *(105)*. Co-expression of the MAPKKK Ask1, along with JNK3 in COS-7 cells, results in very little JNK3 activation. However, when β-arrestin 2 is also expressed, Ask1-stimulated JNK3 activation is dramatically enhanced. β-arrestin 2 forms complexes with Ask1, the MAPKK, MKK4, and JNK3, but not JNK1 or JNK2. Ask1 binds to the β-arrestin 2 N-terminus, whereas JNK3 binding is conferred by a RRSLHL motif in the C-terminal half of β-arrestin 2 *(105,106)*. This motif, which is not present in β-arrestin 1, corresponds to a consensus MAPK binding motif that has been identified in several other MAPK-binding proteins. Binding of the MAPKK MKK4 may be indirect, because its presence in the complex appears to require the presence of at least one of the other kinases.

Interestingly, expression of β-arrestin 2 results in cytosolic retention of JNK3. The physical basis of this phenomenon is the existence of a classical leucine-rich nuclear export sequence (NES) in the β-arrestin 2 C-terminus *(107,108)*. Mutation of this sequence, or replacing it with the corresponding region of β-arrrestin 1 (which lacks the consensus NES), allows nuclear accumulation of both β-arrestin 2 and co-expressed JNK3. Treatment with leptomycin B, which blocks nuclear export, results in active nuclear accumulation of wild-type β-arrestin 2, suggesting that β-arrestin 2 constitutively shuttles in and out of the nucleus and could serve to deliver β-arrestin-binding proteins to the nucleus. Therefore, β-arrestin 2 exhibits all the characteristics of a scaffold for the JNK3 cascade. It assembles the component kinases into a complex with a degree of specificity, increases the efficiency of the sequential phosphorylation steps, and controls the spatial distribution of the active kinase.

6.2.2.2. EXTRACELLULAR SIGNAL-REGULATING KINASE-1/2

β-arrestins also contribute to GPCR stimulation of the ubiquitous ERK1/2 MAPK cascade. Overexpression of β-arrestin 2 paradoxically enhances AT_{1A} receptor-mediated ERK1/2 activation in COS-7 cells while predictably attenuating G protein-mediated phosphatidylinositol hydrolysis *(109)*. Conversely, depletion of β-arrestins in HEK293 cells using RNA interference (RNAi) inhibits AT_{1A} receptor-mediated ERK1/2 activation and receptor sequestration and markedly enhances second-messenger production *(110)*. In each system, GPCR-stimulated ERK1/2 activation correlates with β-arrestin binding, rather than G protein activation. Indeed, recent data indicate that β-arrestins can mediate ERK1/2 activation through a mechanism that is essentially G protein-independent. Stimulation of a G protein-uncoupled mutant AT_{1A} receptor (DRY/AAY) with angiotensin II fails to induce detectable G protein loading but still promotes β-arrestin 2 recruitment, receptor sequestration, and ERK1/2 activation *(82)*. This apparently G protein-independent activation of ERK1/2 is abolished when β-arrestin 2 is selectively depleted by RNAi *(111)*. Identical results were obtained when the wild-type AT_{1A} receptor was treated with the synthetic peptide angiotensin antagonist [Sarcosine1,Ile4,Ile8] AngII. Exposure of the wild-type receptor to [Sarcosine1,Ile4,Ile8] AngII induced β-arrestin 2 recruitment and ERK1/2 activation in the absence of detectable G protein activation, whereas depletion of β-arrestin 2 by RNAi abolished [Sarcosine1,Ile4,Ile8] AngII-stimulated ERK1/2 activation.

Similar paradoxical findings suggesting G protein-independent signaling have been reported for the $β_2$-AR *(112)*. Ligands such as propranolol and ICI118551, which function as inverse agonists for G_s-stimulated adenylyl cyclase activation, act as partial agonists for ERK1/2 activation. The ERK1/2 signal persists in pertussis toxin-treated cells with inactivated $G_{i/o}$ proteins and in S49 cyc⁻ cells that lack functional G_s, but it is inhibited by expression of a dominant-negative β-arrestin and is absent in β-arrestin 1/2 null MEFs. Moreover, activation of ERK1/2 by $β_2$-AR inverse agonists can be conferred upon β-arrestin 1/2 null MEFs by expression of β-arrestin 2, indicating that the G protein-independent signal is transmitted through β-arrestin.

β-arrestin-dependent activation of ERK1/2 apparently results from scaffolding of the MAPK pathway by GPCR-bound β-arrestin. In KNRK cells, stimulation of PAR-2 receptors induces the assembly of a complex containing the internalized receptor, β-arrestin 1, Raf-1, and activated ERK1/2 *(113)*. The complex apparently is required for ERK1/2 activation by the wild-type PAR-2 receptor, because the signal is blocked by expression of a truncated form of β-arrestin that inhibits receptor endocytosis. Qualitatively similar results have been obtained for the AT_{1A} receptor expressed in HEK293 and COS-7 cells *(114)*. AT_{1A} receptor activation results in the formation of complexes containing receptor, β-arrestin 2, and the component kinases of the ERK cascade (cRaf-1, MEK1, and ERK2). Upon receptor internalization, activated ERK2 appears in the same endosomal vesicles that also contain $AT_{1A}R$–β-arrestin complexes. The NK-1 receptor provides a third example. Activation of NK-1 receptors causes the formation of complexes comprised of internalized receptor, β-arrestin, Src, and ERK1/2 *(86)*.

When associated with class B GPCRs such as the PAR-2, AT_{1A} receptors, and NK-1 receptors, β-arrestin–ERK complexes appear to be relatively stable entities that can be isolated by gel filtration or immunoprecipitation *(86,113,114)*. This stability, along with the NES sequence that excludes β-arrestin 2 from the nucleus, may account for data indicating that the nature of the GPCR–β-arrestin interaction affects not only the mechanism of ERK1/2 activation but also the spatial distribution and function of activated ERK1/2.

GPCRs can employ several mechanisms to activate the ERK1/2 pathway *(97,115,116)*. For example, the AT_{1A} receptor activates ERK1/2 not only through β-arrestin-dependent pathways but also through G protein-dependent signals and through the "transactivation" of classical receptor tyrosine kinases such as the epidermal growth factor (EGF) receptor *(117–119)*. Some data suggest that these different ERK1/2 activation pathways are functionally specialized. Crosstalk between GPCRs and EGF receptors, which leads to activation of the Ras pathway, accounts for the proliferative response to GPCR stimulation in several systems *(120,121)*. In contrast, β-arrestin-dependent ERK activation by the PAR-2 and the AT_{1A} receptors does not generate a proliferative signal. Wild-type PAR-2, which mediate β-arrestin-dependent activation of a predominantly cytosolic pool of ERK1/2 in KNRK cells, do not stimulate ^{3}H-thymidine incorporation or cell replication *(113)*. Similarly, overexpression of β-arrestins promotes cytosolic retention of angiotensin II-stimulated ERK1/2 activity and attenuates ERK-dependent transcription of an Elk1–luciferase reporter, a signal that requires nuclear translocation of activated ERK1/2 *(109)*.

By determining the stability of the receptor-β-arrestin interaction, the C-terminal tail of the GPCR appears to control the utilization of β-arrestin scaffolds and, therefore, the physiological consequences of ERK1/2 activation. When expressed in COS-7 cells, the class B AT_{1A} and V2 vasopressin receptors activate β-arrestin-bound ERK2 more efficiently than the class A $α_{1b}$- and $β_2$-ARs *(122)*. The activation of β-arrestin-bound ERK2 correlates with the stability of the GPCR–β-arrestin interaction, because exchanging the C-terminal tails of the V2 and β_2 receptors (which converts the class B receptor to class A and vice versa) reverses the pattern of ERK2 activation. Wild-type V2 vasopressin receptors generate a larger pool of cytosolic phospho-ERK1/2 and less nuclear phospho-ERK1/2 than the comparable chimeric V2–β_2 receptor, whereas the V2–β_2 chimera stimulates Elk1– luciferase reporter expression to a greater extent than the wild-type V2 vasopressin receptor and is capable of eliciting a mitogenic response. Similarly, a mutant PAR-2 that lacks C-terminal GRK phosphorylation sites still activates ERK1/2 but does so through a Ca^{2+}- and Ras-dependent pathway that is mechanistically distinct from the β-arrestin-dependent pathway used by the wild-type receptor. The mutant, unlike the wild-type receptor, induces nuclear translocation of ERK1/2 and stimulates cell proliferation *(113)*.

Although it is clear that β-arrestin-bound ERK1/2 is excluded from the nucleus and does not appear to participate in mitogenic signaling, little is currently known about the functional role of β-arrestin–ERK complexes. In addition to directly phosphorylating nuclear transcription factors, ERK1/2 phosphorylates numerous plasma membrane, cytoplasmic, and cytoskeletal substrates *(100)*, including several proteins involved in GPCR signaling, such as β-arrestin 1 *(123)*, GRK2 *(124,125)*, and $G_{\alpha 1}$-interacting protein (GAIP) *(126)*. Interestingly, phosphorylation of GRK2 by ERK1/2 enhances its rate of degradation, and the process is accelerated by overexpression of β-arrestin 1, suggesting that β-arrestins may target ERK1/2 to GRK2 *(127)*. Therefore, a potential role of β-arrestin–ERK1/2 complex formation could be to specifically target ERK1/2 to non-nuclear substrates involved in the regulation of GPCR signaling or intracellular trafficking. Alternatively, β-arrestin-bound ERK1/2 might phosphorylate other cytosolic proteins involved in transcriptional regulation (such as p90RSK), which in turn relay signals to the nucleus. In such a model, transcriptional events mediated directly by the nuclear pool of ERK1/2 would be attenuated, whereas indirect pathways of ERK-dependent transcription would persist, resulting in an altered pattern of transcription following activation of the GPCR.

In NIH-3T3 cells, PAR-2 stimulate prolonged activation of a plasma-membrane-associated pool of ERK1/2 that is retained in receptor–β-arrestin–ERK1/2 complexes. These complexes are enriched in pseudopodia when exposed to a chemotactic gradient. Furthermore, PAR-2 receptor-mediated cytoskeletal reorganization, polarized pseudopod extension, and chemotaxis are ERK1/2-dependent and inhibited by expression of a dominant-negative mutant of β-arrestin 1. These findings suggest that the formation of β-arrestin–ERK1/2 signaling complexes at the leading edge of a cell may direct localized actin assembly and drive chemotaxis *(128)*. Consistent with this hypothesis, T and B cells from β-arrestin 2 knockout mice are strikingly impaired in their ability to respond to CXCL12 in transwell and in transendothelial migration assays *(129)*.

6.2.2.3. p38 MAPKs

Less information is available regarding the role of β-arrestins in the regulation of p38 MAPKs. In HeLa and HEK293 cells, overexpression of β-arrestin 2 enhances ERK1/2 and p38 MAPK activation as well as the chemotactic response to activation of CXCR4 and CXCR5 *(130)*. Conversely, suppression of endogenous β-arrestin 2 expression by antisense or RNAi attenuates CXCR4-mediated cell migration. In this system, inhibition of p38 MAPK, but not ERK1/2, blocked the effect of β-arrestin 2 on chemotaxis, suggesting that β-arrestin may act as a positive regulator of chemokine receptor-mediated chemotaxis by enhancing activation of the Ask1–p38 MAPK pathway. Similarly, the C-terminal phosphorylation sites and β-arrestin-interacting domain of the human cytomegalovirus GPCR US28 are required for maximal activation of p38 MAPK by this receptor *(54)*.

6.2.3. Arrestins and Ubiquitin Ligases

As previously discussed, the E3 ubiquitin ligase Mdm2 associates directly with β-arrestin 2 and catalyzes ubiquitination of β-arrestin simultaneously with its binding to agonist-occupied GPCRs *(66)*. Ubiquitination of β-arrestin regulates GPCR endocytosis and affects the stability of the GPCR–β-arrestin complex, with important effects on the postendocytic trafficking of GPCRs *(67)*. However, Mdm2 has other cellular roles, most notably as a negative regulator of the p53 tumor suppressor. Some data suggest that β-arrestin binding to Mdm2 may link GPCRs to the p53 signaling pathway *(131)*. Mdm2 catalyzes the ubiquitination of p53, which targets it for degradation. In HEK293 cells, stimulation of δ-opioid receptors leads to β-arrestin-dependent recruitment of Mdm2 to the receptor. In Saos cells, binding of β-arrestin 2 to Mdm2 suppresses Mdm2-catalyzed ubiquitination of p53. As a result of increased p53 abundance, overexpression of β-arrestin 2 enhances p53-mediated apoptosis, whereas suppressing expression of β-arrestin 2 by RNAi has the opposite effect. Although these data do not establish a physiological link between GPCRs and p53 through β-arrestin, they suggest a possible mechanism for β-arrestin effects on cell survival.

6.2.4. Arrestins and cAMP PDEs

β-arrestins bind directly to selected isoforms of cAMP PDE *(132)*. Stimulation of $β_2$-ARs leads to the β-arrestin-dependent recruitment of PDE4D3 and PDE4D5 to the agonist-occupied receptor. The result is accelerated termination of membrane-associated PKA activity, because β-arrestin binding precludes further G_s activation by inducing homologous receptor desensitization, and it enhances the rate of cAMP degradation by recruiting PDE4. In both HEK293 cells and primary cardiac myocytes, inhibiting PDE4 activity either pharmacologically or by expressing a catalytically inactive mutant of PDE4D markedly enhances PKA-mediated phosphorylation of the receptor. Because PKA phosphorylated $β_2$-ARs exhibit enhanced coupling to G_i proteins and stimulate pertussis toxin-sensitive ERK1/2 activation *(37)*, inhibition of PDE activity is also associated with a marked increase in $β_2$-AR stimulation of MAPK *(133)*. By locally controlling cAMP concentration, β-arrestin-dependent PDE recruitment may therefore provide a mechanism for locally dampening PKA activation and modulating the G protein-coupling specificity of β-ARs.

6.2.5. Arrestins and Ral-GDS

The Ral-GDP dissociation stimulator Ral-GDS interacts with β-arrestins in yeast two-hybrid assays and in co-immunoprecipitations from human polymorphonuclear leukocytes *(134)*. Ral-GDS is inactive when bound to cytosolic β-arrestin; however, when β-arrestin is recruited to the membrane in response to formyl-Met-Leu-Phe receptor stimulation, the Ral-GDS dissociates and catalyzes activation of the Ral effector pathway, leading to cytoskeletal re-arrangement. Thus, β-arrestins appear to provide a direct link between GPCR activation and Ral-mediated cytoskeletal reorganization.

6.2.6. Other Signaling Roles for Arrestins

Some data suggest that β-arrestins may regulate cell survival and cell cycle progression through the phosphatidylinositol-3-kinase (PI3K)–Akt pathway. In IIC9 cells, α-thrombin stimulates rapid, PI3K-dependent activation of Akt *(135)*. This response is inhibited by expression of dominant interfering mutants of β-arrestin 1, but not β-arrestin 2.

Alterations in visual arrestin function are associated with retinal disease in flies, mice, and humans. Certain forms of hereditary stationary night blindness, such as Oguchi disease, are attributable to mutations in rhodopsin kinase or arrestin that lead to impaired photoreceptor desensitization *(136–138)*. Many of these patients develop retinitis pigmentosa, with the death of photoreceptor cells. Arrestin knockout mice maintained in continuous or cyclic light, but not in continuous darkness, experienced photoreceptor loss at a rate proportional to the amount of light exposure, consistent with the hypothesis that constitutive signal flow in the absence of arrestin leads to photoreceptor degeneration *(139)*. However, a different mechanism—possibly involving arrestin-dependent signaling—has been demonstrated in retinal degeneration mutants of *Drosophila (140,141)*. In these models, the formation of stable arrestin–rhodopsin complexes leads to apoptotic death of photoreceptor cells, whereas deletion of either rhodopsin or arrestin rescues the degeneration phenotype. This mechanism of retinal degeneration involves the endocytic machinery, suggesting that endocytosis of arrestin–rhodopsin complexes might be involved in triggering the apoptotic pathway *(142)*. In light of the data indicating that arrestins may link GPCRs to the JNK–SAPK, Mdm2–p53, and PI3K–Akt pathways, these results suggest that arrestin scaffolds may play a physiologically relevant role in this form of retinal degeneration.

7. GPCR-INDEPENDENT FUNCTIONS OF ARRESTINS

Whether in the context of negative regulation of receptor–G protein coupling or of signal transduction as adaptor/scaffold proteins, most work on the arrestins has focused on their role in regulating GPCR function. However, several recent reports have suggested that arrestins are involved in the internalization or signaling of some non-GPCRs, indicating that they play a broader role in receptor regulation than previously appreciated.

The receptor for insulin-like growth factor (IGF)-1 is a heterotetrameric receptor tyrosine kinase of the insulin receptor family. Despite its lack of structural relation to the heptahelical GPCRs, some aspects of IGF-1 receptor signaling apparently involve components of the GPCR signalling machinery, including heterotrimeric G proteins and β-arrestins. For example, in cultured Rat 1a fibroblasts, HIRcB cells, and 3T3L1 adipocytes, IGF-1-stimulated ERK1/2 activation and mitogenesis is pertussis toxin-sensitive and inhibited by peptides that sequester free $G_{\beta\gamma}$-subunits *(143)*. β-arrestins also associate with IGF-1 receptors in a ligand-dependent manner. Overexpression of β-arrestin enhances clathrin-mediated endocytosis of the IGF-1 receptor and increases IGF-1-stimulated ERK1/2 phosphorylation and DNA synthesis *(144)*. Conversely, microinjection of antibodies against β-arrestin 1 specifically inhibits IGF-1 stimulated mitogenesis, with no effect on the responses to either insulin or EGF *(145)*. Prolonged exposure to insulin leads to a marked decrease in cellular β-arrrestin 1 content by stimulating β-arrestin ubiquitination and proteosomal degradation, resulting in profound changes in subsequent cellular responses to both GPCR and IGF-1 receptor stimulation. Predictably, β_2-AR-mediated cAMP production is enhanced and receptor sequestration is attenuated *(146)*. Consistent with its signaling role, the downregulaton of β-arrestin 1 reduces IGF-1-stimulated ERK1/2 activation and abolishes ERK1/2 activation by lysophosphatidic acid (LPA) and β_2-ARs. Ectopic expression of β-arrestin 1 in cells where endogenous β-arrestin 1 has been downregulated by insulin exposure rescues IGF-1- and LPA-stimulated ERK1/2 activation *(147)*.

β-arrestin 1 also appears to be involved in IGF-1 activation of a PI3K-dependent cell survival pathway that is distinct from the well-characterized pathway initiated by recruitment of p85/PI3Kα to tyrosine phosphorylated insulin receptor substrate proteins. In subconfluent MEF cultures, IGF-1 treatment rapidly activates the PI3K–Akt pathway and inhibits apoptosis through a mechanism that does not involve the IGF-1 receptor tyrosine kinase activity, pertussis toxin-sensitive G proteins, or ERK1/2. This signal is absent in β-arrestin 1/2 null MEFs but can be restored by stable re-expression of β-arrestin 1, suggesting that the IGF-1 receptor can use β-arrestins in signal transduction *(148)*.

Analogous to their role in GPCR trafficking, β-arrestins play a role in internalization and downregulation of the single transmembrane-spanning transforming growth factor (TGF)-β receptor *(149)*. β-arrestin 2 binds to the TGF-β type III receptor after phosphorylation of the cytoplasmic tail of the receptor. The phosphorylation is not catalyzed by a GRK, but by the TGF-β type II receptor, which possesses intrinsic kinase activity. β-arrestin binding leads to internalization and downregulation of both type II and type III TGF-β receptors.

Constitutive clathrin-dependent endocytosis of the low-density lipoprotein (LDL) receptor involves the autosomal recessive hypercholesterolemia clathrin adaptor protein ARH. Interestingly, however, β-arrestin 2 knockout mice that were fed a high-fat diet exhibited significant elevation of LDL and intermediate-density lipoprotein levels compared to littermate controls, suggesting that β-arrestins may play an accessory role in LDL receptor endocytosis. Indeed, β-arrestins have been shown to coprecipitate with the LDL receptor. In HEK293 cells, overexpression of β-arrestin 1 or β-arrestin 2 substantially increases LDL uptake, whereas RNAi suppression of β-arrestin 2, but not β-arrestin 1, reduces LDL receptor endocytosis. Similarly, LDL receptor endocytosis is impaired in β-arrestin 1/2 null fibroblasts and can be restored by expression of β-arrestin 2, but not β-arrestin 1, at physiological levels. The interaction with β-arrestin 2 involves the LDL receptor C-tail and is enhanced by a phosphoserine-mimetic S833D mutation of the LDL receptor *(150)*.

β-arrestins also appear to be involved in the Wnt signaling pathway, which is a key regulator of development in many organisms. Wnt proteins bind to seven-transmembrane-spanning receptors called Frizzleds. Unlike GPCRs, Frizzleds do not signal via heterotrimeric G proteins. Rather, they recruit the cytoplasmic proteins Dishevelled 1 and Dishevelled 2 (Dvl 1 and Dvl 2), which link the receptor to several signaling cascades, including inhibition of glycogen synthase kinase-3β, stabilization of β-catenin, and activation of lymphoid enhancer factor (LEF). Based on yeast two-hybrid studies, β-arrestin 1 has been identified as a binding partner of both Dvl 1 and Dvl 2 *(151)*. Phosphorylation of Dvl 1 strongly enhances its binding to β-arrestin 1, and overexpression of β-arrestin 1 along with Dvl1 synergistically activates LEF transcription. β-arrestins also participate in the endocytosis of Frizzleds *(152)*. In HEK293 cells, endocytosis of Frizzled 4 in response to the Wnt5A protein is mediated by β-arrestin 2 that is recruited to the receptor by binding to phosphorylated Dvl2. Similarly to GPCRs, these data suggest that β-arrestins function both as regulators of endocytosis and as signaling adaptor proteins coupling Frizzled receptors to transcriptional activation.

8. CONCLUSIONS

The arrestin family of proteins was initially discovered through their involvement in the termination of GPCR coupling to heterotrimeric G proteins. Subsequent research has markedly expanded our appreciation of the diverse roles played by these proteins in both negative and positive receptor regulation. In addition to mediating homologous desensitization of GPCRs, we now appreciate that the arrestins are central regulators of GPCR internalization and intracellular trafficking. Indeed, their role in receptor trafficking appears to extend beyond GPCRs to include other classes of membrane receptor. As receptor-binding proteins that can interact with a host of signaling proteins, the arrestins are also able to confer enzymatic activity upon GPCRs in a ligand-dependent manner. This model of GPCR signaling, in which receptor-bound arrestins act as adaptors or scaffolds that recruit signaling proteins to "desensitized" GPCRs, intimately links the processes of receptor trafficking and signaling. Rather than simply being a mechanism for ending GPCR signaling, arrestin binding may mark the transition point between heterotrimeric G protein activation and the initiation of a second wave of β-arrestin-dependent signaling.

REFERENCES

1. Filipek S, Stenkamp RE, Teller DC, Palczewski K. G protein-coupled receptor rhodopsin: a prospectus. Ann Rev Physiol 2003;65:851–879.
2. Arshavsky VY, Lamb TD, Pugh EN Jr. G proteins and phototransduction. Ann Rev Physiol 2002;64:153–187.
3. Freedman NJ, Lefkowitz R J. Desensitization of G protein-coupled receptors. Recent Prog Horm Res 1996;51:319–351

4. Ferguson SS. Evolving concepts in G protein-coupled receptor endocytosis: the role in receptor desensitization and signaling. Pharm Rev 2001;53:1–24.
5. Smith DP, Shieh BH, Zuker CS. Isolation and structure of an arrestin gene from *Drosophila*. Proc Natl Acad Sci USA 1990;87:1003–1007.
6. Hyde DR, Mecklenburg KL, Pollock JA, Vihtelic TS, Benzer S. Twenty *Drosophila* visual system clones: one is a homolog of human arrestin. Proc Natl Acad Sci USA 1990;87:1008–1012.
7. Yamada T, Takeuchi Y, Komori N, et al. A 49-kilodalton phosphoprotein in the *Drosophila* photoreceptor is an arrestin homolog. Science 1990;248:483–486.
8. Mayeenuddin LH, Mitchell J. Squid visual arrestin; cDNA cloning and calcium-dependent phosphorylation by rhodopsin kinase (SQRK). J Neurochem 2003;85:592–600.
9. Nakagawa M, Orii H, Yoshida N, et al. Ascidian arrestin (Ci-arr), the origin of the visual and nonvisual arrestins of vertebrates. Eur J Biochem 2002;269:5112–5118.
10. Shinohara T, Dietzschold B, Craft CM, et al. Primary and secondary structure of bovine retinal S antigen (48-kDa protein). Proc Nat Acad Sci USA 1987;84:6975–6979.
11. Yamaki K, Takahashi Y, Sakuragi S, Matsubara K. Molecular cloning of the S-antigen cDNA from bovine retina. Biochem Biophys Res Commun 1987;142:904–910.
12. Murakami A, Yajima T, Sakuma H, McClaren MJ, Inana G. X-arrestin: a new retinal arrestin mapping to the X chromosome. FEBS Lett 1993;334:203–209.
13. Craft CM, Whitmore DH, Weichmann AF. Cone arrestin identified by targeting expression of a functional family. J Biol Chem 1994;269:4613–4619.
14. Lohse MJ, Benovic JL, Codina J, Caron MG, Lefkowitz RJ. β-arrestin: a protein that regulates β-adrenergic receptor function. Science 1990;248:1547–1550.
15. Attramadal H, Arriza JL, Aoki C, et al. β-arrestin 2, a novel member of the arrestin/β-arrestin gene family. J Biol Chem 1992;267:17,882–17,890.
16. Goodman OB Jr, Krupnick JG, Santini F, et al. Beta-arrestin acts as a clathrin adaptor in endocytosis of the beta2-adrenergic receptor. Nature 1996;383:447–450.
17. Goodman OB Jr, Krupnick JG, Gurevich VV, Benovic JL, Keen JH. Arrestin/clathrin interaction. Localization of the arrestin binding locus to the clathrin terminal domain. J Biol Chem 1997;272:15,017–15,022.
18. Laporte SA, Oakley RH, Zhang J, et al. The beta2-adrenergic receptor/beta-arrestin complex recruits the clathrin adaptor AP-2 during endocytosis. Proc Natl Acad Sci USA 1999;96:3712–3717.
19. Laporte SA, Oakley RH, Holt JA, Barak LS, Caron MG. The interaction of beta-arrestin with the AP-2 adaptor is required for the clustering of beta 2-adrenergic receptor into clathrin-coated pits. J Biol Chem 2000;275:23,120–23,126.
20. Laporte SA, Miller WE, Kim KM, Caron MG. Beta-Arrestin/AP-2 interaction in G protein-coupled receptor internalization: Identification of a beta-arrestin binding site in beta 2-adaptin. J Biol Chem 2002;277:9247–9254.
21. Graznin J, Wilden U, Choe HW, Labahn J, Krafft B, Buldt G. X-ray crystal structure of arrestin from bovine rod outer segments. Nature 1998;391:918–921.

22. Hirsch JA, Schubert C, Gurevich VV, Sigler PB. The 2.8 A crystal structure of visual arrestin: a model for arrestin's regulation. Cell 1999;97:257–269.
23. Han M, Gurevich VV, Vishnivetsky SA, Sigler PB, Schubert C. Crystal structure of beta-arrestin at 1.9 A: possible mechanism of receptor binding and membrane translocation. Structure 2001;9:869–880.
24. Gurevich VV, Dion SB, Onorato JJ, et al. Arrestin interactions with G protein-coupled receptors. Direct binding studies of wild type and mutant arrestins with rhodopsin, β2-adrenergic, and m2 muscarinic cholinergic receptors. J Biol Chem 1995;270:720–731.
25. Vishnivetskiy SA, Hosey MM, Benovic JL, Gurevich VV. Mapping the arrestin–receptor interface: structural elements responsible for receptor specificity of arrestin proteins. J Biol Chem 2004;279:1262–1268.
26. Shilton BH, McDowell JH, Smith WC, Hargrave PA. The solution structure and activation of visual arrestin studied by small angle X-ray scattering. Eur J Biochem 2002;269:3801–3809.
27. Kovoor A, Celver J, Abdryashitov RI, Chavkin C, Gurevich VV. Targeted construction of phosphorylation-independent beta-arrestin mutants with constitutive activity in cells. J Biol Chem 1999;274:6831–6834.
28. Celver J, Vishnivetskiy SA, Chavkin C, Gurevich VV. Conservation of the phosphate-sensitive elements in the arrestin family of proteins. J Biol Chem 2002;277:9043–9048.
29. Kirchhausen T Adapters for clathrin-mediated traffic. Ann Rev Cell Dev Biol. 1999;15:705–732.
30. Kim YM, Benovic JL. Differential roles of arrestin-2 interaction with clathrin and adaptor protein 2 in G protein-coupled receptor trafficking. J Biol Chem 2002;277:30,760–30,768.
31. Gaidarov I, Krupnick JG, Falck JR, Benovic JL, Keen JH. Arrestin function in G protein-coupled receptor endocytosis requires phosphoinositide binding. EMBO J 1999;18:871–881.
32. McDonald PH, Cote NL, Lin F-T, Premont RT, Pitcher JA, Lefkowitz RJ. Identification of NSF as a beta-arrestin1-binding protein. Implications for beta2-adrenergic receptor regulation. J Biol Chem 1999;274:10,677–10,680.
33. Claing A, Chen W, Miller WE, et al. Beta-Arrestin-mediated ADP-ribosylation factor 6 activation and beta 2-adrenergic receptor endocytosis. J Biol Chem 2001;276:42,509–42,513.
34. Benovic JL, Pike LJ, Cerione RA, et al. Phosphorylation of the mammalian beta-adrenergic receptor by cyclic AMP-dependent protein kinase. Regulation of the rate of receptor phosphorylation and dephosphorylation by agonist occupancy and effects on coupling of the receptor to the stimulatory guanine nucleotide regulatory protein. J Biol Chem 1985;260:7094–7101.
35. Bouvier M, Hausdorff WP, De Blasi A, et al. Removal of phosphorylation sites from the beta 2-adrenergic receptor delays onset of agonist-promoted desensitization. Nature 1988;333:370–373.
36. Lefkowitz RJ. G protein-coupled receptor kinases. Cell 1993; 74: 409–412.
37. Daaka Y, Luttrell LM, Lefkowitz RJ. Switching of the coupling of the beta2-adrenergic receptor to different G proteins by protein kinase A. Nature 1997;390:88–91.

38. Zamah AM, Delahunty M, Luttrell LM, Lefkowitz RJ. Protein kinase A-mediated phosphorylation of the beta2-adrenergic receptor regulates its coupling to Gs and Gi. Demonstration in a reconstituted system. J Biol Chem 2002;277:31,249–31,256.
39. Lawler OA, Miggin SM, Kinsella BT. Protein kinase A-mediated phosphorylation of serine 357 of the mouse prostacyclin receptor regulates its coupling to Gs-, to Gi- and to Gq-coupled effector signaling. J Biol Chem 2001;276:33,596–33,607.
40. Lefkowitz RJ, Pierce KL, Luttrell LM. Dancing with different partners: Protein kinase A phosphorylation of seven membrane-spanning receptors regulates their G protein-coupling specificity. Mol Pharmacol 2002;62:971–974.
41. Benovic JL, Kuhn H, Weyand I, Codina J, Caron MG, Lefkowitz RJ. Functional desensitization of the isolated beta-adrenergic receptor by the beta-adrenergic receptor kinase. Potential role of an analog of the retinal protein arrestin (48-kDa protein). Proc Natl Acad Sci USA 1987;84:8879–8882.
42. Lohse MJ, Andexinger S, Pitcher J, et al. Receptor specific desensitization with purified proteins. Kinase dependence and receptor specificity of β-arrestin and arrestin in the β2-adrenergic receptor and rhodopsin systems. J Biol Chem 1993;267:8558–8564.
43. Weiss ER, Raman D, Shirakawa S, et al. The cloning of GRK7, a candidate cone opsin kinase, from cone- and rod-dominant mammalian retinas. Mol Vis 1998;4:27.
44. Stoffel RH, Pitcher JA, Lefkowitz RJ. Targeting G protein-coupled receptor kinases to their membrane substrates. J Memb Biol 1997;157:1–8.
45. Conner, DA, Mathier MA, Mortensen RM, et al. Beta-Arrestin 1 knockout mice appear normal but demonstrate altered cardiac responses to beta-adrenergic stimulation. Circ Res 1997;81:1021–1026.
46. Bohn LM, Lefkowitz RJ, Gainetdinov RR, Peppel K, Caron MG, Lin F- T. Enhanced morphine analgesia in mice lacking beta-arrestin 2. Science. 1999;286:2495–2498.
47. Bohn LM, Gainetdinov RR, Lin F-T, Lefkowitz RJ, Caron MG. Mu-opioid receptor desensitization by beta-arrestin-2 determines morphine tolerance but not dependence. Nature 2002;408:720–723.
48. Zhang J, Ferguson SS, Barak LS, Menard L, Caron MG. Dynamin and beta-arrestin reveal distinct mechanisms for G protein-coupled receptor internalization. J Biol Chem 1996;271:18,302–18,305.
49. Cao TT, Mays RW, von Zastrow M. Regulated endocytosis of G-protein-coupled receptors by a biochemically and functionally distinct subpopulation of clathrin-coated pits. J Biol Chem 1998;273:24,592–24,602.
50. Barak LS, Oakley RH, Laporte SA, Caron MG. Constitutive arrestin-mediated desensitization of a human vasopressin receptor mutant associated with nephrogenic diabetes insipidus. Proc Natl Acad Sci USA 2001;98:93–98.
51. Paing MM, Stutts AB, Kohout TA, Lefkowitz RJ, Trejo J. beta-Arrestins regulate protease-activated receptor-1 desensitization but not internalization or down-regulation. J Biol Chem 2002;277:1292–1300.

52. Vines CM, Revankar CM, Maestas DC, et al. *N*-formyl peptide receptors internalize but do not recycle in the absence of arrestins. J Biol Chem 2003;278:41,581–41,584.
53. Brasselet S, Guillen S, Vincent JP, Mazella J. Beta-arrestin is involved in the desensitization but not in the internalization of the somatostatin receptor 2A expressed in CHO cells. FEBS Lett 2002;10:124–128.
54. Miller WE, Houtz DA, Nelson CD, Kolattukudy PE, Lefkowitz RJ. G-protein-coupled receptor (GPCR) kinase phosphorylation and beta-arrestin recruitment regulate the constitutive signaling activity of the human cytomegalovirus US28 GPCR. J Biol Chem 2003;278:21,663–21,671.
55. Fraile-Ramos A, Kohout TA, Waldhoer M, Marsh M. Endocytosis of the viral chemokine receptor US28 does not require beta-arrestins but is dependent on the clathrin-mediatted pathway. Traffic 2003;4:243–253.
56. Vogler O, Nolte B, Voss M, Schmidt M, Jakobs KH, van Koppen CJ. Regulation of muscarinic acetylcholine receptor sequestration and function by β-arrestin. J Biol Chem 1999;274:12,333–12,338.
57. Rapacciuolo A, Suvarna S, Barki-Harrington L, et al. Phosphorylation sites of the β-1 adrenergic receptor determine the internalization pathway. J Biol Chem 2003;278:35,403–35,411.
58. Barak LS, Ferguson SS, Zhang J, Caron MG. A beta-arrestin/green fluorescent protein biosensor for detecting G protein-coupled receptor activation. J Biol Chem 1997;272:27,497–27,500.
59. Oakley RH, Laporte SA, Holt JA, Barak LS, Caron MG. Molecular determinants underlying the formation of stable intracellular G protein-coupled receptor-beta-arrestin complexes after receptor endocytosis. J Biol Chem 2001;276:19,452–19,460.
60. Zhang J, Barak LS, Anborgh PH, Laporte SA, Caron MG, Ferguson SS. Cellular trafficking of G protein-coupled receptor/beta-arrestin endocytic complexes. J Biol Chem 1999;274:10,999–11,006.
61. Oakley RH, Laporte SA, Holt JA, Caron MG, Barak LS. Differential affinities of visual arrestin, beta-arrestin1, and beta-arrestin2 for G protein-coupled receptors delineate two major classes of receptors. J Biol Chem 2000;275:17,201–17,210.
62. Kohout TA, Lin F-T, Perry SJ, Conner DA, Lefkowitz RJ. Beta-Arrestin 1 and 2 differentially regulate heptahelical receptor signaling and trafficking. Proc Natl Acad Sci USA 2001;98:1601–1606.
63. Lin F-T, Krueger KM, Kendall HE, et al. Clathrin-mediated endocytosis of the beta-adrenergic receptor is regulated by phosphorylation/dephosphorylation of beta-arrestin1. J Biol Chem 1997;272:31,051–31,057.
64. Lin F-T, Chen W, Shenoy S, Cong M, Exum ST, Lefkowitz RJ. Phosphorylation of beta-arrestin2 regulates it function in internalization of beta(2)-adrenergic receptors. Biochemistry 2002;41:10,692–10,699.
65. Shenoy SK, McDonald PH, Kohout TA, Lefkowitz RJ. Regulation of receptor fate by ubiquitination of activated β2-adrenergic receptor and β-arrestin. Science 2001;294:1307–1313.

66. Martin NP, Lefkowitz RJ, Shenoy SK. Regulation of V2 vasopressin receptor degradation by agonist-promoted ubiquitination. J Biol Chem 2003; 278:45,954–45,959.
67. Shenoy SK, Lefkowitz RJ. Trafficking pattern of beta-arrestin and G protein-coupled receptors determined by the kinetics of beta-arrestin deubiquitination. J Biol Chem 2003;278:14,498–14,506.
68. Sibley DR, Strasser RH, Benovic JL, Daniel K, Lefkowitz RJ. Phosphorylation/dephosphorylation of the beta-adrenergic receptor regulates its functional coupling to adenylate cyclase and subcellular distribution. Proc Natl Acad Sci USA 1986;83:9408–9412.
69. Zhang J, Barak LS, Winkler KE, Caron MG, Ferguson SS. A central role for β-arrestins and clathrin-coated vesicle-mediated endocytosis in β2-adrenergic receptor resensitization. J Biol Chem 1997;272:27,005–27,014.
70. Pitcher JA, Payne ES, Csortos C, DePaoli-Roach AA, Lefkowitz RJ. The G-protein-coupled receptor phosphatase: A protein phosphatase type 2A with a distinct subcellular distribution and substrate specificity. Proc Natl Acad Sci USA 1995;92:8343–8347.
71. Krueger KM, Daaka Y, Pitcher JA, Lefkowitz RJ. The role of sequestration in G protein-coupled receptor resensitization. Regulation of beta2-adrenergic receptor dephosphorylation by vesicular acidification. J Biol Chem 1997;272:5–8.
72. Pan L, Gurevich EV, Gurevich VV. The nature of the arrestin x receptor complex determines the ultimate fate of the internalized receptor. J Biol Chem 2003;278:11,623–11,632.
73. Oakley RH, Laporte SA, Holt JA, Barak LS, Caron MG. Association of beta-arrestin with G protein-coupled receptors during clathrin-mediated endocytosis dictates the profile of receptor resensitization. J Biol Chem 1999; 274:32,248–32,257.
74. Gage RM, Kim KA, Cao TT, von Zastrow M. A transplantable sorting signal that is sufficient to mediate rapid recycling of G protein-coupled receptors. J Biol Chem 2001;276:44,712–44,720.
75. Cao TT, Deacon HW, Reczek D, Bretscher A, von Zastrow M. A kinase-regulated PDZ-domain interaction controls endocytic sorting of the beta 2-adrenergic receptor. Nature 1999;401:286–290.
76. Whistler JL, Enquist J, Marley A, et al. Modulation of postendocytic sorting of G protein-coupled receptors. Science 2002;297:529–531.
77. Hall RA, Lefkowitz RJ. Regulation of G protein-coupled receptor signaling by scaffold proteins. Circ Res 2002;91:672–668.
78. Gurevich VV, Pals-Rylaarsdam R, Benovic JL, Hosey MM, Onorato JJ. Agonist–receptor–arrestin, an alternative ternary complex with high agonist affinity. J Biol Chem 1997;272:28,849–28,852.
79. Key TA, Bennett TA, Foutz TD, Gurevich VV, Sklar LA, Prossnitz ER. Regulation of formyl peptide receptor agonist affinity by reconstitution with arrestins and heterotrimeric G proteins. J Biol Chem 2001;276:49,204–49,212.
80. Martini L, Hastrup H, Holst B, Fraile-Ramos A, Marsh M, Schwartz TW. NK1 receptor fused to beta-arrestin displays a single-component, high-affinity molecular phenotype. Mol Pharmacol 2002;62:30–37.

81. Holloway AC, Qian H, Pipolo L, et al. Side-chain substitutions within angiotensin II reveal different requirements for signaling, internalization, and phosphorylation of type 1a angiotensin receptors. Mol Pharmacol 2002;61:768–777.
82. Gaborik Z, Jagadeesh G, Zhang M, Spat A, Catt KJ, Hunyady L. The role of a conserved region of the second intracellular loop in AT1 angiotensin receptor activation and signaling. Endocrinology. 2003;144:2220–2228.
83. Miller WE, Lefkowitz RJ. Expanding roles for beta-arrestins as scaffolds and adapters in GPCR signaling and trafficking. Curr Opin Cell Biol 2001;13:139–145.
84. Perry SJ, Lefkowitz RJ. Arresting developments in heptahelical receptor signaling and regulation. Trends Cell Biol 2002;12:130–138.
85. Luttrell LM, Ferguson SSG, Daaka Y, et al. β-Arrestin-dependent formation of β2 adrenergic receptor/Src protein kinase complexes. Science 1999;283:655–661.
86. DeFea KA, Vaughn ZD, O'Bryan EM, Nishijima D, Dery O, Bunnett NW. The proliferative and antiapoptotic effects of substance P are facilitated by formation of a β-arrestin-dependent scaffolding complex. Proc Natl Acad Sci USA 2000;97:11,086–11,091.
87. Barlic J, Andrews JD, Kelvin AA, et al. Regulation of tyrosine kinase activation and granule release through β-arrestin by CXCRI. Nature Immunol 2000;1:227–233.
88. Ghalayini AJ, Desai N, Smith KR, Holbrook RM, Elliott MH, Kawakatsu. Light-dependent association of Src with photoreceptor rod outer segment membrane proteins in vivo. J Biol Chem 2002;277:1469–1476.
89. Milano SK, Pace HC, Kim YM, Brenner C, Benovic JL. Scaffolding functions of arrestin-2 revealed by crystal structure and mutagenesis. Biochemistry 2002;41:3321–3328.
90. Miller WE, Maudsley S, Ahn S, Kahn KD, Luttrell, LM, Lefkowitz RJ. β-Arrestin1 interacts with the catalytic domain of the tyrosine kinase c-SRC. J Biol Chem 2000;275:11,312–11,319.
91. Ahn S, Maudsley S, Luttrell LM, Lefkowitz RJ, Daaka Y. Src-mediated tyrosine phosphorylation of dynamin is required for beta2-adrenergic receptor internalization and mitogen-activated protein kinase signaling. J Biol Chem 1999;274:1185–1188.
92. Ahn S, Kim J, Lucaveche CL, et al. Src-dependent tyrosine phosphorylation regulates dynamin self-assembly and ligand-induced endocytosis of the epidermal growth factor receptor. J Biol Chem. 2002;277:26,642–26,651.
93. Penela P, Elorza A, Sarnage S, Mayor F Jr. Beta-arrestin and c-Src-dependent degradation of G-protein-coupled receptor kinase 2. EMBO J 2001;20:5129–5138.
94. Imamura T, Huang J, Dalle S, et al. Beta-Arrestin-mediated recruitment of the Src family kinase Yes mediates endothelin-1-stimulated glucose transport. J Biol Chem 2001;276:43,663–43,667.
95. Luttrell LM, Daaka Y, Lefkowitz RJ. Regulation of tyrosine kinase cascades by G-protein coupled receptors. Curr Opin Cell Biol 1999;11:177–183.

96. Luttrell LM. Activation and targeting of MAP kinases by G protein-coupled receptors. Can J Physiol Pharm 2002;80:375–382.
97. Luttrell LM. Location, location, location. Spatial and temporal regulation of MAP kinases by G protein-coupled receptors. J Mol Endo 2003;30:117–126.
98. Yang M, Zhang H, Voyno-Yasenetskaya T, Ye RD. Requirement of G beta-gamma and c-Src in D2 dopamine receptor-mediated nuclear factor-kappa B activation. Mol Pharmacol 2003;64:447–455.
99. Kryiakis JM, Avruch J. Sounding the alarm: protein kinase cascades activated by stress and inflammation. J Biol Chem 1996;271:24,313–24,316.
100. Pearson G, Robinson F, Beers Gibson T, et al. Mitogen-activated protein (MAP) kinase pathways: Regulation and physiologic functions. Endocr Rev 2001;22:153–183.
101. Burack WR, Shaw AS. Signal transduction: hanging on a scaffold. Curr Opin Cell Biol 2000;12:211–216.
102. Elion EA. The STE5p scaffold. J Cell Sci 2001;114:3967–3978.
103. Whitmarsh AJ, Cavanagh J, Tournier C, Yasuda J, Davis RJ. A mammalian scaffold complex that selectively mediates MAP kinase activation. Science 1998;281:1671–1674.
104. Yasuda J, Whitmarsh AJ, Cavanagh J, Sharma M, Davis RJ. The JIP group of mitogen-activated protein kinase scaffold proteins. Mol Cell Biol 1999;19:7245–7254.
105. McDonald PH, Chow C-W, Miller WE, et al. β-Arrestin 2: a receptor-regulated MAPK scaffold for the activation of JNK3. Science 2000;290:1574–1577.
106. Miller WE, McDonald PH, Cai SF, Field MF, Davis RJ, Lefkowitz RJ. Identification of a motif in the carboxy terminus of β–arrestin2 responsible for activation of JNK3. J Biol Chem 2001;276:27,770–27,777.
107. Scott MG, Le Rouzic E, Perianin A, et al. Differential nucleocytoplasmic shuttling of beta-arrestins. Characterization of a leucine-rich nuclear export sequence in beta-arrestin2. J Biol Chem 2002;277:37,693–37,701.
108. Wang P, Wu Y, Ge X, Ma L, Pei G. Subcellular localization of beta-arrestins is determined by their intact N domain and the nuclear export signal at the C terminus. J Biol Chem 2003;278:11,648–11,653.
109. Tohgo A, Pierce KL, Choy EW, Lefkowitz RJ, Luttrell LM. Beta-Arrestin scaffolding of the ERK cascade enhances cytosolic ERK activity but inhibits ERK-mediated transcription following angiotensin AT_{1A} receptor stimulation. J Biol Chem 2002;277:9429–9436.
110. Ahn S, Nelson CD, Garrison TR, Miller WE, Lefkowitz RJ. Desensitization, internalization, and signaling functions of beta-arrestins demonstrated by RNA interference. Proc Natl Acad Sci USA 2003;100:1740–1744.
111. Wei H, Ahn S, Shenoy SK, et al. Independent beta-arrestin 2 and G protein-mediated pathways for angiotensin II activation of extracellular signal-regulated kinases 1 and 2. Proc Natl Acad Sci USA 2003;100:10,782–10,787.
112. Azzi M, Charest PG, Angers S, et al. Beta-arrestin-mediated activation of MAPK by inverse agonists reveals distinct active conformations for G protein-coupled receptors. Proc Natl Acad Sci USA 2003;100:11,406–11,411.

113. DeFea KA, Zalevsky J, Thoma MS, Dery O, Mullins RD, Bunnett NW. β-Arrestin-dependent endocytosis of proteinase-activated receptor 2 is required for intracellular targeting of activated ERK1/2. J Cell Biol 2000;148:1267–1281.
114. Luttrell LM, Roudabush FL, Choy EW, et al. Activation and targeting of extracellular signal-regulated kinases by β-arrestin scaffolds. Proc Natl Acad Sci USA 2001:98;2449–2454.
115. Gutkind JS. The pathways connecting G protein-coupled receptors to the nucleus through divergent mitogen-activated protein kinase cascades. J Biol Chem 1998;273:1839–1842.
116. Pierce KL, Luttrell LM, Lefkowitz RJ. New mechanisms in heptahelical receptor signaling to mitogen activated protein kinase cascades. Oncogene 2001;20:1532–1539.
117. Eguchi S, Numaguchi K, Iwasaki H, et al. Calcium-dependent epidermal growth factor receptor transactivation mediates the angiotensin II-induced mitogen-activated protein kinase activation in vascular smooth muscle cells. J Biol Chem 1998;273:8890–8896.
118. Heeneman S, Haendeler J, Saito Y, Ishida M, Berk BC. Angiotensin II induces transactivation of two different populations of the platelet-derived growth factor beta receptor. Key role for the p66 adaptor protein Shc. J Biol Chem 2000;275;15,926–15,932.
119. Gschwind A, Zwick E, Prenzel N, Leserer M, Ullrich A. Cell communication networks: epidermal growth factor receptor transactivation as the paradigm for interreceptor signal transmission. Oncogene 2001;20:1594–1600.
120. Murasawa S, Mori Y, Nozawa Y, et al. Angiotensin II type 1 receptor-induced extracellular signal-regulated protein kinase activation is mediated by Ca^{2+}/calmodulin-dependent transactivation of epidermal growth factor receptor. Circ Res 1998;82:1338–1348.
121. Castagliuolo I, Valenick L, Liu J, Pothoulakis C. Epidermal growth factor receptor transactivation mediates substance P-induced mitogenic responses in U-373 MG cells. J Biol Chem 2000;275:26,545–26,550.
122. Tohgo A, Choy EW, Gesty-Palmer D, et al. The stability of the G protein-coupled receptor–beta-arrestin interaction determines the mechanism and functional consequence of ERK activation. J Biol Chem 2003;278:6258–6267.
123. Lin F-T, Miller WE, Luttrell LM, Lefkowitz RJ. Feedback regulation of beta-arrestin1 function by extracellular signal-regulated kinases. J Biol Chem 1999;274;15,971–15,974.
124. Pitcher JA, Tesmer JJ, Freeman JL, Capel WD, Stone WC, Lefkowitz RJ. Feedback inhibition of G protein-coupled receptor kinase 2 (GRK2) activity by extracellular signal-regulated kinases. J Biol Chem 1999;274:34,531–34,534.
125. Elorza A, Sarnago S, Mayor F Jr. Agonist-dependent modulation of G protein-coupled receptor kinase 2 by mitogen-activated protein kinases. Mol Pharm 2000;57:778–783.
126. Ogier-Denis E, Pattingre S, El Benna J, Codogno P. Erk1/2-dependent phosphorylation of Galpha-interacting protein stimulates its GTPase accelerating activity and autophagy in human colon cancer cells. J Biol Chem 2000;275:39,090–39,095.

127. Elorza A, Penela P, Sarnago S, Mayor F Jr. MAPK-dependent degradation of G protein-coupled receptor kinase 2. J Biol Chem 2003;278:29,164–29,173.
128. Ge L, Ly Y, Hollenberg M, DeFea K. A beta-arrestin-dependent scaffold is associated with prolonged MAPK activation in pseudopodia during protease-activated receptor-2-induced chemotaxis. J Biol Chem 2003;278:34,418–34,426.
129. Fong AM, Premont RT, Richardson RM, Yu YR, Lefkowitz RJ, Patel DD. Defective lymphocyte chemotaxis in beta-arrestin2- and GRK6-deficient mice. Proc Natl Acad Sci USA 2002;99:7478–7483.
130. Sun Y, Cheng Z, Ma L, Pei G. Beta-arrestin 2 is critically involved in CXCR4-mediateed chemotaxis, and this is mediated by its enhancement of p38 MAPK activation. J Biol Chem 2002;277:49,212–49,219.
131. Wang P, Gao H, Ni Y, et al. Beta-arrestin 2 functions as a G-protein-coupled receptor-activated regulator of oncoprotein Mdm2. J Biol Chem 2003;278:6363–6370.
132. Perry SJ, Baillie GS, Kohout TA, et al. Targeting of cyclic AMP degradation to beta 2-adrenergic receptors by beta-arrestins. Science 2002;298:834–836.
133. Baillie GS, Sood A, McPhee I, et al Beta-Arrestin-mediated PDE4 cAMP phosphodiesterase recruitment regulates beta-adrenoceptor switching from Gs to Gi. Proc Natl Acad Sci USA 2003;100:940–945.
134. Bhattacharya M, Anborgh PH, Babwah AV, et al. Beta-arrestins regulate a Ral-GDS Ral effector pathway that mediates cytoskeletal reorganization. Nat Cell Biol 2002;4:547–555.
135. Goel R, Baldassare JJ. beta-Arrestin 1 couples thrombin to the rapid activation of the Akt pathway. Ann NY Acad Sci 2002;973:138–141.
136. Nakazawa M, Wada Y, Tamai M. Arrestin gene mutations in autosomal recessive retinitis pigmentosa. Arch Ophthalmol 1998;116:498–501.
137. Yamada T, Matsumoto M, Kadoi C, Nagaki Y, Hayasaka Y, Hayasaka S. 1147 del A mutation in the arrestin gene in Japanese patients with Oguchi disease. Ophthal Genetics 1999;20:117–120.
138. Dryja TP. Molecular genetics of Oguchi disease, fundus albipunctatus, and other forms of stationary night blindness. Am J Ophthal 2000;130:547–563.
139. Chen J, Simon MI, Matthes MT, Yasumura D, LaVail MM. Increased susceptibility to light damage in an arrestin knockout mouse model of Oguchi disease (stationary night blindness). Invest Ophthalmol Vis Sci 1999;40:2978–2982.
140. Alloway PG, Howard L, Dolph PJ. The formation of stable rhodopsin-arrestin complexes induces apoptsis and photoreceptor cell degeneration. Neuron 2000;28:129–138.
141. Kiselev A, Socolich M, Vinos J, Hardy RW, Zuker CS, Ranganathan R. A molecular pathway for light-dependent photoreceptor apoptosis in *Drosophila*. Neuron 2000;28:139–152.
142. Miller WE, Lefkowitz RJ. Arrestins as signaling molecules involved in apoptotic pathways: a real eye opener. Sci STKE 2001;2001:PE1.
143. Luttrell LM, van Biesen T, Hawes BE, Koch WJ, Touhara K, Lefkowitz RJ. G beta gamma subunits mediate mitogen-activated protein kinase activation by

the tyrosine kinase insulin-like growth factor 1 receptor. J Biol Chem 1995;270:16,495–16,498.
144. Lin FT, Daaka Y, Lefkowitz RJ. Beta-Arrestins regulate mitogenic signaling and clathrin-mediated endocytosis of the insulin-like growth factor I receptor. J Biol Chem 1998;273:31,640–31,643.
145. Dalle S, Ricketts W, Imamura T, Vollenweider P, Olefsky JM. Insulin and insulin-like growth factor I receptors utilize different G protein signaling components. J Biol Chem 2001;276:15,688–15,695.
146. Hupfeld CJ, Dalle S, Olefsky JM. Beta-Arrestin 1 down-regulation after insulin treatment is associated with supersensitization of beta 2 adrenergic receptor G alpha s signaling in 3T3-L1 adipocytes. Proc Natl Acad Sci USA 2003;100:161–166.
147. Dalle S, Imamura T, Rose DW, et al. Insulin induces heterologous desensitization of G-protein-coupled receptor and insulin-like growth factor I signaling by down-regulating beta-arrestin-1. Mol Cell Biol 2002;22:6272–6285.
148. Povsic TJ, Kohout TA, Lefkowitz RJ. Beta-Arrestin1 mediates IGF-1 activation of PI-3-K and anti-apoptosis. J Biol Chem 2003;278:51,334–51,339.
149. Chen W, Kirkbride KC, How T, et al. Beta-arrestin 2 mediates endocytosis of type III TGF-beta receptor and down-regulation of its signaling. Science 2003;301:1394–1397.
150. Wu JH, Peppel K, Nelson CD, et al. The adaptor protein beta-arrestin2 enhances endocytosis of the low density lipoprotein receptor. J Biol Chem 2003;278:44,238–44,245.
151. Chen W, Hu LA, Semenov MV, et al. Beta-Arrestin1 modulates lymphoid enhancer factor transcriptional activity through interaction with phosphorylated dishevelled proteins. Proc Natl Acad Sci USA 2001;98:14,889–14,894.
152. Chen W, ten Berge D, Brown J, et al. Dishevelled 2 recruits beta-arrestin 2 to mediate Wnt5A-stimulated endocytosis of Frizzled 4. Science 2003;301:1391–1394.

9
GPCR Interacting Proteins

Classes, Assembly, and Functions

Hongyan Wang, Catherine B. Willmore, Jia Bei Wang

1. INTRODUCTION

The complex transduction of ligand stimulation events at G-protein coupled receptors (GPCRs) by heterotrimeric G proteins has long been appreciated. In addition to this, recent data shows that other protein interactions assist and can fine-tune cellular signals. Scientists have identified other membrane and intracellular proteins that interact, directly or indirectly, with GPCRs. In fact, 50 or more proteins are described in current literature as GPCR interactive proteins. GPCR interacting proteins act as modulators of ligand-evoked signals. Membrane associated or intracellular GPCR interacting proteins have critical roles in mediating: ligand recognition, optimization of signal transduction, trafficking, receptor clustering, and/or compartmentalization. This chapter reviews four aspects of the GPCR interacting protein literature: (a) methods for identifying GPCR interacting proteins; (b) interaction domains on the GPCR; (c) facilitation and fine-tuning of GPCR signaling events by interacting proteins; and (d) particular analysis of proteins that are µ opioid receptor (µOR) interactive. Although GPCR dimerization is viewed by many as a type of protein interaction between the GPCRs, dimer-related protein interactions will not be discussed; an alternate section of this book is devoted to dimerization.

2. GPCR INTERACTING PROTEINS: DETECTION ASSAYS

2.1. Yeast Two-Hybrid Screening

Yeast hybridization techniques capitalize on structural flexibilities that are typical for yeast transcription modulators. For example, the transcription factor Gal4 has both a DNA binding domain and a transcriptional activation domain. These domains do not need to be attached for transcription to occur, but they must be positioned in close proximity. If, by genetic engineering, the Gal4 domains are fused to two functionally unrelated—but interactive—proteins, transcription progresses as a result of protein–protein interaction (1). A further benefit that results from choosing the yeast two-hybrid assay is that this method provides simultaneous access to the genes that encode interacting proteins.

Yeast two-hybrid assays proceed as a complementary DNA (cDNA) library screening, with bait corresponding to either a C-terminal or a third intracellular loop sequence in the favored receptor. Yeast two-hybrid systems are the most commonly used assay systems for identifying G protein-coupled receptor (GPCR) interactive proteins, and these assays have enhanced scientific understanding of protein–protein interplays. γ-aminobutyric acid $_B$ R1 (GABA$_B$R1) and R2 (GABA$_B$R2) proteins in association were detected by the yeast two-hybrid method (2). Similarly, a yeast assay permitted the detection of β_2-adrenergic receptors (β_2-ARs) in association with Na$^+$/H$^+$ exchanger regulatory factor/Ezrin/Radixin/Moesin (ERM)-binding phosphoprotein-50 proteins (NHERF/EBP50) (3). As further examples, the association between somatostatin receptor 2 (SSTR2) and somatostatin receptor interacting protein (SSTRIP) was elucidated by a yeast screening method (4), as was the association between the D$_2$ dopamine receptor (D$_2$R) and spinophilin (5). Clearly, the yeast two-hybrid system provides an excellent method for detecting protein–protein interactions; however, this system does have limitations.

One shortcoming of yeast assaying methods is that a protein can only be identified by yeast assays if the fished protein has direct contact with a GPCR. A second negative factor for yeast hybridization is the low number of proteins identified per library screen. One or, at most, two proteins are typically fished from the cDNA library with each bait. Another downfall of the yeast two-hybrid assaying method is its inability to detect protein–protein interactions that follow posttranslational modification(s). Additionally, as a final caution, investigators contemplating tests in a yeast assaying system should acknowledge that these assays have rendered false-negative as well as false-positive results.

2.2. Expression Cloning

Protein–protein interactions can be characterized by laboratory methods that involve expression cloning. The power of carefully applied expression cloning strategies is demonstrated by work done to characterize calcitonin receptor-like receptor (CRLR)-interacting proteins. CRLR, which is a member of the calcitonin gene-related peptide (CGRP)/CT superfamily of peptides, was proposed to encode the receptor for CGRP. However, several attempts to demonstrate functional expression met with failure *(6)*. Subsequently, an expression cloning effort by McLatchie et al. yielded data suggesting that an alternate protein, receptor-activity modifying protein (RAMP), could affect ligand attachment to CRLR and could alter intracellular signals after CRLR activation *(7)*. The following was the strategy used for this expression cloning effort: (a) genetically engineered human neuroblastoma SK-N-MC cells with good binding of CGRP were generated; (b) the cellular response to bound CGRP was noted as an increase in cytoplasmic cAMP; (c) SK-N-MC cDNA was transcribed in vitro, and pools of complementary RNA were injected into *Xenopus oocytes*, with cRNA encoding the cystic fibrosis transmembrane regulator (CFTR); (d) because CFTR contains a cyclic adenosine monophosphate (cAMP)-activated chloride channel, which was used as a sensitive read-out for indicating the receptors that positively coupled to adenylyl cyclase; and (e) the pool of clones that showed robust response to CGRP was repeatedly subdivided. When this work was completed, McLatchie and colleagues had isolated a single cDNA that encoded RAMP. The study by McLatchie et al., which designated RAMP as a CRLR-interacting protein, was followed-up in other laboratories, and RAMP became the protein credited with facilitating the formation of functional CGRP receptors through its association with CRLR *(8)*. Although this method is not frequently chosen to fish GPCR-associated proteins, it is still a good strategy to search for an interacting protein that might be a functional component of the favored receptor.

2.3. Application of Proteomic Approaches

Recently, a proteomic approach based on peptide affinity chromatography, two-dimensional (2D) electrophoresis, and mass spectrometry was attemtped *(9)*; this combination of methods can be used to identify proteins in a multiprotein complex that is GPCR interactive. Becamel and colleagues examined proteins interacting with the C-terminal tail of 5-HT$_{2C}$ receptors purified from mouse whole-brain extracts by peptide affinity chromatography using the entire C-terminal tail fused to glutathione *S*-transferase (GST), which was immobilized onto glutathione sepharose beads. After the bound proteins were eluted, separated by 2D electrophoresis, and stained with silver, a differential analysis of 2D gel protein patterns from the test sample and from two control samples was performed using image software. Proteins of interest, which were either not detected in the controls or were not distinct from control proteins, were further characterized by matrix-assisted laser desorption/ionization-time of flight mass spectrometry (MALDI-TOF MS; *see* http://www-microbiol.kun.nl/tech/malditof.html) or tandem mass spectrometry after excision from the gel and trypsin digestion. By this proteomic approach, 15 proteins were identified for direct or indirect binding to the 5-HT$_{2C}$ C-terminal; these were synapse-enriched multidomain proteins, some of which contained PSD-95, Dlg, and ZO-1 (PDZ) domains *(10)*. Thus, Becamel et al. delineated a proteomic approach that permits an investigator to globally characterize physiologically relevant protein networks, and this method circumvents known limitations of two-hybrid assays. Protein microarray assays and proteome chip assays are powerful, and, therefore, these methods are used to comprehensively analyze protein–protein interactions. During interaction analyses, a library of immobilized proteins is arrayed on slides, and each slide is probed with fluorescently labeled proteins. For example, a research group recently constructed yeast proteome chips containing 5800 yeast proteins, which were then probed using biotinylated calmodulin (CaM). As follow-up to the CaM probing, a Cy3-labeled streptavidin treatment was used to identify 33 new CaM-interacting proteins *(11)*. Protein microarray assays are ideal for research aiming to detect GPCR-associated proteins in large scale. However, to achieve such a comprehensive analysis, the analyzer might need to label thousands of proteins, which is an arduous task. Although such a process has been simplified by an improved labeling method *(12)*, the assay continues to be viewed as expensive and labor-intensive. Additionally, discrepancies have been observed between protein microarrays and genome mining approaches (such as two-hybrid screening) in studying the complexity of protein interaction networks *(13–15)*. These discrepancies clearly indicate a need to refine and optimize the protein microarray approach.

3. INTERACTING PROTEINS SELECTIVELY TARGET DOMAINS WITHIN A GPCR

Many interacting proteins change cell function by binding to the C-terminus of a GPCR; protein interactions also localize to specific regions of the C-terminus. By direct or indirect binding to C-terminal motifs, the interactive proteins fine-tune GPCR activities. For example, the C-terminal tail of metabotropic glutamate receptor type 7 (mGluR7) has three binding regions *(16)*: proximal, central, and distal.

Proximal C-terminal domains are important for modulating intracellular signals when mGluR7 is stimulated. Agonist stimulation of mGluR7 decreases the formation of cAMP, K^+ channel opening, and voltage-gated Ca^{2+} channel inhibition. The three signals are transduced partly by Gβγ-subunits and partly by CaM. CaM and Gβγ bind to distinct proximal regions of the C-terminus. It is interesting to note that CaM binding promotes the dissociation of Gβγ from mGluR7. The mGluR7 C-terminal signaling system is further regulated by protein kinase C (PKC) *(16)*. In this regard, PKC is capable of phosphorylating a CaM binding domain, which blocks CaM binding. In contrast, CaM binding inhibits the phosphorylation of mGluR7. Collectively, these results indicate that CaM and PKC regulate the activities of mGluR7 by binding to proximal regions of the C-terminal tail.

A more central region of the C-terminal tail on mGluR7 is regarded as an axonal targeting or guiding domain. This region is understood to include residues 883 through 912, which lie between the proximal CaM–Gβγ–PKC recognition sites and a more distally located PDZ domain. Although no interacting proteins have been identified for specific activity centered in this domain, a study performed by Stowell and Craig indicates that this region is functionally linked to axon targeting signals *(17)*.

The distal region of the C-terminus is believed to guide the formation of presynaptic clusters, and PDZ binding motifs are common in distal regions of the C-terminus. Many PDZ domain-containing proteins have been identified as interacting proteins of GPCRs. PDZ is an acronym derived from the names of three proteins in which PDZ coding sequences were originally recognized: PSD-95, Dlg, and ZO-1 proteins. PDZ modules foster protein–protein interactions, and these interactions generate protein layers in scaffold *(18,19)*. Protein interacting with PKC (PICK1) is a single PDZ domain-containing protein originally isolated as a binding protein for PKC-α and is also a substrate for PKC phosphorylation. It was reported that PICK1 interacts with a PDZ binding motif located at the distal region of the C-terminus for mGluR7. Boudin and colleagues demonstrated that the PDZ binding motif critically mediates synaptic aggregation, presumably by interaction with PICK1 *(20)*. Without this structure, PICK1 had no binding affinity at mGluR7, and mGluR7 also failed to cluster in synapses.

3.1. GPCR–PDZ Domain Interactions

In addition to pre- and postsynaptic clustering (as indicated from the study of mGluR7), interactions that follow from PDZ domains cause receptors and ion channels to re-organize into complexes *(21)*. PDZ cassettes also mediate the grouping of signaling components into macromolecular complexes within a microcompartment *(22)*; many interactions between GPCR and PDZ domain-containing proteins depend on PDZ domains *(23)*. The following paragraph provides a few examples of GPCR–PDZ domain interactions.

NHERF is a 55-KDa multidomain protein that contains two tandem PDZ domains at its amino terminus and an ERM-binding domain at its carboxyl terminus *(24)*. EBP50 is a human homolog of rabbit NHERF. It binds to the cytoplasmic tail of β_2-AR through the first PDZ domain and to the cortical actin cytoskeleton through an ERM-binding domain *(25)*. Finer details of the β_2-AR cytoplasmic tail and PDZ domain interaction have also been explained. The specific sequence "D-S/T-x-L" is required for optimal protein–protein interaction, and Serine 411 in the β_2-AR tail is required for both interaction with EBP50 and proper receptor recycling *(25)*.

The importance of PDZ domain-orchestrated protein–protein interactions can also be appreciated by examining details of the interplay between intracellular C-terminal regions of SSTR2 and SSTRIP. The SSTR2–SSTRIP interaction can be viewed as prototypic, because it represents a single example among many multidomain cytoskeletal anchoring protein interactions that enrich the postsynaptic density fractions *(26)*. Postsynaptic density enrichment by protein–protein interaction is important for both maintenance and proper functioning of central nervous system synapses.

3.2. GPCR–Non-PDZ Domain Interactions

Some non-PDZ domain-mediated interactions take place in the intracellular domains of GPCRs. For instance, spinophilin is an 817-amino acid protein enriched in brain tissues; this protein was identified as a D_2R-interacting protein *(5)*. The D_2R–spinophilin protein–protein interaction occurs at the third intracellular loop of receptors (as determined by yeast two-hybrid screening when the receptor's third intracellular loop was used as bait) *(5)*. Spinophilin contains a putative actin-binding domain at the amino terminus, a single PDZ domain, and a region predicted to form a coiled-coil structure at the carboxyl-terminal. Interestingly, the interaction is through a non-PDZ-mediated mechanism. The portion of spinophilin responsible for interacting with the third intracellular loop of the D_2R has been narrowed to a region between its actin-binding domain and its only PDZ domain. Spinophilin can simultaneously interact with both D_2R and a ubiquitously expressed protein phosphatase-1, although these interactions are centered in different binding pockets. Therefore, spinophilin is recognized as an important scaffold or adaptor protein that links receptors to downstream signaling molecules and to cytoskeletal elements, thereby establishing a protein complex that is necessary in dopaminergic neurotransmission. Other GPCR interacting proteins that have been reported to target the intracellular domains include 14-3-3 proteins *(27)* and actin-binding protein-280 *(28)*.

Accessory proteins have been demonstrated to facilitate interactions by binding to both intracellular and C-terminal domains. Contemporary literature reveals that non-PDZ sequences can participate in accessory protein regulation of GPCR signaling *(29,30)*. Non-PDZ sequences have been implicated in partial regulation of chaperone activities, vesicular trafficking, and signal refining after stimulation of a GPCR. Neither inactivation nor afterpotential A (nina A), a 26-KDa integral membrane protein containing a membrane-permeating signal sequence and a single transmembrane domain, and its mammalian homolog RanBP2 *(31)* are the first identified protein chaperones that mediate cell surface expression of two sensory GPCRs—rhodopsin *(32)* and opsin *(31)*. Odr4, which encodes 445 amino acids with a C-terminal transmembrane domain, shares no sequence nor structural similarity to nina A or RanBP2 but interacts with another sensory GPCR, Odr10 (the *Caenorhabditis elegans* odorant receptor), and this interaction localizes the receptor to cilia on olfactory neurons *(33)*.

The association of accessory proteins with GPCRs is not restricted to sensory GPCRs. Thus far, yeast two-hybrid and co-immunoprecipitation studies have demonstrated a stable interaction between angiotensin II type 1 receptor-associated protein (ATRAP) and angiotensin II type 1 receptor (AT_1R) *(34)* as well as associations between filamin A and either D_2R or D_3R *(35)* and between gravin and β_2-AR *(36)*. The mapping studies convincingly show that the C-terminal and/or the third intracellular loop of the receptors are crucial for receptor–protein interaction, whereas the interaction sequences in accessory proteins are much more varied. For example, gravin binds the receptor through β_2-AR C-terminus (Arg329 to Leu413), and the interaction is maintained as the receptor is internalized *(36)*. Filamin A, a ubiquitously expressed actin-crosslinking phosphoprotein, interacts with dopamine receptor through the N-terminal segment on the third intracellular loop of the receptor. Filamin is composed of an N-terminal actin-binding domain, a C-terminal homodimerization domain, and a central rod-like backbone that comprises 23 tandem repeats (each approx 96 amino acids in length) *(37)*. The sequence within repeat 19 of filamin A has been demonstrated to contribute to its association with dopamine receptors *(35)*.

4. FUNCTIONAL PROTEIN NETWORKS: FINE-TUNING OF GPCR ACTIVITY AND MODIFIED LIGAND-BINDING AFFINITY

4.1. Modified Ligand Recognition and Signal Transduction

All receptor proteins, including GPCRs, transduce signals after binding selective ligands. A ligand's affinity for the GPCR is partly derived from its physiochemical properties *(38)*. However, the configuration of corresponding GPCR will also determine affinity *(39)*. Receptor dimerization particularly affects the affinity of a ligand. In fact, once dimerized, receptors can exhibit a novel binding pattern *(40)*. These data (compiled as scientific contemporaries came to understand receptor dimerization) could illustrate the critical activities of interacting proteins and their capacity to modify receptor-binding affinities.

One example involves κ- and δOR, which can assemble into heterodimers. The κ-δOR heterodimer does not significantly bind κ- or δOR-preferring full agonists, nor does it significantly bind κ- or δOR antagonists; however, the κ-δOR heterodimer has strong affinity for partially selective ligands *(41)*. To follow-up the concept that dimerized receptors select among potential ligands, the phenomena of the synergistic binding of a dimerized receptor is also interesting. Gomes et al. reported opioid treatments in cells expressing μ-δOR heterodimers and proved that the dimerization state could influence subsequent affinity measurements *(42)*. In this study, Gomes et al. treated cells isolated for μ-δOR heterodimer expression with δOR-selective ligands at low concentrations. This brought about a significant increase in the cell's binding of μOR agonists. Similarly, treatment with low concentrations of μOR-selective ligands resulted in a significant increase in the binding of δOR agonists.

Several newly identified accessory proteins have been demonstrated to function as regulators of GPCR function *(43)*. Research is warranted to enhance scientific understanding of accessory proteins and accessory protein-provoked modification of GPCR function. For example, it is instructive to consider RAMP isoform-specific modulations of CRLR. CRLR can be co-expressed with RAMP1 to form functional CGRP receptors, and CRLR co-expression with RAMP2 or -3 promoted the formation of a receptor with pharmacological properties of an adrenomedullin receptor *(7,44)*.

From a traditional stance, GPCRs couple only to heterotrimeric G proteins for signal transduction. However, a more recent view of GPCR signaling recognizes that many additional proteins are often required for optimal coupling to downstream effectors. As mentioned earlier, RAMP proteins interact directly with CRLR. They are necessary, but not sufficient, for conveying a full functional CGRP receptor. Another accessory protein, the receptor component protein (RCP), was also found to directly associate with CRLR. Unlike RAMP, RCP does not function as chaperone. Instead, it couples CRLR to cellular signal transduction machinery. Therefore, it is suggested that a functional CGRP receptor is composed of at least three proteins in a complex: the ligand-binding and membrane-spanning protein (CRLR), a chaperone (RAMP), and a coupling protein for signal transduction (RCP) *(45)*.

4.2. Modified Receptor Trafficking, Sorting, and Intracellular Compartmentalization

Small accessory or chaperone proteins also play important roles in promoting receptor delivery to plasma membrane regions of a cell. The CGRP/CRLR system is again instructive. A specific RAMP protein, RAMP1, is critical for terminal glycosylation events in the posttranslational processing of CRLR. A mature and fully glycosylated CRLR is inserted at plasma membranes only if RAMP1 interactions ensue. The D_2R-filamin A interplay is another example that illustrates the criticality of protein–protein interactions for proper cell surface expression. Lin et al. tried to express D_2R in a filamin A-deficient cell line and found that in lieu of surface expression, D_2R was detected mainly within intracellular compartments. In contrast, when a filamin A-reconstituted cell line was used, the D_2R became localized in cell membranes *(35)*.

Another phenomena that is believed to involve protein–protein interactions is cell-compartment—or, more specifically, endoplasmic reticulum (ER)-compartment—retention of imperfect proteins. Although protein–protein interactions enhance the efficiency of protein folding *(46)*, which fosters the insertion of viable cell surface receptors, folding and insertion processes are corruptible. Improperly folded proteins are generally retained within the ER compartment of a cell *(47)*. New data indicate that protein–protein interactions are instrumental to bring about such ER retentions. For example, the chaperone protein calnexin targets improperly folded receptor proteins to the ER compartment. This was demonstrated by work performed by Morello et al., who found a greater interaction between calnexin and an ER-retained R337X mutant receptor than with the wild-type V_2 vasopressin receptor (AVPR2). This demonstrated that calnexin played a role in increasing the ER retention of misfolded GPCRs in addition to its general role in protein folding *(48)*.

Recent work in many laboratories has contributed to our understanding of the process of receptor endocytosis. Once internalized, receptors are processed by one of the following two sorting options *(49,50)*: (a) receptors can be dephosphorylated, recycled, and re-inserted in plasma membrane or (b) receptors can be taken up by cellular lysosomes and degraded. GPCR-associated and interacting proteins may participate, and ultimately control, receptor-sorting pathways *(51)*. Consider, for example, that the fate of an internalized β_2-AR depends on interactions between the accessory phosphoprotein EBP50 and the distal cytoplasmic tail of the β_2-AR. The β_2-AR–EBP50 interplay was elucidated as pivotal after results from a site-directed mutagenesis study were published *(25)*. In this study, the Ser411 residue of native β_2-AR was replaced with an aspartic acid (S411D), thus blocking the β_2-AR–EBP50 interaction and detracting from the efficiency of β_2-AR recycling, which ultimately provoked ligand-induced degradation of the mutant receptor.

Homer proteins, which contain a PDZ-like domain at the N-terminus, are designated as 1a, 1b, or 1c variant proteins. Homer proteins are interesting both because they exist as a part of the postsynaptic density in excitatory brain synapses and because they interact with each other to induce plasma membrane clustering of receptors. To illustrate, Homer-1b has been demonstrated to retain Group I mGluRs in the ER *(52)*. If there is subsequent synaptic activity, then Homer-1a competes with Homer-1b for the receptor in the ER and efficiently chaperones the receptor to the cell surface. Then, once the receptor is inserted in cell membrane, Homer-1c replaces Homer-1a in its interaction with the receptor, forming large clusters of receptors on cell surface *(53,54)*.

5. INTERACTING PROTEINS THAT SELECTIVELY MODULATE µOR SIGNALING PATHWAYS

5.1. PKC Interacting Protein (PKCI)

An interacting protein's fine-tuning of the receptor signal might occur by "dampening" a second messenger pathway. An example of dampened receptor signals exists in the µOR–mPKCI interaction (reported at the 2002 INRC meeting). The mPKCI protein contains 126 amino acids and is a ubiquitous member of the histidine triad (HIT) protein family. The association between mPKCI and HIT proteins is evident in a conserved HIT (His-X-His-X-His, X is a hydrophobic amino acid) motif *(55)*. mPKCI was originally identified as an in vitro inhibitor of PKC isoforms *(59)*. With further investigation, inconsistencies were noted between mPKCI and a well-established PKC inhibitor in vivo *(55)*. In our laboratory, a two-hybrid screening method was used to determine that mPKCI underwent specific interactions with µOR and that this interaction localized to the C-terminus of µOR. These findings were also confirmed by co-immunoprecipitating full-length μOR sequences with mPKCI in CHO cells. The affinity of μOR for opioid ligands and its ability to mediate G protein activation were not changed by μOR–mPKCI interactions, but its ability to inhibit adenylyl cyclase activity was moderately reduced. The association of mPKCI and μOR also induced suppression of receptor desensitization at the adenylyl cyclase level. In these interaction studies, phorbol 12-myristate 3-acetate (PMA)-induced, but not [D-Ala^2, MePhe^4, Glyol^5]enkephalin (DAMGO)-induced, μOR phosphorylation was partly inhibited.

In addition to in vitro analysis of μOR–mPKCI interactions, some behavioral tests were performed. We observed an enhanced morphine-induced analgesia in mPKCI knockout mice and noted a faster development of tolerance to morphine-induced analgesia in knockout mice than in wild-type controls. Therefore, studies from our laboratory revealed some roles for mPKCI in fine-tuning the signaling of the stimulated μOR. The sensitization and phosphorylation states of μOR appeared to depend on μOR–mPKCI interactions in vitro. In experiments that quantified morphine-induced antinociception, our results indicate that μOR–mPKCI interactions have potential to alter pain perceptions in a living organism. Our results in experiments that measured morphine-induced dependence and tolerance also indicate that μOR–mPKCI interactions participate in manifestations of tolerance and dependence to an opiate drug (data unpublished but reported in the 2002 INRC meeting). Taken together, these data indicate that mPKCI is a direct and specific modulator of μOR.

5.2. Phospholipase D2

Koch et al. used the yeast two-hybrid technique to screen a rat cDNA library, and the selected bait was rat μOR C-terminus. Results from this library screening indicated specific interactions between μOR C-terminus and an N-terminal coding sequence for phospholipase D2 (PLD2). This μOR–PLD2 interaction was confirmed by co-immunoprecipitation in an HEK293 cell line with stable expression of μOR and PLD2, and the interaction was shown to be constitutive. The Phox homologous domain in the N-terminus of PLD2 was further determined as an important site for interaction with the C-terminus of μOR *(56)*. PLD2 is a widely distributed phospholipid-specific diesterase that hydrolyzes phosphatidylcholine to phosphatidic acid and choline and is believed to play important roles in cell regulation *(57)*. Functional studies have revealed that the μOR agonist DAMGO activated PLD2 and induced receptor internalization, whereas morphine, which did not induce receptor endocytosis, failed to activate PLD2. DAMGO-mediated PLD2 activation was dependent on adenosine diphosphate-ribosylation factor (ARF) but not PKC.

Heterologous stimulation of PLD2 accelerated μOR internalization, whereas inhibition of PLD2 prevented agonist-mediated receptor endocytosis. Taken together, the findings of Koch et al. indicate that ARF-dependent PLD2 activation is required for agonist-induced μOR endocytosis *(56)*.

6. CONCLUSIONS

Research to discern the functional relevance of GPCR-associated protein interaction is progressing rapidly. The picture that emerges from this research is that interacting proteins are structurally and functionally diverse, "partnering" easily with alternate interacting proteins. This diversity may help individual GPCRs to form a physically and functionally distinct unit that is important for carrying specificity and selectivity along unique signaling pathways *(58)*. Although specific GPCR-interacting proteins have not yet been designated as promising drug targets, the potential to use these interacting proteins as novel drug targets and to develop cell type-specific disease intervention should not be underestimated.

REFERENCES

1. Ma J, Ptashne M. Deletion analysis of GAL4 defines two transcriptional activating segments. Cell 1987; 48:847–853.
2. Kuner R, Kohr G, Grunewald S, Eisenhardt G, Bach A, Kornau HC. Role of heteromer formation in GABAB receptor function. Science 1999; 283:74–77.
3. Hall RA, Premont RT, Chow CW, et al. The beta2-adrenergic receptor interacts with the Na^+/H^+-exchanger regulatory factor to control Na^+/H^+ exchange. Nature 1998; 392:626–630.
4. Zitzer H, Honck HH, Bachner D, Richter D, Kreienkamp HJ. Somatostatin receptor interacting protein defines a novel family of multidomain proteins present in human and rodent brain. J Biol Chem 1999; 274:32,997–33,001.
5. Smith FD, Oxford GS, Milgram SL. Association of the D2 dopamine receptor third cytoplasmic loop with spinophilin, a protein phosphatase-1-interacting protein. J Biol Chem 1999; 274:19,894–19,900.
6. Fluhmann B, Muff R, Hunziker W, Fischer JA, Born W. A human orphan calcitonin receptor-like structure. Biochem Biophys Res Commun 1995; 206:341–347.
7. McLatchie LM, Fraser NJ, Main MJ, et al. RAMPs regulate the transport and ligand specificity of the calcitonin-receptor-like receptor. Nature 1998; 393:333–339.
8. Christopoulos G, Perry KJ, Morfis M, et al. Multiple amylin receptors arise from receptor activity-modifying protein interaction with the calcitonin receptor gene product. Mol Pharmacol 1999; 56:235–242.
9. Becamel C, Galeotti N, Poncet J, et al. A proteomic approach based on peptide affinity chromatography, 2-dimensional electrophoresis and mass spectrom-

etry to identify multiprotein complexes interacting with membrane-bound receptors. Biol Proced Online 2002; 4:94–104.
10. Becamel C, Alonso G, Galeotti N, et al. Synaptic multiprotein complexes associated with 5-HT$_{2C}$ receptors: a proteomic approach. EMBO J 2002; 21:2332–2342.
11. Zhu H, Bilgin M, Bangham R, et al. Global analysis of protein activities using proteome chips. Science 2001; 293:2101–2105.
12. Kawahashi Y, Doi N, Takashima H, et al. In vitro protein microarrays for detecting protein-protein interactions: application of a new method for fluorescence labeling of proteins. Proteomics 2003; 3:1236–1243.
13. Marcotte EM, Pellegrini M, Ng HL, Rice DW, Yeates TO, Eisenberg D. Detecting protein function and protein-protein interactions from genome sequences. Science 1999; 285:751–753.
14. Uetz P, Giot L, Cagney G, et al. A comprehensive analysis of protein–protein interactions in *Saccharomyces cerevisiae*. Nature 2000; 403:623–627.
15. Ito T, Chiba T, Ozawa R, Yoshida M, Hattori M, Sakaki Y. A comprehensive two-hybrid analysis to explore the yeast protein interactome. Proc Natl Acad Sci USA 2001; 98:4569–4574.
16. Dev KK, Nakanishi S, Henley JM. Regulation of mglu(7) receptors by proteins that interact with the intracellular C-terminus. Trends Pharmacol Sci 2001; 22:355–361.
17. Stowell JN, Craig AM. Axon/dendrite targeting of metabotropic glutamate receptors by their cytoplasmic carboxy-terminal domains. Neuron 1999; 22:525–536.
18. Zhang M, Wang W. Organization of signaling complexes by PDZ-domain scaffold proteins. Acc Chem Res 2003; 36:530–538.
19. Garner CC, Nash J, Huganir RL. PDZ domains in synapse assembly and signalling. Trends Cell Biol 2000; 10:274–280.
20. Boudin H, Doan A, Xia J, et al. Presynaptic clustering of mGluR7a requires the PICK1 PDZ domain binding site. Neuron 2000; 28:485–497.
21. Zhang Q, Fan JS, Zhang M. Interdomain chaperoning between PSD-95, Dlg, and Zo-1 (PDZ) domains of glutamate receptor-interacting proteins. J Biol Chem 2001; 276:43,216–43,220.
22. Fanning AS, Anderson JM. Protein modules as organizers of membrane structure. Curr Opin Cell Biol 1999; 11:432–439.
23. Xu XZ, Choudhury A, Li X, Montell C. Coordination of an array of signaling proteins through homo- and heteromeric interactions between PDZ domains and target proteins. J Cell Biol 1998; 142:545–555.
24. Shenolikar S, Weinman EJ. NHERF: targeting and trafficking membrane proteins. Am J Physiol Renal Physiol 2001; 280:F389–F395.
25. Cao TT, Deacon HW, Reczek D, Bretscher A, von Zastrow M. A kinase-regulated PDZ-domain interaction controls endocytic sorting of the beta2-adrenergic receptor. Nature 1999; 401:286–290.
26. Kreienkamp HJ, Zitzer H, Richter D. Identification of proteins interacting with the rat somatostatin receptor subtype 2. J Physiol Paris 2000; 94:193–198.
27. Prezeau L, Richman JG, Edwards SW, Limbird LE. The zeta isoform of 14-3-

3 proteins interacts with the third intracellular loop of different alpha$_2$-adrenergic receptor subtypes. J Biol Chem 1999; 274:13,462–13,469.
28. Li M, Bermak JC, Wang ZW, Zhou QY. Modulation of dopamine D(2) receptor signaling by actin-binding protein (ABP-280). Mol Pharmacol 2000; 57:446–452.
29. Ishii M, Kurachi Y. Physiological actions of regulators of G-protein signaling (RGS) proteins. Life Sci 2003; 74:163–171.
30. Kovoor A, Chen CK, He W, Wensel TG, Simon MI, Lester HA. Co-expression of Gbeta5 enhances the function of two Ggamma subunit-like domain-containing regulators of G protein signaling proteins. J Biol Chem 2000; 275:3397–3402.
31. Ferreira PA, Nakayama TA, Pak WL, Travis GH. Cyclophilin-related protein RanBP2 acts as chaperone for red/green opsin. Nature 1996; 383:637–640.
32. Schneuwly S, Shortridge RD, Larrivee DC, Ono T, Ozaki M, Pak WL. Drosophila ninaA gene encodes an eye-specific cyclophilin (cyclosporine A binding protein). Proc Natl Acad Sci USA 1989; 86:5390–5394.
33. Dwyer ND, Troemel ER, Sengupta P, Bargmann CI. Odorant receptor localization to olfactory cilia is mediated by ODR-4, a novel membrane-associated protein. Cell 1998; 93:455–466.
34. Daviet L, Lehtonen JY, Tamura K, Griese DP, Horiuchi M, Dzau VJ. Cloning and characterization of ATRAP, a novel protein that interacts with the angiotensin II type 1 receptor. J Biol Chem 1999; 274:17,058–17,062.
35. Lin R, Karpa K, Kabbani N, Goldman-Rakic P, Levenson R. Dopamine D2 and D3 receptors are linked to the actin cytoskeleton via interaction with filamin A. Proc Natl Acad Sci USA 2001; 98:5258–5263.
36. Fan G, Shumay E, Wang H, Malbon CC. The scaffold protein gravin (cAMP-dependent protein kinase-anchoring protein 250) binds the beta2-adrenergic receptor via the receptor cytoplasmic Arg-329 to Leu-413 domain and provides a mobile scaffold during desensitization. J Biol Chem 2001; 276:24,005–24,014.
37. Gorlin JB, Yamin R, Egan S, et al. Human endothelial actin-binding protein (ABP-280, nonmuscle filamin): a molecular leaf spring. J Cell Biol 1990; 111:1089–1105.
38. Zartler ER, Yan J, Mo H, Kline AD, Shapiro MJ. ID NMR Methods in ligand-receptor interactions. Curr Top Med Chem 2003; 3:25–37.
39. Hulme EC. Muscarinic acetylcholine receptors: typical G-coupled receptors. Symp Soc Exp Biol 1990; 44:39–54.
40. Baneres JL, Parello J. Structure-based analysis of GPCR function: evidence for a novel pentameric assembly between the dimeric leukotriene B4 receptor BLT1 and the G-protein. J Mol Biol 2003; 329:815–829.
41. Jordan BA, Devi LA. G-protein-coupled receptor heterodimerization modulates receptor function. Nature 1999; 399:697–700.
42. Gomes I, Jordan BA, Gupta A, Trapaidze N, Nagy V, Devi LA. Heterodimerization of mu and delta opioid receptors: A role in opiate synergy. J Neurosci 2000; 20:RC110.
43. Brown D, Breton S. Sorting proteins to their target membranes. Kidney Int 2000; 57:816–824.
44. Fraser NJ, Wise A, Brown J, McLatchie LM, Main MJ, Foord SM. The amino terminus of receptor activity modifying proteins is a critical determinant of

glycosylation state and ligand binding of calcitonin receptor-like receptor. Mol Pharmacol 1999; 55:1054–1059.
45. Evans BN, Rosenblatt MI, Mnayer LO, Oliver KR, Dickerson IM. CGRP-RCP, a novel protein required for signal transduction at calcitonin gene-related peptide and adrenomedullin receptors. J Biol Chem 2000; 275:31,438–31,443.
46. Haynes RL, Zheng T, Nicchitta CV. Structure and folding of nascent polypeptide chains during protein translocation in the endoplasmic reticulum. J Biol Chem 1997; 272:17,126–17,133.
47. Trombetta ES, Parodi AJ. Quality control and protein folding in the secretory pathway. Annu Rev Cell Dev Biol 2003; 19:649–676.
48. Morello JP, Salahpour A, Petaja-Repo UE, et al. Association of calnexin with wild type and mutant AVPR2 that causes nephrogenic diabetes insipidus. Biochemistry 2001; 40:6766–6775.
49. Rosenfeld JL, Knoll BJ, Moore RH. Regulation of G-protein-coupled receptor activity by rab GTPases. Receptors Channels 2002; 8:87–97.
50. Grimes ML, Miettinen HM. Receptor tyrosine kinase and G-protein coupled receptor signaling and sorting within endosomes. J Neurochem 2003; 84:905–918.
51. Shenoy SK, Lefkowitz RJ. Trafficking patterns of beta-arrestin and G protein-coupled receptors determined by the kinetics of beta-arrestin deubiquitination. J Biol Chem 2003; 278:14,498–14,506.
52. Roche KW, Tu JC, Petralia RS, Xiao B, Wenthold RJ, Worley PF. Homer 1b regulates the trafficking of group I metabotropic glutamate receptors. J Biol Chem 1999; 274:25,953–25,957.
53. Ciruela F, Soloviev MM, Chan WY, McIlhinney RA. Homer-1c/Vesl-1L modulates the cell surface targeting of metabotropic glutamate receptor type 1alpha: evidence for an anchoring function. Mol Cell Neurosci 2000; 15:36–50.
54. Ciruela F, Soloviev MM, McIlhinney RA. Co-expression of metabotropic glutamate receptor type 1alpha with homer-1a/Vesl-1S increases the cell surface expression of the receptor. Biochem J 1999; 341:795–803.
55. Klein MG, Yao Y, Slosberg ED, Lima CD, Doki Y, Weinstein IB. Characterization of PKCI and comparative studies with FHIT, related members of the HIT protein family. Exp Cell Res 1998; 244:26–32.
56. Koch T, Brandenburg LO, Schulz S, Liang Y, Klein J, Hollt V. ADP-ribosylation factor-dependent phospholipase D2 activation is required for agonist-induced mu-opioid receptor endocytosis. J Biol Chem 2003; 278:9979–9985.
57. Morris AJ, Frohman MA, Engebrecht J. Measurement of phospholipase D activity. Anal Biochem 1997; 252:1–9.
58. Bockaert J, Marin P, Dumuis A, Fagni L. The 'magic tail' of G protein-coupled receptors: an anchorage for functional protein networks. FEBS Lett 2003; 546:65–72.

III
GPCR Dimerization/Oligomerization

10
Biophysical and Biochemical Methods to Study GPCR Oligomerization

Karen M. Kroeger, Kevin D. G. Pfleger, and Karin A. Eidne

1. INTRODUCTION

Traditionally, G protein-coupled receptors (GPCRs) were believed to exist and function as single monomeric entities that interacted with only G proteins to produce an intracellular signal. However, intensive research in the field now clearly indicates that receptors exist in a multiprotein complex, interacting with other GPCRs and intracellular regulatory proteins to form homo- or hetero-oligomeric signaling units (reviewed in refs. *1–5*). The existence of direct receptor–receptor interactions adds an additional level of complexity to the regulation of GPCR function in cells co-expressing various GPCRs. Furthermore, the discovery that GPCRs can interact to form hetero-oligomeric complexes, often with novel pharmacological and functional properties, has shed much light on the previously unexplained behavior of many agonists in vivo and on the mechanisms by which different pathways and receptor systems can intersect and crossreact to produce an integrated signal and cellular response.

Investigations into GPCR oligomerization have involved several different approaches. Early studies used such techniques as receptor complementation to provide indirect evidence for oligomerization. However, the notion of GPCR oligomerization was not generally accepted until a large body of evidence accumulated using biochemical methods and, more recently, biophysical methods to demonstrate and monitor receptor–receptor interactions. This chapter discusses techniques that have been applied to the study of GPCR oligomerization, focusing both on the more conventional biochemical approaches and the newer biophysical approaches and the mechanisms by which they have been used to demonstrate the existence of GPCR homo- and hetero-oligomers.

2. HISTORICAL PERSPECTIVE OF GPCR OLIGOMERIZATION

2.1. Earlier Studies

Even prior to the understanding of the genetic structure of GPCRs, there was evidence that pointed toward the phenomenon of oligomerization within this receptor superfamily. However, the significance of many of these studies was not fully appreciated until much later, when more direct evidence for GPCR oligomerization emerged. In retrospect, early studies employing radioligand binding, radiation inactivation, and receptor crosslinking—all reported data that could be interpreted or explained by the presence of receptor homo- or hetero-oligomers.

Many groups have observed complex binding of combinations of agonists and antagonists to receptors with positive and negative cooperativity that, with hindsight, could be explained by the presence of more than one ligand binding site resulting from the formation of a dimeric or oligomeric receptor complex *(6–8)*. Crosslinking experiments using crosslinked agonists, antibodies, and cell-surface crosslinking reagents also provided early evidence that GPCRs could function as dimers *(9–13)*.

Radiation inactivation is a technique based on the inverse relationship between the size of a macromolecule and the dose-dependent inactivation of that molecule by ionizing radiation. It has been used to demonstrate that the functional receptor has a mass higher than that predicted from the monomeric structure for several GPCRs, including the α_2-adrenergic receptor (AR; ref. *14*), D2 dopamine receptor *(15)*, and the gonadotropin-releasing hormone (GnRH) receptor *(16)*. However, it was unclear whether these higher-molecular-weight complexes represented receptor–receptor complexes or merely receptor–protein complexes.

2.2. Trans-Complementation Studies

Although earlier pharmacological and biochemical studies suggested that GPCRs function as oligomers, mechanisms of receptor function were still modeled on a monomeric unit. *Trans*-complementation studies provided a resurgence in interest in the concept of oligomerization by reporting functional *trans*-complementation upon co-expression of various chimeric and/or mutant receptor constructs. In initial *trans*-complementation studies, two nonfunctional chimeric α_2-ARs/m3 muscarinic receptors (each containing transmembranes (TMs) 1 to 5 of one receptor and TMs 6 to 7 of the other) were nonfunctional when expressed alone. However, upon co-expression of the two chimeras, binding of both adrenergic and muscarinic ligands was restored, providing evidence for dimerization *(17)*. A similar *trans*-complementation was also observed for two nonfunctional mutant angiotensin type II AT1 receptors (AT1R) *(18)* and truncated D2 and D3 dopamine receptors *(19)*. The dominant-negative and dominant-positive effects observed for certain mutants on wild-type receptor function also provides evidence for direct receptor–receptor interactions *(20–24)*.

3. BIOCHEMICAL METHODS TO STUDY GPCR OLIGOMERIZATION

Because of the cloning of many GPCR complementary DNAs (cDNAs) and the availability of antibodies toward several GPCRs and epitope tags, it has been possible for biochemical studies to be conducted, providing strong evidence for GPCR oligomerization. The observation (using immunoprecipitation and Western blotting) of molecular species corresponding to twice (or more) the molecular weight of the receptor provided support for the notion that GPCRs could form dimers or oligomers. Co-immunoprecipitation of differentially epitope-tagged receptors provided much stronger evidence for the existence of GPCR homodimers and has also been used to demonstrate the occurrence of heterodimers (reviewed in refs. *1–3,5*) (Table 1).

3.1. Detection of Higher Order Receptor Complexes by Immunoprecipitation and Co-Immunoprecipitation

To detect receptor–receptor interactions biochemically, tissues or cells endogenously expressing the receptor(s) or, more typically, cells heterologously expressing epitope-tagged receptors are employed. Following cell lysis and the solubilization of membranes, receptors are immunoprecipitated using receptor- or epitope-specific antibodies. Immunoprecipitates are then analyzed by sodium dodecyl sulfate-polyacrylamide gel electrophoresis (SDS-PAGE) and detected by Western blotting.

Immunoprecipitation has allowed the detection of higher order complexes of several GPCRs, including the thyroid-stimulating hormone (TSH; ref. *25*) and D2 dopamine receptor *(26)*, suggesting the existence of homodimers and oligomers. Furthermore, studies using native tissues have strengthened the concept that homodimers and oligomers occur in vivo and do not simply result from heterologous overexpression of receptor. Receptor-specific antibodies detected higher order complexes of D3 dopamine receptors in immunoprecipitates from monkey and rat brain *(27)*. Similarly, higher order complexes have been detected in immunoprecipitates prepared from native tissues for the calcium-sensing receptor *(28)*, adenosine A_1 receptor *(29)*,

Table 1
Summary of Studies Applying Biochemical and Biophysical Techniques to Investigate Homo- and Hetero-Oligomerization Within the GPCR Superfamily

GPCR	Technique	References
Homo-oligomerization class A		
α_{1a}-AR	Immunoprecipitation, FRET	60
α_{1b}-AR	Immunoprecipitation, FRET	60
β_1-AR	BRET	82,96
β_2-AR	Immunoprecipitation	30
	BRET	11,82,96
TRH receptor 1	BRET	71,73
	Immunoprecipitation	32
TRH receptor 2	BRET	73
GnRH receptor	pbFRET	61,90
	BRET	71
D2 dopamine receptor	Immunoprecipitation	97
	FRET	56
D1 dopamine receptor	Immunoprecipitation	98
δ-opioid receptor	Immunoprecipitation	31,40,99
	BRET, TR-FRET	63
κ-opioid receptor	Immunoprecipitation	39
μ-opioid receptor	Immunoprecipitation	40
	BRET	99
SSTR 5	Immunoprecipitation, FRET	65
Cholecystokinin type A receptor	BRET	72
LH receptor	pbFRET	100
α-Factor receptor	FRET	57
CXCR2	Immunoprecipitation	34
CXCR4	BRET	78
Chemokine receptor CCR5	Immunoprecipitation	101
	BRET	91
Chemokine receptor CCR2	Immunoprecipitation	10
Complement C5A receptor	FRET	59
Melatonin M1 receptor	Immunoprecipitation, BRET	86
Melatonin M2 receptor	Immunoprecipitation, BRET	86
TSH receptor	FRET	62
Neuropeptide Y Y1 receptor	FRET	58
Adenosine A_1 receptor	BRET	50
Adenosine A_2 receptor	BRET	81
Oxytocin receptor	BRET	74
V1a vasopressin receptor	BRET	74
V2 vasopressin receptor	BRET	74
Neuropeptide Y Y4 receptor	Immunoprecipitation, BRET	80
Class C		
Calcium-sensing receptor	Immunoprecipitation	9
	BRET	79

(continued)

mGluR5	Immunoprecipitation	*33*
mGluR1	Immunoprecipitation	*52*
Hetero-oligomerization class A and class A		
SSTRs 1 and 5	pbFRET	*65*
SSTRs 2A and 3	Immunoprecipitation	*42*
SSTR2A and μ-opioid receptors	Immunoprecipitation	*102*
D2 dopamine receptors and SSTR5	pbFRET	*66*
Adenosine A_1 and D1 dopamine receptors	Immunoprecipitation	*46*
Adenosine A_{2A} and D2 dopamine receptors	Immunoprecipitation	*103*
	BRET	*75,81*
AT1A and AT2 receptors	Immunoprecipitation	*43*
AT1A receptors and B_2Rs	Immunoprecipitation	*48*
κ- and δ-opioid receptors	Immunoprecipitation	*39*
	BRET	*99*
μ- and δ-opioid receptors	Immunoprecipitation	*40,41*
μ-opioid and substance P (NK1) receptor	Immunoprecipitation, BRET	*47*
CCR2 and CCR5	Immunoprecipitation, FRET	*104*
CCKA and CCKB receptors	Immunoprecipitation, BRET	*76*
TRHR1 and TRHR2	BRET	*73*
5-HT_{1B} and 5-HT_{1D} receptors	Immunoprecipitation	*105*
Melatonin M1 and M2 receptors	Immunoprecipitation, BRET	*86*
$β_1$- and $β_2$-ARs	Immunoprecipitation, BRET	*82,96*
$β_2$-ARs and angiotensin AT1Rs	Immunoprecipitation	*106*
Adenosine A_1 and P_2Y_1 receptors	Immunoprecipitation, BRET	*50*
$β_2$-AR and δ-opioid receptors	BRET, TR-FRET	*63*
	Immunoprecipitation	*44*
$β_2$-ARs and κ-opioid receptors	Immunoprecipitation	*44*
$α_{2a}$-adrenergic and $β_1$-ARs	Immunoprecipitation	*107*
$α_{1a}$-adrenergic and $α_{1b}$-ARs	FRET	*60*
Oxytocin and V1a vasopressin receptors	BRET	*74*
Oxytocin and V2 vasopressin receptors	BRET	*74*
V1a and V2 vasopressin receptors	BRET	*74*
Class C and class C		
$GABA_BR1$ and $GABA_BR2$	Immunoprecipitation	*36–38*
Calcium-sensing receptor and mGluR1	Immunoprecipitation	*51*
Calcium-sensing receptor and mGluR5	Immunoprecipitation	*51*
Class A and class C		
5HT1A and $GABA_BR2$	Immunoprecipitation	*94*
Adenosine A_1 receptor and mGluR1	Immunoprecipitation	*108*
Adenosine A_{2A} receptor and mGluR5	Immunoprecipitation	*45*

Abbreviations: GPCR, G protein-coupled receptor; FRET, fluorescence resonance energy transfer; BRET, bioluminescence resonance energy transfer; AR, adrenergic receptor; mGluR, metabotropic glutamate receptor; GABA, γ-aminobutyric acid; B_2R, bradykinin 2 receptor; TRH, thyrotropin-releasing hormone; GnRH, gonadotropin-releasing hormone; TR-FRET, time-resolved FRET; pbFRET, photobleaching FRET; SSTR, somatostatin receptor.

and the chemokine CCR5 receptor *(29)*. However, whether these higher-molecular-weight complexes represent receptor dimers or oligomers or merely represent receptor–protein complexes often is not addressed.

Through the heterologous expression of differentially epitope-tagged receptors, co-immunoprecipitation has been used to detect homo- and hetero-oligomerization of many GPCRs; a signal is obtained on Western blot only if there is an association between the two tagged receptors. For example, the immunoprecipitation of *myc*-tagged β_2-AR and the subsequent immunoblotting of co-expressed HA-tagged β_2-AR demonstrated the co-immunoprecipitation of the differentially tagged receptors and indicated homo-oligomerization of the β_2-AR *(30)*. Similar studies have demonstrated the homo-oligomerization of many GPCRs, including the δ-opioid receptor *(31)*, thyrotropin-releasing hormone receptor (TRHR) *(32)*, metabotropic glutamate receptor 5 (mGluR5) *(33)*, calcium-sensing receptor *(9)*, CXCR2 receptor *(34)*, and lutropin receptor *(35)*. More recently, co-immunoprecipitation experiments have been used to demonstrate hetero-oligomerization between closely related receptors, such as γ-aminobutyric acid (GABA)$_B$R1 and GABA$_B$R2 *(36–38)*, opioid receptor subtypes δ and κ *(39)* as well as δ and μ *(40,41)*, somatostatin receptor (SSTR) subtypes SSTR2A and SSTR3 *(42)*, and AT1A and AT2 receptors *(43)*. Hetero-oligomerization between more distantly related receptors, such as the β_2-ARs and δ-opioid receptors *(44)*, adenosine A_{2A} and mGluR5 receptors *(45)*, adenosine A_1 and D1 dopamine receptors *(46)*, and μ-opioid and substance P (neurokinin [NK]1) receptors *(47)*, has also been observed using co-immunoprecipitation techniques.

Ideally, studies should investigate receptor hetero-oligomers in cell lines or tissues in which they are endogenously expressed. Hetero-oligomeric receptor complexes of endogenously expressed receptors have been detected for the AT1R and bradykinin 2 receptors (B$_2$R) in rat smooth muscle cells *(48)* and human platelets and omental vessels *(49)*. Direct associations between the adenosine A_1 receptor and P$_2$ adenosine triphosphate (ATP) purinoceptor (P$_2$Y$_1$) *(50)* as well as between the mGluR1 and calcium-sensing receptor *(51)* have also been demonstrated in co-immunoprecipitation experiments using membrane extracts from rat brain. This provides evidence that dimerization is not simply an artifact of receptor overexpression in heterologous expression systems. However, demonstrating heterodimerization in native tissues is often problematic because of the low level of endogenous receptor expression observed for certain GPCRs and the lack of receptor-specific antibodies.

3.2. Concerns and Controls

Although immunoprecipitation is commonly used to study GPCR oligomerization, it can be problematic because a solubilization step with detergents is usually required, which, as a result of the highly hydrophobic nature of GPCRs, can lead to artifactual aggregation of receptor. However, many studies validate immunoprecipitation data by demonstrating that complexes are obtained only upon co-expression of receptors within the same cell and not merely upon mixing of cells individually expressing each receptor. Furthermore, the use of an additional GPCR to demonstrate the specificity of the receptor–receptor interaction represents another potentially valuable control for immunoprecipitation experiments. The finding that whole-cell crosslinking prior to solubilization leads to an increase in the proportion of dimers formed for several receptors (suggesting that preformed complexes are present at the cell surface) further supports that dimers or oligomers do not merely result from aggregation *(9,30,31,33)*.

Various immunoprecipitation conditions, including the use of different combinations of detergents, have also been used to address the concerns regarding artificial aggregation of GPCRs. It is possible that the detergent used could induce receptor aggregation. However, a study investigating opioid receptor heterodimerization detected dimers under a wide range of conditions *(39)*. Reducing agents have also been used to address the concerns regarding the potential nonspecific aggregation of GPCRs during the extraction and immunoprecipitation process. Dimers were detected on SDS-PAGE for the calcium-sensing receptor *(9)* and mGluR1*(52)* in nonreducing conditions, but in reducing conditions (presence of dithiothreitol [DTT]), only monomers were observed, suggesting that dimers/oligomers were not just a result of artifactual aggregation but, rather, resulted from covalent (disulfide) intermolecular interactions. Capping agents that carboxy-methylate free sulfhydryl groups on cysteine residues are often used in conjunction with reducing agents to reduce the chance of artifactual receptor associations occurring as a result of nonspecific disulfide bond formation during solubilization and immunoprecipitation procedures *(39,53)*. Therefore, provided that the appropriate controls are performed, immunoprecipitation can represent a valid technique for detecting GPCR oligomerization.

4. BIOPHYSICAL METHODS TO STUDY GPCR OLIGOMERIZATION

Biophysical techniques, which overcome many of the problems associated with the use of biochemical methods, have recently been used to provide strong evidence for the existence of GPCR oligomerization in intact cells (reviewed in refs. *2,3,5,54,* and *55*) (Table 1). Fluorescence resonance energy transfer (FRET) and the newly emerging derivative, bioluminescence resonance energy transfer (BRET), have the distinct advantage of being able to monitor receptor–receptor interactions in live cells in real time.

4.1. Fluorescence Resonance Energy Transfer

4.1.1. Principle of FRET

FRET is a strictly distance-dependent energy transfer technique that has been used for several years to detect and monitor protein–protein interactions in live cells, both temporally and spatially. It involves the transfer of energy from a fluorescent energy donor, following its excitation, to a fluorescent energy acceptor when in close enough proximity (<100 Å). Measuring this transfer of energy relies on a sufficient overlap between the emission spectrum of the donor and the excitation spectrum of the acceptor to allow energy transfer. Ideally, there should be little overlap between the excitation spectra of the donor and acceptor molecules to minimize the degree of direct excitation of the acceptor fluorophore as well as sufficient separation between the emission spectra of the donor and acceptor molecules to allow measurement of acceptor fluorescence without contaminating bleedthrough from donor fluorescence.

4.1.2. Fluorescent Proteins and Dyes

The fluorescent donor and acceptor molecules can be either fluorescent proteins genetically added to the proteins of interest to produce fusion proteins or fluorescent dyes, which are used to label the proteins in combination with receptor- or epitope-specific antibodies.

Green fluorescent protein (GFP) and its color variants have a wide range of spectral properties and thus have been used as FRET pairs to detect many GPCR oligomers. Cyan fluorescent protein (CFP) and yellow fluorescent protein (YFP), have been commonly used as donor and acceptor fluorescent molecules, respectively, to detect GPCR interactions using FRET. The homo-oligomerization of the D2 dopamine receptors *(56)*, the yeast α-factor receptor (Ste2)*(57)*, the neuropeptide Y receptors (Y1, Y2, and Y5), *(58)* and the complement C5A receptors *(59)* all have been demonstrated using CFP and YFP receptor fusion in FRET-based assays (Fig. 1A). Additionally, CFP and GFP *(60)*, GFP and red fluorescent protein (RFP) *(58,61)*, and YFP and RFP *(62)* have been used as donor and acceptor pairs in FRET.

Fluorescent dyes conjugated to receptor- and/or epitope-specific antibodies have also been used to demonstrate homo-oligomerization of heterologously expressed δ-opioid receptors *(63)*, D2 dopamine receptors *(64)*, and SSTR5 *(65)* as well as hetero-oligomerization of SSTR5 and D2 dopamine receptor *(66)* (Fig. 1B). Using fluorescent dyes conjugated to receptor-specific antibodies, it should be possible to detect FRET between endogenous receptors and image these interactions in a single cell, as has been done for other membrane receptors, including the epidermal growth factor receptor *(67)*. An increase in the availability of receptor-specific antibodies for GPCRs would make it more feasible to monitor endogenous receptor–receptor interactions, thus providing strong evidence for the existence of GPCR oligomers in vivo.

4.1.3. Time-Resolved FRET

The simplest FRET-based approaches to monitor GPCR oligomers involve measuring the ratio of acceptor versus donor emission (or intensity). As a result of FRET, donor emission decreases and acceptor emission increases, and the acceptor/donor fluorescence intensity ratio in a cell co-expressing both fusion proteins increases compared to the ratio obtained from cells expressing only the donor-tagged protein. The homo-oligomerization of the yeast α-factor receptor (Fig. 1A) *(57,68)* and the complement C5A receptor *(59)* as well as the hetero-oligomerization of the α_{1a}- and α_{1b}-ARs *(60)* have all been demonstrated by employing scanning spectrofluorometry on cell suspensions to measure donor and acceptor emissions to detect FRET. Alternatively, donor and acceptor emissions have been measured and FRET has been imaged in single cells for the TSH receptor *(62,69)* and neuropeptide Y homodimers *(58)*.

Ratiometric, intensity-based FRET approaches are at a disadvantage because they require a donor that only expresses cell samples with expression levels similar to those in the donor-plus-acceptor cell population. Furthermore, in a typical FRET experiment, the emitted fluorescent light is a combination of donor emission, acceptor emission caused by direct excitation, and acceptor emission caused by FRET. Therefore, to calculate the FRET efficiency, the proportion of emission resulting from contaminating fluorescence from the donor and direct excitation of the acceptor fluorophore must be accounted for (Fig. 1A). Often, this correction can result in a very low actual increase in acceptor emission, making this FRET-based technique relatively insensitive for detecting weak or low-level protein interactions.

Time-resolved FRET (TR-FRET) can be performed to reduce the background fluorescence resulting from cellular autofluorescence and to improve the signal/noise ratio. Fluorophores with long-lived fluorescence are used in TR-FRET, enabling delayed FRET measurements to be taken after the back-

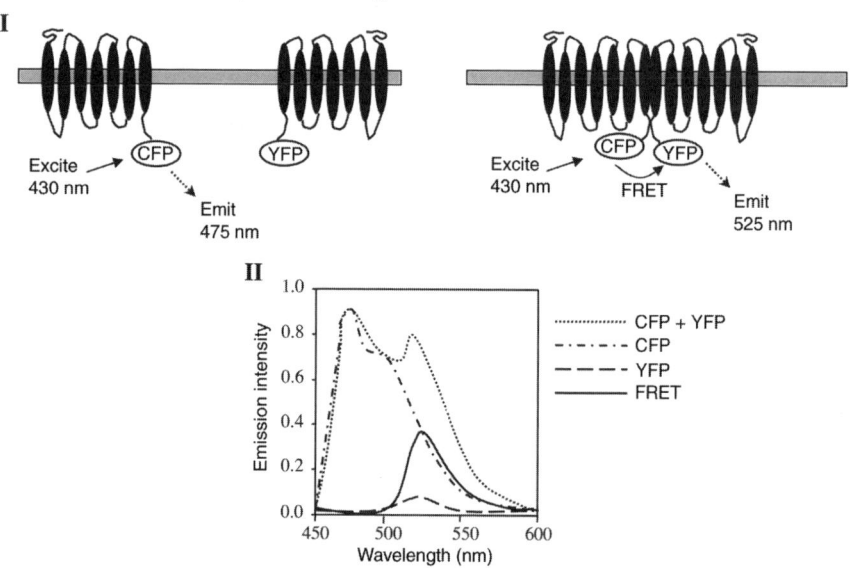
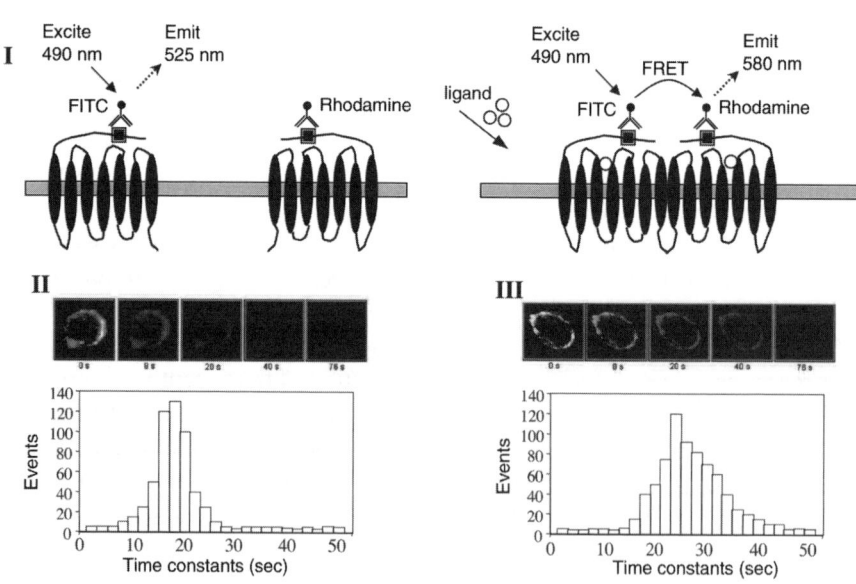

Fig. 1

ground fluorescence has decayed. Donor fluorophore europium^{3+} has been used with acceptor fluorophore allophycocyanin to detect δ-opioid receptor *(63)* and D2 dopamine receptor *(64)* homo-oligomers.

4.1.4. Photobleaching FRET

Photobleaching FRET (pbFRET) is an alternative to intensity-based FRET measurements. Unlike intensity-based or ratiometric FRET assays (which measure the ratio of acceptor vs donor emission), pbFRET measures the rate of fluorescence decay of the donor fluorophore. Fluorescent proteins naturally decay when continually excited, but as a result of FRET, the rate of photobleaching decay of the donor fluorophore is reduced. This derivative of FRET has the advantage of being independent of donor and/or acceptor fluorophore concentrations. However, it has the disadvantage that repeated measurements cannot be performed on the same sample, because the fluorophores are destroyed by continual exposure to the excitation light source. Homo-oligomerization of the SSTR5 (Fig. 1B) *(65)* and hetero-oli-

---

Fig. 1. The application of FRET to study GPCR oligomerization. (A) Monitoring yeast α-factor homo-oligomerization in vivo using FRET. (I) Schematic diagram illustrating the application of FRET to detect oligomerization. Yeast cells co-expressing CFP-tagged (donor) and YFP-tagged (acceptor) receptors are irradiated at 425 nm, and in the absence of an interaction, light is emitted at 475 nm. If a receptor–receptor interaction occurs, energy is transferred via FRET to YFP and re-emitted at 525 nm. (II) Schematic diagram demonstrating how FRET is detected by scanning spectroscopy with fluorescence emission recorded between 450 and 600 nm. The emission from YFP caused by FRET is calculated as (CFP + YFP emission) − (CFP only emission) − (YFP only emission). (B) Use of pbFRET to detect SSTR5 ligand-induced homo-oligomerization. *(I)* Schematic diagram demonstrating how ligand-induced oligomerization is detected using epitope-specific antibodies with FRET occurring between antibody conjugated FITC and rhodamine dyes. *(II)* Photobleaching of FITC (donor) in the absence of rhodamine (acceptor), and the presence of ligand, with images taken every 4 s (selection of images shown) illustrating the decay. For analysis of photobleaching emission, only the high-intensity membrane region is used, with background and intracellular regions masked. The decrease in fluorescence intensity of each pixel in the unmasked (membrane) region is then plotted to show the average rate of decay of donor fluorescence caused by photobleaching. *(III)* Photobleaching of FITC in the presence of rhodamine and ligand leads to a reduction in the rate of photobleaching decay of the donor resulting from FRET. (Images in Fig. 1B parts II and III are from ref. 65. Copyright 2000 by American Society for Biochemistry & Molecular Biology. Reproduced with permission from the authors and American Society for Biochemistry & Molecular Biology in textbook format via the Copyright Clearance Center.)

gomerization between the D2 dopamine receptor and SSTR5 *(66)* were demonstrated using pbFRET. Although pbFRET is a powerful way to monitor GPCR oligomerization, its use may be limited by the low availability of the instrumentation and technology required to perform the single-cell FRET microscopy.

4.2. Bioluminescence Resonance Energy Transfer

BRET involves the distance-dependent transfer of energy between a bioluminescent energy donor and a fluorescent acceptor molecule. It is a naturally occurring phenomenon observed in *Renilla reniformis*, with energy resulting from the degradation of coelenterazine by Renilla luciferase (Rluc) transferred to GFP *(70)*. Similarly, BRET is observed between aequorin and GFP in the jellyfish *Aequora victoria*. By generating fusion proteins with either Rluc or *Aequora* GFP or its derivatives, BRET can be used to monitor protein–protein interactions in live cells *(70)*.

Performing a BRET assay initially involves genetically fusing one protein with the Rluc and fusing the second protein to the red shifted variant of GFP, enhanced YFP (EYFP). Following the addition of the cell-permeable substrate coelenterazine to cells co-expressing the fusion proteins, energy is transferred from Rluc (peak emission: 480 nm) to EYFP if the proteins are in close enough proximity (<100 Å). Energy is then re-emitted at a wavelength characteristic of EYFP (peak emission: 530 nm) (Fig. 2A). Similarly to FRET, BRET involves a ratiometric measurement (BRET ratio = the ratio of light emitted at 530 nm over that emitted at 480 nm). The extent of BRET is then determined by subtracting the BRET ratio for cells expressing only the Rluc-fused protein from the BRET ratio for cells co-expressing both fusion proteins; an increase in the BRET ratio is indicative of an interaction.

BRET was first developed to enable the dimerization of the light-sensitive circadian clock protein KaiB from cyanobacteria to be studied *(70)*. It has now been used to demonstrate the oligomerization of many GPCRs in living cells (Table 1), including the homo-oligomerization of the β_2-AR *(11)*, TRHR1 *(71)*, GnRH receptor *(71)*, and type A cholecystokinin (CCKA) receptor *(72)* as well as hetero-oligomerization between the type 1 and 2 TRHRs *(73)*, oxytocin and V1 or V2 vasopressin receptors *(74)*, adenosine A_{2A} and D2 dopamine receptors *(75)*, and CCKA and CCKB receptors *(76)*.

More recently, BRET was performed using Rluc and GFP as donor and acceptor molecules, respectively. In this modified version of the original BRET assay, marketed as BRET^2, light is emitted from Rluc as a result of the degradation of the coelenterazine "DeepBlueC" at a peak wavelength of 410 nm. It is then transferred to GFP, resulting in its excitation and subse-

Detection of GPCR Oligomerization

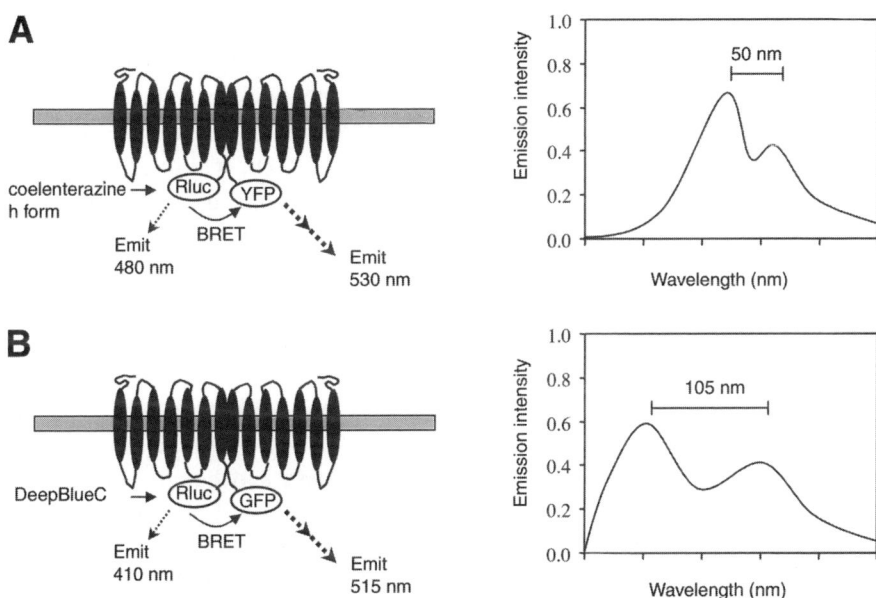

Fig. 2. Schematic diagram illustrating the application of BRET and BRET^2 for the detection of GPCR oligomers. (**A**) Receptors C-terminally tagged with either the energy donor, Renilla luciferase (Rluc) or the energy acceptor, EYFP are coexpressed in cells. In the absence of dimerization or oligomerization, no energy transfer is observed following addition of the cell permeable Rluc substrate, coelenterazine (h form), and light is emitted from Rluc at a peak wavelength of 480 nm. If an interaction occurs, then the tags are in close enough proximity (<100 Å), allowing energy transfer from Rluc to EYFP and additional emission of fluorescent light at a peak wavelength of 530 nm. (**B**) In BRET^2, energy is emitted from the donor, Rluc, at a peak of 410 nm, as a result of degradation of its substrate DeepBlueC. It is then transferred to the acceptor, GFP, leading to the appearance of an emission peak at 515 nm characteristic of the GFP used. BRET^2 provides an advantage over the original BRET system, with greater spectral resolution (105 nm) between donor and acceptor emissions.

quent fluorescence at a peak wavelength of 515 nm (Fig. 2B). Therefore, BRET^2 offers an improved spectral resolution between donor and acceptor emissions (105 nm) compared to the original BRET pair (50 nm), theoretically increasing range and sensitivity. However, the quantum yield of Rluc using DeepBlueC in BRET^2 is less than that with coelenterazine h, commonly used in the original BRET assay system. BRET^2 has been used to demonstrate constitutive homo-oligomerization of the δ-opioid receptor *(77)*, human immunodeficiency virus coreceptor CXCR4 *(78)*, calcium-sensing receptor *(79)*, and neuropeptide Y Y4 receptor *(80)* as well as hetero-oligomerization between the adenosine A_{2A} and D2 dopamine receptors *(81)* and adenosine A_1 and P_2Y_1 receptors *(50)*.

BRET has also been applied quantitatively to gain insight into the relative affinities of various homo-oligomers and hetero-oligomers *(82)*. BRET saturation curves can be generated by expressing an increasing concentration of acceptor fusion protein with a constant amount of donor fusion protein. The BRET signal should rise with increasing concentration of acceptor fusion protein until a maximum signal is reached, with the maximum signal being a function of the relative orientation and distance between the donor and acceptor tags and the quantity of dimerized receptors. However, the concentration of acceptor fusion producing 50% of the maximum BRET signal ($BRET_{50}$) is a measure of the relative affinity of the interaction. This approach was used to show that the β_1- and β_2-ARs formed homodimers and heterodimers with similar relative affinities, suggesting that homo-dimers and hetero-dimers co-expressed at equivalent levels should form in equal proportions *(82)*. The V1a, V2, and oxytocin vasopressin receptors were also demonstrated to have a similar propensity to interact with themselves as they were to each other *(74)*. This may suggest that in vivo, the degree of homo- versus hetero-oligomerization may be determined by the relative expression levels of the receptors rather than by interaction affinities. Alternatively, accessory proteins may be involved in modulating homo- and hetero-oligomer formation.

5. DIMER OR OLIGOMER?

Although the interaction between receptors has been demonstrated for many GPCRs, it is still unclear whether such complexes are dimeric or oligomeric and what proportion of the receptor exists in the monomeric, dimeric, or oligomeric state. Furthermore, the issue of whether the receptor monomer is functional has yet to be addressed for most GPCRs. The monomeric receptors are nonfunctional for at least the $GABA_B$ receptors *(36–38,83)* and the T1R1 *(84)* and T1R2 *(85)* taste receptors, with hetero-oligomerization an obligate requirement for function.

Many studies do not make a clear distinction between dimers and oligomers, with dimerization and oligomerization often used interchangeably when describing the formation of the GPCR complex. Immunoprecipitation experiments often reveal the presence of monomers, dimers, and oligomers. However, the relative quantities of each molecular species observed on immunoblots is often influenced by the experimental conditions employed. Although biophysical techniques have the advantage because the receptor–receptor interactions are monitored in intact cells, they do not easily reveal

Detection of GPCR Oligomerization

the stoichiometry of the receptor complex. However, a recent study used BRET competition in conjunction with theoretical models to predict that the melatonin MT1 and MT2 receptors exist predominantly as homodimers, rather than oligomers or monomers *(86)*. The BRET signal between either MT1 or MT2 receptor homodimers was observed to reduce linearly with increasing concentrations of untagged MT1 or MT2 receptor, respectively. A similar approach was applied to demonstrate that more than 80% of β_2-ARs existed as dimers *(82)*. However, such studies were performed in heterologous expression systems, and the size of GPCR oligomers in native tissues still remains largely unclear. A recent study using atomic force microscopy on mouse eye-disc membranes showed that, at least for rhodopsin, the receptors exist as dimers arranged in an oligomeric array *(87,88)*.

6. LIGAND MODULATION OF OLIGOMERS DETECTED BY BIOCHEMICAL AND BIOPHYSICAL METHODOLOGIES

To better understand the functional relevance of GPCR oligomerization, the effect of ligand binding on oligomerization has been investigated by many researchers in the field. However, it is still not clear whether dimers or oligomers are preformed during synthesis and traffic to the cell surface as an oligomer or whether they oligomerize in response to agonist binding. Many studies have detected preformed receptor oligomers, with ligands causing an increase, decrease, or no change in oligomerization. However, for certain GPCRs, such as the SSTR1 and SSTR5 *(66,89)*, SSTR5 and D2 dopamine receptor *(66)*, and the GnRH receptor *(61,71,90)*, oligomerization appears to be agonist-dependent.

The observed diversity of ligand-induced effects on GPCR oligomerization may reflect differences between GPCRs in how oligomerization is regulated. However, a partial source of the conflicting results may be differences in methodologies used to detect GPCR oligomers. Several studies have examined the effects of ligands on dimer or oligomer formation using co-immunoprecipitations, reporting an increase *(30,32,65)*, decrease *(31)*, or no change *(34,53)* in the extent of dimer formation observed. However, ligand-induced changes in the amount of dimer or oligomer observed in co-immunprecipitations may be caused by conformational changes that result in a change in the accessibility of the antibody and, hence, in detection. Ligand binding may also affect the stability of the dimer or oligomer and, therefore, the quantity of higher order species that appear on Western blots.

Similarly, changes in FRET and BRET signals have been observed in response to ligand binding. Using FRET, the yeast α-factor receptor *(57)*, D2 dopamine receptor *(64)*, and complement C5A receptor *(59)* were all found to exist as preformed oligomers unaffected by ligand binding. However, in another study using FRET, the D2 dopamine receptor was found to undergo constitutive and agonist-induced oligomerization *(56)*. In contrast, oligomerization of the TSH receptor was inhibited by receptor occupation by the TSH ligand *(62)*.

Similarly, investigations using BRET have revealed differences in the effects of ligand on oligomerization. Agonist-induced BRET was observed for homo-oligomerization of the β_2-ARs *(11)* and TRH receptors *(71)* as well as hetero-oligomerization between the adenosine A_1 and P_2Y_1 receptors *(50)*. In contrast, agonist-induced decreases in the BRET signal between CCKA *(72)* and neuropeptide Y Y4 receptor *(80)* homo-oligomers were interpreted as indicative of receptor dissociation. However, many studies have demonstrated no change in BRET signals in the presence of agonists *(74,76,79,91)*. This may suggest that the oligomerization state of the receptor is not affected by ligand and is consequently unrelated to the activation state of the receptor. Detection of changes in BRET between melatonin MT2 receptors following the binding of agonists and antagonists supports the idea that changes in resonance energy transfer efficiencies following ligand binding result from conformational changes that occur in the receptor that may not necessarily be related to activation *(86)*. Indeed, ligand-induced changes in BRET may not necessarily reflect an increase in oligomer formation or a dissociation but, rather, reflect conformational changes in the dimer that bring the donor and acceptor pairs into a more or less favorable orientation for energy transfer. These conformational changes may or may not be detected, depending on the sensitivity of the assay and instrumentation.

7. ADVANTAGES AND LIMITATIONS OF BIOCHEMICAL AND BIOPHYSICAL APPROACHES

The oligomerization of GPCRs has been studied using many different approaches. Biochemical and biophysical approaches both have provided strong evidence to support the existence of GPCR dimers or oligomers. However, the different methodologies have advantages and disadvantages that need to be considered when deciding which technique is to be used in the investigation of receptor–receptor interactions and also when interpreting the findings obtained from applying different approaches.

Biophysical techniques, such as FRET and BRET, provide a powerful means to study GPCR oligomerization. Unlike biochemical techniques such as co-immunoprecipitation, they are performed in intact cells and do not require receptors to be extracted and solubilized from cell membranes, which can often lead to artifactual receptor aggregation.

Resonance energy transfer techniques are strictly distance-dependent, with energy transfer only occurring between donor and acceptor molecules less than 100 Å apart. Because the structure of rhodopsin predicts that the center-to-center distance between monomers in a dimer would be approx 40 to 50 Å *(92,93)*, such biophysical techniques are ideal for monitoring these interactions and their modulation in living cells.

Unlike FRET, BRET does not require an excitation light source, with donor energy derived from a bioluminescent reaction. This eliminates the associated problems of autofluorescence, photobleaching, cell damage, and signal loss. Furthermore, the subsequent reduced background fluorescence of BRET makes it a highly sensitive technique for detecting weak or low-level protein–protein interactions. This may be an important consideration, because many studies of GPCR oligomerization have involved the overexpression of receptors in heterologous expression systems. It has been suggested that GPCRs have a natural tendency to form dimers or oligomers upon co-expression in heterologous expression systems, and studies need to be performed in cells that endogenously co-express the receptors *(94)*. However, this is not always possible, thus it is advantageous to perform high-sensitivity assays with receptor expression levels at near physiological levels, as has been done recently using BRET *(77,86,91)*. The problem of nonspecific receptor–receptor interactions caused by heterologous receptor overexpression can also be reduced through the use of receptor controls. An interaction between two receptors has often been shown to be specific by using additional receptors that show a lack of interaction. In the case of BRET, nontagged receptors can be used to specifically compete out receptor–receptor BRET signals *(71)*.

A potential limitation of BRET is the inability to clearly localize the receptor interaction. Imaging GPCR dimers through FRET microscopy has been performed *(58,62)*. Furthermore, through the use of receptor-specific antibodies, dimers and oligomers could be monitored in native tissues. Imaging of BRET has been performed on *Escherichia coli* colonies *(70)* and in Chinese hamster ovary (CHO) cell extracts *(95)*. Single-cell BRET was performed on the melatonin MT1R and MT2R expressed in HEK293 cells; however, subcellular resolution was not observed *(86)*. The development of BRET imaging to detect and localize GPCR dimers and oligomers within a single cell would represent a significant advance in this technology and allow the fate of dimers and oligomers to be monitored following agonist activation, signaling, and receptor trafficking to and from the cell surface. The development of alternative BRET donor and acceptor molecules that could be conjugated to receptor-specific antibodies would also be extremely useful, enabling this relatively simple technique to detect endogenously expressed GPCR dimers and oligomers.

8. CONCLUSIONS

Evidence for the existence of GPCR dimers and oligomers has rapidly accumulated over the recent years. A range of techniques has been applied to detect and monitor receptor–receptor interactions. The use of biophysical techniques, such as FRET and BRET, produced a significant advance, providing the ability to detect GPCR oligomerization in live cells and eliminating the problems associated with artifactual receptor aggregation that can potentially occur in co-immunoprecipitation procedures. Continuing advancements in the reagents and instrumentation required to perform FRET- and BRET-based techniques will only widen their application and increase our understanding of GPCR oligomerization, the mechanisms involved, its regulation, and, importantly, its role in receptor and cellular function.

REFERENCES

1. Gomes I, Jordan BA, Gupta A, Rios C, Trapaidze N, Devi LA. G protein coupled receptor dimerization: implications in modulating receptor function. J Mol Med 2001;79:226–242.
2. Rios CD, Jordan BA, Gomes I, Devi LA. G-protein-coupled receptor dimerization: modulation of receptor function. Pharmacol Ther 2001;92:71–87.
3. Angers S, Salahpour A, Bouvier M. Dimerization: an emerging concept for G protein-coupled receptor ontogeny and function. Annu Rev Pharmacol Toxicol 2002;42:409–435.
4. George SR, O'Dowd BF, Lee SP. G-protein-coupled receptor oligomerization and its potential for drug discovery. Nat Rev Drug Discov 2002;1:808–820.
5. Kroeger KM, Pfleger KDG, Eidne KA. G-protein coupled receptor oligomerization in neuroendocrine pathways. Front Neuroendocrinol 2003;24:254–278.
6. Limbird LE, Meyts PD, Lefkowitz RJ. Beta-adrenergic receptors: evidence for negative cooperativity. Biochem Biophys Res Commun 1975;64:1160–1168.
7. Mattera R, Pitts BJ, Entman ML, Birnbaumer L. Guanine nucleotide regulation of a mammalian myocardial muscarinic receptor system. Evidence for homo- and heterotropic cooperativity in ligand binding analyzed by computer-assisted curve fitting. J Biol Chem 1985;260:7410–7421.
8. Potter LT, Ballesteros LA, Bichajian LH, et al. Evidence of paired M2 muscarinic receptors. Mol Pharmacol 1991;39:211–221.
9. Bai M, Trivedi S, Brown EM. Dimerization of the extracellular calcium-sensing receptor (CaR) on the cell surface of CaR-transfected HEK293 cells. J Biol Chem 1998;273:23,605–23,610.
10. Rodriguez-Frade JM, Vila-Coro AJ, de Ana AM, Albar JP, Martinez AC, Mellado M. The chemokine monocyte chemoattractant protein-1 induces functional responses through dimerization of its receptor CCR2. Proc Natl Acad Sci USA 1999;96:3628–3633.

11. Angers S, Salahpour A, Joly E, Hilairet S, Chelsky D, Dennis M, Bouvier M. Detection of beta 2-adrenergic receptor dimerization in living cells using bioluminescence resonance energy transfer (BRET). Proc Natl Acad Sci USA 2000;97:3684–3689.
12. Rogers TB. High affinity angiotensin II receptors in myocardial sarcolemmal membranes. Characterization of receptors and covalent linkage of 125I-angiotensin II to a membrane component of 116,000 daltons. J Biol Chem 1984;259:8106–8114.
13. AbdAlla S, Zaki E, Lother H, Quitterer U. Involvement of the amino terminus of the B(2) receptor in agonist-induced receptor dimerization. J Biol Chem 1999;274:26,079–26,084.
14. Venter JC, Schaber JS, U'Prichard DC, Fraser CM. Molecular size of the human platelet alpha 2-adrenergic receptor as determined by radiation inactivation. Biochem Biophys Res Commun 1983;116:1070–1075.
15. Lilly L, Fraser CM, Jung CY, Seeman P, Venter JC. Molecular size of the canine and human brain D2 dopamine receptor as determined by radiation inactivation. Mol Pharmacol 1983;24:10–14.
16. Conn PM, Venter JC. Radiation inactivation (target size analysis) of the gonadotropin-releasing hormone receptor: evidence for a high molecular weight complex. Endocrinology 1985;116:1324–1326.
17. Maggio R, Vogel Z, Wess J. Coexpression studies with mutant muscarinic/adrenergic receptors provide evidence for intermolecular "cross-talk" between G-protein-linked receptors. Proc Natl Acad Sci USA 1993;90:3103–3107.
18. Monnot C, Bihoreau C, Conchon S, Curnow KM, Corvol P, Clauser E. Polar residues in the transmembrane domains of the type 1 angiotensin II receptor are required for binding and coupling. Reconstitution of the binding site by co-expression of two deficient mutants. J Biol Chem 1996;271:1507–1513.
19. Scarselli M, Novi F, Schallmach E, et al. D2/D3 dopamine receptor heterodimers exhibit unique functional properties. J Biol Chem 2001;276:30,308–30,314.
20. Grosse R, Schoneberg T, Schultz G, Gudermann T. Inhibition of gonadotropin-releasing hormone receptor signaling by expression of a splice variant of the human receptor. Mol Endocrinol 1997;11:1305–1318.
21. Zhu X, Wess J. Truncated V2 vasopressin receptors as negative regulators of wild-type V2 receptor function. Biochemistry 1998;37:15,773–15,784.
22. Benkirane M, Jin DY, Chun RF, Koup RA, Jeang KT. Mechanism of transdominant inhibition of CCR5-mediated HIV-1 infection by ccr5delta32. J Biol Chem 1997;272:30,603–30,606.
23. Hebert TE, Loisel TP, Adam L, Ethier N, Onge SS, Bouvier M. Functional rescue of a constitutively desensitized beta2AR through receptor dimerization. Biochem J 1998;330:287–293.
24. Leanos-Miranda A, Ulloa-Aguirre A, Ji TH, Janovick JA, Conn PM. Dominant-negative action of disease-causing gonadotropin-releasing hormone receptor (GnRHR) mutants: a trait that potentially coevolved with decreased plasma membrane expression of GnRHR in humans. J Clin Endocrinol Metab 2003;88:3360–3367.

25. Ban T, Kosugi S, Kohn LD. Specific antibody to the thyrotropin receptor identifies multiple receptor forms in membranes of cells transfected with wild-type receptor complementary deoxyribonucleic acid: characterization of their relevance to receptor synthesis, processing, structure, and function. Endocrinology 1992;131:815–829.
26. Ng GY, O'Dowd BF, Caron M, Dennis M, Brann MR, George SR. Phosphorylation and palmitoylation of the human D2L dopamine receptor in Sf9 cells. J Neurochem 1994;63:1589–1595.
27. Nimchinsky EA, Hof PR, Janssen WG, Morrison JH, Schmauss C. Expression of dopamine D3 receptor dimers and tetramers in brain and in transfected cells. J Biol Chem 1997;272:29,229–29,237.
28. Ward DT, Brown EM, Harris HW. Disulfide bonds in the extracellular calcium-polyvalent cation-sensing receptor correlate with dimer formation and its response to divalent cations in vitro. J Biol Chem 1998;273:14,476–14,483.
29. Ciruela F, Casado V, Mallol J, Canela EI, Lluis C, Franco R. Immunological identification of A1 adenosine receptors in brain cortex. J Neurosci Res 1995;42:818–828.
30. Hebert TE, Moffett S, Morello JP, et al. A peptide derived from a beta2-adrenergic receptor transmembrane domain inhibits both receptor dimerization and activation. J Biol Chem 1996;271:16,384–16,392.
31. Cvejic S, Devi LA. Dimerization of the delta opioid receptor: implication for a role in receptor internalization. J Biol Chem 1997;272:26,959–26,964.
32. Zhu CC, Cook LB, Hinkle PM. Dimerization and phosphorylation of thyrotropin-releasing hormone receptors are modulated by agonist stimulation. J Biol Chem 2002;277:28,228–28,237.
33. Romano C, Yang WL, O'Malley KL. Metabotropic glutamate receptor 5 is a disulfide-linked dimer. J Biol Chem 1996;271:28,612–28,616.
34. Trettel F, Di Bartolomeo S, Lauro C, Catalano M, Ciotti TM, Limatola C. Ligand-independent CXCR2 dimerization. J Biol Chem 2003;278:40,980–40,988.
35. Tao YX, Johnson NB, Segaloff DL. Constitutive and agonist-dependent self-association of the cell surface human lutropin receptor. J Biol Chem 2004;279:5904–5914.
36. Jones KA, Borowsky B, Tamm JA, et al. GABA(B) receptors function as a heteromeric assembly of the subunits GABA(B)R1 and GABA(B)R2. Nature 1998;396:674–679.
37. Kaupmann K, Malitschek B, Schuler V, et al. GABA(B)-receptor subtypes assemble into functional heteromeric complexes. Nature 1998;396:683–687.
38. White JH, Wise A, Main MJ, et al. Heterodimerization is required for the formation of a functional GABA(B) receptor. Nature 1998;396:679–682.
39. Jordan BA, Devi LA. G-protein-coupled receptor heterodimerization modulates receptor function. Nature 1999;399:697–700.
40. George SR, Fan T, Xie Z, et al. Oligomerization of mu- and delta-opioid receptors. Generation of novel functional properties. J Biol Chem 2000;275:26,128–26,135.

41. Gomes I, Jordan BA, Gupta A, Trapaidze N, Nagy V, Devi LA. Heterodimerization of mu and delta opioid receptors: A role in opiate synergy. J Neurosci 2000;20:RC110.
42. Pfeiffer M, Koch T, Schroder H, et al. Homo- and heterodimerization of somatostatin receptor subtypes. Inactivation of sst(3) receptor function by heterodimerization with sst(2A). J Biol Chem 2001;276:14,027–14,036.
43. AbdAlla S, Lother H, Abdel-tawab AM, Quitterer U. The angiotensin II AT2 receptor is an AT1 receptor antagonist. J Biol Chem 2001;276:39,721–39,726.
44. Jordan BA, Trapaidze N, Gomes I, Nivarthi R, Devi LA. Oligomerization of opioid receptors with beta 2-adrenergic receptors: a role in trafficking and mitogen-activated protein kinase activation. Proc Natl Acad Sci USA 2001;98:343–348.
45. Ferre S, Karcz-Kubicha M, Hope BT, et al. Synergistic interaction between adenosine A2A and glutamate mGlu5 receptors: implications for striatal neuronal function. Proc Natl Acad Sci USA 2002;99:11,940–11,945.
46. Gines S, Hillion J, Torvinen M, Le Crom S, Casado V, Canela EI, Rondin S, Lew JY, Watson S, Zoli M, Agnati LF, et al. Dopamine D1 and adenosine A1 receptors form functionally interacting heteromeric complexes. Proc Natl Acad Sci USA 2000;97:8606–8611.
47. Pfeiffer M, Kirscht S, Stumm R, et al. Heterodimerization of substance P and mu-opioid receptors regulates receptor trafficking and resensitization. J Biol Chem 2003;278:51,630–51,637.
48. AbdAlla S, Lother H, Quitterer U, et al. AT1-receptor heterodimers show enhanced G-protein activation and altered receptor sequestration. Nature 2000;407:94–98.
49. AbdAlla S, Lother H, el Massiery A, Quitterer U. Increased AT(1) receptor heterodimers in preeclampsia mediate enhanced angiotensin II responsiveness. Nat Med 2001;7:1003–1009.
50. Yoshioka K, Hosoda R, Kuroda Y, Nakata H. Hetero-oligomerization of adenosine A1 receptors with P2Y1 receptors in rat brains. FEBS Lett 2002;531:299–303.
51. Gama L, Wilt SG, Breitwieser GE. Heterodimerization of calcium sensing receptors with metabotropic glutamate receptors in neurons. J Biol Chem 2001;276:39,053–39,059.
52. Robbins MJ, Ciruela F, Rhodes A, McIlhinney RA. Characterization of the dimerization of metabotropic glutamate receptors using an N-terminal truncation of mGluR1alpha. J Neurochem 1999;72:2539–2547.
53. Zeng FY, Wess J. Identification and molecular characterization of m3 muscarinic receptor dimers. J Biol Chem 1999;274:19,487–19,497.
54. Eidne KA, Kroeger KM, Hanyaloglu AC. Applications of novel resonance energy transfer techniques to study dynamic hormone receptor interactions in living cells. Trends Endocrinol Metab 2002;13:415–421.
55. Pfleger KDG, Eidne KA. New technologies: bioluminescence resonance energy transfer (BRET) for the detection of real time interactions involving G-protein coupled receptors. Pituitary 2003;6:141–151.

56. Wurch T, Matsumoto A, Pauwels PJ. Agonist-independent and -dependent oligomerization of dopamine D(2) receptors by fusion to fluorescent proteins. FEBS Lett 2001;507:109–113.
57. Overton MC, Blumer KJ. G-protein-coupled receptors function as oligomers in vivo. Curr Biol 2000;10:341–344.
58. Dinger MC, Bader JE, Kobor AD, Kretzschmar AK, Beck-Sickinger AG. Homodimerization of neuropeptide y receptors investigated by fluorescence resonance energy transfer in living cells. J Biol Chem 2003;278:10,562–10,571.
59. Floyd DH, Geva A, Bruinsma SP, Overton MC, Blumer KJ, Baranski TJ. C5a receptor oligomerization II: fluorescence resonance energy transfer studies of a human G protein-coupled receptor expressed in yeast. J Biol Chem 2003;278:35,354–35,361.
60. Stanasila L, Perez JB, Vogel H, Cotecchia S. Oligomerization of the alpha1a and alpha1b-adrenergic receptor subtypes:potential implications in receptor internalization. J Biol Chem 2003;278:40,239–40,251.
61. Cornea A, Janovick JA, Maya-Nunez G, Conn PM. Gonadotropin-releasing hormone receptor microaggregration: rate monitored by fluorescence resonance energy transfer. J Biol Chem 2001;276:2153–2158.
62. Latif R, Graves P, Davies TF. Ligand-dependent inhibition of oligomerization at the human thyrotropin receptor. J Biol Chem 2002;277:45,059–45,067.
63. McVey M, Ramsay D, Kellett E, et al. Monitoring receptor oligomerization using time-resolved fluorescence resonance energy transfer and bioluminescence resonance energy transfer. The human delta-opioid receptor displays constitutive oligomerization at the cell surface, which is not regulated by receptor occupancy. J Biol Chem 2001;276:14,092–14,099.
64. Gazi L, Lopez-Gimenez JF, Rudiger MP, Strange PG. Constitutive oligomerization of human D2 dopamine receptors expressed in Spodoptera frugiperda 9 (Sf9) and in HEK293 cells. Analysis using co-immunoprecipitation and time-resolved fluorescence resonance energy transfer. Eur J Biochem 2003;270:3928–3938.
65. Rocheville M, Lange DC, Kumar U, Sasi R, Patel RC, Patel YC. Subtypes of the somatostatin receptor assemble as functional homo- and heterodimers. J Biol Chem 2000;275:7862–7869.
66. Rocheville M, Lange DC, Kumar U, Patel SC, Patel RC, Patel YC. Receptors for dopamine and somatostatin: formation of hetero-oligomers with enhanced functional activity. Science 2000;288:154–157.
67. Gadella TW Jr., Jovin TM. Oligomerization of epidermal growth factor receptors on A431 cells studied by time-resolved fluorescence imaging microscopy. A stereochemical model for tyrosine kinase receptor activation. J Cell Biol 1995;129:1543–1558.
68. Overton MC, Blumer KJ. Use of fluorescence resonance energy transfer to analyze oligomerization of G-protein-coupled receptors expressed in yeast. Methods 2002;27:324–332.

69. Latif R, Graves P, Davies TF. Oligomerization of the human thyrotropin receptor: fluorescent protein-tagged hTSHR reveals post-translational complexes. J Biol Chem 2001;276:45,217–45,224.
70. Xu Y, Piston DW, Johnson CH. A bioluminescence resonance energy transfer (BRET) system: application to interacting circadian clock proteins. Proc Natl Acad Sci USA 1999;96:151–156.
71. Kroeger KM, Hanyaloglu AC, Seeber RM, Miles LE, Eidne KA. Constitutive and agonist-dependent homo-oligomerization of the thyrotropin-releasing hormone receptor. Detection in living cells using bioluminescence resonance energy transfer. J Biol Chem 2001;276:12,736–12,743.
72. Cheng ZJ, Miller LJ. Agonist-dependent dissociation of oligomeric complexes of G protein-coupled cholecystokinin receptors demonstrated in living cells using bioluminescence resonance energy transfer. J Biol Chem 2001;276:48,040–48,047.
73. Hanyaloglu AC, Seeber RM, Kohout TA, Lefkowitz RJ, Eidne KA. Homo- and hetero-oligomerization of thyrotropin-releasing hormone (TRH) receptor subtypes. Differential regulation of beta-arrestins 1 and 2. J Biol Chem 2002;277:50,422–50,430.
74. Terrillon S, Durroux T, Mouillac B, et al. Oxytocin and vasopressin V1a and V2 receptors form constitutive homo- and heterodimers during biosynthesis. Mol Endocrinol 2003;17:677–691.
75. Canals M, Marcellino D, Fanelli F, et al. Adenosine A2A-dopamine D2 receptor-receptor heteromerization. Qualitative and quantitative assessment by fluorescence and bioluminescence energy transfer. J Biol Chem 2003;278:46,741–46,749.
76. Cheng ZJ, Harikumar KG, Holicky EL, Miller LJ. Heterodimerization of type A and B cholecystokinin receptors enhance signaling and promote cell growth. J Biol Chem 2003;278:52,972–52,979.
77. Ramsay D, Kellett E, McVey M, et al. Homo- and hetero-oligomeric interactions between G protein-coupled receptors in living cells monitored by two variants of bioluminescence resonance energy transfer. Biochem J 2002;365:429–440.
78. Babcock GJ, Farzan M, Sodroski J. Ligand-independent dimerization of CXCR4, a principal HIV-1 coreceptor. J Biol Chem 2003;278:3378–3385.
79. Jensen AA, Hansen JL, Sheikh SP, Brauner-Osborne H. Probing intermolecular protein-protein interactions in the calcium-sensing receptor homodimer using bioluminescence resonance energy transfer (BRET). Eur J Biochem 2002;269:5076–5087.
80. Berglund MM, Schober DA, Esterman MA, Gehlert DR. Neuropeptide Y Y4 receptor homodimers dissociate upon agonist stimulation. J Pharmacol Exp Ther 2003;307:1120–1126.
81. Kamiya T, Saitoh O, Yoshioka K, Nakata H. Oligomerization of adenosine A2A and dopamine D2 receptors in living cells. Biochem Biophys Res Commun 2003;306:544–549.

82. Mercier JF, Salahpour A, Angers S, Breit A, Bouvier M. Quantitative assessment of beta 1- and beta 2-adrenergic receptor homo- and heterodimerization by bioluminescence resonance energy transfer. J Biol Chem 2002;277: 44,925–44,931.
83. Galvez T, Duthey B, Kniazeff J, et al. Allosteric interactions between GB1 and GB2 subunits are required for optimal GABA(B) receptor function. Embo J 2001;20:2152–2159.
84. Nelson G, Chandrashekar J, Hoon MA, et al. An amino-acid taste receptor. Nature 2002;416:199–202.
85. Nelson G, Hoon MA, Chandrashekar J, Zhang Y, Ryba NJ, Zuker CS. Mammalian sweet taste receptors. Cell 2001;106:381–390.
86. Ayoub MA, Couturier C, Lucas-Meunier E, et al. Monitoring of ligand-independent dimerization and ligand-induced conformational changes of melatonin receptors in living cells by bioluminescence resonance energy transfer. J Biol Chem 2002;277:21,522–21,528.
87. Fotiadis D, Liang Y, Filipek S, Saperstein DA, Engel A, Palczewski K. Rhodopsin dimers in native disc membranes. Nature 2003;421:127–128.
88. Liang Y, Fotiadis D, Filipek S, Saperstein DA, Palczewski K, Engel A. Organization of the G protein-coupled receptors rhodopsin and opsin in native membranes. J Biol Chem 2003;278:21,655–21,662.
89. Patel RC, Kumar U, Lamb DC, et al. Ligand binding to somatostatin receptors induces receptor-specific oligomer formation in live cells. Proc Natl Acad Sci USA 2002;99:3294–3299.
90. Horvat RD, Roess DA, Nelson SE, Barisas BG, Clay CM. Binding of agonist but not antagonist leads to fluorescence resonance energy transfer between intrinsically fluorescent gonadotropin-releasing hormone receptors. Mol Endocrinol 2001;15:695–703.
91. Issafras H, Angers S, Bulenger S, et al. Constitutive agonist-independent CCR5 oligomerization and antibody-mediated clustering occurring at physiological levels of receptors. J Biol Chem 2002;277:34,666–34,673.
92. Palczewski K, Kumasaka T, Hori T, et al. Crystal structure of rhodopsin: a G protein-coupled receptor. Science 2000;289:739–745.
93. Okada T, Le Trong I, Fox BA, Behnke CA, Stenkamp RE, Palczewski K. X-Ray diffraction analysis of three-dimensional crystals of bovine rhodopsin obtained from mixed micelles. J Struct Biol 2000;130:73–80.
94. Salim K, Fenton T, Bacha J, et al. Oligomerization of G-protein-coupled receptors shown by selective co-immunoprecipitation. J Biol Chem 2002;277:15,482–15,485.
95. Wang Y, Wang G, O'Kane DJ, Szalay AA. A study of protein–protein interactions in living cells using luminescence resonance energy transfer (LRET) from *Renilla luciferase* to Aequorea GFP. Mol Gen Genet 2001;264:578–587.
96. Lavoie C, Mercier JF, Salahpour A, et al. Beta 1/beta 2-adrenergic receptor heterodimerization regulates beta 2-adrenergic receptor internalization and ERK signaling efficacy. J Biol Chem 2002;277:35,402–35,410.

97. Lee SP, O'Dowd BF, Ng GY, et al. Inhibition of cell surface expression by mutant receptors demonstrates that D2 dopamine receptors exist as oligomers in the cell. Mol Pharmacol 2000;58:120–128.
98. George SR, Lee SP, Varghese G, et al. A transmembrane domain-derived peptide inhibits D1 dopamine receptor function without affecting receptor oligomerization. J Biol Chem 1998;273:30,244–30,248.
99. Gomes I, Filipovska J, Jordan BA, Devi LA. Oligomerization of opioid receptors. Methods 2002;27:358–365.
100. Roess DA, Horvat RD, Munnelly H, et al. Luteinizing hormone receptors are self-associated in the plasma membrane. Endocrinology 2000;141:4518–4523.
101. Vila-Coro AJ, Mellado M, Martin de Ana A, et al. HIV-1 infection through the CCR5 receptor is blocked by receptor dimerization. Proc Natl Acad Sci USA 2000;97:3388–3393.
102. Pfeiffer M, Koch T, Schroder H, Laugsch M, Hollt V, Schulz S. Heterodimerization of somatostatin and opioid receptors cross-modulates phosphorylation, internalization, and desensitization. J Biol Chem 2002; 277:19,762–19,772.
103. Hillion J, Canals M, Torvinen M, et al. Coaggregation, cointernalization, and codesensitization of adenosine A2A receptors and dopamine D2 receptors. J Biol Chem 2002;277:18,091–18,097.
104. Mellado M, Rodriguez-Frade JM, Vila-Coro AJ, et al. Chemokine receptor homo- or heterodimerization activates distinct signaling pathways. EMBO J 2001;20:2497–2507.
105. Xie Z, Lee SP, O'Dowd BF, George SR. Serotonin 5-HT1B and 5-HT1D receptors form homodimers when expressed alone and heterodimers when co-expressed. FEBS Lett 1999;456:63–67.
106. Barki-Harrington L, Luttrell LM, Rockman HA. Dual inhibition of beta-adrenergic and angiotensin II receptors by a single antagonist: a functional role for receptor-receptor interaction in vivo. Circulation 2003;108:1611–1618.
107. Xu J, He J, Castleberry AM, Balasubramanian S, Lau AG, Hall RA. Heterodimerization of alpha 2A- and beta 1-adrenergic receptors. J Biol Chem 2003;278:10,770–10,777.
108. Ciruela F, Escriche M, Burgueno J, et al. Metabotropic glutamate 1alpha and adenosine A1 receptors assemble into functionally interacting complexes. J Biol Chem 2001;276:18,345–18,351.

11
Oligomerization Domains of G Protein-Coupled Receptors

Insights Into the Structural Basis of GPCR Association

Marta Filizola, Wen Guo, Jonathan A. Javitch, and Harel Weinstein

1. INTRODUCTION

Many recent reviews have thoroughly described the ability of a wide range of G protein-coupled receptors (GPCRs) to exist and to potentially function as oligomers *(1–6)*. This chapter summarizes the computational and experimental studies that have provided insight into the understanding of the structural basis of GPCR association. Particular emphasis is placed on the combined computational and experimental approach that we have recently developed to characterize the homodimerization interface of rhodopsin-like GPCRs.

Traditional models of GPCR activation have been based on the assumption that single agonists induce conformational changes in single receptors, which in turn stimulate heterotrimeric G proteins and produce signal amplification. However, recent reports on GPCR oligomerization have suggested that the ligands bind to and may activate an oligomeric complex *(7–10)*, giving rise to various signaling events in the ensuing cascade. The requirement for GPCR dimerization in signaling has been demonstrated explicitly for the γ-aminobutyric acid $(GABA)_{B1}$–$GABA_{B2}$ heterodimer. In this case, the subunit $GABA_{B1}$ binds GABA but does not appear to be capable of G protein coupling, whereas the subunit $GABA_{B2}$ cannot bind GABA but does appear to couple to G protein *(8)*. Therefore, heterodimerization is a prereq-

From: *Contemporary Clinical Neuroscience: The G Protein-Coupled Receptors Handbook*
Edited by: L. A. Devi © Humana Press Inc., Totowa, NJ

uisite for receptor activation, which seems to occur through transactivation *(8)*. It is also possible that for other GPCRs, the relevant unit defining their pharmacological characteristics may not be the monomer. Consequently, it is necessary to consider the structural details and modes of oligomerization to achieve an accurate understanding of receptor function.

Both computational and experimental efforts have been made to understand the basis of protein–protein interaction in GPCR oligomerization, but the specific molecular determinants required for receptor–receptor dimerization are still unknown. Additionally, there is still an issue regarding whether dimerization interfaces differ among highly related GPCRs. We have recently developed a combined experimental *(11)* and computational *(12–15)* approach to identify the molecular determinants responsible for GPCR oligomerization, with the goal of discovering mutations that disrupt the interface between protomers and, therefore, may interfere with receptor function. The computational approach produces putative three-dimensional (3D) models of oligomers based on the structural information contained in the crystal structure of rhodopsin *(16)* as well as on correlated mutation analysis (CMA) *(12–15,17–21)* that serves to significantly limit the number of different packing modes of the transmembrane (TM) bundles of GPCRs that must be considered in the modeling of oligomers. This approach has recently been applied to the three-cloned opioid receptor subtypes to identify their likely interfaces in both homo- *(15)* and heterodimers *(13)*. Once the likely oligomerization interfaces were identified with the CMA-based approach, molecular modeling served in the construction of 3D models of GPCR dimers to maximize the number of interactions between the correlated residues that were predicted from CMA on the appropriate lipid-facing surface of the TMs in each protomer.

The computational procedure has recently been described for the opioid receptors *(15)*. This chapter presents the results of CMA calculations performed on the other rhodopsin-like GPCR subtypes for which homodimerization has been experimentally demonstrated. To test these predictions, we have developed an experimental strategy that uses cysteine crosslinking to map the dimer interface of GPCRs. By applying this approach to the D2 dopamine receptor, we recently showed that TM4 forms part of a symmetric homodimer interface for this receptor *(11)*. Interestingly, our CMA-based approach predicted TM4 as a likely dimerization interface for the D2 dopamine receptor. We propose to use the interdisciplinary approach described in this chapter as a tool to provide new insights into the understanding of the structural basis of GPCR oligomerization.

2. DOMAINS OF GPCR OLIGOMERIZATION INFERRED FROM EXPERIMENTS

To date, oligomerization of GPCRs has been suggested to be mediated by direct protein–protein interaction involving all the structural regions—extracellular, intracellular, and/or TM. A summary of the GPCR oligomerization domains that are experimentally suggested is provided later.

2.1. Extracellular Amino Terminus Domain

The involvement of the N-terminus in GPCR oligomerization has been clearly demonstrated for class C receptor subtypes. Direct evidence exists for the metabotropic glutamate receptor (mGluR)1 based on X-ray crystallographical data *(22)*. Three different high-resolution crystal structures of the N-terminal ligand-binding region of this receptor, with and without the ligand, appeared as disulfide-linked homodimers. Several conformers of this amino terminus were identified by combining crystallographical data with modeling studies *(22)*. Specifically, "active" and "resting" conformations resulted from interdomain movement and relocation of the dimer interface. Binding of glutamate to the extracellular ligand-binding domain of mGluR1 stabilized both the "active" dimer and a "closed" conformation of the protomer in dynamic equilibrium. The ligand-induced interdomain movements in the mGluR1 dimeric complex were suggested to produce an allosteric effect on the TM or intracellular regions of the receptor, leading to its activation.

A similar activation mechanism was recently proposed for the class C $GABA_B$ receptor *(23)*. Introduction of two cysteines, which were expected to stabilize the N-terminal ligand-binding domain of $GABA_{B1}$ in a closed state by a disulfide bridge, locked the receptor into an almost fully active state.

Additional, although indirect, evidence for the participation of the N-terminal domain in GPCR oligomerization exists for other members of class C, as well as members of classes A, B, and D. Particularly, mutagenesis studies showed that cysteines within the N-terminal domain of the class C human extracellular calcium sensing receptor were critical for the formation of intermolecular disulfide bond(s) formed between receptor protomers *(24,25)*.

The involvement of the amino terminus has also been implicated in the dimerization of the class A bradykinin 2 receptor (B_2R) *(26)*. In the wild-type B_2R, the fraction of B_2R crosslinked in a dimeric or oligomeric form was increased by the binding of the agonist bradykinin, whereas this crosslinking was greatly reduced in a mutant that lacked the wild-type amino terminus and that started at amino acid 65, just before TM1 *(26)*. Furthermore, the addition of a peptide corresponding to the amino terminus of the receptor reduced the amount of crosslinked B_2R dimers observed after bradykinin treatment, whereas peptides derived from the extracellular loops had no effect.

Western blot analysis of Ig-Hepta (a novel member of the GPCR superfamily defining a new subfamily of class B with a large extracellular amino terminus domain) indicated that this protein exists as a disulfide-linked dimer *(27)*. The same analysis performed on the amino terminus domain of Ig-Hepta alone indicated that by itself, this receptor region lacks the ability to dimerize by forming disulfide bond(s). Nonetheless, experiments carried out on the mutant Ig-Hepta truncated after the first TM span indicated dimer formation, suggesting that the disulfide-linked dimer is formed through the cysteine residues in the extracellular domain, rather than in the seven-TM (7TM) helices. Assuming that the truncated Ig-Hepta mutants adopt a conformation similar to the extracellular domain of the wild-type receptor, these results suggest that although the covalent dimer is formed through intermolecular disulfide bond(s) in the amino terminus, TM1 is necessary for the molecular association of Ig-Hepta receptor.

Fluorescence resonance energy transfer (FRET) and endocytosis-based assays (which detect the ability of green fluorescent protein (GFP)-tagged endocytosis-defective receptors to interact with and be rescued by co-expressed untagged wild-type receptors) were used to analyze receptor deletion mutants of the class D yeast α-factor receptor sex pheromone exporter (STE)2 *(28)*. These studies suggested that the α-factor receptor STE2 amino terminus, as well as TM1 and TM2, mediate receptor dimerization.

2.2. Intracellular C-Terminal Domain

Early indirect evidence suggested the involvement of the C-terminal region in the heterodimerization process of $GABA_{B1}$–$GABA_{B2}$ *(29,30)* and in the homodimerization of δ-opioid receptor *(31)*. In particular, yeast two-hybrid screening *(29,30)* showed that $GABA_{B1}$ and $GABA_{B2}$ subunits interact via a stretch of approx 30 amino acid residues within their intracellular C-terminal domains. Additionally, circular dischroism spectroscopy of polypeptide chain fragments containing the heterodimerization site of $GABA_B$ receptor showed that these peptides preferentially form parallel coiled-coil heterodimers in a physiological buffer *(32)*, suggesting that the functional $GABA_B$ receptor is a heterodimer assembled by parallel coiled-coil α-helices contained in the intracellular C-terminal domain. Subsequent experiments with a series of $GABA_{B1}$ receptor C-terminal truncation

Domains of GPCR Oligomerization

mutants identified a sequence of four amino acids (RSRR) within the proposed coiled-coil interaction domain of the $GABA_{B1}$ receptor that function as an endoplasmic reticulum (ER) retention signal in this receptor subunit *(33)*. Therefore, the coiled-coil interaction with $GABA_{B2}$ masks this ER retention signal and allows $GABA_{B1}$ to come to the cell surface as a heterodimer with $GABA_{B2}$. Disruption of both the ER retention signal and the coiled-coil interaction domain allowed the $GABA_{B1}$ receptor mutant to reach the cell surface, where it retained the ability to bind agonist but did not function. However, co-expression of this mutant with $GABA_{B2}$ produced a heterodimer capable of inducing GABA-evoked G protein-coupled inwardly rectifying potassium (GIRK) current *(33)*, demonstrating that regions (maybe TM helices) other than the intracellular C-terminal domain can mediate appropriate heterodimerization of $GABA_B$ receptor subunits.

2.3. TM Domains

Dimer interfaces involving TM regions have been suggested for several GPCR subtypes, including rhodopsin receptors *(34,35)*, β_2-adrenergic receptors (ARs *[36]*), D1 *(37)* and D2 *(5,11,38)* dopamine receptors, C5a *(39)*, α_{1b}-adrenoreceptor *(40)*, and yeast α-factor receptors *(41)*. As detailed below, the experimental approaches used atomic force microscopy and molecular modeling, synthetic peptides corresponding to various TMs, co-expression and FRET, and a strategy of disulfide-trapping of endogenous or substituted cysteine residues.

Specific oligomerization interfaces were suggested from molecular modeling based on the recently published atomic force microscopy analysis of rhodopsin. These pointed to the involvement of TM4 and TM5 in intradimeric contacts and to TM1 and TM2, as well as the cytoplasmic loop connecting helices TM5 and TM6, in the formation of dimeric rows *(34,35)*. A 3D model of the rhodopsin homodimer was derived from these studies (Protein Data Bank identification code 1N3M *[34,35]*), offering predictions of specific interaction sites. Inhibition studies with a synthetic peptide corresponding to TM6 of the β_2-AR suggested the involvement of TM6 in the dimerization of this family A GPCR *(36)*. Specifically, a glycophorin A-like dimerization motif (^{272}LKTLGIIMGTFTL^{284}) in TM6 of the β_2-AR was hypothesized to play a role in the dimerization of this receptor. The use of synthetic peptides also identified TM6 and TM7 as possible dimerization interfaces of D2 dopamine receptors *(38)*. In contrast, studies using the substituted cysteine accessibility method *(42)* and cysteine crosslinking experiments *(11)* suggested that TM4 forms part of a symmetric homodimer interface in the D2 dopamine receptor.

In contrast to the early findings for the β_2-ARs *(36)* and D2 dopamine *(38)* receptors, a peptide based on the TM6 sequence of the D1 dopamine receptor did not affect the extent of dimerization *(37)*, suggesting that the dimerization interfaces of closely related GPCRs could differ. Although the findings discussed earlier were shown to be specific for the sequences of these particular synthetic peptides, these data do not necessarily establish TM6 and/or TM7 as the dimer interface in β_2-ARs or D2 dopamine receptors, because a specific peptide–receptor interaction at one site may modulate the ability of the receptor to form dimers at a different interface.

Specificity of oligomerization was also observed for GPCR heterodimerization *(7,43,44)*, because some GPCRs were found to interact with one type of receptor but not another. For example, the μ-opioid receptor is known to heterodimerize with the δ-opioid receptor but not with the κ-opioid receptor. However, the κ-opioid receptor may form heterodimers with the δ-opioid receptor *(43)*. Interestingly, the notion of selectivity in heterodimerization is also supported by computational analysis with the subtractive correlated mutation (SCM) method that we recently developed to identify likely heterodimerization interfaces among GPCR subtypes *(13)*. Similarly to the opioid receptors, the somatostatin (SST)5 receptor has been reported to heterodimerize selectively, with SST1 but not with SST4 subtypes *(7)*, whereas the chemokine receptor (CCR)2 has been demonstrated to associate with CCR5 but not with CXCR4 subtypes *(44)*.

An interface involving TM1 has been proposed for C5a *(39)*, α_{1b}-adrenoreceptor *(40)*, and yeast α-factor *(41)* receptors. Thus, co-expression of the α_{1b}-adrenoreceptor with a fusion protein incorporating the N-terminal domain and TM1 of the α_{1b}-adrenoreceptor and $G_{11}\alpha$ was interpreted to indicate a role for TM1 in dimerization. Both TM1 and TM2 were suggested to form the interface of yeast α-factor receptor oligomers *(28)* based on the result of FRET experiments and assays showing that GFP-tagged endocytosis-defective receptors are recruited into the endocytic pathway through interaction with untagged wild-type receptors. Additionally, symmetric dimer interfaces involving TM1 and TM2 or TM4 of the C5 receptor were inferred based on disulfide trapping *(39)*.

Taken together, the experimental results for various GPCR types suggest that TM1 and TM4 are the most likely interfaces for GPCR dimerization. Assuming a rhodopsin-like packing of the TM bundle for all GPCRs, TM1 and TM4 could not simultaneously participate in a single symmetric dimerization interface. Therefore, these results suggest either different dimerization interfaces for highly related GPCRs or the possible formation of higher order oligomers, as discussed below.

3. MODES OF INTERACTION BETWEEN TM REGIONS OF GPCR MONOMERS

Two modes of interaction between the TM helices of GPCRs have been proposed. First, "domain swapping" occurs when the TM bundles interpenetrate, and the interacting TMs from two different polypeptides appear as interlaced units *(18,20,45,46)*. Second, "contact dimerization" occurs when each protomer TM bundle presents a separate binding site that is packed against that of another protomer through interactions at interfaces that would otherwise face the lipid environment. The domain-swapping mode of interaction was first suggested based on co-expression studies with mutant muscarinic and adrenergic receptors *(47)*. Two chimeric constructs were used for this experimental study: α_{2C}–m3 and m3–α_{2C}, in which the TM6 and TM7 segments had been exchanged between the α_{2C}-adrenergic and the m3 muscarinic receptors. Although transfection with either of the two chimeric receptors alone did not result in detectable binding activity for muscarinic or adrenergic ligands, cotransfection with both α_{2C}–m3 and m3–α_{2C} restored binding of both ligands. This prompted the explanation that the TM1 to TM5 part of the receptor chimera formed a ligand-binding site using TM6 and TM7 from the second receptor chimera, thereby reconstituting a normal binding site comprised of TMs incorporated in two different polypeptide chains.

Although other examples of domain swapping have been proposed *(19)*, more recent experimental results *(28,34,35,48–51)* have not supported this mode of interaction as a dominant form of GPCR dimerization but propose it only as a mechanism of functional rescue in heterologously expressed GPCRs. For example, nonfunctional point mutants and truncation mutants of D2 dopamine receptor were used to examine TM domain swapping in the oligomerization process of this receptor. Specifically, it was demonstrated that receptor function was antagonized when D2 dopamine mutant receptors that were incapable of ligand binding were expressed with the wild-type receptor *(48)*, which is contrary to the expectation of reconstitution of intact binding pockets as a result of TM domain swapping. Additionally, no specific binding was detected upon co-expression of an Asp^{114}Asn D2 dopamine receptor-defective mutant with a truncation mutant containing TM1 through TM5, suggesting that TM domains 1 through 5 do not participate in swapping with TM domains 6 and 7 in the D2 dopamine receptor. Similarly, nonfunctional constructs of the V2 vasopressin receptor with mutations in the N-terminal folding domain (TM1–TM5) could not be rescued by co-expressing a nonmutated N-terminal receptor fragment *(51)*. Because all attempts to restore function of V2 vasopressin receptor mutants failed, but a noncovalent interaction between monomers was detected by co-immunoprecipitation studies, oligomers of this receptor were also suggested to form by contact rather than by domain swapping. A recent study on the yeast α-factor receptor *(28)* also favored a dimer contact model over domain swapping: small fragments of this receptor (as simple as the N-terminal region plus TM1) could self-associate, unlike other receptor fragments lacking TM1. Intramolecular crosslinking between the TM1 and TM7 domains of m3 muscarinic acetylcholine receptor using an *in situ* disulfide crosslinking strategy did not produce dimers *(49)*. In any domain-swapped dimer, regardless of the crossover point between the polypeptides, TM1 and TM7 (the first and last TM helices) must be from different protomers (unless multiple swaps are proposed). Therefore, crosslinking between TM1 and TM7 without the formation of a disulfide-bonded dimeric complex on nonreducing sodium dodecyl sulfate-polyacrylamide gel electrophoresis (SDS-PAGE) and immunoblotting is strong evidence against domain swapping. Similarly, photo-affinity label experiments using a peptide agonist of the cholecystokinin receptor showed binding of this peptide to TM1 and TM7 of the same receptor *(50)*, also arguing against the mechanism of domain swapping in the dimerization of this GPCR.

Finally, the recent atomic force microscopy map of rhodopsin *(34,35)*, which shows receptor monomers organized into two-dimensional arrays of dimers, provides direct evidence for the formation of contact dimers, rather than domain-swapped dimers, in GPCRs.

4. AN INTERDISCIPLINARY APPROACH TO CHARACTERIZE THE OLIGOMERIZATION INTERFACE OF RHODOPSIN-LIKE GPCRS

Based on the preponderance of experimental data suggesting that the contact-dimer geometry is the most likely form of oligomerization among rhodopsin-like GPCRs, we recently developed a combined computational and experimental strategy to identify the molecular determinants for the oligomerization of rhodopsin-like GPCRs. The number of possibilities in which the 7TM regions of two GPCR monomers can be packed together is extremely large (at least 49 [=7 × 7] for heterodimers and 28 [=7{7 + 1}/2] for homodimers). To reduce these possibilities, we recently developed two different computational approaches *(6,13,15)* based on a combination of CMA with the structural information of GPCR monomers derived from homology modeling, using the rhodopsin crystal structure *(16)* as a template. The computational method was used to identify likely interfaces of homodimerization *(15)* for all the rhodopsin-like GPCRs that are known to

Domains of GPCR Oligomerization 251

dimerize to investigate whether oligomerization interfaces in different rhodopsin-like GPCRs are the same or different. The results from this study are reported below with a summary of the experimental method that we developed to test the computational predictions and characterize, in detail, the oligomerization interface(s) of GPCRs.

4.1. Dimerization Interfaces of Rhodopsin-Like GPCRs Predicted Computationally

Based on the principle that residues involved in a common function tend to mutate together, we recently searched for lipid-exposed correlated mutations within multiple sequence alignments of rhodopsin-like GPCRs for which homodimerization had been experimentally demonstrated. Specifically, calculations were performed for the following GPCRs: adenosine 1 receptors; angiotensin II type 1 receptors; α_{1b}-adrenoreceptors; β_1-ARs; β_2-ARs; B$_2$Rs; cannabinoid 1 receptors; CCR2s, CCR5s, and CXCR4s; D1, D2, D3 dopamine receptors; H1, H2 , and H4 histamine receptors; 5-HT$_{1B}$ and 5-HT$_{1D}$ serotonin receptors; leukotriene B$_4$-1 receptors; luteinizing hormone receptors; m2 and m3 muscarinic acetylcholine receptors; MT1 and MT2 melatonin receptors; neuropeptide Y type 1, type 2, and type 5 receptors; δ-, μ-, and κ-opioid receptors; SST2A, SST5, SST3, and SST1 receptors; thyrotropin-releasing hormone receptors; and V2 vasopressin receptors *(3,40,52–55)*. Several receptors were excluded from analysis, including rhodopsin-like GPCRs with fewer than five full-length sequences from different species (CCR2, D3DR, H4R, L4R1, MT2, SST2A, SST5, SST3, and SST1) and GPCR subtypes for which the structural similarity with rhodopsin has been questioned (CB1, CCR5, and CXCR4) *(56)*.

Rhodopsin-like GPCR sequences were retrieved from the GPCR database *(57)*. Human receptor sequences were used as reference sequences for the multiple alignments of GPCRs, which were performed using the CLUSTALW program version 1.81 *(58)*. TM regions were assigned based on the multiple sequence alignment of the entire rhodopsin family reported in the GPCR database *(57)*. The 2.8 Å crystallographical structure of bovine rhodopsin *(16)* was used as a structural template in the definition of surface-exposed residues from 3D models *(59,60)* of the GPCR TM domains. Models of all receptors were constructed using the homology modeling approach implemented in the program MODELLER *(61)*. Any residual steric repulsions between atoms of the side-chains in the resulting models were eliminated with mild energy minimization using version 27 of CHARMM *(62)*. Specifically, 200 cycles of steepest descent followed by 200 cycles of conjugate gradient minimization were performed using a distance-dependent dielectric constant of 4r and keeping all the backbone atoms restrained by harmonic potentials.

The sequence alignments and the 3D models were used in a computer program that identified lipid-exposed correlated mutations in GPCRs *(6,15)*. This program builds on concepts embodied in general algorithms for the identification of correlated mutations *(63,64)*. Lists of lipid-exposed correlated mutations for a given GPCR were extracted from the program outputs using the solvent accessibility values calculated from the atomic coordinates of each residue in the GPCR 3D models as criteria. Based on these values, pairs of correlated residues in which either one or both residues were inaccessible to lipid were eliminated from the lists. This pruning was performed to eliminate intramolecular contacts from the initial list of correlated mutations, which may have included predictions of both intra- and intermolecular contacts. Additional filtering criteria were applied to reduce the number of false-positives, although this procedure may produce some false-negatives by eliminating an actual interface from the resulting predictions. Therefore, the number of correlated pairs was first reduced to L/2, where L was the length of the GPCR sequence used as a reference in each multiple alignment. This filtering was performed because a list of L/2 was demonstrated to contain more correct predictions *(64)*. Further eliminations from the L/2 list included any correlated pairs with a correlation index of 0.7 or less, which was done to reduce the number of false-positives. However, all correlated pairs with a correlation index equal to 1 were considered, even if the total exceeded the number of L/2 correlated pairs (*see* above). The residues that remained after filtering by these criteria were considered putative candidates for the interface of homodimerization of each GPCR under study. However, among these identified residues, an interface was considered to be predicted only if at least three were within seven residues from one another.

The constituent residues of the predicted structural neighborhood at dimerization interfaces (i. e., at least three lipid-exposed correlated mutations within seven residues from one another) of the rhodopsin-like GPCRs we studied are shown in Table 1. Notably, although calculations were performed for α_{1b}-adrenoreceptors, m2 muscarinic acetylcholine receptors, m3 muscarinic acetylcholine receptors, neuropeptide type 2 receptors, neuropeptide type 5 receptors, and κ-opioid receptors, no likely interface is reported because no residues satisfied the interface prediction criterion of at least three lipid-exposed correlated mutations close in sequence (within seven residues from one another). Because of the stringency of the criterion chosen to define dimerization interfaces and the filtering used to eliminate

Domains of GPCR Oligomerization 253

most of the false-positives (*see* above), it is likely that some actual interfaces were not detected computationally.

Gouldson et al. *(17,18,20)* also carried out CMA calculations to identify likely dimerization interfaces of GPCRs. The main differences between those calculations and the protocol discussed earlier are: (a) the present method uses separate sequence alignments for each GPCR subtype, whereas Gouldson et al. did not separate subtypes and considered, for example, the entire family of biogenic amine receptors in a single sequence alignment and (b) additional criteria and filtering methods are incorporated into the present protocol to prune the original list of correlated mutations and to identify the dimerization interface neighborhood (e.g., the definition of likely dimerization interfaces based on the presence of at least three correlated mutations within seven residues from one another). Consequently, predictions from our CMA-based approach can focus on a small number of strongly predicted interfaces and involve only a few TMs, whereas the results from the larger alignments by Gouldson et al. *(17,18,20)* predict nearly every TM as a putative interface, most likely because they miss any putative subtype differences.

As demonstrated in Table 1, residues in TM1 and TM4 appear most often as putative interfaces among the studied GPCRs, making these TMs the most likely segments of rhodopsin-like GPCRs to be involved in oligomerization interfaces. This finding is intriguing, given the recent experimental data summarized in Section 1, which have suggested a role for these two segments in the dimerization/oligomerization of rhodopsin-like GPCRs, including rhodopsin receptors *(34,35)*, D2 dopamine receptors *(11)*, α1b-adrenoreceptors *(40)*, and C5a receptors *(39)*. In particular, structural inferences from the recently published atomic force microscopy analysis of rhodopsin *(34,35)* have suggested that TM1 and TM4 form distinct symmetrical interfaces. The TM4 segment was specifically implicated in intradimeric contact between monomers, whereas TM1 was suggested to facilitate the formation of rhodopsin dimer rows. Despite the refinement and stringency of the CMA-based protocol described earlier, we believe that application of the CMA approach alone is not sufficient to exactly determine the entire correct interface for each GPCR, and additional computational efforts (such as the analysis of 3D models of GPCR dimers using the CMA predictions as a starting point) must be added for this purpose. Nevertheless, the agreement found thus far between CMA-based predictions with the stringent protocol and the experimental evidence regarding the likely interfaces (e.g., the preponderance of TM4 and TM1 in rhodopsin-like GPCRs, the subtype variability) underscores the usefulness and predictive

Table 1
Multi-Subtype Correlated Mutation Analysis of Rhodopsin-Like GPCRs

Res[a]	GPCR subtype																	
	AA1R	AG2R	B1AR	B2AR	BRB2	D1DR	D2DR	H1R	H2R	5H1B	5H1D	LHR	MT1	NY1R	OPRD	OPRM	TRHR	V2R
1.30									A16									
1.31	F8								C17				L26					
1.32	Q9								K18				A27					
1.33									I19		I39							
1.34	A11																	
1.36																		
1.37											V43							
1.38	I15										V44		V33					
1.39									V24									
1.40	V17										S46							
1.41	L18								V27			L368						
1.44									L30									
1.45	V22								I31							V83		
1.47									V33			I374	I42					
1.48	P25											M375	L43			F86		
1.51	V28															F89	I44	
1.52												T379						
1.55												F382	L50					
1.58	K35											L385				V96	M51	
1.59	V36																R52	
2.55														I91				
2.62														F98				
2.63						V83								V99				
2.64						A84												
2.66						I86												
2.67						A87												

Res[a]	GPCR subtype																			
	AA1R	AG2R	B1AR	B2AR	BRB2	D1DR	D2DR	H1R	H2R	5H1B	5H1D	LHR	MT1	NY1R	OPRD	OPRM	TRHR	V2R
3.23																I140		
3.24										V120						L141		
3.27										V121								
3.29										F124						I146		
4.40		L143				K138			V133				K142		A163			A154
4.41		V144				A139												H155
4.43		K146				F141							L145					
4.44								A146	I137									
4.45	V126														L167			
4.47													L149		I170		I147	
4.48						V146	I158		L141	L172				V164	V174		A151	V162
4.51	I133							F153				L492		L165				
4.52	L134							L154						V167				
4.54										I178			L156	A168				
4.55	V137						T165						A157					
4.57																		
4.58							C168			L182		M499		L171			M158	
4.59	T141																L159	
4.60							L170							F173				
4.61										F185								
4.62												V503	R164					
5.35		P192																
5.36			R222															
5.37			A223											D190				
5.40											I199			K193			L193	

255

Res[a]	GPCR subtype																
	AA1R	AG2R	B1AR	B2AR	BRB2	D1DR	D2DR	H1R	H2R	5H1B	5H1D	LHR	MT1	NY1R	OPRD	OPRM	TRHR	V2R
5.41		T198	A227									I533		D194			M194	
5.43														Y196				
5.44		I201									C203		V193		L219			
5.45												V537					V198	
5.47														D200				
5.48		F206											L197				V201	
5.49											I208	I541	V198					
5.51														S204	I226			
5.52	L194								I197									
5.53													I202	S206				
5.55	V197								I200				I204					
5.56	L198												F205		V231			
5.58									Y203									
5.59	L201																	
6.30		N235																
6.31					R266													
6.32					R267													
6.34																		V266
6.35																		V270
																		R271

Res[a]	GPCR subtype																
	AA1R	AG2R	B1AR	B2AR	BRB2	D1DR	D2DR	H1R	H2R	5H1B	5H1D	LHR	MT1	NY1R	OPRD	OPRM	TRHR	V2R
6.36		I241																
6.39					V274								V242					
6.42		L247			L277													
6.45					I280													
6.46													I249					
6.49													F256					
6.53											S321							
6.55																		
6.57																		
6.58											L324							
6.59											P325							
6.60													S263					
7.33									V268									
7.34				V307	I313				L269									
7.36																		
7.37					I316				I272									
7.41				I314														
7.40					F319													
7.41					M320													
7.44				V317														

[a] Residue numbers refer to human receptor sequences. The generic numbering scheme for GPCR sequences (73) is also used to make possible comparisons among the different GPCRs.

power of the approach. Notably, the computational method is also capable of providing useful information about conserved interfaces in oligomeric assemblies larger than dimers.

4.2. Testing and Validation Using Cysteine Crosslinking Experiments

Numerous studies using crosslinking have demonstrated that GPCRs in the membrane are dimeric or oligomeric complexes *(26,31,36,65–67)*. These studies used relatively long, lysine-reactive bifunctional crosslinking reagents, which made it impossible to infer the specific residues or regions that were crosslinked. To directly identify the dimer interface, we designed a strategy to use cysteine crosslinking *(11)* of our collection of D2 dopamine receptor-substituted-cysteine mutants *(59,68)*. It was essential to develop a system that would allow a non-crosslinked receptor to run as a monomer on nonreducing SDS-PAGE. Our background construct FLAG-D2 dopamine receptor *(11)* ran almost exclusively as a heterogeneously glycosylated monomer of approx 65 kDa on nonreducing SDS-PAGE (Fig. 1A). Therefore, if this D2 dopamine receptor is oligomeric, the oligomer dissociates in SDS. Additionally, unlike some class C receptors, the D2 dopamine receptor is not an obligatory disulfide-linked dimer in the plasma membrane *(69,70)*.

As a control before introducing engineered cysteines into FLAG-D2 dopamine receptor for disulfide crosslinking experiments, we reacted FLAG-D2 dopamine receptor in intact cells with copper phenanthroline (CuP), an oxidizing reagent that promotes the formation of disulfide bonds directly between cysteines *(71,72)*. Reaction with CuP produced a new band of approx 133 kDa (Fig. 1A), which is approximately twice the size of monomer *(11)*. The fraction of total density that was present in the approx 133-kDa band was plotted against increasing CuP concentrations (Fig. 1), providing half-maximal crosslinking at 60 ± 10 μM CuP and maximal crosslinking of $80 \pm 14\%$ ($n = 3$).

The apparent mass of the crosslinked species was consistent with it being a homodimer of D2 dopamine receptor; however, because it was possible that it might represent D2 dopamine receptor crosslinked to another protein of similar size, the partners in the crosslinked species were definitively identified by co-immunoprecipitation of *myc*-D2 dopamine receptor stably coexpressed with FLAG-D2 dopamine receptor *(11)*. These results established that the approx 133-kDa band is a D2 dopamine receptor homodimer that is disulfide crosslinked via one of the remaining endogenous cysteines. Mutation of Cys168$^{4.58}$, but not Cys56$^{1.54}$, Cys126$^{3.44}$, or Cys356$^{6.47}$, in TM4 to Ser completely prevented CuP-induced crosslinking (Fig. 2; ref. *11*), dem-

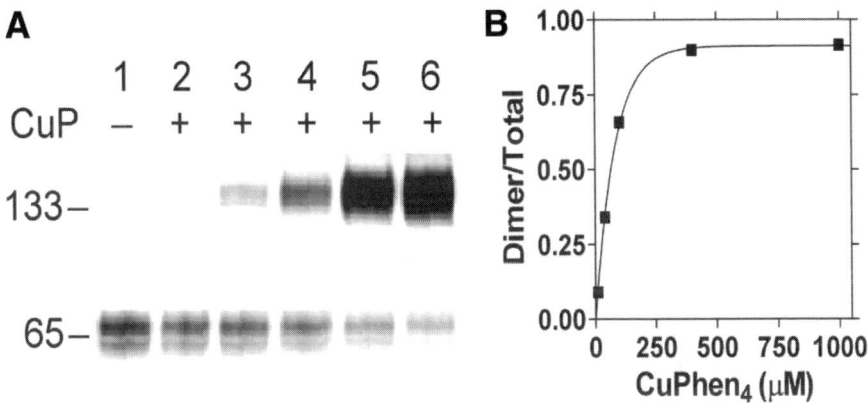

Fig.1. Crosslinking of D2 dopamine receptor to a homodimer by copper phenanthroline. (A) Treatment of FLAG–D2 dopamine receptor with 0, 10/40, 40/160, 100/400, 400/1600, 1000/4000 mM CuP (lanes 1–6, respectively). (B) Exponential association fit of dimer/total density plotted against CuP from panel a. The molecular masses of protein standards are given in kDa. Representative data from $n = 3$ experiments are shown. (Adapted from ref. *11*.)

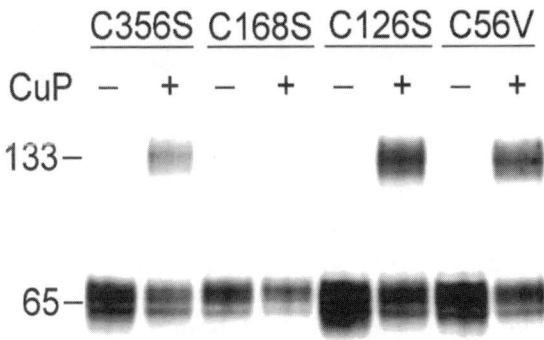

Fig. 2. Crosslinking of Cys mutants by 100/400 mM CuP in FLAG–D2 dopamine receptor. Representative data from $n = 3$ experiments are shown. (Adapted from ref. *11*.)

onstrating that this Cys at the extracellular end of TM4 forms the CuP-induced disulfide crosslink at a symmetrical homodimer interface. However, mutation of Cys168$^{4.58}$ to Ser does not prevent interaction at this D2 dopamine receptor dimer interface, based on the crosslinking of this construct observed when another residue at the same interface is simultaneously mutated to cysteine (unpublished observations).

Because crosslinking requires that only one of the two cysteines involved is modified initially by the reagent, and the derivatized cysteine then reacts by collision with the second unmodified cysteine, the rate of collision must be much faster than the rate of initial modification. This is consistent with the cysteines being very close initially. The very high fraction of receptor that can be crosslinked, the apparent specificity of the crosslinking (based on the appearance of a single homodimer band), and the lack of crosslinking of $Cys56^{1.54}$ (which, based on the bovine rhodopsin structure, has a lipid accessibility similar to $Cys168^{4.58}$) all argue for the proximity of the TM4 cysteines in the native state. Therefore, in the membrane, D2 dopamine receptor, untreated with CuP, very likely exists as a homodimer, but this dimer does not survive detergent solubilization.

Our finding that the site of crosslinking in D2 dopamine receptor is $Cys168^{4.58}$ at the extracellular end of TM4 is consistent with the hypothesis that TM4 forms a symmetrical dimer interface. Notably, the computational method we described earlier predicted $C^{4.58}$ as a possible dimeric contact in D2 dopamine receptor, together with $I^{4.48}$, $T^{4.55}$, and $L^{4.60}$. To reduce the number of false-positives among the predicted correlated mutations and to reveal additional loci of interaction between monomeric interfaces as discussed in detail earlier, we recently built geometrically feasible configurations of D2 dopamine receptor homodimers (in preparation). The information available from these monomer-based models is currently guiding experiments that are serving to map the entire interface of D2 dopamine receptor dimerization, with the goal of understanding the role of the interface in ligand binding and receptor activation and of discovering mutations that disrupt the interface between monomers, which therefore interfere with receptor function. Because of the success in the application of our interdisciplinary approach to D2 dopamine receptors, we propose to use our methodology as a tool to provide new insights into the understanding of the structural basis and functional consequences of GPCR oligomerization.

ACKNOWLEDGMENTS

Support from National Institutes of Health grants DA12923 and DA00060 (to H. Weinstein) and from MH57324 and MH54137 (to J. A. Javitch). M. Filizola is supported by NRSA T32 DA07135 from National Institute on Drug Abuse.

REFERENCES

1. Angers S, Salahpour A, Bouvier M. Dimerization: an emerging concept for G-protein coupled receptor ontogeny and function. Annu Rev Pharmacol Toxicol 2002;42:409–435.

2. Brady AE, Limbird LE. G-protein coupled receptor interacting proteins: emerging roles in localization and signal transduction. Cell Signal 2002;14:297–309.
3. George SR, O'Dowd BF, Lee SP. Oligomerization and its potential for drug discovery. Nature Rev Drug Disc 2002;1:808–820.
4. Milligan G, Ramsay D, Pascal G, Carrillo JJ. GPCR dimerisation. Life Sci 2003;74:181–188.
5. Lee SP, O'Dowd BF, Rajaram RD, Nguyen T, George SR. D2 dopamine receptor homodimerization is mediated by multiple sites of interaction, including an intermolecular interaction involving transmembrane domain 4. Biochemistry 2003;42:11,023–11,031.
6. Filizola M, Visiers, I., Skrabanek, L., Campagne, F. & Weinstein, H. Functional mechanisms of GPCRs in a structural context. In ed.^eds. Schousboe, A. & Bräuner-Osborne, H. Strategies in Molecular Neuropharmacology: Humana Press, Chapter 13, 2003, pp. 235–266.
7. Rocheville M, Lange DC, Kumar U, Sasi R, Patel RC, Patel YC. Subtypes of the somatostatin receptor assemble as functional homo- and heterodimers. J Biol Chem 2000;275:7862–7869.
8. Robbins MJ, Calver AR, Filippov AK, Hirst WD, Russell RB, Wood MD, Nasir S, Couve A, Brown DA, Moss SJ, Pangalos MN. GABA(B2) is essential for g-protein coupling of the GABA(B) receptor heterodimer J Neurosci 2001;21:8043–8052.
9. Rocheville M, Lange DC, Kumar U, Patel SC, Patel RC, Patel Y C. Receptors for dopamine and somatostatin: formation of hetero-oligomers with enhanced functional activity. Science 2000;288:154–157.
10. Pfeiffer M, Koch T, Schroder H, Laugsch M, Hollt V, Schulz S. Heterodimerization of somatostatin and opioid receptors cross-modulates phosphorylation, internalization, and desensitization. J Biol Chem 2002;277:19,762–19,772.
11. Guo W, Shi L, Javitch JA. The fourth transmembrane segment forms the interface of the dopamine D2 receptor homodimer. J Biol Chem 2003;278:4385–4388.
12. Filizola M, Guo W, Javitch JA, Weinstein H. Dimerization in G-protein coupled receptors: Correlation analysis and electron density maps of rhodopsin from different species suggest subtype-specific interfaces. Biophys J 2003;84(pt 2):2235.
13. Filizola M., Olmea O, Weinstein H. Prediction of heterodimerization interfaces of G-protein coupled receptors with a new subtractive correlated mutation method. Prot Eng 2002;15:881–885.
14. Filizola M, Olmea O, Weinstein H. Using correlated mutation analysis to predict the heterodimerization interface of GPCRs. Biophys J 2002;82(pt 2):2307.
15. Filizola M, Weinstein, H. Structural Models for Dimerization of G-Protein Coupled Receptors: The Opioid Receptor Homodimers. Biopolymers (Peptide Science) 2002;66:317–325.
16. Palczewski K, Kumasaka T, Hori T, Behnke CA, Motoshima H, Fox BA, LeTrong I, Teller DC, Okada T, Stenkamp RE, Yamamoto M, Miyano M. Crystal structure of rhodopsin: A G protein-coupled receptor Science 2000;289:739–745.

17. Gouldson PR, Snell CR, Reynolds, C. A. A new approach to docking in the beta 2-adrenergic receptor that exploits the domain structure of G-protein-coupled receptors. J Med Chem 1997;40:3871–3886.
18. Gouldson PR, Snell CR, Bywater RP, Higgs C, Reynolds CA. Domain swapping in G-protein coupled receptor dimers. Prot Eng 1998;11:1181–1193.
19. Dean MK, Higgs C, Smith RE, Bywater RP, Snell CR, Scott PD, Upton GJ, Howe TJ, Reynolds CA. Dimerization of G-protein-coupled receptors J Med Chem 2001;44:4595–4614.
20. Gouldson PR, Dean MK, Snell CR, Bywater RP, Gkoutos G, Reynolds CA. Lipid-facing correlated mutations and dimerization in G-protein coupled receptors. Prot Eng 2001;14:759–767.
21. Gkoutos GV, Higgs C, Bywater RP, Gouldson, PR, Reynolds CA. Evidence for dimerization in the b2-adrenergic receptor from the evolutionary trace method Intl J Quantum Chem Biophys Q 1999;74:371–379.
22. Kunishima N, Shimada Y, Tsuji Y, Sato T, Yamamoto M, Kumasaka T, Nakanishi S, Jingami H, Morikawa K. Structural basis of glutamate recognition by a dimeric metabotropic glutamate receptor Nature 2000;407:971–977.
23. Kniazeff J, Saintot PP, Goudet C, Liu J, Charnet A, Guillon G, Pin JP. Locking the dimeric GABA(B) G-protein-coupled receptor in its active state. J Neurosci 2004;24:370–377.
24. Ray K, Hauschild BC, Steinbach PJ Goldsmith PK, Hauache O, Spiegel AM. Identification of the cysteine residues in the amino-terminal extracellular domain of the human Ca(2+) receptor critical for dimerization. Implications for function of monomeric Ca(2+) receptor. J Biol Chem 1999;274:27,642–27,650.
25. Pace AJ, Gama L, Breitwieser GE. Dimerization of the calcium-sensing receptor occurs within the extracellular domain and is eliminated by Cys —> Ser mutations at Cys101 and Cys236 J Biol Chem 1999;274:11,629–11,634.
26. AbdAlla S, Zaki E, Lother H, Quitterer, U. Involvement of the amino terminus of the B(2) receptor in agonist-induced receptor dimerization. J Biol Chem 1999;274:26,079–26,084.
27. Abe J, Suzuki H, Notoya M, Yamamoto T, Hirose S. Ig-hepta, a novel member of the G protein-coupled hepta-helical receptor (GPCR) family that has immunoglobulin-like repeats in a long N-terminal extracellular domain and defines a new subfamily of GPCRs. J Biol Chem 1999;274:19,957–19,964.
28. Overton MC, Blumer, K. J. The extracellular N-terminal domain and transmembrane domains 1 and 2 mediate oligomerization of a yeast G protein-coupled receptor. J Biol Chem 2002;277:41,463–41,472.
29. Kuner R, Kohr G, Grunewald S, Eisenhardt G, Bach A, Kornau HC. Role of heteromer formation in GABAB receptor function. Science 1999;283:74–77.
30. White JH, Wise A, Main MJ, Green A, Fraser NJ, Disney GH, Barnes AA, Emson P, Foord SM,Marshall FH. Heterodimerization is required for the formation of a functional GABA(B) receptor. Nature 1998;396:679–682.
31. Cvejic S, Devi LA. Dimerization of the delta opioid receptor: implication for a role in receptor internalization. J Biol Chem 1997;272:26,959–26,964.
32. Kammerer RA, Frank S, Schulthess T, Landwehr R, Lustig A, Engel J. Heterodimerization of a functional GABAB receptor is mediated by parallel coiled-coil alpha-helices Biochemistry 1999;38:13,263–13,269.

33. Margeta-Mitrovic M, Jan YN, Jan LY. A trafficking checkpoint controls GABA(B) receptor heterodimerization. Neuron 2000;27:97–106.
34. Liang Y, Fotiadis D, Filipek S, Saperstein DA, Palczewski K, Engel A. Organization of the G protein-coupled receptors rhodopsin and opsin in native membranes. J Biol Chem 2003;278:21,655–21,662.
35. Fotiadis D, Liang Y, Filipek S, Saperstein DA, Engel A, Palczewski K. Atomic-force microscopy: Rhodopsin dimers in native disc membranes Nature 2003;421:127,128.
36. Hebert TE, Moffett S, Morello JP, Loisel TP, Bichet DG, Barret C, Bouvier M. A peptide derived from a beta2-adrenergic receptor transmembrane domain inhibits both receptor dimerization and activation. J Biol Chem 1996;271:16,384–16,392.
37. George SR, Lee SP, Varghese, G, Zeman PR, Seeman P, Ng GY, O'Dowd BF. A transmembrane domain-derived peptide inhibits D1 dopamine receptor function without affecting receptor oligomerization. J Biol Chem 1998;273:30,244–30,248.
38. Ng GY, O'Dowd BF, Lee SP, Chung HT, Brann MR, Seeman P, George SR. Dopamine D2 receptor dimers and receptor-blocking peptides. Biochem. Biophys Res Commun 1996;227:200–204.
39. Klco JM, Lassere TB, Baranski TJ. C5a receptor oligomerization. I. Disulfide trapping reveals oligomers and potential contact surfaces in a G protein-coupled receptor J Biol Chem 2003;278:35,345–35,353.
40. Carrillo JJ, Pediani J, Milligan G. Dimers of class A G protein-coupled receptors function via agonist-mediated trans-activation of associated G proteins J Biol Chem 2003;278:42,578–42,587.
41. Overton MC, Chinault SL, Blumer, K. J. Oligomerization, biogenesis, and signaling is promoted by a glycophorin A-like dimerization motif in transmembrane domain 1 of a yeast G protein-coupled receptor J Biol Chem 2003;278:49,369–49,377.
42. Javitch JA, Shi L, Simpson MM, Chen J, Chiappa V, Visiers I, Weinstein H, Ballesteros JA. The fourth transmembrane segment of the dopamine D2 receptor: accessibility in the binding-site crevice and position in the transmembrane bundle Biochemistry 2000;39:12,190–12,199.
43. Jordan BA, Devi LA. G-protein-coupled receptor heterodimerization modulates receptor function. Nature 1999;399:697–700.
44. Mellado M, Rodriguez-Frade JM, Vila-Coro AJ, Fernandez S, Martin De AnaA, Jones DR, Toran JL, Martinez, A. C. Chemokine receptor homo- or heterodimerization activates distinct signaling pathways EMBO J 2001;20:2497–2507.
45. Gouldson PR, Higgs C, Smith RE, Dean MK, Gkoutos GV, Reynolds CA. Dimerization and domain swapping in G-protein-coupled receptors: a computational study. Neuropsychopharmacology 2000;23:S60–S77.
46. Gouldson PR, Reynolds CA. Simulations on dimeric peptides: evidence for domain swapping in G-protein-coupled receptors? Biochem Soc Trans 1997;25:1066–1071.

47. Maggio R, Vogel Z, Wess J. Coexpression studies with mutant muscarinic/ adrenergic receptors provide evidence for intermolecular "cross-talk" between G-protein-linked receptors Proc Natl Acad Sci USA 1993;90:3103–3107.
48. Lee SP, O'Dowd BF, Ng GY, Varghese, G, Akil H, Mansour A, Nguyen T, George SR. Inhibition of cell surface expression by mutant receptors demonstrates that D2 dopamine receptors exist as oligomers in the cell. Mol Pharmacol 2000;58:120–128.
49. Hamdan FF, Ward SD, Siddiqui N A, Bloodworth L M, Wess J. Use of an *in situ* disulfide cross-linking strategy to map proximities between amino acid residues in transmembrane domains I and VII of the M3 muscarinic acetylcholine receptor. Biochemistry 2002;41:7647–7658.
50. Hadac EM, Ji Z, Pinon DI, Henne RM, Lybrand TP, Miller LJ. A peptide agonist acts by occupation of a monomeric G protein-coupled receptor: dual sites of covalent attachment to domains near TM1 and TM7 of the same molecule make biologically significant domain-swapped dimerization unlikely. J Med Chem 1999;42:2105–2111.
51. Schulz A, Grosse R, Schultz G, Gudermann T, Schoneberg T. Structural implication for receptor oligomerization from functional reconstitution studies of mutant V2 vasopressin receptors. J Biol Chem 2000;275:2381–2389.
52. Mercier JF, Salahpour A, Angers S, Breit A, Bouvier M. Quantitative assessment of beta 1 and beta 2-adrenergic receptor homo and hetero-dimerization by bioluminescence resonance energy transfer. J Biol Chem 2002;277:44,925–44,931.
53. Mukhopadhyay S, McIntosh HH, Houston DB, Howlett AC. The CB(1) cannabinoid receptor juxtamembrane C-terminal peptide confers activation to specific G proteins in brain. Mol Pharmacol 2000;57:162–170.
54. Dinger MC, Bader JE, Kobor AD, Kretzschmar AK, Beck-Sickinger AG. Homodimerization of neuropeptide Y receptors investigated by fluorescence resonance energy transfer in living cells J Biol Chem 2003;10:10.
55. Baneres JL, Parello J. Structure-based analysis of GPCR function: evidence for a novel pentameric assembly between the dimeric leukotriene B4 receptor BLT1 and the G-protein J Mol Biol 2003;329:815–829.
56. Singh R, Hurst DP, Barnett-Norris J, Lynch DL, Reggio PH, Guarnieri F. Activation of the cannabinoid CB1 receptor may involve a W6.48/F3.36 rotamer toggle switch J Peptide Res 2002;60:357–370.
57. Horn F, Weare J, Beukers MW, Horsch S, Bairoch A, Chen W, Edvardsen O, Campagne F, Vriend G. GPCRDB: an information system for G protein-coupled receptors Nucleic Acids Res 1998;26:275–279.
58. Thompson JD, Higgins DG, Gibson TJ. CLUSTAL W: improving the sensitivity of progressive multiple sequence alignment through sequence weighting, positions-specific gap penalties and weight matrix choice. Nucleic Acids Res 1994;22:4673–4680.
59. Ballesteros JA, Shi L, Javitch JA. Structural Mimicry in G Protein-Coupled Receptors: Implications of the High-Resolution Structure of Rhodopsin for Structure-Function Analysis of Rhodopsin-Like Receptors. Mol Pharmacol 2001;60:1–19.

60. Visiers I, Ballesteros JA, Weinstein H. Three-dimensional representations of G protein-coupled receptor structures and mechanisms. Meth Enzymol 2002;343:329–371.
61. Sali A, Potterton L, Yuan F, van Vlijmen H, Karplus M. Evaluation of comparative protein modeling. Proteins Struct Funct Genet 1995;23:318–326.
62. Brooks BR., Bruccoleri RE, Olafson BD, States DJ, Swaminathan S , Karplus M. CHARMM: A program for macromolecular energy, minimization and dynamics calculations. J Comput Chem 1983;4:187–217.
63. Gobel U, Sander C, Schneider R, Valencia A. Correlated mutations and residue contacts in proteins Proteins 1994;18:309–317.
64. Olmea O, Valencia A. Improving contact predictions by the combination of correlated mutations and other sources of sequence information Fold Des 1997; 2: S25–S32.
65. Rodriguez-Frade JM, Vila-Coro AJ, de Ana AM, Albar JP, Martinez-A C, Mellado M. The chemokine monocyte chemoattractant protein-1 induces functional responses through dimerization of its receptor CCR2. Proc Natl Acad Sci USA 1999;96:3628–3633.
66. Vila-Coro AJ, Mellado M, Martin de Ana A, Lucas P, del Real G, Martinez-A C, Rodriguez-Frade JM. HIV-1 infection through the CCR5 receptor is blocked by receptor dimerization. Proc Natl Acad Sci USA 2000;97:3388–3393.
67. Vila-Coro AJ, Rodriguez-Frade JM, Martin De Ana A, Moreno-Ortiz MC, Martinez-A C, Mellado M. The chemokine SDF-1alpha triggers CXCR4 receptor dimerization and activates the JAK/STAT pathway. FASEB J 1999;13:1699–1710.
68. Shi L, Javitch JA. The binding site of aminergic G protein-coupled receptors. Annu Rev Pharmacol Toxicol 2002;42:437–467.
69. Bai M, Trivedi S, Brown EM. Dimerization of the extracellular calcium-sensing receptor (CaR) on the cell surface of CaR-transfected HEK293 cells. J Biol Chem 1998;273:23,605–23,610.
70. Romano C, Yang WL, O'Malley KL. Metabotropic glutamate receptor 5 is a disulfide-linked dimer. J Biol Chem 1996;271:28,612–28,616.
71. Wu J, Kaback HR. A general method for determining helix packing in membrane proteins in situ: helices I and II are close to helix VII in the lactose permease of Escherichia coli. Proc Natl Acad Sci USA 1996;93:14,498–14,502.
72. Careaga CL, Falke JJ. Thermal motions of surface alpha-helices in the D-galactose chemosensory receptor. Detection by disulfide trapping. J Mol Biol 1992;226:1219–1235.
73. Ballesteros JA, Weinstein H. Integrated methods for the construction of three-dimensional models and computational probing of structure-function relations in G protein-coupled receptors. Methods Neurosci 1995;25:366–428.

12
Functional Complementation and the Analysis of GPCR Dimerization

Graeme Milligan, Juan J. Carrillo, and Geraldine Pascal

1. INTRODUCTION

Cloning of complementary DNAs (cDNAs), which are predicted to encode G protein-coupled receptors (GPCRs), required a mechanism to ascertain if the single polypeptide encoded by such cDNAs was sufficient to generate the pharmacology and function anticipated for the receptor in question. Because this generally was the case—and despite evidence of greater complexity (reviewed in ref. *1*)—it became axiomatic that GPCRs were single, seven-transmembrane-span polypeptides. However, a series of immunoblotting studies has suggested that a fraction of cellular GPCRs might exist as dimers or higher order species in both transfected cell lines and native tissues *(2–4)*. The known propensity of hydrophobic proteins to aggregate—particularly when samples were heated prior to separation by sodium dodecyl sulfate-polyacrylamide gel electrophoresis (SDS-PAGE)—meant that it was possible to disregard these data. However, studies in which differentially epitope-tagged forms of a single GPCR could be co-immunoprecipitated following their co-expression in heterologous cell lines provided considerable evidence for the presence of dimeric or, indeed, higher oligomeric complexes *(2,5)*. This was generally not observed when the two forms of the GPCR were expressed in separate cell populations that were mixed prior to membrane solubilization and immunoprecipitation. These results indicate that co-immunoprecipitation did not result simply from aggregation of the hydrophobic transmembrane elements of these polypeptides following removal of lipid by treatment with detergents. Equivalent studies then began to examine the proclivity of different, co-expressed GPCRs to be co-immunoprecipitated. These studies have produced a large body of evidence sup-

From: *Contemporary Clinical Neuroscience: The G Protein-Coupled Receptors Handbook*
Edited by: L. A. Devi © Humana Press Inc., Totowa, NJ

porting the capacity of GPCRs to exist as heterodimers/-oligomers (for review, *see* refs. 6 and 7). However, studies that have reported a very broad capacity of particular GPCRs to allow co-immunoprecipitation of co-expressed GPCRs *(8)* have questioned the relevance of GPCR co-immunoprecipitation data when this is not accompanied by similar data produced by alternative strategies.

There is also a concern that many "co-immunoprecipitation" studies are poorly performed and controlled. For example, many studies in this area centrifuge samples to remove particulate material after detergent treatment to generate a "soluble" preparation for the immunoprecipitation steps. Often, however, the centrifugation step is performed for only short periods, and the centrifugal force is too low to achieve this end. This inevitably results in small membrane fragments being present in the "soluble" fraction and the possibility that "co-immunoprecipitation" represents nothing more than both polypeptides presenting within the same membrane fragment.

These types of issues have resulted in the widespread adoption and use of techniques that can monitor protein–protein interactions in either membrane preparations or intact living cells. Various forms of resonance energy transfer techniques have been employed to explore GPCR dimerization/oligomerization because of their relative simplicity and exquisite dependence on distance between the energy donor and acceptor species *(9–14)*. These issues and the utility of the systems have recently been reviewed *(15,16)* and are discussed further within this volume *(17)*. An alternative strategy that can be used instead of, or in parallel with, resonance energy transfer techniques is functional complementation produced following co-expression of pairs of distinct GPCR or GPCR–G protein fusion mutants.

2. COMPLEMENTATION OF PAIRS OF MUTANT GPCRS

GPCRs can be assembled from co-expressed fragments *(12,18)*. This has led to the hypothesis that segments of GPCRs can be considered distinct domains capable of independent folding. Extensions of this idea have also allowed the envisaging of models of GPCR dimers that include explicit requirement for domain swapping *(19,20)*. Although direct experimental evidence in favor of such models is somewhat limited, such evidence would potentially explain marked variations in ligand pharmacology that are sometimes observed when two GPCRs capable of heterodimer formation are co-expressed *(21)* or when chimeric GPCRs are constructed *(21,22)*. Reciprocal chimeric GPCRs between the α_{2C}-adrenoceptor and the m3 muscarinic acetylcholine receptor that contained the N-terminus and transmembrane segments I–V of one receptor and transmembrane segments VI–VII and the

C-terminal tail of the second bound neither muscarinic or α_2-adrenergic ligands. However, the co-expression of the two chimeras resulted in regeneration of some level of high-affinity binding for both muscarinic and α_2-adrenergic antagonists *(22)*. These workers were also able to reconstitute agonist-mediated signal transduction when pairs of distinct, functionally impaired mutant muscarinic receptors were co-expressed *(22)*. Related studies on the angiotensin II type 1 receptor showed that forms of this GPCR with specific mutations in either transmembrane region III or transmembrane segment V were unable to bind ligands. However, ligand binding was observed upon co-expression *(23)*. These studies were central in providing evidence in favor of molecular interactions between GPCRs. Despite these conclusions, the number of ligand-binding sites produced following co-expression was low, and it is unclear if this simply reflects that the individual mutations are very poorly expressed compared to the wild-type protein, or if the efficiency of "reconstitution" is poor. One potential explanation is that if the reconstitution of high-affinity ligand binding requires a domain exchange mechanism, then this may be thermodynamically feasible but uncommon compared with dimerization based on linear packing, which would not be anticipated to regenerate ligand binding on co-expression of the two mutants.

The first definitive evidence of GPCR hetero-dimerization was the recognition that the γ-aminobutyric acid $(GABA)_BR1$ polypeptide could bind agonist (although with lower than the anticipated affinity) and antagonist ligands but could not generate signals or be trafficked effectively to the cell surface unless co-expressed with the $GABA_BR2$ polypeptide *(24,25)*. Similarly to other family C GPCRs, the long extracellular N-terminal extension binds the ligand, but the architecture of the prototypic seven-transmembrane core defines the signal transduction unit. Because such GPCRs can be considered bifunctional polypeptides, chimeras between the extracellular region and the transmembrane segments and intracellular elements have been generated to aid understanding of the mechanisms of signal propagation, from ligand binding to G protein activation *(26)*.

Several family A GPCRs that respond to large glycoprotein hormones also have long extracellular N-termini and, therefore, are amenable to related approaches. One well-studied example is the luteinizing hormone receptor. This GPCR has been used to generate differentially defective mutant pairs—one in the extracellular exodomain that lacks binding of the ligand human chorionic gonadotropin and a second that can bind the ligand but is unable to signal because of a mutation in the endodomain *(27,28)*. Their co-expression resulted in the restoration of ligand generation of the second messenger cyclic adenosine monophosphate (cAMP). Again, although this provided a persuasive case for intermolecular interactions among GPCRs, the amplification mechanisms resulting in cAMP generation (which also often result in the characteristic of "spare receptors") do not allow for easy assessment of the effectiveness of the complementation process.

It has been suggested that unlike many other GPCRs, the luteinizing hormone receptor does not form stable dimers. Such functional complementation studies can provide evidence for both *cis*-activation (i.e., ligand bound to the exodomain, causing activation on the same endodomain) or *trans*-activation, where the ligand activates the partner endodomain within a dimer *(29)*. It is noteworthy that such a model may also explain the complex interactions between activated protease-activated receptor (PAR)3 and PAR4 GPCRs *(30)* and, potentially, other complex interactions between PAR subtypes *(31)*. The glycoprotein hormones and PARs are unusual examples of family A, rhodopsin-like GPCRs. For the bulk of these, the agonist ligands bind at least partially within the cavity created by the seven-transmembrane helix architecture. We wished to develop a generic functional complementation strategy that would be suitable to study both homo- and heterodimerization of family A GPCRs and that would provide quantitative assays for the effectiveness of complementation. Therefore, we employed pairs of nonfunctional, but potentially complementary, GPCR–G protein α-subunit fusion proteins.

3. FUNCTIONAL COMPLEMENTATION OF PAIRS OF GPCR–G PROTEIN FUSIONS

3.1. GPCR–G Protein Fusions

The first GPCR–G protein fusion was constructed between the β_2-adrenoceptor and the α-subunit of the long isoform of the G protein G_s. When this was expressed in S49 cyc⁻ cells lacking expression of $G\alpha_s$, the β-adrenoceptor agonist isoprenaline was able to stimulate adenylyl cyclase *(32)*, thus confirming the functionality of both elements of the fusion construct. Although the fusion construct had several interesting characteristics *(32)*, it was initially viewed as little more than a curiosity. However, over the intervening period, a wide range of GPCRs and G proteins have been used to generate similar fusions *(33,34)*. These have been used to address issues such as the selectivity of GPCRs in activating different G proteins *(35)*, the role and regulation of posttranslational acylation in the function and cellular location of GPCRs and G proteins *(36–38)*, as reagents to screen for agonists at "orphan" GPCRs *(39,40)* and, most recently, to explore the basis and selectivity of GPCR dimerization *(41)*.

GPCR–G protein fusions can also be considered as bifunctional polypeptides because, although they are generated from a single open frame, they contain the sequence and functionality of both GPCR and G protein. Therefore, mutants that eliminate the ability of agonists to generate a signal corresponding to activation of the G protein can be produced by alterations in the sequence of either the GPCR or the G protein element. This allows for the production of distinct pairs of fusions that are individually inactive but have the potential to complement function if they interact. Two obvious regions can be targeted for the G protein segment of the fusion. All G protein α-subunits have a conserved Val-Gly-Gly-Gln-Arg sequence (Table 1), where mutation of the second Gly to Ala generates a form of the G protein that is unable to exchange guanosine diphosphate (GDP) for guanosine triphosphate (GTP) and hence become activated. This provides a simple strategy to generate GPCR–G protein fusion proteins that are unable to respond to agonist ligands even though the GPCRs linked to such mutants are wild-type and able to bind ligands. Equally, because it is well established that the extreme C-terminal region of most G protein α-subunits is a key contact domain for GPCR-mediated activation *(42)*, judicious mutation in this region can produce forms of the G protein that are not responsive to GPCRs. For example, because pertussis toxin-catalyzed adenosine diphosphate (ADP)-ribosylation of the Cys residue four amino acids from the C-terminus of all widely expressed G_i-family G protein α-subunits prevents GPCR-mediated activation of these G proteins, Bahia et al. *(43)* replaced this residue in $G\alpha_{i1}$ with each of the other naturally occurring amino acids and then assessed the ability of each of the mutants to be activated by agonist occupancy of the α_{2A}-adrenoceptor. Substitution of the Cys with more hydrophilic amino acids reduced coupling effectiveness, and there was no significant activation with either positively or negatively charged amino acids at this position. Similar data have been produced for activation of equivalently mutated forms of $G\alpha_{i3}$ by the 5-HT_{1A} receptor *(44)*. Several of the $G\alpha_{i1}$ mutations have been constructed into fusion proteins with the α_{2A}-adrenoceptor and have been used to monitor the effects on these alterations on information transfer from GPCR to G protein, which was measured as alterations in agonist potency and relative efficacy *(45,46)*. Similarly, mutation of the Tyr four amino acids from the C-terminal of $G\alpha_{11}$ resulted in reduction of its ability to be activated by the α_{1b}-adrenoceptor *(47)*. With Asp at this position, no activation was observed by the agonist-occupied α_{1b}-adrenoceptor, both in co-expression studies and when the modified G protein was constructed into a fusion with this GPCR *(47)*. Although it has not been explored in such detail, it is well known that the lack of function of $G\alpha_s$ in the S49

Table 1
A Generic Strategy to Generate Inactive G Protein α-Subunits

G α-subunit	Species	Sequence
G_s	Human	VGGQRDERRK
G_{olf}	Human	VGGQRDERRK
G_{i1}	Human	VGGQRSERKK
	Rat	VGGQRSERKK
G_{i2}	Human	VGGQRSERKK
	Mouse	VGGQRSERKK
G_{i3}	Human	VGGQRSERKK
	Rat	VGGQRSERKK
G_{o1}	Human	VGGQRSERKK
	Bovine	VGGQRSERKK
G_{o2}	Human	VGGQRSERKK
G_z	Rat	VGGQRSERKK
	Bovine	VGGQRSERKK
Transducin$_1$	Human	VGGQRSERKK
Transducin$_2$	Rat	VGGQRSERKK
Gustducin	Human	VGGQRSERKK
G_q	Mouse	VGGQRSERKK
	Human	VGGQRSERKK
G_{11}	Rat	VGGQRSERKK
	Human	VGGQRSQRQK
G_{12}	Rat	VGGQRSQRQK
	Human	VGGQRSERKR
G_{13}	Human	VGGQRSERRK
G_{14}	Mouse	VGGQRSERRK
G_{15}	Human	VGGQRSERKK
G_{16}		

G protein α-subunits are highly conserved between mammalian species, particularly in core domains involved in the binding and hydrolysis of guanine nucleotides. Alteration of the highlighted glycine residue to alanine results in forms of the G protein that are unable to bind GTP. This provides a generic means to generate G proteins that are unable to be activated by G protein-coupled receptors (GPCRs) and thus to produce one of the pair of inactive GPCR–G protein fusion proteins used in the functional complementation strategy.

unc cell line reflects an Arg to Pro substitution six amino acids from the C-terminus of this G protein *(48)*. Therefore, although the details of contacts between different GPCRs and a specific G protein differ subtly *(49)*, alterations in the C-terminal tail of the G protein can produce forms than cannot be activated by the GPCR of choice.

A range of potential strategies are available to produce inactive GPCR–G protein fusions via mutation of the GPCR. However, the most generic (at least, for many rhodopsin-like class A GPCRs) appears to be via mutation of one or more highly conserved hydrophobic residues in the second intracellular loop of the GPCR to acidic residues. The first of these is frequently three amino acids downstream of the highly conserved DRY domain, and the second is four amino acids further downstream (Table 2).

3.2. Complementation of GPCR–G Protein Fusions in Cell Membranes

A useful feature of GPCR–G protein fusions is that the defined 1:1 stoichiometry ensures that saturation ligand-binding studies define the expression levels of the G protein as well as the GPCR. Agonist stimulation of the binding of [^{35}S]GTPγS *(50)* to such fusion proteins can then provide a proximal, and potentially quantitative, assay of G protein activation which direct experiments have shown to be linear over a good range of membrane amounts *(38)*. It is important to demonstrate that agonist-mediated binding of [^{35}S]GTPγS is truly to the G protein element of the fusion rather than to endogenously expressed G proteins. This is an important issue because—at least, for some fusion constructs—it is clear that activation of endogenously expressed G proteins can occur *(51,52)*. However, generally, this seems to be a serious concern only with high-level expression of the GPCR–G protein fusion or when a free G protein is also co-expressed at high levels *(53)*. For fusions incorporating members of the G_i family of G proteins, mutation of the pertussis toxin-sensitive Cys residue to a hydrophobic amino acid, such as Ile, prevents pertussis toxin-catalyzed ADP-ribosylation without inhibiting GPCR-mediated activation of the G protein *(45,51)*. Therefore, following transfection of cells to express a GPCR–G protein fusion containing such a pertussis toxin-resistant mutant, treatment of the cells with pertussis toxin results in ADP-ribosylation of only the endogenously expressed forms of G_i. Binding of [^{35}S]GTPγS to these expressed forms of G_i hence cannot be stimulated by agonist. Therefore, following immunoprecipitation with an anti-$G\alpha_i$ antiserum, agonist-enhanced binding of the nucleotide must reflect loading of the G protein element of the fusion protein. For example, membranes prepared from pertussis toxin-treated HEK293 cells expressing a δ-opioid peptide (DOP)-opioid receptor–Cys^{351}Ile$G\alpha_{i1}$ fusion protein (15 fmol of [^{3}H]antagonist binding sites) were subjected to a [^{35}S]GTPγS binding assay in the absence or presence of the agonist D-ala^2, D-leu^5-enkephalin (DADLE) and were then subjected to immunoprecipitation. Radioactivity in the presence of the agonist was approximately sixfold higher than in its absence, con-

Table 2
A Generic Strategy to Generate GPCR Variants Unable to Activate G Proteins

GPCR	Specie	G protein	Sequence
5-HT$_{1A}$ receptor	Human	G$_i$/G$_o$	DRYWAITDPID
5-HT$_{1B}$ receptor	Mouse	G$_i$/G$_o$	DRYWAITDAVE
5-HT$_{1D}$ receptor	Rabbit	G$_i$/G$_o$	DRYWAITDALE
5-HT$_{2A}$ receptor	Rat	G$_q$/G$_{11}$	DRYVAIQNPIH
5-HT$_{2C}$ receptor	Rat	G$_q$/G$_{11}$	DRYVAIRNPIE
5-HT$_4$ receptor	Mouse	G$_s$	DRYYAICCQPL
5-HT$_6$ receptor	Rat	G$_s$	DRYLLILSPLR
α$_{1a}$-adrenoceptor	Bovine	G$_q$/G$_{11}$	DRYIGVSYPLR
α$_{1b}$-adrenoceptor	Hamster	G$_q$/G$_{11}$	DRYIGVRYSLQ
α$_{2b}$-adrenoceptor	Rat	G$_i$/G$_o$	DRYWAVSRALE
β$_1$-adrenoceptor	Human	G$_s$	DRYLAITSPFR
β$_2$-adrenoceptor	Bovine	G$_s$	DRYLAITSPFK
β$_3$-adrenoceptor	Mouse	G$_s$	DRYLAVTNPLR
A1 adenosine receptor	Human	G$_i$/G$_o$	DRYLRVKIPLR
A3 adenosine receptor	Human	G$_i$/G$_o$	DRYLRVKLTVR
Acetylcholine M1 receptor	Mouse	G$_q$/G$_{11}$	DRYFSVTRPLS
Acetylcholine M2 receptor	Human	G$_i$/G$_o$	DRYFCVTKPLT
Acetylcholine M3 receptor	Mouse	G$_q$/G$_{11}$	DRYFSITRPLT
Angiotensin AT$_{1A}$ receptor	Human	G$_q$/G$_{11}$	DRYLAIVHPMK
Angiotensin AT$_{1B}$ receptor	Rat	G$_q$/G$_{11}$	DRYLAIVHPMK
Bradykinin B2 receptor	Human	G$_q$/G$_{11}$	DRYLALVKTMS
Chemokine CXCR3	Mouse	G$_i$/G$_o$	DRYLSIVHATQ
Chemokine CXCR4	Human	G$_i$/G$_o$	DRYLAIVHATN
Dopamine D2 receptor	Mouse	G$_i$/G$_o$	DRYTAVAMPML
Dopamine D3 receptor	Rat	G$_i$/G$_o$	DRYTAVVMPVH
FSH receptor	Bovine	G$_s$	ERWHTITHAMQ
GnRH receptor	Mouse	G$_q$/G$_{11}$	DRSLAITQPLA
Histamine H1 receptor	Mouse	G$_q$/G$_{11}$	DRYRSVQQPLR
Histamine H2 receptor	Human	G$_s$	DRYCAVMDPLR
LH receptor	Mouse	G$_s$	ERWHTITYAVQ
Melanocortin 2 receptor	Human	G$_s$	DRYITIFHALR
DOP-opioid receptor	Rat	G$_i$/G$_o$, G$_z$	DRYIAVCHPVK
KOP-opioid receptor 1	Mouse	G$_i$/G$_o$	DRYIAVCHPVK
MOP-opioid receptor 1	Rat	G$_i$/G$_o$, G$_z$	DRYIAVCHPVK
Rhodopsin	Bovine	Gt	ERYVVVCKPMS
Oxytocin receptor	Rat	G$_q$/G$_{11}$	DRCLAICQPLR
P2U purinoceptor 1	Rat	G$_q$/G$_{11}$	HRCLGVLRPLH
Prostaglandin D2 receptor	Mouse	G$_s$	ECWLSLGHPFF
Prostaglandin E2 receptor	Rat	G$_q$/G$_{11}$	ERCVGVTQPLI
Somatostatin receptor 2	Human	G$_i$/G$_o$ G$_q$	DRYLAVVHPIK
Thyrotropin-releasing hormone receptor-1	Rat	G$_q$/G$_{11}$	ERYIAICHPIK

The vast majority of class A, rhodopsin-like G protein-coupled receptors (GPCRs) possess a pa[ir] of hydrophobic residues (highlighted) in the second intracellular loop, downstream of the highly

continued

firming the functionality of this construct. When similar experiments were performed with a version of this fusion protein that also incorporated a Gly^{203}Ala mutation in the G protein, DADLE did not enhance incorporation of the nucleotide. This was also true for a version of the fusion that incorporated Val^{150}Glu and Val^{154}Asp mutations into the second intracellular loop of the GPCR element but that was wild-type for Gα_{i1}, other than the pertussis toxin-resistant Cys^{351}Ile mutation. However, when Val^{150}Glu,Val^{154}Asp DOP–Cys^{351}IleGα_{i1} was co-expressed with DOP–Gly^{203}Ala,Cys^{351}IleGα_{i1} membranes prepared after treatment with pertussis toxin and containing 15 fmol of DOP-opioid receptor binding sites displayed marked reconstitution of [^{35}S]GTPγS binding in response to DADLE (Fig. 1). Assuming that the intracellular loop 2 mutations do not interfere with DOP–DOP interactions, that GPCRs function as dimers, that dimers bind two ligands, and that Val^{150}Glu,Val^{154}Asp DOP–Cys^{351}IleGα_{i1} and DOP–Gly^{203}Ala,Cys^{351}IleGα_{i1} express to equal extents in the cotransfections, simple arithmetic defines that the heterodimer (containing one copy of each of Val^{150}Glu,Val^{154}Asp DOP–Cys^{351}IleGα_{i1} and DOP–Gly^{203}Ala,Cys^{351}IleGα_{i1}) should represent only 50% of the ligand-binding sites. Of the other 50%, 25% would be expected to be Val^{150}Glu,Val^{154}Asp DOP–Cys^{351}IleGα_{i1} homodimers, and 25% would be expected to be DOP-Gly^{203}Ala,Cys^{351}IleGα_{i1} homodimers.

As demonstrated earlier, neither the Val^{150}Glu,Val^{154}Asp DOP–Cys^{351}IleGα_{i1} nor the DOP–Gly^{203}Ala,Cys^{351}IleGα_{i1} homodimers bind [^{35}S]GTPγS in response to agonists. Therefore, conceptually, following coexpression, membranes containing 30 fmol of antagonist binding sites could be anticipated to bind the same level of [^{35}S]GTPγS in response to agonist as membranes expressing 15 fmol of the DOP–Cys^{351}Ile Gα_{i1} fusion protein. Reconstitution was not this effective (Fig. 1). This may imply that the quaternary structure of the GPCR–G protein fusion is a higher order oligomer rather than a dimer or that there is a complex mixture of species that may also include monomers. Further analyses are required to test these ideas.

Table 2 *(From opposite page)* conserved 'DRY' domain at the bottom of transmembrane region III. In the examples we tested, alteration of one or both of these to acidic residues resulted in forms of the GPCR that are unable to activate G protein. This provides a potentially generic means to form an inactive GPCR–G protein fusion protein. Such mutations do not affect the binding of antagonist ligands. The effect on the binding of agonist ligands is more complex and generally reflects whether the GPCR in question shows a significant alteration in agonist binding in response to addition of guanine nucleotides. For example, many G$_q$/G$_{11}$-coupled GPCRs display only a marginal effect on agonist binding affinity; however, for G$_s$ and G$_i$/G$_o$-coupled receptors, this alteration in affinity can be substantial.

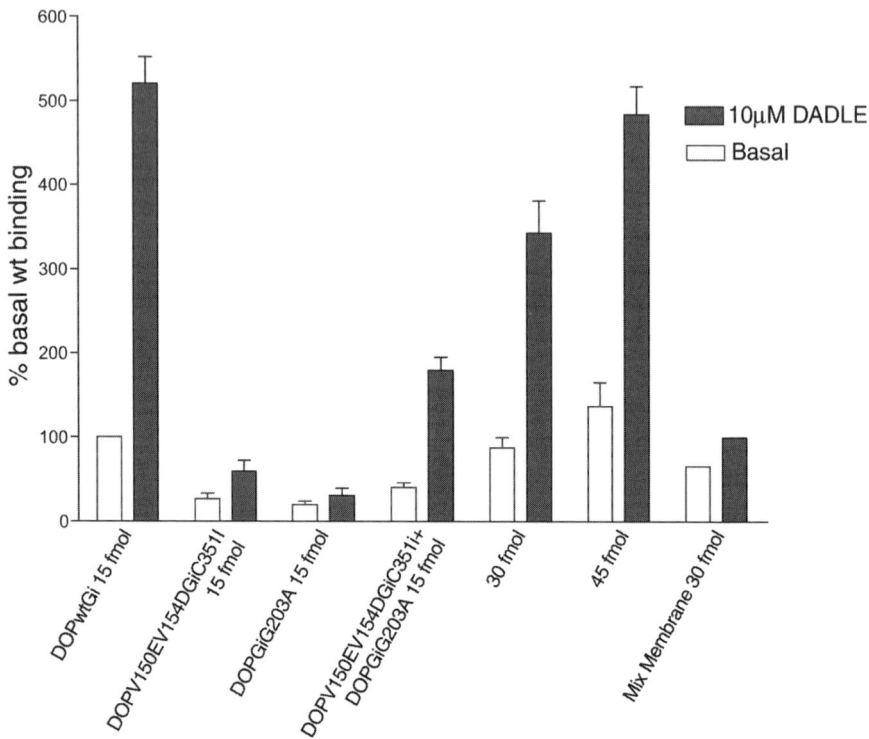

Fig. 1. Functional complementation of agonist-mediated [^{35}S]GTPγS binding by co-expression of pairs of inactive DOP-opioid receptor–Gα$_{i1}$ fusion proteins. A fusion protein was created between the human DOP-opioid receptor and a pertussis toxin-resistant Cys^{351}Ile mutant of Gα$_{i1}$. Further modifications to this fusion introduced either Val^{150}Glu,Val^{154}Asp mutations into the receptor element or Gly^{203}Ala into the G protein. Each of these three fusion proteins was transfected into HEK293 cells that were then treated with pertussis toxin to cause ADP-ribosylation of endogenously expressed forms of Gα$_i$/Gα$_o$. Expression levels of the fusion proteins were measured by the specific binding of the opioid receptor antagonist [^{3}H]diprenorphine and membrane amounts containing 15 fmol of each fusion protein used in [^{35}S]GTPγS binding studies, which were conducted in the absence or presence of the synthetic enkephalin DADLE (10 µM). At the termination of assay, samples were immunoprecipitated with an antiserum directed to the C-terminal decapeptide of Gα$_{i1}$ and were then counted. Equivalent studies were also performed on membranes of cells into which Val^{150}Glu,Val^{154}Asp DOP-opioid receptor–Cys^{351}IleGα$_{i1}$ and DOP-opioid receptor–Gly^{203}Ala,Cys^{351}IleGα$_{i1}$ were co-expressed. Co-expression reconstituted DADLE-stimulated [^{35}S]GTPγS binding, but this was not produced when membrane preparations individually expressing either Val^{150}Glu,Val^{154}Asp DOP-opioid receptor–Cys^{351}IleGα$_{i1}$ or DOP-opioid receptor–Gly^{203}Ala,Cys^{351}IleGα$_{i1}$ were mixed.

Importantly, however, co-expression is required to reconstitute function. Membranes prepared from cells each expressing one of the inactive fusion constructs did not allow agonist-stimulated binding of [^{35}S]GTPγS when mixed. Equivalent fusion constructs incorporating the μ-opioid peptide (MOP)-opioid receptor have similar characteristics, and reconstitution of agonist function is again achieved following co-expression of the pair of individually inactive fusion constructs. Co-expression of one inactive fusion containing the MOP-opioid receptor, with the partner containing the DOP-opioid receptor, also results in reconstitution of function and is consistent with the formation of a MOP–DOP heterodimer *(54,55)*.

Other G proteins are not targets for pertussis toxin-catalyzed ADP-ribosylation, but are widely expressed. Therefore, other strategies are required to ensure that enhanced binding of [^{35}S]GTPγS truly reflects activation of the G protein element of the fusion, rather than endogenous G protein. The α_1-adrenoceptor agonist phenylephrine elevates binding of [^{35}S]GTPγS following expression of a fusion protein between the α_{1b}-adrenoceptor and $G\alpha_{11}$ in HEK293 cells and membrane preparation *(53)*. However, nucleotide binding in the absence of agonist is high, thus limiting the agonist-induced signal to background. The vast majority of the agonist-independent binding is not to either the G protein within the fusion construct or to endogenously expressed $G\alpha_{11}$ and/or $G\alpha_q$. When samples were immunoprecipitated with an anti-$G\alpha_{11}$/$G\alpha_q$ antiserum, virtually all the agonist-independent binding was lost, but because the agonist-mediated signal was maintained, the signal to background ratio was increased greatly *(53)*. This did not establish whether all (or any) of the [^{35}S]GTPγS was bound to the G protein element of the fusion protein, because both the fusion construct and the endogenously expressed G proteins were pulled down. However, following expression of an α_{1b}-adrenoceptor–Gly^{208}Ala$G\alpha_{11}$ fusion protein, virtually no agonist-stimulated binding of [^{35}S]GTPγS was observed in such immunoprecipitates *(53)*. These results imply that at the levels of expression achieved, the receptor in this fusion cannot access and activate the endogenous pools of $G\alpha_{11}/G\alpha_q$ to any significant degree. Therefore, the agonist-stimulation of [^{35}S]GTPγS binding produced in membranes expressing the wild-type α_{1b}-adrenoceptor–$G\alpha_{11}$ fusion is a direct measure of activation of the fusion protein. A fusion between Leu^{151}Aspα_{1b}-adrenoceptor and wild-type $G\alpha_{11}$ also generated very little agonist-dependent [^{35}S]GTPγS binding *(41)*, defining the poor ability of this mutated receptor to activate G protein. Similarly to the opioid receptor constructs discussed earlier, co-expression of Leu^{151}Aspα_{1b}-adrenoceptor–$G\alpha_{11}$ and α_{1b}-adrenoceptor–Gly^{208}Ala$G\alpha_{11}$ resulted in reconstitution of agonist function consistent with

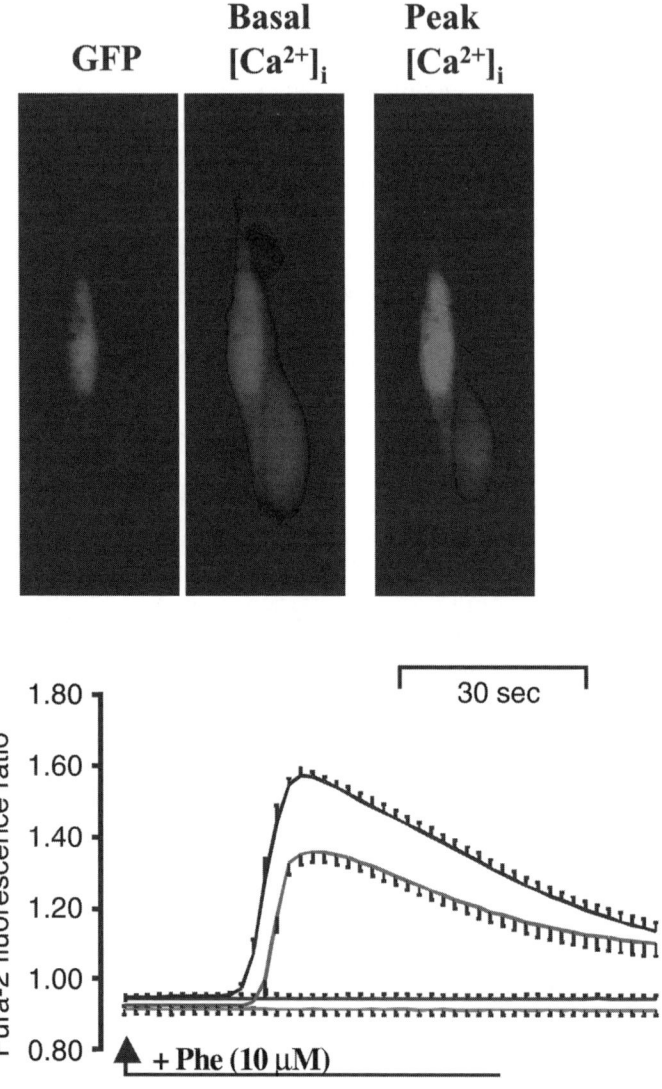

Fig. 2. Detection of α_{1b}-adrenoceptor dimerization by functional complementation of a pair of inactive α_{1b}-adrenoceptor–$G\alpha_{11}$ fusion proteins using single-cell Ca^{2+} imaging studies. A fusion protein was constructed between the hamster α_{1b}-adrenoceptor and $G\alpha_{11}$. This construct was transfected into a mouse embryo fibroblast cell line (EF88 cells) derived from a double $G\alpha_q/G\alpha_{11}$ knockout mouse along with green fluorescent protein (GFP). Imaging of these cells (top panel left) shows only a single positive cell in the field, and this was the only cell that responded to phenylephrine (basal vs peak $[Ca^{2+}]$). Because these cells do not express any G protein able to induce Ca^{2+} signaling, these data demonstrate the functionality of

dimerization of the α_{1b}-adrenoceptor bringing the wild-type G protein and wild-type receptor into contact *(41)*. Similar data have been obtained for fusion proteins between the histamine H1 receptor and $G\alpha_{11}$.

As mentioned earlier, the binding affinity of both antagonist and agonist ligands for the α_{1b}-adrenoceptor–$G\alpha_{11}$ fusion protein was unaffected by introduction of the $Leu^{151}Asp$ mutation that blocked agonist activation. This may seem rather surprising, because the affinity of agonist ligands is often modulated by the interaction status of GPCR and G protein, and such effects have been observed for other GPCR–G protein fusions *(56)*. However, guanine nucleotide-induced shifts in agonist affinity are well established to be very small for α_1-adrenoceptors. Significant alterations in agonist affinity might be observed with the introduction of equivalent mutations in other GPCR–G protein fusions.

3.3. Complementation Between GPCR–G Protein Fusions in Intact Cells

When the α_{1b}-adrenoceptor–$G\alpha_{11}$ fusion protein was expressed in EF88 cells, addition of phenylephrine resulted in elevation of $[Ca^{2+}]i$ *(41,47)*. This reflects functionality of the fusion protein because EF88 cells are a line of mouse embryo fibroblasts derived from an animal in which the genes for both of the widely expressed Ca^{2+} mobilizing G proteins, $G\alpha_{11}$ and $G\alpha_q$ had been inactivated *(57)*. Earlier studies demonstrated the requirement for co-expression of both a suitable GPCR and an appropriate G protein in these cells to generate function *(38)*. Furthermore, only cells that are positively transfected respond to the agonist. The poor transfection efficiency of these cells when using cationic lipid-based methods means that analysis has generally required the use of single-cell Ca^{2+} imaging *(38,41)*. However, retrovirally based infection or the development of other effective means of transfection would allow the use of standard, cuvet-based measurements of Ca^{2+} levels and easier generation of concentration–response curves. As suggested under Subheading 3.2., transfection of these cells with a fusion protein between the wild-type α_{1b}-adrenoceptor and $Gly^{208}AlaG\alpha_{11}$ failed to generate an elevation in

---

Fig. 2 *(From opposite page)* the fusion protein. The detailed time-course of $[Ca^{2+}]$ regulation in response to phenylephrine is shown in the bottom panel (top-most line). The fusion was modified to introduce a $Leu^{151}Asp$ mutation into the receptor element (bottom-most line) or a $Gly^{208}Ala$ mutation into the G protein (second line from bottom). When expressed in EF88 cells, neither of these mutants responded to phenylephrine; however, agonist-mediated signaling was reconstituted when they were co-expressed (second line from top).

[Ca^{2+}]i in response to phenylephrine *(41)*. This was also the case for Leu^{151}Aspα$_{1b}$-adrenoceptor–Gα$_{11}$. Similarly to the membrane-based assays, co-expression of Leu^{151}Aspα$_{1b}$-adrenoceptor–Gα$_{11}$ and α$_{1b}$-adrenoceptor–Gly^{208}AlaGα$_{11}$11 resulted in a restoration of phenylephrine-mediated elevation of [Ca^{2+}]i (Fig. 2). Because there are no appropriate endogenous G proteins in these cells, such data require interactions between the complementary fusions. Similar results were obtained when equivalent fusion proteins that incorporated the histamine H1 receptor rather than the α$_{1b}$-adrenoceptor were used *(41)*. When Leu^{151}Aspα$_{1b}$-adrenoceptor–Gα$_{11}$ and histamine H1–Gly^{208}AlaGα$_{11}$ were co-expressed, histamine, but not phenylephrine, was able to elevate [Ca^{2+}]i. The reverse was true when α$_{1b}$-adrenoceptor–Gly^{208}AlaGα$_{11}$ was co-expressed with Leu^{133}Asp histamine H1–Gα$_{11}$ *(41)*.

A weakness of single-cell imaging studies is the impossibility of monitoring expression levels of each construct. However, when the wild-type histamine H1 receptor–Gα$_{11}$ fusion was co-expressed with the isolated Leu^{151}Aspα$_{1b}$-adrenoceptor, the capacity of histamine to elevate [Ca^{2+}]i was reduced. Although these studies are more difficult to interpret than the parallel experiments performed in membranes of HEK293 cells (in which the expression levels of both the histamine H1 receptor–Gα$_{11}$ fusion and Leu^{151}Aspα$_{1b}$-adrenoceptor could be measured directly by [^{3}H]ligand-binding studies; *see* Subheading 3.2.), their results are consistent with the Leu^{151}Aspα$_{1b}$-adrenoceptor forming a nonfunctional heterodimer with the histamine H1 receptor–Gα$_{11}$ fusion protein and thus reducing levels of functional histamine H1 receptor–Gα$_{11}$/histamine H1 receptor–Gα$_{11}$ homodimer. This type of strategy should be generally applicable for GPCRs that couple to the Ca^{2+} mobilizing G proteins G$_q$ and G$_{11}$, and may be appropriate to measure the relative interaction affinities among GPCRs.

4. FUTURE PERSPECTIVES

Potentially, although the subject has yet to be directly explored, studies akin to those described earlier could be expanded. Many GPCRs can interact with the so-called "promiscuous" G proteins G$_{15}$ and G$_{16}$ *(58)*. Fusion proteins that incorporate either G$_{15}$ or G$_{16}$ with several GPCRs that do not routinely elevate [Ca^{2+}]i have been generated and have been used to allow ligand screening and detailed pharmacological characterization using platforms such as the fluorescence imaging plate reader system, which is widely employed by the pharmaceutical industry *(59–61)*. Incorporation of a Gly^{211}Ala form of Gα$_{16}$ into a fusion should render it unresponsive to agonists and allow equivalent single-cell reconstitution assays, as detailed under Subheading 3.3. Equally, chimeric G proteins containing the backbone of a

Ca^{2+}-mobilizing G protein (with the extreme C-terminal tail modified to provide interaction with GPCRs that couple selectively to G_i-family or G_s G proteins) have been widely employed *(62–64)* to channel signal to a common and easy-to-measure assay endpoint. GPCR–G protein fusions containing chimeric G proteins have also been generated and have been shown to be functional *(65,66)*. Selectivity of GPCR interactions can be quantitatively measured from the extent of reconstitution of [^{35}S]GTPγS binding, with expression of known amounts of two GPCR fusions monitored by saturation ligand-binding studies. Equally, the domains and amino acids that provide the interfaces of GPCR dimerization should be amenable to analysis via mutagenesis studies followed by reconstitution.

ACKNOWLEDGMENTS

Work in the authors' laboratory in this area is supported by the Medical Research Council, the Biotechnology and Biological Sciences Research Council, and the Wellcome Trust.

REFERENCES

1. Salahpour A, Angers S, Bouvier M. Functional significance of oligomerization of G-protein-coupled receptors. Trends Endocrinol Metab 2000;11:163–168.
2. Hebert TE, Moffett S, Morello JP, et al. A peptide derived from a beta2-adrenergic receptor transmembrane domain inhibits both receptor dimerization and activation. J Biol Chem 1996;271:16,384–16,392.
3. Ng GYK, O'Dowd BF, Lee SP, et al. Dopamine D2 receptor dimers and receptor-blocking peptides. Biochem Biophys Res Comm 1996;227:200–204.
4. Nimchinsky EA, Hof PR, Janssen WG, Morrison JH, Schmauss C. Expression of dopamine D3 receptor dimers and tetramers in brain and in transfected cells. J Biol Chem 1997;272:29,229–29,237.
5. Cvejic S, Devi LA. Dimerization of the delta opioid receptor: implication for a role in receptor internalization. J Biol Chem 1997;272:26,959–26,964.
6. George SR, O'Dowd BF, Lee SP. G-protein-coupled receptor oligomerization and its potential for drug discovery. Nat Rev Drug Discov 2002;1:808–820.
7. Devi LA. Heterodimerization of G-protein-coupled receptors: pharmacology, signaling and trafficking. Trends Pharmacol Sci 2001;22:532–537.
8. Salim K, Fenton T, Bacha J, et al. Oligomerization of G-protein-coupled receptors shown by selective co-immunoprecipitation. J Biol Chem 2002;277:15,482–15,485.
9. Angers S, Salahpour A, Joly E, et al. Detection of beta 2-adrenergic receptor dimerization in living cells using bioluminescence resonance energy transfer (BRET). Proc Natl Acad Sci USA 2000;97:3684–3689.
10. McVey M, Ramsay D, Kellett E, et al. Monitoring receptor oligomerization using time-resolved fluorescence resonance energy transfer and bioluminescence resonance energy transfer. The human delta-opioid receptor displays

constitutive oligomerization at the cell surface, which is not regulated by receptor occupancy. J Biol Chem 2001;276:14,092–14,099.
11. Mercier JF, Salahpour A, Angers S, Breit A, Bouvier M. Quantitative assessment of beta 1- and beta 2-adrenergic receptor homo- and heterodimerization by bioluminescence resonance energy transfer. J Biol Chem 2002;277:44,925–44,931.
12. Overton MC, Blumer KJ. The extracellular N-terminal domain and transmembrane domains 1 and 2 mediate oligomerization of a yeast G protein-coupled receptor. J Biol Chem 2002;277:41,463–41,472.
13. Ramsay D, Kellett E, McVey M, Rees S, Milligan G. Homo- and hetero-oligomeric interactions between G-protein-coupled receptors in living cells monitored by two variants of bioluminescence resonance energy transfer (BRET): hetero-oligomers between receptor subtypes form more efficiently than between less closely related sequences. Biochem J 2002;365:429–440.
14. Stanasila L, Perez J-B, Vogel H, Cotecchia S. Oligomerization of the α_{1a}- and $\alpha_{1\beta}$-adrenergic receptor subtypes. J Biol Chem 2003;278:40,239–40,251.
15. Eidne KA, Kroeger KM, Hanyaloglu AC. Applications of novel resonance energy transfer techniques to study dynamic hormone receptor interactions in living cells. Trends Endocrinol Metab 2002;13:415–421.
16. Milligan G. Applications of bioluminescence and fluorescence resonance energy transfer to drug discovery at G protein-coupled receptors. Eur J Pharm Sci 2004;21:397–405.
17. Kroeger KM, Pfleger KDG, Eidne KA. Biophysical and biochemical methods to study G protein-coupled receptor oligomerization. In Devi LA, ed., *The G Protein-Coupled Receptors Handbook*. Humana, Totowa, NJ: 2005; pp. 217–241.
18. Martin NP, Leavitt LM, Sommers CM, Dumont ME. Assembly of G protein-coupled receptors from fragments: identification of functional receptors with discontinuities in each of the loops connecting transmembrane segments. Biochemistry 1999;38:682–695.
19. Dean MK, Higgs C, Smith RE, et al. Dimerization of G-protein-coupled receptors. J Med Chem 2001;44:4595–4614.
20. Gouldson PR, Dean MK, Snell CR, Bywater RP, Gkoutos G, Reynolds CA. Lipid-facing correlated mutations and dimerization in G-protein coupled receptors. Protein Eng 2001;14:759–767.
21. Maggio R, Scarselli M, Novi F, Millan MJ, Corsini GU. Potent activation of dopamine D3/D2 heterodimers by the antiparkinsonian agents, S32504, pramipexole and ropinirole. J Neurochem 2003;87:631–641.
22. Maggio R, Vogel Z, Wess J. Coexpression studies with mutant muscarinic/ adrenergic receptors provide evidence for intermolecular "cross-talk" between G-protein-linked receptors. Proc Natl Acad Sci USA 1993;90:3103–3107.
23. Monnot C, Bihoreau C, Conchon S, Curnow KM, Corvol P, Clauser E. Polar residues in the transmembrane domains of the type 1 angiotensin II receptor are required for binding and coupling. Reconstitution of the binding site by coexpression of two deficient mutants. J Biol Chem 1996;271:1507–1513.
24. Mohler H, Fritschy JM. $GABA_B$ receptors make it to the top—as dimers. Trends Pharmacol Sci 1999;20:87–89.

25. Marshall FH, Jones KA, Kaupmann K, Bettler B. $GABA_B$ receptors—the first 7TM heterodimers. Trends Pharmacol Sci 1999;20:396–399.
26. Parmentier ML, Prezeau L, Bockaert J, Pin JP. A model for the functioning of family 3 GPCRs. Trends Pharmacol Sci 2002;23:268–274.
27. Lee C, Ji I, Ryu K, Song Y, Conn PM, Ji TH. Two defective heterozygous luteinizing hormone receptors can rescue hormone action. J Biol Chem. 2002;277:15,795–15,800.
28. Lee C, Ji IJ, Ji TH. Use of defined-function mutants to access receptor–receptor interactions. Methods 2002;27:318–323.
29. Ji I, Lee C, Song Y, Conn PM, Ji TH. Cis- and trans-activation of hormone receptors: the LH receptor. Mol Endocrinol 2002;16:1299–1308.
30. Nakanishi-Matsui M, Zheng YW, Sulciner DJ, Weiss EJ, Ludeman MJ, Coughlin SR. PAR3 is a cofactor for PAR4 activation by thrombin. Nature 2000;404:609–613.
31. O'Brien PJ, Prevost N, Molino M, et al. Thrombin responses in human endothelial cells. Contributions from receptors other than PAR1 include the transactivation of PAR2 by thrombin-cleaved PAR1. J Biol Chem 2000;275:13,502–13,509.
32. Bertin B, Freissmuth M, Jockers R, Strosberg AD, Marullo S. Cellular signaling by an agonist-activated receptor/Gs alpha fusion protein. Proc Natl Acad Sci USA 1994;91:8827–8831.
33. Milligan G. Insights into ligand pharmacology using receptor–G protein fusion proteins. Trends Pharmacol Sci 2000;21:24–28.
34. Milligan G. Construction and analysis of function of GPCR–G protein fusion proteins. Methods Enzymol 2002;343:260–273.
35. Moon HE, Cavalli A, Bahia DS, Hoffmann M, Massotte D, Milligan G. The human δ opioid receptor activates $G_{i1}\alpha$ more efficiently than $G_{o1}\alpha$. J Neurochem 2001;76:1805–1813.
36. Ugur O, Onaran HO, Jones TL. Partial rescue of functional interactions of a nonpalmitoylated mutant of the G-protein G alpha s by fusion to the beta-adrenergic receptor. Biochemistry 2003;42:2607–2615.
37. Loisel TP, Ansanay H, Adam L, et al. Activation of the beta(2)-adrenergic receptor-Galpha(s) complex leads to rapid depalmitoylation and inhibition of repalmitoylation of both the receptor and Galpha(s). J Biol Chem 1999;274:31,014–31,019.
38. Stevens PA, Pediani J, Carrillo JJ, Milligan G. Coordinated agonist regulation of receptor and G protein palmitoylation and functional rescue of palmitoylation-deficient mutants of the G protein G_{11}alpha following fusion to the alpha$_{1b}$-adrenoreceptor: palmitoylation of G_{11}alpha is not required for interaction with beta/gamma complex. J Biol Chem 2001;276:35,883–35,890.
39. Hosoi T, Koguchi Y, Sugikawa E, et al. Identification of a novel human eicosanoid receptor coupled to G(i/o). J Biol Chem 2002;277:31,459–31,465.
40. Takeda S, Yamamoto A, Okada T, et al. Identification of surrogate ligands for orphan G protein-coupled receptors. Life Sci 2003;74:367–377.
41. Carrillo JJ, Pediani J, Milligan G. Dimers of class A G protein-coupled receptors function via agonist-mediated trans-activation of associated G proteins. J Biol Chem 2003;278:42,578–42,587.

42. Gether U. Uncovering molecular mechanisms involved in activation of G protein-coupled receptors. Endocr Rev 2000;21:90–113.
43. Bahia DS, Wise A, Fanelli F, Lee M, Rees S, Milligan G. Hydrophobicity of residue^{351} of the G-protein $G_{i1}\alpha$ determines the extent of activation by the α_{2A}-adrenoceptor. Biochemistry 1998;37:11,555–11,562.
44. Dupuis DS, Wurch T, Tardif S, Colpaert FC, Pauwels PJ. Modulation of 5-HT(1A) receptor activation by its interaction with wild-type and mutant g(alphai3) proteins. Neuropharmacology 2001;40:36–47.
45. Jackson VN, Bahia DS, Milligan G. Modulation of the relative intrinsic activity of agonists at the α_{2A}-adrenoceptor by mutation of residue^{351} of the G protein $G_{i1}\alpha$. Mol Pharmacol 1999;55:195–201.
46. Milligan G. The use of receptor-G protein fusion proteins for the study of ligand activity. Receptors Channels 2002;8:309–317.
47. Liu S, Carrillo JJ, Pediani J, Milligan G. Effective information transfer from the α_{1b}-adrenoceptor to $G_{11}\alpha$ requires both β/γ interactions and an aromatic group 4 amino acid from the C-terminus of the G protein. J Biol Chem 2002;277:25,707–25,714.
48. Sullivan KA, Miller RT, Masters SB, Beiderman B, Heideman W, Bourne HR. Identification of receptor contact site involved in receptor–G protein coupling. Nature 1987;330:758–760.
49. Waldhoer M, Wise A, Milligan G, Freissmuth M, Nanoff C. Kinetics of ternary complex formation with fusion proteins composed of the A1-adenosine receptor and G protein α-subunits. J Biol Chem 1999;274:30,571–30,579.
50. Milligan G. Principles: extending the utility of [^{35}S]GTPγS binding assays. Trends Pharmacol Sci 2003;24:87–90.
51. Burt AR, Sautel M, Wilson MA, Rees S, Wise A, Milligan G. Agonist-occupation of an α_{2A}-adrenoceptor–$G_{i1}\alpha$ fusion protein results in activation of both receptor -linked and endogenous G proteins: comparisons of their contributions to GTPase activity and signal transduction and analysis of receptor–G protein activation stoichiometry. J Biol Chem 1998;273:10,367–10,375.
52. Molinari P, Ambrosio C, Riitano D, Sbraccia M, Gro MC, Costa T. Promiscuous coupling at receptor–Galpha fusion proteins. The receptor of one covalent complex interacts with the alpha-subunit of another. J Biol Chem 2003;278:15,778–15,788.
53. Carrillo JJ, Stevens PA, Milligan G. Measurement of agonist-dependent and - independent signal initiation of alpha(1b)-adrenoceptor mutants by direct analysis of guanine nucleotide exchange on the G protein γalpha(11). J Pharmacol Exp Ther 2002;302:1080–1088.
54. George SR, Fan T, Xie Z, et al. Oligomerization of μ- and δ-opioid receptors. J Biol Chem 2000;275:26,128–26,135.
55. Jordan BA, Devi LA. G-protein-coupled receptor heterodimerization modulates receptor function. Nature 1999;399:697–700.
56. Wenzel-Seifert K, Seifert R. Molecular analysis of beta(2)-adrenoceptor coupling to G(s)-, G(i)-, and G(q)-proteins. Mol Pharmacol 2000;58:954–966.
57. Yu R, Hinkle PM. Signal transduction and hormone-dependent internalization of the thyrotropin-releasing hormone receptor in cells lacking Gq and G11. J Biol Chem 1999;274:15,745–15,750.

58. Milligan G, Marshall F, Rees S. G16 as a universal G protein adapter: implications for agonist screening strategies. Trends Pharmacol Sci 1996;17:235–237.
59. Pauwels PJ, Colpaert FC. Disparate ligand-mediated Ca(2+) responses by wild-type, mutant Ser(200)Ala and Ser(204)Ala alpha(2A)-adrenoceptor: G(alpha15) fusion proteins: evidence for multiple ligand-activation binding sites. Br J Pharmacol 2000;130:1505–1512.
60. Pauwels PJ, Tardif S, Finana F, Wurch T, Colpaert FC. Ligand-receptor interactions as controlled by wild-type and mutant Thr(370)Lys alpha2B-adrenoceptor–Galpha15 fusion proteins. J Neurochem 2000;74:375–384.
61. Martin RS, Reynen PH, Calixto JJ, et al. Pharmacological comparison of a recombinant CB1 cannabinoid receptor with its G(alpha 16) fusion product. J Biomol Screen 2002;7:281–289.
62. Milligan G, Rees S. Chimaeric G alpha proteins: their potential use in drug discovery. Trends Pharmacol Sci 1999;20:118–124.
63. Coward P, Chan SD, Wada HG, Humphries GM, Conklin BR. Chimeric G proteins allow a high-throughput signaling assay of Gi-coupled receptors. Anal Biochem 1999;270:242–248.
64. Liu AM, Ho MK, Wong CS, Chan JH, Pau AH, Wong YH. Galpha(16/z) chimeras efficiently link a wide range of G protein-coupled receptors to calcium mobilization. J Biomol Screen 2003;8:39–49.
65. Feng GJ, Cavalli A, Milligan G. Engineering a V(2) vasopressin receptor agonist- and regulator of G-protein-signaling-sensitive G protein. Anal Biochem 2002;300:212–220.
66. Fong CW, Milligan G. Analysis of agonist function at fusion proteins between the IP prostanoid receptor and cognate, unnatural and chimaeric G-proteins. Biochem J 1999;342:457–463.-

13
The Role of Oligomerization in G Protein-Coupled Receptor Maturation

Michael M.C. Kong, Christopher H. So, Brian F. O'Dowd, and Susan R. George

1. INTRODUCTION

A large body of evidence now shows that the basic functional unit of seven transmembrane-spanning G protein-coupled receptors (GPCRs) is a dimer, with the possibility of the existence of higher order oligomeric species. GPCR oligomerization has been demonstrated to be a physiological process that defines receptor pharmacology and function *(1,2)*. There is substantial evidence indicating that these receptors are assembled as dimers and, possibly, oligomers prior to cell-surface expression (Fig. 1). Although it has generally been accepted that constitutive GPCR oligomers exist at the plasma membrane, there is evidence demonstrating that the extent of oligomerization at the plasma membrane may be altered by ligand induction *(3–5)* (Fig. 1). For many other classes of cell-surface receptors, oligomerization has been found to be a prerequisite for activation and signaling. For example, the epidermal growth factor receptor, a prototypical member of the tyrosine kinase (TK) family, requires a ligand-induced dimeric configuration for the auto-phosphorylation of tyrosine residues on the cytoplasmic domain and subsequent recruitment of various signaling proteins *(6)*. With the exception of the insulin receptor, agonist-induced dimerization appears to be the rule of thumb for TKs. Conversely, a large proportion of receptors belonging to the cytokine receptor superfamily have been reported as intracellularly derived dimers at the plasma membrane *(7–10)*. Ligand binding triggers a conformational change in these receptors, facilitating Janus kinase-

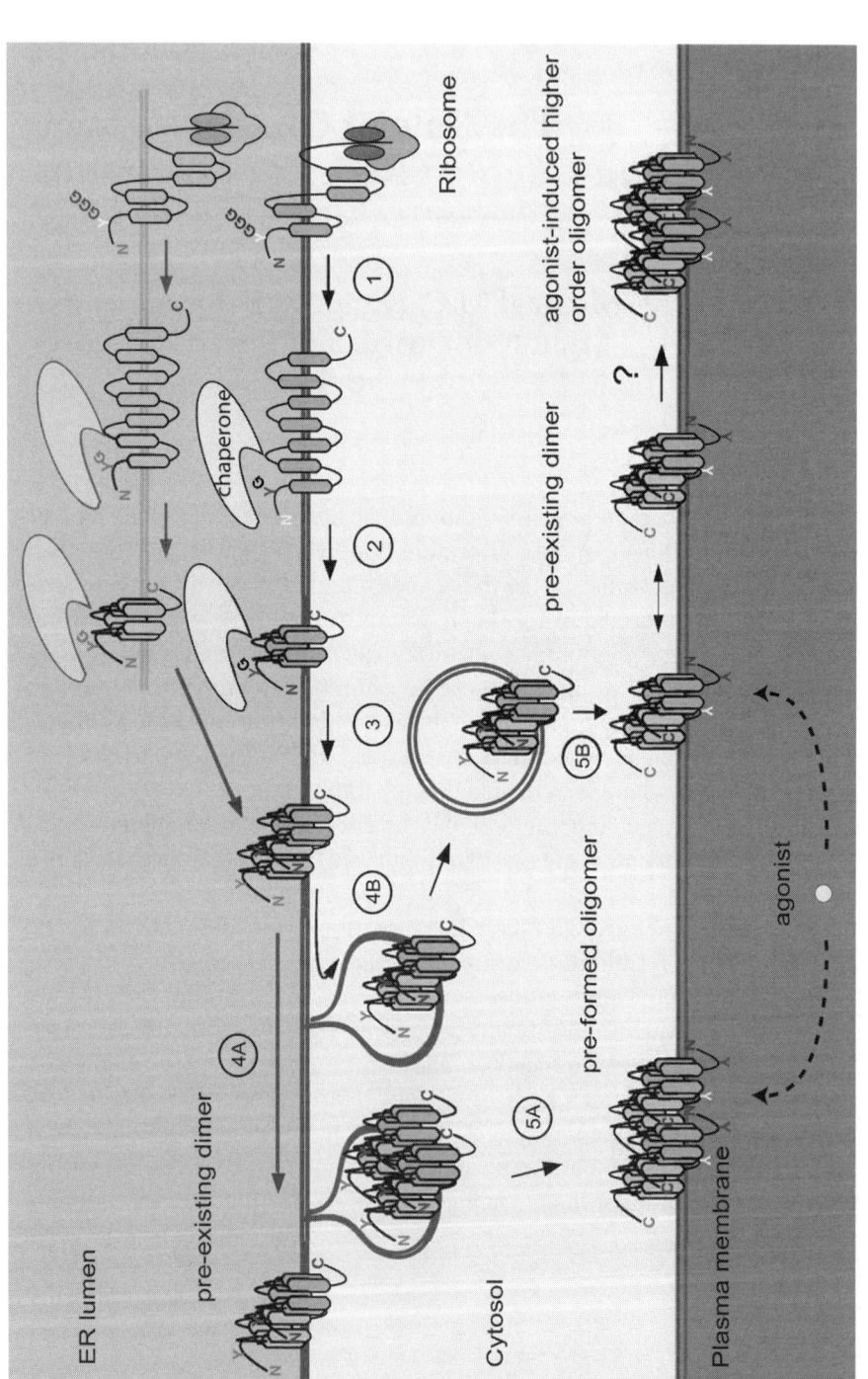

Fig. 1

mediated phosphorylation of various cytosolic substrates *(11)*. Although there is a wealth of knowledge regarding the formation and functional significance of oligomerization in these other receptor families, progress is still being made to determine the cellular implications of the relatively novel concept of GPCR oligomerization.

A key step in dissecting the functional consequences of GPCR oligomerization involves understanding how receptors are formed in the cell. The current understanding of the folding and maturation process of a GPCR (or any other α-helical transmembrane protein) assumes an initial monomeric configuration. The formation of a membrane-spanning receptor begins in the endoplasmic reticulum (ER) and occurs in two stages. The first stage involves the sequential pair-wise insertion of transmembrane α-helices into the ER membrane. Several landmark studies on single transmembrane fragments of the seven-transmembrane-domain protein opsin were among the first to demonstrate that translocation of the nascent transmembrane domains through the membrane requires signal sequences and stop–transfer sequences *(12–16)*.

The maturation of polytopic integral membrane proteins such as GPCRs begins with the insertion of two α-helical peptide segments into the membrane as a hairpin loop (Fig. 1). Translocation of each hairpin loop involves coincident insertion of two transmembrane domains, with intrinsic alternating signal–anchor and stop–transfer sequences. Asparagine-linked (*N*-linked) glycosylation can occur cotranslationally as the translocation mechanism proceeds. This concept of membrane insertion of integral proteins was first demonstrated in a multitransmembrane repeat mutant of the single-membrane-spanning asialoglycoprotein receptor H *(17)*.

The second stage of receptor formation involves assembly of the transmembrane segments into a heptahelical bundle that yields the receptor's tertiary structure. This is driven by a number of factors, including helix–helix interactions and structural constraints imposed by the connecting loops *(18)*. This model was first proposed in studies involving bacteriorhodopsin frag-

---

Fig. 1. *(From opposite page)* Maturation process of a GPCR oligomer. GPCR monomers are synthesized in the endoplasmic reticulum (ER) and inserted in the membrane sequentially as transmembrane domain pairs *(1)*. Folding of the polypeptide is mediated by specific ER-resident molecular chaperones, which may also function to mediate dimeric assembly *(2,3)*. Higher order oligomeric assembly may occur with other dimers in the ER (4A) and these complexes will then be trafficked to the cell surface as constitutively formed GPCR oligomers (5A). Alternatively, ER-formed dimers may traffic to the plasma membrane (4B, 5B) and form higher order oligomeric units upon agonist induction.

ments that contained multiple transmembrane domains. These fragments were demonstrated to insert separately into lipid vesicles, and subsequent assembly between complementary domains was found to result in reconstitution of the native receptor *(19,20)*. This principle has been shown with other GPCRs, including rhodopsin *(21,22)* and muscarinic receptors *(23)*.

To date, there is little information regarding the mechanism by which GPCR oligomers are actually synthesized and which factors are involved in oligomer trafficking through the secretory pathway. However, there are clues from other receptor families and emerging evidence from many rigorous studies on GPCR oligomerization to suggest exactly where and how GPCR oligomers are made and processed.

2. BIOSYNTHESIS OF OLIGOMERS OF GPCRS

2.1. Intracellular Formation of GPCR Oligomers

The study of specific GPCR mutants, both naturally occurring and genetically modified, has provided a useful tool in locating the intracellular site of GPCR oligomer formation *(24–29)*. An increasing number of reports demonstrate that co-expression of various intracellularly sequestered GPCR mutants with the corresponding wild-type receptors results in intracellular retention of the wild-type receptor. These dominant negative effects are a consequence of receptor oligomerization and provide evidence for constitutively formed GPCR oligomers. A physiologically relevant example of dominant negative inhibition of GPCR function is provided by the naturally occurring ccr5Δ32 deletion mutant of the CCR5 chemokine receptor, a coreceptor for human immunodeficiency virus (HIV) infection. This truncated nonfunctional variant of the CCR5 receptor is localized in the endoplasmic reticulum and reduces cell surface expression of the wild-type CCR5 by oligomerization, rendering it aberrantly trapped and unable to support HIV1 infection *(26)*. Other naturally occurring examples of dominant inhibition can be drawn from splice variants of certain GPCRs such as the gonadotropin-releasing hormone receptor *(27)* and the photoreceptor rhodopsin *(29)*. Each of these truncated receptors sequester their respective wild-type receptor in an intracellular compartment, likely by oligomerization in the ER. There are also examples of genetically derived receptor mutants that yield dominant inhibition of native receptors as a consequence of receptor–receptor interactions. Truncation mutants of the V2 vasopressin receptor have been shown to negatively regulate wild-type receptor function by forming a hetero-oligomer that is intracellularly retained *(25)*. Similarly, point mutants of the human platelet-activating factor receptor *(30)* and the D2 dopamine receptor *(31)* have been shown to decrease binding and cell-surface expression of the cognate wild-type receptor. Although the precise site of intracellular retention has not been conclusively determined for most of these sequestered oligomers, it indicates that oligomerization occurs prior to cell-surface expression, lending support for the constitutive oligomeric assembly of these receptors.

Several methods have been used to determine where receptor oligomers are formed in the biosynthetic pathway. Sucrose density gradient fractionation has provided a reliable means of isolating various subcellular compartments. Immunoblot analysis of these fractions has provided information regarding where GPCR oligomers are formed and how they are processed as they make their way to the plasma membrane. The advent of biophysical techniques in the study of GPCR oligomerization has provided a unique strategy for assessing the proximity of two receptors in the cell. Bioluminescence resonance energy transfer (BRET) and fluorescence resonance energy transfer (FRET) have enabled the measurement of receptor–receptor proximity within a range of 50 to 100 angstroms, a distance that would permit receptor oligomerization. The combination of BRET or FRET and subcellular fractionation has provided a powerful tool for determining the presence of GPCR oligomers in specific organelles. BRET signals have been reported to be the highest in ER and plasma membrane-rich fractions of cells expressing oxytocin, vasopressin, and CCR5 chemokine receptor oligomers *(32,33)*. This indicates that the earliest site of oligomer formation is the ER and that oligomeric stability is maintained during transit through the secretory pathway to the cell surface. Similar expression studies of the human complement C5a anaphylatoxin receptor have used FRET to demonstrate that these receptors also exist as oligomers in the ER, Golgi, and cell surface *(34)*. C5a receptor FRET signals are not affected by ligand induction, implying that GPCR oligomerization is insensitive to ligand treatment and favoring the view that oligomers are assembled in the ER. These reports corroborate well with the studies involving dominant negative receptor mutants that imply that GPCR oligomers are constitutively formed in the ER.

2.2. The Role of Glycosylation in Oligomer Formation

Most GPCRs have been shown to possess *N*-linked glycosylation sites in extracellular regions that serve as sites for cotranslational addition of high-mannose oligosaccharides. Mutation of these sites can result in reduced cell-surface expression of certain GPCRs, including the D5 dopamine receptor *(35)* and the AT1 angiotensin receptor *(36)*, implicating a role for glycosylation in intracellular trafficking. It has been suggested that various elements, such as Rab GTPases, vesicular composition, and posttranslational modifications (like glycosylation), differentially modulate exocytosis-mediated transport *(37)*. The role of glycosylation (if a role exists) in GPCR oligomer formation has not been clearly established. There is evidence to suggest that N-linked glycosylation-deficient mutants of the V2 vasopressin receptor *(25)* and the D1 dopamine receptor (Fig. 2) can form oligomeric complexes on sodium dodecyl sulfate-polyacylamide gel electrophoresis. Similar results were found in cells expressing metabotropic glutamate receptor 1α that were pre-incubated with glycosylation inhibitors such as tunicamycin *(38)*. Although these examples suggest that glycosylation has no role in GPCR oligomerization, studies of adrenergic receptor (AR) oligomers challenge this notion and implicate receptor-specific modulation by glycosylation. The decreased ability to co-immunoprecipitate differentially tagged glycosylation mutants of the β_1-AR compared to the wild-type receptor provides evidence that in this case, glycosylation may actually be required for receptor homo-oligomerization *(39)*. Conversely, it was demonstrated that the same glycosylation mutant of the β_1-AR could heterodimerize more efficiently with wild-type α_{2A}- AR than with wild-type β_1-AR *(40)*. The reciprocal experiment with a glycosylation-impaired α_{2A}-AR yielded similar results, suggesting that glycosylation may sterically hinder the efficiency of these ARs to hetero-oligomerize. Thus, it appears that abolishing glycosylation in the β_1-AR has differential effects on its propensity to homo- and hetero-oligomerize.

2.3. Resident ER Chaperones Aid in Receptor Oligomerization

The processing of proteins in the ER involves rigorous quality control mechanisms to ensure that the proteins adopt a conformation compatible for proper trafficking through the distal secretory pathway *(41)*. Because of the hydrophobic nature of many nascent proteins, the cell employs ER-resident chaperone proteins that function within the framework of a quality control mechanism to monitor the folding of functional oligomeric proteins, thus ensuring that they do not aggregate or misfold.

Constitutive oligomeric assembly of glycoproteins in the cell, including receptors and ion channels, is tightly regulated by ER-resident proteins known as molecular chaperones *(42–45)*. These proteins function by binding to and assisting the folding kinetics of polypeptides as they are extruded from the ER (Fig. 1). Molecular chaperones can be classified into four main families: the heat shock proteins (including Hsp40, Hsp70, and Hsp90), the lectin family of chaperones (including calnexin and calreticulin), the peptidyl-prolyl isomerases, and the thiol-disulphide-oxidoreductases *(46)*. An elegant

Biosynthesis of G Protein-Coupled Receptor Oligomers

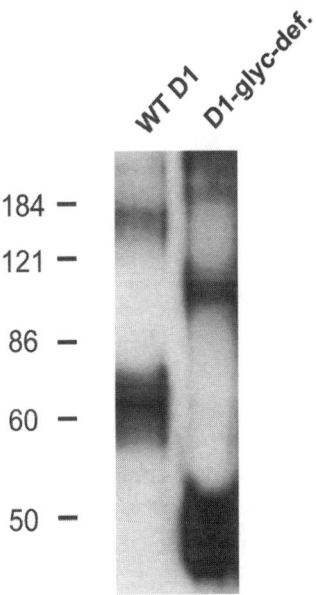

Fig. 2. The wild-type D1 dopamine receptor (WT-D1) exists as dimeric and higher order oligomeric forms. The glycosylation-deficient mutant (D1-glyc def.) has alanine mutations at N5 and N175 and exhibits a similar expression pattern, with a reduction in size of all species corresponding to the expected size of the unglycosylated D1 receptor. Monomeric species are dissociation products resulting from treatment with reducing agents.

study exemplifying the role of chaperones in oligomeric receptor assembly involves the single-transmembrane-spanning human insulin receptor (HIR), which is expressed at the cell surface as a functional ER-derived homodimer *(42)*. HIR maturation involves the cotranslational trimming of three glucose residues by glucosidase I and II to a single, terminal glucose on high-mannose-type oligosaccharides. The resulting monoglucosylated core glycan serves as a substrate for binding to calnexin and calreticulin, which is required for proper folding and dimerization of nascent receptor monomers. The addition of glucose trimming inhibitors such as castanospermine prevents the binding of these chaperones to the HIR, resulting in premature processing manifested as accelerated dimerization and misfolded oligomeric assembly *(42)*. Therefore, the HIR requires chaperone association to maintain oligomer fidelity, possibly by sterically masking hydrophobic interfaces that otherwise would cause aggregation of the nascent monomeric protein.

Similar detailed studies with GPCRs have not been reported, but there is evidence suggesting that molecular chaperones may participate in GPCR oligomeric assembly in the ER. Both the V2 vasopressin receptor and the gonadotropin-releasing hormone receptor have been shown to form oligomers constitutively *(27,32,47)* and to interact with calnexin *(48,49)*. The thyrotropin receptor (TSHR) has been reported to interact with BiP (a prototypic Hsp70), calnexin, and calreticulin in the ER *(50)*, and each interaction has unique effects on receptor synthesis and folding. Calnexin and calreticulin appear to stabilize the TSHR and blunt degradation of newly synthesized receptors, whereas association with BiP destabilizes the receptor and promotes proteasomal degradation. As a result, the maturation of TSHR and its ultimate cellular fate is highly dependent on which chaperone system participates in folding after protein synthesis. The TSHR has also been shown to form constitutive oligomers, as detected by FRET and by immunodetection of oligomers in detergent-solubilized thyroid membranes *(51,52)*. Therefore, because of the role of lectin chaperones in insulin receptor oligomer maturation, it is conceivable that calnexin and calreticulin promote maturation of TSHR by helping to mediate proper oligomeric assembly in the ER. Evidence for the self-dimerization of calreticulin *(53)*, calnexin *(54)*, and specific HSP90 chaperones *(55,56)* suggests a mechanism by which chaperone dimers bind to and facilitate the folding of GPCR oligomers. Thus, the ubiquitous role of chaperones in general oligomeric assembly of proteins implicates a functional role for molecular chaperones in GPCR oligomer formation.

3. THE ROLE OF GPCR OLIGOMERIZATION IN RECEPTOR TRAFFICKING TO THE CELL SURFACE

The dominant negative effect of receptor mutants on the ability of the wild-type receptor to traffic to the cell surface when co-expressed indicates that a GPCR oligomer must be in an appropriate conformation within the ER to permit cell-surface expression. For other membrane-spanning proteins such as the potassium ion channels, oligomerization is well-established to be important for proper cell-surface expression *(57)*. In this case, correct oligomerization of ion channel subunits results in the formation of a fully functional ion channel within the ER that is trafficked to the cell surface. Proper oligomeric formations may facilitate cell-surface expression by masking ER retention signals (such as the RXR motif *[58]*) and exposing export motifs (such as the DXE motif *[59]*) found on ion channel subunits. These retention and retrieval motifs may also play a role in the alignment of GPCR oligomers, because many of these motifs are found within the primary sequences of several GPCRs.

3.1. γ-Aminobutyric Acid BR1–γ-Aminobutyric Acid BR2 Hetero-Oligomerization

The γ-aminobutyric acid B (GABA$_B$) receptors provide a significant example of the importance of GPCR oligomerization in receptor trafficking, as illustrated by the interaction between GABA$_B$R1 and GABA$_B$R2, which share a 35% overall amino acid homology, that results in the formation of a fully functional GABA$_B$ receptor *(60–64)*. When individually expressed, GABA$_B$R1 is mostly intracellularly localized within the ER *(65)*, whereas the GABA$_B$R2 is expressed at the cell surface and within the cytoplasm *(62,66)*. Co-expression of both receptors in heterologous cell lines forms GABA$_B$R1–GABA$_B$R2 hetero-oligomers, which display functional characteristics similar to endogenous GABA$_B$ receptors *(60–62,64)*, and increases GABA$_B$R1 cell-surface expression *(61)*. Interactions between the coiled-coil domains within the carboxyl termini of the GABA$_B$R1 and GABA$_B$R2 *(60,67)* and their transmembrane domains *(68,69)* participate in hetero-oligomer formation *(60–64)*. The coiled-coil domain also masks the ER retention motif RXR on the carboxyl terminus of GABA$_B$R1, allowing the receptor to exit the ER to be further processed within the Golgi into a mature, glycosylated receptor before cell-surface localization *(61–63)*. The resulting glycosylation of the GABA$_B$R1 becomes resistant to the activity of the enzyme endoglycosidase H (Endo H), which specifically cleaves glycoproteins that have high-mannose oligosaccharides attached at their *N*-linked glycosylation sites—a characteristic of glycoproteins that are intracellularly retained in the ER *(61–63)*.

3.2. Hetero-Oligomerization Between α$_1$-AR Subtypes

A further example demonstrating the importance of GPCR oligomerization in receptor cell-surface trafficking is the interaction of the α$_{1B}$-AR with the α$_{1A}$- or α$_{1D}$-ARs, forming α$_{1A}$–α$_{1B}$ and α$_{1B}$–α$_{1D}$ hetero-oligomers, respectively *(70)*. Expressed alone, α$_{1A}$-AR and α$_{1D}$-AR *(71,72)* are poorly expressed at the cell surface, whereas the α$_{1B}$-AR is predominantly expressed at the cell surface *(73)*. The formation of hetero-oligomers resulted in no change in the pharmacology of either receptor but increased cell-surface expression of the α$_{1A}$- and α$_{1D}$-ARs, compared to the α$_{1A}$- and α$_{1D}$-ARs expressed alone *(70)*. These observations suggest that the α$_{1B}$-AR facilitates transport of α$_{1A}$- and α$_{1D}$-ARs to the cell surface *(70)*. Physical interactions forming hetero-oligomers are not mediated by interactions between the carboxyl or amino terminus of the receptors and are specific, because co-expression of α$_{1A}$- and α$_{1D}$-ARs did not lead to formation of hetero-oligomers or change in cell-surface expression of either receptor.

4. INTERACTIONS BETWEEN GPCRS AND ACCESSORY PROTEINS REGULATE GPCR EXPORT TO THE CELL SURFACE

Some accessory proteins, both cytosolic and membrane-bound, have been observed to enhance cell-surface expression of GPCRs such as rhodopsin and olfactory receptors, whereas others have been observed to inhibit cell-surface expression of certain GPCRs, such as the group I metabotropic glutamate receptors and the D1 dopamine receptor.

4.1. Calcitonin Receptor-Like Receptor and Receptor Activity-Modifying Proteins

The observation that the calcitonin receptor-like receptor (CRLR) did not express at the cell surface when transfected into heterologous cell lines suggests that it may require a specific accessory protein for proper cell-surface trafficking *(74–76)*. The accessory protein for CRLR was found to be the receptor activity-modifying protein (RAMP), of which there are three subtypes that are 31% identical to each other *(75)*. The RAMPs are ubiquitously expressed and have a single-transmembrane domain, a large extracellular domain, and a short cytoplasmic domain. Expressed alone, RAMPs are intracellularly retained, cycling between the Golgi and the ER *(74,75,77)*, and exist as dimers *(74)*. The intracellular retention of RAMPs results from an ER retention motif on its carboxyl terminus *(77)*.

When RAMP1 is cotransfected with the CRLR, surface localization of both CRLR and RAMP1 are observed *(74,75)* and CRLR is differentially processed into a mature, glycosylated receptor. This glycosylation of CRLR is resistant to the activity of Endo H, indicating that the RAMP–CRLR complex has exited the ER *(74,75)*. A similar interaction exists between RAMP2 or RAMP3 and CRLR, which also facilitates transport of the receptor to the cell surface *(75,78–80)*. The RAMP–CRLR complex, formed by transmembrane domain interactions between CRLR and RAMP *(77,79,81,82)*, is maintained at the cell surface and during agonist-induced internalization *(74,76)*. RAMPs also interact with other group 2 GPCRs, such as the glucagon and parathyroid hormone receptor 1 and 2; however, this occurs without change in receptor pharmacology or surface localization *(83)*. However, these receptors aid RAMP cell-surface expression *(83)*. Interactions between RAMP1 or RAMP3 and the calcitonin receptor (CTR) are also observed, resulting in decreased receptor cell-surface expression and increased affinity to amylin *(84)*.

4.2. Rhodopsin and NinaA

In *Drosophila*, rhodopsin processing is mediated by the protein neither inactivation nor afterpotential A (ninaA), which escorts the receptor through the secretory pathway to the cell surface *(85,86)*. This protein, which is homologous to cyclophilins that are cytosolic isomerases involved in the folding of proteins *(87,88)*, is colocalized with rhodopsin within the ER and at the cell surface in a certain subset of photoreceptor cells *(86,89)*. NinaA is a membrane-bound protein with the carboxyl terminus anchored in the membrane and the active cyclophilin homologous domain protruding into the lumen of the ER *(88,89)*.

The ninaA protein performs two functions. First, because it contains a cyclophilin homologous domain, it re-arranges peptide bonds within rhodopsin in the ER to ensure proper receptor folding *(88)*. Second, it forms a stable complex with rhodopsin and acts as a chaperone to allow rhodopsin to exit the ER for further processing within the Golgi *(85)*. This has been demonstrated in experiments involving *drosophila* mutants that lacked ninaA expression, where increasing the expression of ninaA protein increased the amount of Endo H-resistant receptors, indicating that ninaA facilitated the exit of the receptor from the ER *(85,86,88,90)*. The interaction between ninaA and rhodopsin is mediated through the carboxyl terminus of ninaA *(85)* and is maintained despite deglycosylation of rhodopsin *(90)*.

A mammalian homolog of ninaA has not been discovered. However, the Ran binding protein 2 (RBP2) is a mammalian cyclophilin that functions similarly to ninaA to target cell-surface expression of mammalian red and green opsin. This protein is prenylated in membranes and binds red and green opsin, but not blue opsin or rhodopsin *(91,92)*. The cyclophilin domain re-arranges peptide bonds on the opsin to stabilize the RBP2–opsin complex *(92)*.

4.3. Odorant Response Mutant-10 and Odorant Response Mutant-4

Odorant receptors are predominantly found on the plasma membrane when transfected into olfactory neurons *(93)* and cells of olfactory lineage *(94)*, but they are retained in intracellular compartments when expressed in heterologous cell lines *(95)*. Because these receptors are retained within the ER as a result of receptor misfolding *(96)*, accessory proteins, which are present only in olfactory cells, may be required to ensure proper cell-surface expression. One such accessory protein identified from *Caenorhabditis elegans* is the single-transmembrane protein odorant response mutant-4 (ODR-4), which promotes cell-surface expression of the *C. elegans* GPCR

ODR-10. ODR-4 is localized within the ER, Golgi, and transport vesicles and acts as a chaperone protein, stabilizing ODR-10 receptor processing throughout the secretory pathway *(94,97)*. ODR-4 also interacts with the odorant receptor STR-2 in *C. elegans* by mediating receptor cell-surface expression *(97)*. Accessory proteins like ODR-4 may be required for proper cell-surface localization of mammalian odorant receptors, because cell-surface localization of the rat odorant receptor U131 is increased when it is coexpressed with ODR-4 *(94)*.

4.4. Homer and Group 1 Metabotropic Glutamate Receptors

Metabotropic glutamate receptors (mGluRs) are involved in synaptic activity *(98)* and are classified into three groups depending on their structural, pharmacological, and functional similarities. Group I mGluRs, comprised of mGluR1 (splice variants mGluR1α and mGluR1β) and mGluR5, have been demonstrated to interact with the Homer family of cytosolic proteins. There are three classes of Homer proteins; class 1 is made up of three alternatively spliced homer proteins: Homer 1a, 1b, and 1c *(99)*. Unlike Homer 1a, Homer 1b and 1c contain a large coiled-coil domain that aids in self-oligomerization *(100)*. Interactions between group I mGluRs and Homer proteins are mediated by an interaction between the P-P-X-X-F motif on the mGluR carboxyl terminus and a distinct domain on the Homer proteins *(101)*. These interactions result in changes in receptor cell-surface trafficking. A specific interaction between Homer 1b and mGluR5 retained the receptor within the ER, as determined by immunofluorescence and increased susceptibility of the receptor species to Endo H activity *(102,103)*. Interactions between mGluR1α and Homer 1c have also been observed, but the effect is unclear because reports suggest enhanced cell-surface expression, with Homer 1c anchoring the receptor at the cell surface *(104,105)*. However, another study observed increased receptor sequestration within the ER *(106)*.

4.5. D1 Dopamine Receptor and Dopamine Receptor Interacting Protein-78

ER-resident chaperone proteins can not only mediate formation of GPCR oligomers but can also mediate cellular trafficking. An interaction between the D1 dopamine receptor and an ER-resident transport protein dopamine receptor interacting protein (DRIP)78 results in ER retention of the D1 receptor, as demonstrated by a shift of receptor expression from the plasma membrane to the ER *(107)*. DRIP78 is a transmembrane structure that associates, through two potential zinc finger domains, with the D1 receptor at the ER-export motif F-X-X-X-F-X-X-X-F in the proximal carboxyl terminus *(107)*. Once the interaction between the D1 receptor and DRIP78 is lost, the D1 receptor is released from the ER for further modifications within the Golgi. Other receptors that have also been observed to interact with DRIP78 are the M2 muscarinic receptor *(107)* and the type 1 receptor for angiotensin II *(108)*.

4.6. $α_{1B}$-AR and gC1q-R

The $α_{1B}$-AR is expressed at the cell surface, but upon co-expression with gC1q-R (a regulatory protein of the complement pathway), intracellular retention of the receptor is observed by immunofluorescence microscopy *(109)*, cell-surface cell-flow cytometry analysis, and radioligand binding *(110)*. gC1q-R binds to the carboxyl terminus of the $α_{1B}$- and $α_{1D}$-ARs, as determined by yeast two-hybrid assays and co-immunoprecipitation studies *(109–111)*.

5. SIGNIFICANCE OF INTRACELLULARLY RETAINED GPCRS

Immunofluorescence microscopy has revealed that by default, certain members of the GPCR family are intracellularly retained when expressed in heterologous cell lines. Some of these receptors include the $α_{2c}$-AR *(112)* and the rat trace amine receptor 1 *(113)*. As described earlier, intracellular localization has also been observed for the CRLR, ODR-10, and $GABA_BR1$ receptors when expressed alone; these receptors have also been demonstrated to require a protein partner for proper cell-surface expression. Therefore, it is possible that these other receptors require an unidentified protein partner, possibly another receptor or accessory protein, to mediate cell-surface expression.

6. CONCLUSIONS

To date, the evidence for GPCR oligomerization indicates that it is an early event in receptor maturation. The intracellular processing of a GPCR, such as glycosylation, may participate in determining whether a specific receptor is subject to oligomerization. Although there are some reports regarding agonist-induced oligomerization at the plasma membrane, most of the current evidence suggests that GPCR oligomerization occurs in the ER. There, the receptor may be required to achieve a certain oligomeric configuration to exit the ER for further processing in the Golgi—possibly with the assistance of ER-resident chaperone proteins. Some GPCRs may require accessory proteins, which regulate receptor folding and enhance or impede cell-surface expression. In some cases, these accessory proteins can also modulate receptor function at the cell surface. Other GPCRs (e. g., $GABA_B$

receptor) may require GPCR oligomerization to mask intrinsic trafficking motifs that direct the cellular fate of the receptor.

Considerable progress has been made in elucidating the stages involved in GPCR oligomeric assembly. Further studies must be conducted to determine how oligomerization occurs in the ER and the maturation steps subsequent to oligomerization. This involves defining the molecular chaperones that orchestrate the folding kinetics of the specific oligomer, the role of posttranslational modifications in oligomerization, and the other cellular factors involved in ensuring that receptors interact in a specific manner.

ACKNOWLEDGMENTS

Work in the laboratory is supported by grants from the National Institute on Drug Abuse and the Canadian Institutes of Health Research. S. R. George holds a Canada Research Chair in Molecular Neuroscience.

REFERENCES

1. George SR, O'Dowd BF, Lee SP. G-protein-coupled receptor oligomerization and its potential for drug discovery. Nat Rev Drug Discov 2002;1:808–820.
2. Bouvier M. Oligomerization of G-protein-coupled transmitter receptors. Nat Rev Neurosci 2001;2:274–286.
3. Rocheville M, Lange DC, Kumar U, Patel SC, Patel RC, Patel YC. Receptors for dopamine and somatostatin: formation of hetero-oligomers with enhanced functional activity. Science 2000;288:154–157.
4. Cornea A, Janovick JA, Maya-Nunez G, Conn PM. Gonadotropin-releasing hormone receptor microaggregation. Rate monitored by fluorescence resonance energy transfer. J Biol Chem 2001;276:2153–2158.
5. Cheng ZJ, Miller LJ. Agonist-dependent dissociation of oligomeric complexes of G protein-coupled cholecystokinin receptors demonstrated in living cells using bioluminescence resonance energy transfer. J Biol Chem 2001;276:48,040–48,047.
6. Schlessinger J. Ligand-induced, receptor-mediated dimerization and activation of EGF receptor. Cell 2002;110:669–672.
7. Constantinescu SN, Keren T, Socolovsky M, Nam H, Henis YI, Lodish HF. Ligand-independent oligomerization of cell-surface erythropoietin receptor is mediated by the transmembrane domain. Proc Natl Acad Sci USA 2001;98:4379–4384.
8. Devos R, Guisez Y, Van der Heyden J, et al. Ligand-independent dimerization of the extracellular domain of the leptin receptor and determination of the stoichiometry of leptin binding. J Biol Chem 1997;272:18,304–18,310.
9. Livnah O, Stura EA, Middleton SA, Johnson DL, Jolliffe LK, Wilson IA. Crystallographic evidence for preformed dimers of erythropoietin receptor before ligand activation. Science 1999;283:987–990.

10. Gent J, van Kerkhof P, Roza M, Bu G, Strous GJ. Ligand-independent growth hormone receptor dimerization occurs in the endoplasmic reticulum and is required for ubiquitin system-dependent endocytosis. Proc Natl Acad Sci USA 2002;99:9858–9863.
11. Heldin CH. Dimerization of cell surface receptors in signal transduction. Cell 1995;80:213–223.
12. Audigier Y, Friedlander M, Blobel G. Multiple topogenic sequences in bovine opsin. Proc Natl Acad Sci USA 1987;84:5783–5787.
13. Blobel G. Intracellular protein topogenesis. Proc Natl Acad Sci USA 1980;77:1496–500.
14. Friedlander M, Blobel G. Bovine opsin has more than one signal sequence. Nature 1985;318:338–343.
15. Sabatini DD, Kreibich G, Morimoto T, Adesnik M. Mechanisms for the incorporation of proteins in membranes and organelles. J Cell Biol 1982;92:1–22.
16. Singer SJ, Maher PA, Yaffe MP. On the transfer of integral proteins into membranes. Proc Natl Acad Sci USA 1987;84:1960–1964.
17. Wessels HP, Spiess M. Insertion of a multispanning membrane protein occurs sequentially and requires only one signal sequence. Cell 1988;55:61–70.
18. Lemmon MA, Engelman DM. Specificity and promiscuity in membrane helix interactions. FEBS Lett 1994;346:17–20.
19. Popot JL, Gerchman SE, Engelman DM. Refolding of bacteriorhodopsin in lipid bilayers. A thermodynamically controlled two-stage process. J Mol Biol 1987;198:655–676.
20. Marti T. Refolding of bacteriorhodopsin from expressed polypeptide fragments. J Biol Chem 1998;273:9312–9322.
21. Ridge KD, Lee SS, Abdulaev NG. Examining rhodopsin folding and assembly through expression of polypeptide fragments. J Biol Chem 1996;271:7860–7867.
22. Ridge KD, Lee SS, Yao LL. In vivo assembly of rhodopsin from expressed polypeptide fragments. Proc Natl Acad Sci USA 1995;92:3204–3208.
23. Schoneberg T, Liu J, Wess J. Plasma membrane localization and functional rescue of truncated forms of a G protein-coupled receptor. J Biol Chem 1995;270:18,000–18,006.
24. Karpa KD, Lin R, Kabbani N, Levenson R. The dopamine D3 receptor interacts with itself and the truncated D3 splice variant d3nf: D3–D3nf interaction causes mislocalization of D3 receptors. Mol Pharmacol 2000;58:677–683.
25. Zhu X, Wess J. Truncated V2 vasopressin receptors as negative regulators of wild-type V2 receptor function. Biochemistry 1998;37:15,773–15,784.
26. Benkirane M, Jin DY, Chun RF, Koup RA, Jeang KT. Mechanism of transdominant inhibition of CCR5-mediated HIV-1 infection by ccr5delta32. J Biol Chem 1997;272:30,603–30,606.
27. Grosse R, Schoneberg T, Schultz G, Gudermann T. Inhibition of gonadotropin-releasing hormone receptor signaling by expression of a splice variant of the human receptor. Mol Endocrinol 1997;11:1305–1318.

28. Coge F, Guenin SP, Renouard-Try A, et al. Truncated isoforms inhibit [3H]prazosin binding and cellular trafficking of native human alpha1A-adrenoceptors. Biochem J 1999;343(Pt 1):231–239.
29. Colley NJ, Cassill JA, Baker EK, Zuker CS. Defective intracellular transport is the molecular basis of rhodopsin-dependent dominant retinal degeneration. Proc Natl Acad Sci USA 1995;92:3070–3074.
30. Le Gouill C, Parent JL, Caron CA, et al. Selective modulation of wild type receptor functions by mutants of G-protein-coupled receptors. J Biol Chem 1999;274:12,548–12,554.
31. Lee SP, O'Dowd BF, Ng GY, et al. Inhibition of cell surface expression by mutant receptors demonstrates that D2 dopamine receptors exist as oligomers in the cell. Mol Pharmacol 2000;58:120–128.
32. Terrillon S, Durroux T, Mouillac B, et al. Oxytocin and vasopressin V1a and V2 receptors form constitutive homo- and heterodimers during biosynthesis. Mol Endocrinol 2003;17:677–691.
33. Issafras H, Angers S, Bulenger S, et al. Constitutive agonist-independent CCR5 oligomerization and antibody-mediated clustering occurring at physiological levels of receptors. J Biol Chem 2002;277:34,666–34,673.
34. Floyd DH, Geva A, Bruinsma SP, Overton MC, Blumer KJ, Baranski TJ. C5a receptor oligomerization. II. Fluorescence resonance energy transfer studies of a human G protein-coupled receptor expressed in yeast. J Biol Chem 2003;278:35,354–35,361.
35. Karpa KD, Lidow MS, Pickering MT, Levenson R, Bergson C. *N*-linked glycosylation is required for plasma membrane localization of D5, but not D1, dopamine receptors in transfected mammalian cells. Mol Pharmacol 1999;56:1071–1078.
36. Lanctot PM, Leclerc PC, Escher E, Leduc R, Guillemette G. Role of *N*-glycosylation in the expression and functional properties of human AT1 receptor. Biochemistry 1999;38:8621–8627.
37. Wu G, Zhao G, He Y. Distinct pathways for the trafficking of angiotensin II and adrenergic receptors from the endoplasmic reticulum to the cell surface: Rab1-independent transport of a G protein-coupled receptor. J Biol Chem 2003;278:47,062–47,069.
38. Robbins MJ, Ciruela F, Rhodes A, McIlhinney RA. Characterization of the dimerization of metabotropic glutamate receptors using an N-terminal truncation of mGluR1alpha. J Neurochem 1999;72:2539–2547.
39. He J, Xu J, Castleberry AM, Lau AG, Hall RA. Glycosylation of beta(1)-adrenergic receptors regulates receptor surface expression and dimerization. Biochem Biophys Res Commun 2002;297:565–572.
40. Xu J, He J, Castleberry AM, Balasubramanian S, Lau AG, Hall RA. Heterodimerization of alpha 2A- and beta 1-adrenergic receptors. J Biol Chem 2003;278:10,770–10,777.
41. Aridor M, Balch WE. Membrane fusion: timing is everything. Nature 1996;383:220–221.

42. Bass J, Chiu G, Argon Y, Steiner DF. Folding of insulin receptor monomers is facilitated by the molecular chaperones calnexin and calreticulin and impaired by rapid dimerization. J Cell Biol 1998;141:637–646.
43. Boyd GW, Low P, Dunlop JI, et al. Assembly and cell surface expression of homomeric and heteromeric 5-HT3 receptors: the role of oligomerization and chaperone proteins. Mol Cell Neurosci 2002;21:38–50.
44. Vassilakos A, Cohen-Doyle MF, Peterson PA, Jackson MR, Williams DB. The molecular chaperone calnexin facilitates folding and assembly of class I histocompatibility molecules. EMBO J 1996;15:1495–1506.
45. Hebert DN, Foellmer B, Helenius A. Calnexin and calreticulin promote folding, delay oligomerization and suppress degradation of influenza hemagglutinin in microsomes. EMBO J 1996;15:2961–2968.
46. Ellgaard L, Helenius A. Quality control in the endoplasmic reticulum. Nat Rev Mol Cell Biol 2003;4:181–191.
47. Schulz A, Grosse R, Schultz G, Gudermann T, Schoneberg T. Structural implication for receptor oligomerization from functional reconstitution studies of mutant V2 vasopressin receptors. J Biol Chem 2000;275:2381–2389.
48. Morello JP, Salahpour A, Laperriere A, et al. Pharmacological chaperones rescue cell-surface expression and function of misfolded V2 vasopressin receptor mutants. J Clin Invest 2000;105:887–895.
49. Rozell TG, Davis DP, Chai Y, Segaloff DL. Association of gonadotropin receptor precursors with the protein folding chaperone calnexin. Endocrinology 1998;139:1588–1593.
50. Siffroi-Fernandez S, Giraud A, Lanet J, Franc JL. Association of the thyrotropin receptor with calnexin, calreticulin and BiP. Effects on the maturation of the receptor. Eur J Biochem 2002;269:4930–4937.
51. Graves PN, Vlase H, Bobovnikova Y, Davies TF. Multimeric complex formation by the thyrotropin receptor in solubilized thyroid membranes. Endocrinology 1996;137:3915–3920.
52. Latif R, Graves P, Davies TF. Ligand-dependent inhibition of oligomerization at the human thyrotropin receptor. J Biol Chem 2002;277:45,059–45,067.
53. Jorgensen CS, Ryder LR, Steino A, et al. Dimerization and oligomerization of the chaperone calreticulin. Eur J Biochem 2003;270:4140–4148.
54. Ou WJ, Bergeron JJ, Li Y, Kang CY, Thomas DY. Conformational changes induced in the endoplasmic reticulum luminal domain of calnexin by Mg-ATP and Ca2+. J Biol Chem 1995;270:18,051–18,059.
55. Chadli A, Ladjimi MM, Baulieu EE, Catelli MG. Heat-induced oligomerization of the molecular chaperone Hsp90. Inhibition by ATP and geldanamycin and activation by transition metal oxyanions. J Biol Chem 1999;274:4133–4139.
56. Wearsch PA, Nicchitta CV. Endoplasmic reticulum chaperone GRP94 subunit assembly is regulated through a defined oligomerization domain. Biochemistry 1996;35:16,760–16,769.
57. Papazian DM. Potassium channels: some assembly required. Neuron 1999;23:7–10.

58. Zerangue N, Schwappach B, Jan YN, Jan LY. A new ER trafficking signal regulates the subunit stoichiometry of plasma membrane K(ATP) channels. Neuron 1999;22:537–548.
59. Ma D, Zerangue N, Lin YF, et al. Role of ER export signals in controlling surface potassium channel numbers. Science 2001;291:316–319.
60. Kuner R, Kohr G, Grunewald S, Eisenhardt G, Bach A, Kornau HC. Role of heteromer formation in GABAB receptor function. Science 1999;283:74–77.
61. White JH, Wise A, Main MJ, et al. Heterodimerization is required for the formation of a functional GABA(B) receptor. Nature 1998;396:679–682.
62. Jones KA, Borowsky B, Tamm JA, et al. GABA(B) receptors function as a heteromeric assembly of the subunits GABA(B)R1 and GABA(B)R2. Nature 1998;396:674–679.
63. Kaupmann K, Malitschek B, Schuler V, et al. GABA(B)-receptor subtypes assemble into functional heteromeric complexes. Nature 1998;396:683–687.
64. Ng GY, Clark J, Coulombe N, et al. Identification of a GABAB receptor subunit, gb2, required for functional GABAB receptor activity. J Biol Chem 1999;274:7607–7610.
65. Couve A, Filippov AK, Connolly CN, Bettler B, Brown DA, Moss SJ. Intracellular retention of recombinant GABAB receptors. J Biol Chem 1998;273:26,361–26,367.
66. Martin SC, Russek SJ, Farb DH. Molecular identification of the human GABABR2: cell surface expression and coupling to adenylyl cyclase in the absence of GABABR1. Mol Cell Neurosci 1999;13:180–191.
67. Kammerer RA, Frank S, Schulthess T, Landwehr R, Lustig A, Engel J. Heterodimerization of a functional GABAB receptor is mediated by parallel coiled-coil alpha-helices. Biochemistry 1999;38:13,263–13,269.
68. Calver AR, Robbins MJ, Cosio C, et al. The C-terminal domains of the GABA(b) receptor subunits mediate intracellular trafficking but are not required for receptor signaling. J Neurosci 2001;21:1203–1210.
69. Pagano A, Rovelli G, Mosbacher J, et al. C-terminal interaction is essential for surface trafficking but not for heteromeric assembly of GABA(b) receptors. J Neurosci 2001;21:1189–1202.
70. Uberti MA, Hall RA, Minneman KP. Subtype-specific dimerization of alpha 1-adrenoceptors: effects on receptor expression and pharmacological properties. Mol Pharmacol 2003;64:1379–1390.
71. Taguchi K, Yang M, Goepel M, Michel MC. Comparison of human alpha1-adrenoceptor subtype coupling to protein kinase C activation and related signalling pathways. Naunyn Schmiedebergs Arch Pharmacol 1998;357:100–110.
72. Theroux TL, Esbenshade TA, Peavy RD, Minneman KP. Coupling efficiencies of human alpha 1-adrenergic receptor subtypes: titration of receptor density and responsiveness with inducible and repressible expression vectors. Mol Pharmacol 1996;50:1376–1387.
73. Hirasawa A, Sugawara T, Awaji T, Tsumaya K, Ito H, Tsujimoto G. Subtype-specific differences in subcellular localization of alpha1-adrenoceptors:

chlorethylclonidine preferentially alkylates the accessible cell surface alpha1-adrenoceptors irrespective of the subtype. Mol Pharmacol 1997;52:764–770.
74. Hilairet S, Belanger C, Bertrand J, Laperriere A, Foord SM, Bouvier M. Agonist-promoted internalization of a ternary complex between calcitonin receptor-like receptor, receptor activity-modifying protein 1 (RAMP1), and beta-arrestin. J Biol Chem 2001;276:42,182–42,190.
75. McLatchie LM, Fraser NJ, Main MJ, et al. RAMPs regulate the transport and ligand specificity of the calcitonin-receptor-like receptor. Nature 1998;393:333–339.
76. Kuwasako K, Shimekake Y, Masuda M, et al. Visualization of the calcitonin receptor-like receptor and its receptor activity-modifying proteins during internalization and recycling. J Biol Chem 2000;275:29,602–29,609.
77. Steiner S, Muff R, Gujer R, Fischer JA, Born W. The transmembrane domain of receptor-activity-modifying protein 1 is essential for the functional expression of a calcitonin gene-related peptide receptor. Biochemistry 2002;41:11,398–11,404.
78. Kamitani S, Asakawa M, Shimekake Y, Kuwasako K, Nakahara K, Sakata T. The RAMP2/CRLR complex is a functional adrenomedullin receptor in human endothelial and vascular smooth muscle cells. FEBS Lett 1999;448:111–114.
79. Hilairet S, Foord SM, Marshall FH, Bouvier M. Protein–protein interaction and not glycosylation determines the binding selectivity of heterodimers between the calcitonin receptor-like receptor and the receptor activity-modifying proteins. J Biol Chem 2001;276:29,575–29,581.
80. Fraser NJ, Wise A, Brown J, McLatchie LM, Main MJ, Foord SM. The amino terminus of receptor activity modifying proteins is a critical determinant of glycosylation state and ligand binding of calcitonin receptor-like receptor. Mol Pharmacol 1999;55:1054–1059.
81. Kuwasako K, Kitamura K, Onitsuka H, et al. Rat RAMP domains involved in adrenomedullin binding specificity. FEBS Lett 2002;519:113–116.
82. Miret JJ, Rakhilina L, Silverman L, Oehlen B. Functional expression of heteromeric calcitonin gene-related peptide and adrenomedullin receptors in yeast. J Biol Chem 2002;277:6881–6887.
83. Christopoulos A, Christopoulos G, Morfis M, et al. Novel receptor partners and function of receptor activity-modifying proteins. J Biol Chem 2003;278:3293–3297.
84. Muff R, Buhlmann N, Fischer JA, Born W. An amylin receptor is revealed following co-transfection of a calcitonin receptor with receptor activity modifying proteins-1 or -3. Endocrinology 1999;140:2924–2927.
85. Baker EK, Colley NJ, Zuker CS. The cyclophilin homolog NinaA functions as a chaperone, forming a stable complex in vivo with its protein target rhodopsin. EMBO J 1994;13:4886–4895.
86. Colley NJ, Baker EK, Stamnes MA, Zuker CS. The cyclophilin homolog ninaA is required in the secretory pathway. Cell 1991;67:255–263.
87. Fischer G, Wittmann-Liebold B, Lang K, Kiefhaber T, Schmid FX. Cyclophilin and peptidyl-prolyl *cis-trans* isomerase are probably identical proteins. Nature 1989;337:476–478.

88. Shieh BH, Stamnes MA, Seavello S, Harris GL, Zuker CS. The ninaA gene required for visual transduction in Drosophila encodes a homologue of cyclosporin A-binding protein. Nature 1989;338:67–70.
89. Stamnes MA, Shieh BH, Chuman L, Harris GL, Zuker CS. The cyclophilin homolog ninaA is a tissue-specific integral membrane protein required for the proper synthesis of a subset of Drosophila rhodopsins. Cell 1991;65:219–227.
90. Webel R, Menon I, O'Tousa JE, Colley NJ. Role of asparagine-linked oligosaccharides in rhodopsin maturation and association with its molecular chaperone, NinaA. J Biol Chem 2000;275:24,752–24,759.
91. Ferreira PA, Nakayama TA, Travis GH. Interconversion of red opsin isoforms by the cyclophilin-related chaperone protein Ran-binding protein 2. Proc Natl Acad Sci USA 1997;94:1556–1561.
92. Ferreira PA, Nakayama TA, Pak WL, Travis GH. Cyclophilin-related protein RanBP2 acts as chaperone for red/green opsin. Nature 1996;383:637–640.
93. Zhao H, Ivic L, Otaki JM, Hashimoto M, Mikoshiba K, Firestein S. Functional expression of a mammalian odorant receptor. Science 1998;279:237–242.
94. Gimelbrant AA, Haley SL, McClintock TS. Olfactory receptor trafficking involves conserved regulatory steps. J Biol Chem 2001;276:7285–7290.
95. McClintock TS, Landers TM, Gimelbrant AA, et al. Functional expression of olfactory-adrenergic receptor chimeras and intracellular retention of heterologously expressed olfactory receptors. Brain Res Mol Brain Res 1997;48:270–278.
96. Gimelbrant AA, Stoss TD, Landers TM, McClintock TS. Truncation releases olfactory receptors from the endoplasmic reticulum of heterologous cells. J Neurochem 1999;72:2301–2311.
97. Dwyer ND, Troemel ER, Sengupta P, Bargmann CI. Odorant receptor localization to olfactory cilia is mediated by ODR-4, a novel membrane-associated protein. Cell 1998;93:455–466.
98. Ichise T, Kano M, Hashimoto K, et al. mGluR1 in cerebellar Purkinje cells essential for long-term depression, synapse elimination, and motor coordination. Science 2000;288:1832–1835.
99. Xiao B, Tu JC, Worley PF, et al. Homer regulates the association of group 1 metabotropic glutamate receptors with multivalent complexes of homer-related, synaptic proteins. Curr Opin Neurobiol 2000;10:370–374.
100. Brakeman PR, Lanahan AA, O'Brien R, et al. Homer: a protein that selectively binds metabotropic glutamate receptors. Nature 1997;386:284–288.
101. Tu JC, Xiao B, Yuan JP, et al. Homer binds a novel proline-rich motif and links group 1 metabotropic glutamate receptors with IP3 receptors. Neuron 1998;21:717–726.
102. Roche KW, Tu JC, Petralia RS, Xiao B, Wenthold RJ, Worley PF. Homer 1b regulates the trafficking of group I metabotropic glutamate receptors. J Biol Chem 1999;274:25,953–25,957.
103. Ango F, Robbe D, Tu JC, et al. Homer-dependent cell surface expression of metabotropic glutamate receptor type 5 in neurons. Mol Cell Neurosci 2002;20:323–329.

104. Ciruela F, Soloviev MM, McIlhinney RA. Co-expression of metabotropic glutamate receptor type 1alpha with homer-1a/Vesl-1S increases the cell surface expression of the receptor. Biochem J 1999;341(pt 3):795–803.
105. Ciruela F, Soloviev MM, Chan WY, McIlhinney RA. Homer-1c/Vesl-1L modulates the cell surface targeting of metabotropic glutamate receptor type 1alpha: evidence for an anchoring function. Mol Cell Neurosci 2000;15:36–50.
106. Abe H, Misaka T, Tateyama M, Kubo Y. Effects of coexpression with Homer isoforms on the function of metabotropic glutamate receptor 1alpha. Mol Cell Neurosci 2003;23:157–168.
107. Bermak JC, Li M, Bullock C, Zhou QY. Regulation of transport of the dopamine D1 receptor by a new membrane-associated ER protein. Nat Cell Biol 2001;3:492–498.
108. Leclerc PC, Auger-Messier M, Lanctot PM, Escher E, Leduc R, Guillemette G. A polyaromatic caveolin-binding-like motif in the cytoplasmic tail of the type 1 receptor for angiotensin II plays an important role in receptor trafficking and signaling. Endocrinology 2002;143:4702–4710.
109. Xu Z, Hirasawa A, Shinoura H, Tsujimoto G. Interaction of the alpha(1B)-adrenergic receptor with gC1q-R, a multifunctional protein. J Biol Chem 1999;274:21,149–21,154.
110. Hirasawa A, Awaji T, Xu Z, Shinoura H, Tsujimoto G. Regulation of subcellular localization of alpha1-adrenoceptor subtypes. Life Sci 2001;68:2259–2267.
111. Pupo AS, Minneman KP. Specific interactions between gC1qR and alpha1-adrenoceptor subtypes. J Recept Signal Transduct Res 2003;23:185–195.
112. Daunt DA, Hurt C, Hein L, Kallio J, Feng F, Kobilka BK. Subtype-specific intracellular trafficking of alpha2-adrenergic receptors. Mol Pharmacol 1997;51:711–720.
113. Bunzow JR, Sonders MS, Arttamangkul S, et al. Amphetamine, 3,4-methylenedioxymethamphetamine, lysergic acid diethylamide, and metabolites of the catecholamine neurotransmitters are agonists of a rat trace amine receptor. Mol Pharmacol 2001;60:1181–1188.

14
Receptor Oligomerization and Trafficking

Selena E. Bartlett and Jennifer L. Whistler

1. INTRODUCTION

G protein-coupled receptor (GPCR) dimerization is a mechanism for regulating the signaling from several classes of plasma membrane receptors and has been a particularly well-studied mechanism for the regulation of tyrosine kinase (Trk) receptors (reviewed in ref. *1*). As reported for Trk and cytokine receptors, some GPCRs may also dimerize in response to agonist *(2–6)*. However, GPCRs can also form constitutive dimers—often as early as during their biosynthesis *(7–9)*. Adding to this complexity, GPCRs can form not only homodimers but also heterodimers with altered properties. Ultimately, one main goal is to understand the functional consequences of GPCR dimerization.

GPCR dimerization was first reported for the gonadotropin-releasing hormone receptor *(10)*, then for the β_2-adrenergic receptor (AR) *(11,12)*, and is now considered to be a common theme for many GPCRs (for review, *see* refs. *13–15*). GPCRs can form homo- and heterodimers as well as higher order oligomeric structures such as trimers, tetramers, and pentamers *(12,16–18)*. For the purposes of this chapter, we use the word "dimer" to describe all of these oligomers because it is the smallest possible oligomeric unit. GPCR homodimerization has been shown to occur with several receptors, including the D2 dopamine receptor *(19)*, the µ-opioid peptide receptor (MOP-R; ref. *20*), several of the somatostatin (SST) receptors *(21)*, and the chemokine CCR5 receptors *(2)*, to name just a few. Heterodimerization has been reported not only between receptors within the same GPCR family, such as the µ and δ-opioid peptide receptor (DOP-R) *(22)*, the m2 and m3 muscarinic receptors *(23)*, and the CCR2 and CCR5 chemokine receptors

(2), but also between diverse GPCR families, such as opioid receptors and β_2-ARs *(24,25)*, adenosine A2a receptors and D1 dopamine receptors *(26)*, and SSTR5 and D2 dopamine receptors *(27)*. Although the existence of these diverse heterodimers has yet to be demonstrated in vivo, the potential for their existence has profound implications for the biology of GPCR signaling and for drug design. There is a growing body of literature describing various GPCR dimers and dimerization's effects on receptor maturation through the secretory pathway, ligand binding, and signaling in vitro (for review, *see* ref. *28*). In contrast, there are relatively few studies describing the effects of dimerization on receptor trafficking—particularly postendocytic trafficking. This chapter summarizes the ways that dimerization can alter the trafficking properties of various GPCRs.

2. SECRETION

GPCR dimerization has been demonstrated in the absence of receptor activation by agonists (for review, *see* ref. *14*). Thus, it has been proposed that GPCR dimerization may be the result of a constitutive process that occurs early in the biosynthetic pathway and is necessary for trafficking receptors from the endoplasmic reticulum (ER) to the plasma membrane. For some GPCRs, heterodimerization with highly homologous receptors within the same family is actually a prerequisite for the expression of functional receptor units. The γ-aminobutyric acid-B (GABA$_B$) receptor is composed of two subunits, GABA$_B$R1 and GABA$_B$R2. GABA$_B$R1 is not expressed at the cell surface because it contains an ER retention signal, whereas GABA$_B$R2 is efficiently transported to the plasma membrane but is nonfunctional *(29)*. Co-expression of GABA$_B$R2 with GABA$_B$R1 allows transport of both subunits, presumably because heterodimerization of GABA$_B$R2 to GABA$_B$R1 masks the ER retention signal *(30,31)*. Intriguingly, it has also been demonstrated that the γ2S-subunit of the ionotropic GABA$_A$ receptor interacts with GABA$_B$R1 and can promote its cell-surface expression, thereby substituting for GABA$_B$R2 for trafficking of GABA$_B$R1 to the membrane. It can also interact with the GABA$_B$R1–GABA$_B$R2 heterodimeric complex to significantly enhance GABA$_B$ receptor internalization in response to agonist *(32)*. This is the first demonstration of an interaction between a GPCR and an ion channel affecting receptor trafficking. Although a general role of dimerization in ER export has yet to be established, studies that use fluorescence and/or bioluminescence resonance energy transfer with cell fractionation have demonstrated the existence of receptor dimers in the ER for several receptor classes, including CCR5 chemokine receptors *(7)*, vasopressin receptors and oxytocin receptors *(8)*, the yeast α-factor receptor *(33)*, and the C5a chemotactic receptor *(34)*.

Once the receptors have reached the plasma membrane, the debate continues regarding whether ligands alter the dimerization state of the receptors. To date, there is no clear answer. Ligand binding has been shown to both promote *(6,27,35–40)* or inhibit *(26,41,42)* dimerization, and several groups report no effect of ligand, suggesting instead that dimerization is a constitutive process *(7,8,34,43-50)*.

3. DESENSITIZATION AND ENDOCYTOSIS

Following transport to the plasma membrane and activation by an agonist ligand, GPCRs are regulated by several mechanisms, many of which have been shown to be altered by receptor homo- or heterodimerization, including receptor desensitization, receptor endocytosis, receptor recycling, and receptor degradation. For the purposes of this chapter, we define receptor desensitization as any process that alters the functional coupling of a receptor to its G protein/second messenger-signaling pathway. Endocytosis/internalization is defined as the translocation of receptors from the cell surface to an intracellular compartment. Receptor recycling or receptor resensitization is defined as receptors returning to the cell surface following their endocytosis. Finally, receptor degradation or downregulation is defined as any process that decreases the number of ligand-binding sites. Signaling from GPCRs is rapidly regulated by a well-characterized and highly conserved cascade of events involving G protein coupling and activation, receptor phosphorylation by GPCR kinases (GRKs), and subsequent β-arrestin recruitment that has been extensively reviewed elsewhere *(51)*. These processes contribute directly to receptor desensitization by facilitating the uncoupling of the receptor from its G protein. GRK- and β-arrestin-mediated desensitization is a rapid process that often occurs within minutes of receptor activation. However, receptors can also be desensitized/uncoupled from G proteins by GRK and β-arrestin-independent mechanisms (for review, *see* ref. 52).

It has been reported that some GPCR heterodimeric complexes have altered desensitization properties. For example, co-expression of the SST receptors SSTR2A and SSTR3 results in a slower desensitization of the SSTR2A–SSTR3 heterodimer compared to cells expressing SSTR2 or SSTR3 homodimers *(53)*. Similarly, heterodimerization of the MOP-R and the SST2A receptor leads to cross-desensitization of receptor function *(21)*. This pattern of cross-desensitization is mirrored by co-expression and heterodimerization of the adenosine A2a receptor with the D2 dopamine receptor *(54)* as well as by heterodimerization of the MOP-R with the CCR5 chemokine receptor *(55)*.

Following desensitization by GRKs and β-arrestin, GPCRs are rapidly endocytosed into an intracellular compartment. This process occurs following even brief agonist exposure and is independent of signal transduction *(56,57)*. GPCR internalization is predominantly mediated via the recruitment of β-arrestins 1 and/or 2 to agonist-activated phosphorylated receptors. GPCR dimerization can alter the complement of proteins—particularly arrestins— that are recruited during endocytosis, thereby affecting the endocytic properties of the receptors. In fact, heterodimerization has been shown to both inhibit and facilitate endocytosis, depending on the pair of receptors expressed. Heterodimerization of the $β_1$-AR (a poor internalizer) and the $β_2$-AR (a good internalizer) results in inhibition of agonist-promoted internalization of the $β_2$-AR *(58)*. Similarly, the κ-opioid peptide receptor (KOP-R), which is a poor internalizer, inhibits the endocytosis of both the DOP-R *(12)* and the $β_2$-AR *(24)* when it heterodimerizes with these receptors.

On the other hand, there are several examples of co-internalization of both receptors in a heterodimer pair, even when agonist ligand is present for only one of the receptors. These include the A2a adenosine–D2 dopamine dimer pair *(54)*, in which addition of either the adenosine agonist or the dopamine agonist promotes internalization of both receptors, and the SSTR2a–MOP receptor pair, in which activation with either the SSTR2a agonist or the MOP-R agonist promotes internalization of both receptors *(21)*.

As mentioned earlier, in both cases, co-internalization was associated with cross-desensitization of receptor-mediated signaling. Similarly, $α_{1a}$- and $α_{1b}$-AR heterodimerization results in co-internalization of both receptors following treatment with oxymetazoline, an $α_{1a}$-specific agonist. In contrast, $α_{1b}$ does not co-internalize with agonist-activated neurokinin (NK)1 receptors or agonist-activated CCR5 chemokine receptors, and neither of these receptors heterodimerizes with the $α_{1b}$-AR *(49)*, demonstrating that dimerization appears to be required for co-internalization.

Another example of dimer-facilitated endocytosis occurs with mutant and wild-type MOP-Rs. Wild-type MOP-Rs do not endocytose when activated by morphine *(59)*. However, co-expression of a mutant MOP-R, D-MOP-R (which does internalize in response to morphine *[60]*) leads to dimerization of MOP-R with D-MOP-R and endocytosis of both the wild-type and mutant MOP-R in response to morphine *(20)*.

In some cases, it has been shown that the co-internalization effects are modulated by changes in recruitment of β-arrestin. This has been elegantly demonstrated with the thyrotropin-releasing hormone receptors 1 and 2 (TRHR1 and TRHR2) dimers 961 and also with MOP-R and NK1-R dimers *(61,62)*. TRHR2 interacts preferentially with β-arrestin 2 when expressed alone in HEK293 and simian fibroblast (COS) cells. However, co-expression of TRHR1 leads to interaction of both TRHR1 and TRHR2 with β-arrestin 1. As a consequence, the internalization rates for both TRHR1 and TRHR2 are altered, with no effect on ligand binding or basal inositol phosphate production *(61)*.

Based on the pattern of arrestin recruitment, GPCRs can be divided into two categories: class A and class B *(63)*. Class A receptors have a higher affinity for β-arrestin 2 than β-arrestin 1, and, after initial recruitment, class A receptors rapidly dissociate from β-arrestin. Class B receptors bind β-arrestin 1 and 2 with equal affinities, and, unlike class A receptors, class B receptors remain in a stable complex with arrestins during endocytosis and, therefore, are colocalized with arrestins in endosomes. Alterations in the pattern of β-arrestin 2 recruitment are seen with MOP-R–NK1-R heterodimers. In cells expressing only the MOP-R, β-arrestin 2 is recruited to the activated receptors; however, β-arrestin 2 does not co-internalize with the receptor classifies MOP-Rs as class A. However, when the NK1-R is co-expressed with the MOP-R, the heterodimers recruit β-arrestin 2 in response to either the MOP-R agonist [D-Ala(2)-*N*-Me-Phe(4), Gly(5)-ol]-enkephalin (DAMGO) or the NK1-R agonist substance P, and β-arrestin 2 is then co-internalized with the MOP-R–NK1-R heterodimeric complex, switching the classification of the MOP-R to class B *(62)*. Furthermore, in the case of the MOP-R–NK1-R pair, co-internalization of β-arrestin 2 with the heterodimer leads to a delay in both MOP-R recycling and MOP-R resensitization kinetics *(62)*. This observation leads to the conclusion that heterodimerization may also affect receptor trafficking—more specifically, postendocytic sorting.

4. POSTENDOCYTIC TRAFFICKING

Receptors that have been desensitized and rapidly endocytosed are uniquely poised to make an important decision that has substantial impact on future signal transduction. As mentioned earlier, following endocytosis, receptors can be recycled, thereby restoring the functional complement of receptors. Alternatively, receptors that have been endocytosed can be targeted for degradation, thereby decreasing the functional complement of receptors and ultimately resulting in receptor downregulation. Although endocytosis and subsequent degradation of receptors is not the only means to produce receptor downregulation, it can produce receptor downregulation rapidly, even following brief exposure to agonist *(64)*. Apparent receptor downregulation can also be affected by alterations in rate of receptor synthesis and/or folding or secretion *(65)*. Not all GPCRs are downregulated following their endocytosis. For example, although both the MOP-R and

DOP-R are class A receptors that are endocytosed via clathrin-coated pits following agonist-induced activation, GRK phosphorylation, and association with β-arrestins, they differ in their fate following endocytosis. Whereas MOP-Rs are recycled following their endocytosis, DOP-Rs are transported deeper into the endocytic pathway and are rapidly degraded by the lysosome *(60,66)* and hence downregulated.

As discussed earlier, β-arrestin 2 is recruited and co-internalized with the MOP-R–NK1-R heterodimeric complex in response to either the μ-opioid agonist DAMGO or the NK1 receptor agonist substance P. This leads to a delay in both MOP-R recycling and MOP-R resensitization kinetics *(62)*, suggesting that heterodimerization can alter the post-endocytic trafficking properties of the MOP-R.

Heterodimerization of GPCRs may also result in changes, not only to the rate of postendocytic recycling but also to the actual postendocytic fate of the receptors. This has been beautifully demonstrated using the V1a vasopressin receptor (V1aR), which heterodimerizes with V2 vasopressin receptor (V2R) *(8)*. V1aR activation results in the recruitment of β-arrestin. Following receptor internalization, β-arrestin dissociates from V1aR, and the receptor is rapidly recycled to the plasma membrane, which classifies the V1aR as a class A GPCR. In contrast, V2Rs recruit β-arrestin, and following internalization, β-arrestin fails to dissociate from the receptor, resulting in V2R accumulation in endosomes in the perinuclear compartment *(67)*; this classifies V2R as a class B receptor. The postendocytic fates of V1aR and V2R are altered when the receptors are co-expressed. Nonselective agonist activation of both V1aR and V2R results in the co-internalization of both V1aR and V2R with β-arrestin into endosomes, where the interaction inhibits the recycling of both receptors to the plasma membrane, thereby switching the V1aR from a class A to a class B receptor. Selective activation of only V1aR results in co-internalization of the V2R, without co-internalization of β-arrestin. This results in rapid recycling of both V1aR and V2R, thereby converting V2R to class A receptor *(67)*. These data suggest that for the vasopressin receptors, it is the activated receptor within the heterodimeric complex that determines the overall postendocytic fate of all receptors within the complex. It remains to be determined whether this is true for all reported receptor heterodimers.

5. CLINICAL RELEVANCE AND/OR BENEFITS

When one begins to think of all the combinations and permutations whereby GPCRs could homo- or heterodimerize, the implications for unique drug targets or "landing pads" becomes apparent and really quite overwhelming. Already there are several examples of ways in which heterodimerization, solely by affecting trafficking, affects phenomena important for human disease.

5.1. Human Immunodeficiency Virus Infection

CCR5Δ32, the truncated form of the human immunodeficiency virus (HIV) coreceptor CCR5, conveys resistance to HIV infection even in individuals in the heterozygous state *(68–71)*. Co-expression of this truncated CCR5 with the wild-type CCR5 leads to retention of both receptors in the ER. Because CCR5 forms heterodimers, it has been suggested that by forming a heterodimer with the wild-type receptor, the mutant receptor delays transport of CCR5 coreceptors to the surface of cells, thereby preventing HIV coreceptor function *(72)*. The existence of naturally occurring chemokine receptor mutants that alter CCR5 trafficking may be an explanation for the subpopulations of patients with acquired immune deficiency syndrome that display differential resistance to HIV infection. This may help guide future therapeutic strategies for its treatment.

5.2. Attenuation of Morphine Tolerance and Dependence

The current evidence suggests that receptor oligomerization may open the door to the discovery of new, more effective drug treatments for disease. One example is the amelioration of the development of morphine tolerance and dependence. This section outlines studies that suggest that targeting the MOP-R oligomeric complex may be helpful to reduce the development of morphine tolerance.

Although opioids such as morphine remain the analgesic of choice in many cases, a major limitation to their long-term use is the development of physiological tolerance, which is a profound decrease in analgesic effect observed in all patients during prolonged administration of opioid drug. Despite considerable progress, the molecular and cellular mechanisms mediating the development of tolerance and withdrawal to morphine remain controversial. Multiple hypotheses exist to explain morphine tolerance.

In contrast to other opioids and many other agonists in general, morphine fails to promote endocytosis of the wild-type μ-opioid receptor in cultured cells *(59,73)* and native neurons *(74,75)*, whereas endogenous peptide ligands such as endorphins and several opioid drugs, such as methadone *(76)*, readily drive receptor endocytosis *(74,75,77)*. Morphine-activated MOP-Rs are relatively unique because they are not GRK-phosphorylated nor do they efficiently recruit β-arrestin, although they are in an "active" receptor conformation *(78–80)*. Hence, morphine-activated MOP-Rs generally elude an important, highly conserved regulatory mechanism designed to rapidly modulate receptor-mediated signaling.

Whistler et al. have proposed that the regulation of opioid receptors by endocytosis serves as a protective role in reducing the development of tolerance and dependence to opioid drugs *(20,81)*. According to this model, promoting morphine-induced endocytosis of the MOP-R should lead to an attenuation of the development of morphine tolerance and dependence. This has been shown to be true in a cell culture model, where mutations of the MOP-R that enhance morphine-induced endocytosis ameliorate a cellular hallmark of tolerance *(60)*. Conversely, mutations that prevented endocytosis of the MOP-R in response to ligands (such as methadone) enhanced methadone-mediated tolerance *(60)*. These observations led to the prediction that one might be able to use the oligomeric state of the MOP-R to alter its endocytosis in response to morphine, thereby altering the development of morphine tolerance.

Briefly, could a MOP-R dimer in which one protomer was occupied with DAMGO (which facilitates endocytosis) "drag" the other morphine-occupied protomer into the cell? In fact, co-administration of a small dose of DAMGO facilitated morphine-induced endocytosis of the MOP-R both in cell culture models and in vivo *(20)*. Importantly, facilitation of MOP-R endocytosis also prevented the development of morphine tolerance *(20)*. Pharmaceutical interventions that target heterodimeric MOP-Rs that facilitate the endocytosis of the MOP-R in response to morphine may play important roles in the attenuation of morphine tolerance and dependence.

6. CONCLUSIONS

Determining whether opioid receptor heterodimers or opioid–other GPCR dimers exist in vivo is a true challenge. Advances in bivalent ligands that selectively target heterodimers may provide the necessary tools (for review, *see* refs. *82* and *83*). Additionally, the phenotypes of the opioid receptor knockout mice also provide hints that heterodimerization, at least among the opioid receptors, is a real phenomenon. For example, some μ-opioid receptor-specific analgesia is lost in the δ-opioid receptor-deficient mice *(84)*. One explanation for this observation could be that a μ–δ opioid receptor dimer has also been eliminated in the genetically modified animals. Making the assumption that GPCRs form dimers in vivo, it may be possible to design therapeutics to specifically target these complexes, which therefore provide more specific drug treatments. Signaling from GPCRs will ultimately depend on the cell-type and its specific environment. Different signals may be generated from the same receptor, depending on the intracellular complement of other GPCRs available and/or the environment of the cell. Understanding these differences will be important for the design of drugs and the treatment of diseases.

ACKNOWLEDGMENTS

S. E. Bartlett is supported in part by the Sidney Baer Trust as a NARSAD Young Investigator. J. L. Whistler is supported by a National Institutes of Health grant R01 DA015232 and by funds provided by the State of California for medical research on alcohol and substance abuse through UCSF. We would like to thank Dr. Maria Waldhoer for a critical reading of the manuscript.

REFERENCES

1. Hackel PO, Zwick E, Prenzel N, Ullrich A. Epidermal growth factor receptors: critical mediators of multiple receptor pathways. Curr Opin Cell Biol 1999; 11:184–189.
2. Mellado M, Rodriguez-Frade JM, Vila-Coro AJ, et al. Chemokine receptor homo- or heterodimerization activates distinct signaling pathways. EMBO J 2001;20:2497–2507
3. AbdAlla S, Lother H, Quitterer U. AT1-receptor heterodimers show enhanced G-protein activation and altered receptor sequestration. Nature 2000;407:94–98.
4. AbdAlla S. Zaki E, Lother H, Quitterer U. Involvement of the amino terminus of the B(2) receptor in agonist-induced receptor dimerization. J Biol Chem 1999;274:26,079–26,084
5. Horvat RD, Barisas BG, Roess DA. Luteinizing hormone receptors are self-associated in slowly diffusing complexes during receptor desensitization. Mol Endocrinol 2001;15:534–542.
6. Horvat RD, Roess DA, Nelson SE, Barisas BG, Clay CM. Binding of agonist but not antagonist leads to fluorescence resonance energy transfer between intrinsically fluorescent gonadotropin-releasing hormone receptors. Mol Endocrinol 2001;15:695–703.
7. Issafras H, Angers S, Bulenger S, et al. Constitutive agonist-independent CCR5 oligomerization and antibody-mediated clustering occurring at physiological levels of receptors. J Biol Chem 2002;277:34,666–34,673.
8. Terrillon S, Durroux T, Mouillac B, et al. Oxytocin and vasopressin V1a and V2 receptors form constitutive homo- and heterodimers during biosynthesis. Mol Endocrinol 2003;17:677–691.
9. Romano C, Yang WL, O'Malley KL. Metabotropic glutamate receptor 5 is a disulfide-linked dimer. J Biol Chem 1996;271:28,612–28,616.
10. Conn PM, Rogers DC, Stewart JM, Niedel J, Sheffield T. Conversion of a

gonadotropin-releasing hormone antagonist to an agonist. Nature 1982;296:653–655.
11. Hebert TE, Moffett S, Morello JP, et al. A peptide derived from a beta2-adrenergic receptor transmembrane domain inhibits both receptor dimerization and activation. J Biol Chem 1996;271:16,384–16,392.
12. Jordan BA, Devi LA. G-protein-coupled receptor heterodimerization modulates receptor function. Nature 1999;399:697–700.
13. Terrillon S, Bouvier M. Roles of G-protein-coupled receptor dimerization. EMBO Rep 2004;5:30–34.
14. Bai M. Dimerization of G-protein-coupled receptors: roles in signal transduction. Cell Signal 2004;16:175–186.
15. Gomes I, Filipovska J, Jordan BA, Devi LA. Oligomerization of opioid receptors. Methods 2002;27:358–365.
16. Lee SP, O'Dowd BF, Ng GY, et al. Inhibition of cell surface expression by mutant receptors demonstrates that D2 dopamine receptors exist as oligomers in the cell. Mol Pharmacol 2000;58:120–128.
17. Zawarynski P, Tallerico T, Seeman P, Lee SP, O'Dowd BF, George SR. Dopamine D2 receptor dimers in human and rat brain. FEBS Lett 1998;441:383–386.
18. Klco JM, Lassere TB, Baranski TJ. C5a receptor oligomerization. I. Disulfide trapping reveals oligomers and potential contact surfaces in a G protein-coupled receptor. J Biol Chem 2003;278:35,345–35,353.
19. Lee SP, O'Dowd BF, Rajaram RD, Nguyen T, George SR. D2 dopamine receptor homodimerization is mediated by multiple sites of interaction, including an intermolecular interaction involving transmembrane domain 4. Biochemistry 2003;42:11,023–11,031.
20. He L, Fong J, von Zastrow M, Whistler JL. Regulation of opioid receptor trafficking and morphine tolerance by receptor oligomerization. Cell 2002;108:271–282.
21. Pfeiffer M, Koch T, Schroder H, Laugsch M, Hollt V, Schulz S. Heterodimerization of somatostatin and opioid receptors cross-modulates phosphorylation, internalization, and desensitization. J Biol Chem 2002;277: 19,762–19,772.
22. Gomes I, Jordan BA, Gupta A, Trapaidze N, Nagy V, Devi LA. Heterodimerization of mu and delta opioid receptors: A role in opiate synergy. J Neurosci 2000;20:RC110.
23. Maggio R, Vogel Z, Wess J. Coexpression studies with mutant muscarinic/adrenergic receptors provide evidence for intermolecular "cross-talk" between G-protein-linked receptors. Proc Natl Acad Sci USA 1993;90:3103–3107.
24. Jordan BA, Trapaidze N, Gomes I, Nivarthi R, Devi LA. Oligomerization of opioid receptors with beta 2-adrenergic receptors: a role in trafficking and mitogen-activated protein kinase activation. Proc Natl Acad Sci USA 2001;98:343–348.
25. Rios CD, Jordan BA, Gomes I, Devi LA. G-protein-coupled receptor dimerization: modulation of receptor function. Pharmacol Ther 2001;92:71–87.
26. Gines S, Hillion J, Torvinen M, et al. Dopamine D1 and adenosine A1 recep-

tors form functionally interacting heteromeric complexes. Proc Natl Acad Sci USA 2000;97:8606–8611.
27. Rocheville M, Lange DC, Kumar U, Patel SC, Patel RC, Patel YC. Receptors for dopamine and somatostatin: formation of hetero-oligomers with enhanced functional activity. Science 2000;288:154–157.
28. Devi LA. Heterodimerization of G-protein-coupled receptors: pharmacology, signaling and trafficking. Trends Pharmacol Sci 2001;22:532–537.
29. Margeta-Mitrovic M, Jan YN, Jan LY. A trafficking checkpoint controls GABA(B) receptor heterodimerization. Neuron 2000;27:97–106.
30. Galvez T, Duthey B, Kniazeff J, et al. Allosteric interactions between GB1 and GB2 subunits are required for optimal GABA(B) receptor function. EMBO J 2001;20:2152–2159.
31. Pagano A, Rovelli G, Mosbacher J, et al. C-terminal interaction is essential for surface trafficking but not for heteromeric assembly of GABA(b) receptors. J Neurosci 2001;21:1189–1202.
32. Balasubramanian S, Teissere JA, Raju DV, Hall RA. Hetero-oligomerization between GABA-A and GABA-B receptors regulates GABA-B receptor trafficking. J Biol Chem 2004;279:18,840–18,850.
33. Overton MC, Blumer KJ. The extracellular N-terminal domain and transmembrane domains 1 and 2 mediate oligomerization of a yeast G protein-coupled receptor. J Biol Chem 2002;277:41,463–41,472.
34. Floyd DH, Geva A, Bruinsma SP, Overton MC, Blumer KJ, Baranski TJ. C5a receptor oligomerization. II. Fluorescence resonance energy transfer studies of a human G protein-coupled receptor expressed in yeast. J Biol Chem 2003;278:35,354–35,361.
35. Patel RC, Kumar U, Lamb DC, et al. Ligand binding to somatostatin receptors induces receptor-specific oligomer formation in live cells. Proc Natl Acad Sci USA 2002;99:3294–3299.
36. Rodriguez-Frade JM, Vila-Coro AJ, de Ana AM, Albar JP, Martinez AC, Mellado M. The chemokine monocyte chemoattractant protein-1 induces functional responses through dimerization of its receptor CCR2. Proc Natl Acad Sci USA 1999;96:3628–3633
37. Cornea A, Janovick JA, Maya-Nunez G, Conn PM. Gonadotropin-releasing hormone receptor microaggregation. Rate monitored by fluorescence resonance energy transfer. J Biol Chem 2001;276:2153–2158.
38. Kroeger KM, Hanyaloglu AC, Seeber RM, Miles LE, Eidne KA. Constitutive and agonist-dependent homo-oligomerization of the thyrotropin-releasing hormone receptor. Detection in living cells using bioluminescence resonance energy transfer. J Biol Chem 2001;276:12,736–12,743.
39. Zhu CC, Cook LB, Hinkle PM. Dimerization and phosphorylation of thyrotropin-releasing hormone receptors are modulated by agonist stimulation. J Biol Chem 2002;277:28,228–28,237.
40. Hunzicker-Dunn M, Barisas G, Song J, Roess DA. Membrane organization of luteinizing hormone receptors differs between actively signaling and desensitized receptors. J Biol Chem 2003;278:42,744–42,749.
41. Cheng ZJ, Miller LJ. Agonist-dependent dissociation of oligomeric complexes

of G protein-coupled cholecystokinin receptors demonstrated in living cells using bioluminescence resonance energy transfer. J Biol Chem 2001;276:48,040–48,047.
42. Latif R, Graves P, Davies TF. Ligand-dependent inhibition of oligomerization at the human thyrotropin receptor. J Biol Chem 2002;277:45,059–45,067.
43. Overton MC, Blumer KJ. G-protein-coupled receptors function as oligomers in vivo. Curr Biol 2000;10:341–344.
44. Jensen AA, Hansen JL, Sheikh SP, Brauner-Osborne H. Probing intermolecular protein–protein interactions in the calcium-sensing receptor homodimer using bioluminescence resonance energy transfer (BRET). Eur J Biochem 2002;269:5076–5087.
45. Babcock GJ, Farzan M, Sodroski J. Ligand-independent dimerization of CXCR4, a principal HIV-1 coreceptor. J Biol Chem 2003;278:3378–3385.
46. Canals M, Marcellino D, Fanelli F, et al. Adenosine A2A-dopamine D2 receptor–receptor heteromerization: qualitative and quantitative assessment by fluorescence and bioluminescence energy transfer. J Biol Chem 2003;278:46,741–46,749.
47. Dinger MC, Bader JE, Kobor AD, Kretzschmar AK, Beck-Sickinger AG. Homodimerization of neuropeptide y receptors investigated by fluorescence resonance energy transfer in living cells. J Biol Chem 2003;278:10,562–10,571.
48. Guo W, Shi L, Javitch JA. The fourth transmembrane segment forms the interface of the dopamine D2 receptor homodimer. J Biol Chem 2003;278:4385–4388.
49. Stanasila L, Perez JB, Vogel H, Cotecchia S. Oligomerization of the alpha 1a- and alpha 1b-adrenergic receptor subtypes. Potential implications in receptor internalization. J Biol Chem 2003;278:40,239–40,251.
50. Trettel F, Di Bartolomeo S, Lauro C, Catalano M, Ciotti MT, Limatola C. Ligand-independent CXCR2 dimerization. J Biol Chem 2003;278:40,980–40,988.
51. Ferguson SS. Evolving concepts in G protein-coupled receptor endocytosis: the role in receptor desensitization and signaling. Pharmacol Rev 2001;53:1–24.
52. Liu JG, Anand KJ. Protein kinases modulate the cellular adaptations associated with opioid tolerance and dependence. Brain Res Brain Res Rev 2001;38:1–19.
53. Pfeiffer M, Koch T, Schroder H, et al. Homo- and heterodimerization of somatostatin receptor subtypes. Inactivation of sst(3) receptor function by heterodimerization with sst(2A). J Biol Chem 2001;276:14,027–14,036.
54. Hillion J, Canals M, Torvinen M, et al. Coaggregation, cointernalization, and codesensitization of adenosine A2A receptors and dopamine D2 receptors. J Biol Chem 2002;277:18,091–18,097.
55. Chen C, Li J, Bot G, Szabo I, Rogers TJ, Liu-Chen LY. Heterodimerization and cross-desensitization between the mu-opioid receptor and the chemokine CCR5 receptor. Eur J Pharmacol 2004;483:175–186.
56. Zaki PA, Keith DE, Jr., Thomas JB, Carroll FI, Evans CJ. Agonist-, antagonist-, and inverse agonist-regulated trafficking of the delta-opioid receptor cor-

relates with, but does not require, G protein activation. J Pharmacol Exp Ther 2001;298:1015–1020.
57. Remmers AE, Clark MJ, Liu XY, Medzihradsky F. Delta opioid receptor downregulation is independent of functional G protein yet is dependent on agonist efficacy. J Pharmacol Exp Ther 1998;287:625–632.
58. Lavoie C, Mercier JF, Salahpour A, et al. Beta 1/beta 2-adrenergic receptor heterodimerization regulates beta 2-adrenergic receptor internalization and ERK signaling efficacy. J Biol Chem 2002;277:35,402–35,410.
59. Keith DE, Murray SR, Zaki PA, et al. Morphine activates opioid receptors without causing their rapid internalization. J Biol Chem 1996;271:19,021–19,024.
60. Finn AK, Whistler JL. Endocytosis of the mu opioid receptor reduces tolerance and a cellular hallmark of opiate withdrawal. Neuron 2001;32:829–839.
61. Hanyaloglu AC, Seeber RM, Kohout TA, Lefkowitz RJ, Eidne KA. Homo- and hetero-oligomerization of thyrotropin-releasing hormone (TRH) receptor subtypes. Differential regulation of beta-arrestins 1 and 2. J Biol Chem 2002;277:50,422–50,430.
62. Pfeiffer M, Kirscht S, Stumm R, et al. Heterodimerization of substance P and mu-opioid receptors regulates receptor trafficking and resensitization. J Biol Chem 2003;278:51,630–51,637.
63. Oakley RH, Laporte SA, Holt JA, Caron MG, Barak LS. Differential affinities of visual arrestin, beta arrestin1, and beta arrestin2 for G protein-coupled receptors delineate two major classes of receptors. J Biol Chem 2000;275:17,201–17,210.
64. Tsao PI, von Zastrow M. Type specific sorting of G protein-coupled receptors after endocytosis. J Biol Chem 2000;275:11,130–11,140.
65. Petaja-Repo UE, Hogue M, Bhalla S, Laperriere A, Morello JP, Bouvier M. Ligands act as pharmacological chaperones and increase the efficiency of delta opioid receptor maturation. EMBO J 2002;21:1628–1637.
66. Whistler JL, Enquist J, Marley A, et al. Modulation of postendocytic sorting of G protein-coupled receptors. Science 2002;297:615–620.
67. Terrillon S, Barberis C, Bouvier M. Heterodimerization of V1a and V2 vasopressin receptors determines the interaction with beta-arrestin and their trafficking patterns. Proc Natl Acad Sci USA 2004;101:1548–1553.
68. Dean M, Jacobson LP, McFarlane G, et al. Reduced risk of AIDS lymphoma in individuals heterozygous for the CCR5-delta32 mutation. Cancer Res 1999;59:3561–3564.
69. Samson M, Libert F, Doranz BJ, et al. Resistance to HIV-1 infection in caucasian individuals bearing mutant alleles of the CCR-5 chemokine receptor gene. Nature 1996;382:722–725.
70. Huang Y, Paxton WA, Wolinsky SM, et al. The role of a mutant CCR5 allele in HIV-1 transmission and disease progression. Nat Med 1996;2:1240–1243.
71. Michael NL, Chang G, Louie LG, et al. The role of viral phenotype and CCR-5 gene defects in HIV-1 transmission and disease progression. Nat Med 1997;3:338–340.

72. Benkirane M, Jin DY, Chun RF, Koup RA, Jeang KT. Mechanism of transdominant inhibition of CCR5-mediated HIV-1 infection by ccr5delta32. J Biol Chem 1997;272:30,603–30,606.
73. Arden JR, Segredo V, Wang Z, Lameh J, Sadee W. Phosphorylation and agonist-specific intracellular trafficking of an epitope-tagged mu-opioid receptor expressed in HEK 293 cells. J Neurochem 1995;65:1636–1645.
74. Sternini C, Spann M, Anton B, et al. Agonist-selective endocytosis of mu opioid receptor by neurons in vivo. Proc Natl Acad Sci USA 1996;93:9241–9246.
75. Keith DE, Anton B, Murray SR, et al. mu-Opioid receptor internalization: opiate drugs have differential effects on a conserved endocytic mechanism in vitro and in the mammalian brain. Mol Pharmacol 1998;53:377–384.
76. Garrido MJ, Troconiz IF. Methadone: a review of its pharmacokinetic/pharmacodynamic properties. J Pharmacol Toxicol Methods 1999;42:61–66.
77. Trapaidze N, Gomes I, Cvejic S, Bansinath M, Devi LA. Opioid receptor endocytosis and activation of MAP kinase pathway. Brain Res Mol Brain Res 2000;76:220–228.
78. Blake AD, Bot G, Freeman JC, Reisine T. Differential opioid agonist regulation of the mouse mu opioid receptor. J Biol Chem 1997;272:782–790.
79. Whistler JL, von Zastrow M. Morphine-activated opioid receptors elude desensitization by beta-arrestin. Proc Natl Acad Sci USA 1998;95:9914–9919.
80. Zhang J, Ferguson SS, Barak LS, et al. Role for G protein-coupled receptor kinase in agonist-specific regulation of mu-opioid receptor responsiveness. Proc Natl Acad Sci USA 1998;95:7157–7162.
81. Whistler JL, Chuang HH, Chu P, Jan LY, von Zastrow M. Functional dissociation of mu opioid receptor signaling and endocytosis: implications for the biology of opiate tolerance and addiction. Neuron 1999;23:737–746.
82. Portoghese PS. From models to molecules: opioid receptor dimers, bivalent ligands, and selective opioid receptor probes. J Med Chem 2001;44:2259–2269.
83. Lutz RA, Pfister HP. Opioid receptors and their pharmacological profiles. J Recept Res 1992;12:267–286.
84. Zhu Y, King MA, Schuller AG, et al. Retention of supraspinal delta-like analgesia and loss of morphine tolerance in delta opioid receptor knockout mice. Neuron 1999;24:243–252.

15
Modulation of Receptor Pharmacology by G Protein-Coupled Receptor Dimerization

Noura S. Abul-Husn, Achla Gupta, Lakshmi A. Devi, and Ivone Gomes

1. INTRODUCTION

Although it was classically believed that G protein-coupled receptors (GPCRs) acted as monomeric entities, it is now well-established that they exist and function as dimers (or oligomers) in the plasma membrane. In addition to forming homodimers, GPCRs can associate with closely or distantly related members of the GPCR superfamily to form heterodimers. Dissection of the functional relevance of these associations is currently an area of enormous interest. Additionally, there is mounting evidence that heterodimerization can generate receptors with novel characteristics, leading to altered pharmacological properties. This could at least partially account for pharmacologically defined receptor subtypes for which no gene has been identified. This chapter reviews recent reports of GPCR dimerization and its effect on ligand pharmacology and function. Because current techniques do not readily allow the distinction between the functional effects of GPCR dimers and oligomers, the receptor complexes resulting from GPCR interactions are referred to as "dimers" and the phenomenon as "dimerization." Additionally, interactions between identical proteins are referred to as "homodimers," and interactions between nonidentical proteins are referred to as "heterodimers."

2. HOMODIMERS

Recent biochemical and biophysical evidence has shown that an increasing number of GPCRs exist as dimers/oligomers (*see* refs. *1–5*). X-ray crystallo-

From: *Contemporary Clinical Neuroscience: The G Protein-Coupled Receptors Handbook*
Edited by: L. A. Devi © Humana Press Inc., Totowa, NJ

graphic and atomic force microscopy (AFM) studies have demonstrated the presence of constitutive GPCR homodimers. The finding that GPCRs exist as dimers raises the question: How does this phenomenon modulate receptor pharmacology? This has been studied by investigating the effect of ligand treatment and peptide competition on the levels of receptor dimers as well as by functional complementation studies using mutant receptors (Table 1). These topics are described in the following sections.

2.1. Crystallographic and AFM Studies

X-ray crystallographic studies have revealed that the extracellular region of the metabotropic glutamate receptor (mGluR) exists as a disulfide-linked dimer. Binding of the ligand glutamate induces a movement of the two lobes in each ligand-binding domain, leading to stabilization of the dimer in the active conformation (6). The three-dimensional crystal structure of rhodopsin confirms that this prototype of family A GPCRs exists as a constitutive dimer (7). Palczewski (8) as well as Fotiadis et al. (9) used infrared AFM to demonstrate the native arrangement of rhodopsin in isolated mouse rod outer-segment membranes. High-magnification analysis revealed distinct rows of rhodopsin dimers densely packed in paracrystalline arrays. Based on these findings, a molecular model for the rhodopsin paracrystal has been proposed in which points of contact in the interface between the monomers in the rhodopsin dimer consist of transmembranes (TMs)-IV and -V (10).

Based on geometrical constraints, it appears that a single heterotrimeric G protein covers four rhodopsin molecules (11). According to this model, ligand binding to one rhodopsin monomer induces a conformational change that is transmitted to the second monomer, which then signals through the heterotrimeric G protein. The second rhodopsin dimer appears to serve as a docking platform (11). This model for G protein activation by GPCR oligomers needs to be validated by further studies with other family A GPCRs. Note that a recent study using mass spectrometry after chemical crosslinking and neutron scattering in solution is consistent with this model because it showed that one G protein trimer binds to the dimeric leukotriene B_4 receptor BLT1 to form a pentameric assembly (12).

2.2. Peptide Competition Studies

The domains involved in dimer formation have been examined using peptides directed against certain TM regions of GPCRs. A few studies have shown that ligand binding and signaling can be affected by disruption of dimer formation with such peptides (13–17). In the case of the β_2-adrenergic receptor (AR), adding a peptide corresponding to TM-VI substantially reduced the level of detected β_2-AR dimers as well as isoproterenol-stimu-

Table 1
GPCR Homodimers

Homodimers	References
Crystallographic and AFM studies	
mGluR	*6*
Rhodopsin	*7–11*
Leukotriene B$_4$	*12*
Peptide competition studies	
β$_2$AR	*14*
D1R dopamine	*15*
B$_2$R	*17*
δ-opioid	This chapter
D2 dopamine	*16*
Functional complementation studies	
α$_{1b}$-AR	*20*
Platelet activating factor	*21*
AT$_1$R angiotensin	*18*
H1 histamine	*20*
Calcium-sensing	*22*
CCR2b	*23*
CXCR2	*24*
V2 vasopressin	*25*
α-mating factor	*26*
SSTR	*19*
Melanocortin-4	*27*
Ligand-mediated modulation of receptor dimerization	
m3 muscarinic	*29*
β$_2$AR	*14,30*
B$_2$R bradykinin	*17*
TRHR	*31,32*
MT1 melatonin	*33*
MT2 melatonin	*33*
GnRHR	*35–38*
Calcium-sensing	*39*
Sphingosine-1-phosphate	*40*
Metabotropic glutamate	*41*
CXCR2	*24*
CXCR4	*34*
CCR2b	*23*
CCR5	*43*
Oxytocin	*44*
Neuropeptide NPY1	*42*
Neuropeptide Y Y4	*45*
δ-opioid	*46*

mGluR, metabotropic glutamate receptor; AR, adrenergic receptor; TRHR, thyrotropin-releasing hormone receptor; GnRH, gonadotropin-releasing hormone receptor; CCR, chemokine receptor; SSTR, somatostatin receptor.

lated adenylyl cyclase activity. Furthermore, pretreatment of membranes with isoproterenol protected the dimer from the disruptive effects of the TM-VI peptide, suggesting that agonist treatment resulted in dimer stabilization *(14)*. These studies suggest a role for TM-VI in β_2-AR dimer formation and are consistent with a role for this domain, as proposed by correlation mutational analysis *(13)*. Similarly, the TM-VI domain of the leukotriene B_4 receptor BLT1 was shown to be involved in dimerization because a TM-V1 peptide inhibited the dimerization of this receptor and affected its ability to interact with heterotrimeric G proteins *(12)*.

In contrast, in the case of the D1 dopamine receptor, the peptide corresponding to TM-VI did not disrupt oligomer formation; however, it caused a dose-dependent irreversible inhibition of antagonist binding and attenuation of signaling *(15)*, suggesting a role for additional TMs in the dimerization of this receptor. In the case of D2 dopamine receptor, an elegant set of cysteine crosslinking studies was used to identify TM-IV as a symmetrical dimer interface for this receptor. Additionally, this study showed that crosslinking did not affect ligand binding or receptor activation, suggesting that D2 dopamine receptor forms a constitutive dimer *(16)*.We investigated the role for TM-V to -VII domains in dimerization of δ-opioid receptors using synthetic peptides to the putative TM domains. For this, peptides were incubated with immunoprecipitates obtained from cells co-expressing Flag-δ and *myc*-δ receptors, and the level of monomers released into the supernatant as well as the dimers left behind in the pellet were examined as described (*see* Fig. 1 legend). We discovered that the level of dimers present in the pellet was reduced by greater than 80% upon treatment with either TM-V or TM-VI peptide (Fig. 1A). There was little (approx 25%) and no change with the TM-VII and mutant TM-VI peptide, respectively (Fig. 1A). The decrease in the level of dimeric receptors matched well with the increase in the level of monomers (Fig. 1B) observed in the supernatant. These results suggest a role for TM-V and -VI in the formation of δ-receptor dimers, although the involvement of additional TMs cannot be ruled out.

2.3. Functional Complementation Studies

Modulation of receptor pharmacology by dimerization has been demonstrated using functional complementation studies. In these studies, mutants of the receptor that exhibit decreased or no agonist-mediated activity are cotransfected with the wild-type receptor, and the resulting change in agonist-mediated binding or signaling is examined. For example, co-expression of Lys[102] and Lys[199] mutants of the type 1 angiotensin II receptor (which individually did not bind angiotensin II or related analogs) led to the restora-

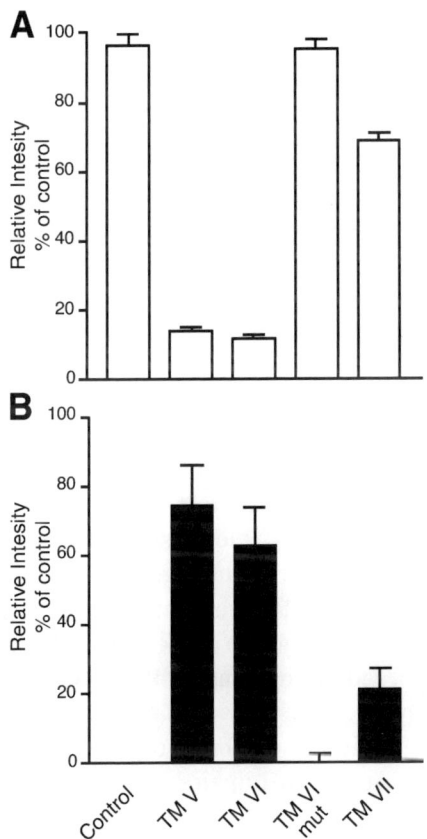

Fig. 1. Effect of various peptides on δ receptor dimerization. Human embryonic kidney (HEK)293 cells were transfected with myc-tagged and Flag-tagged δ-receptors. Immunoprecipitates with anti-myc antibodies were incubated with vehicle (control) or with 1 mM peptides corresponding to TM-V(^{217}VFLFAFVVPILIIT VCYGLML^{237}), TM-VI (^{263}VLVVVGAFVVCW APIHIFVIV^{283}), TM-VII (^{297}VAALHLCIALGYANSSLNPVLYAF^{320}), or TM-VI mutant (^{263}VaVVVaAFVa CWAaIHIaVIV^{283}) in a buffer containing a protease inhibitor cocktail for 1 h, as described in ref. 14. The mixture was centrifuged and the resulting pellet (A) and supernatant (B) fractions were separately subjected to SDS/PAGE and Western blotting with anti-Flag antibody as described in ref. 14. The autoradiograms were analyzed by densitometry and the density of the vehicle band was taken as 100%. Data represent mean ± SEM of three independent experiments.

tion of a normal binding site (18). Co-expression of mutant somatostatin receptors lacking the second extracellular loop (and ligand binding) with mutants lacking the C-terminal tail (and signaling) led to the reconstitution of agonist-mediated signaling *(19)*. Studies with α_{1b}-adrenoreceptors containing a fused $G\alpha_{11}$ at the C-terminus demonstrated that co-expression of two nonfunctional, but complementary, fusion constructs led to reconstitution of agonist-mediated signaling *(20)*. Similar observations were made with mutant fusion constructs of the histamine H1 receptor *(20)*. These and additional studies using G protein fusion constructs are described in detail in Chapter 12. Interestingly, cotransfection of a mutant human platelet-activating factor receptor, which does not couple to G proteins (and therefore does not exhibit agonist-mediated increases in inositol phosphate), with the wild-type receptor led to the formation of a constitutively active receptor that exhibited a higher production of inositol phosphate than the wild-type receptor *(21)*.

Functional complementation studies have also demonstrated a partial reconstitution of signaling. In the case of the calcium-sensing receptor, cells co-expressing cysteine to serine mutants—each with reduced (or absent) agonist-mediated, calcium-dependent signaling—exhibited partial reconstitution of signaling *(22)*. In the case of CCR2b chemokine receptors, cotransfection of a loss-of-function CCR2 (Y139F) mutant receptor with the wild-type receptor led to a decreased affinity and responsiveness to agonists *(23)*. Additionally, co-expression of truncated mutant chemokine CXCR2 receptors that did not exhibit agonist-mediated signaling with the wild-type receptor resulted in impaired agonist-mediated signaling and chemotaxis *(24)*.

Cotransfection of mutant vasopressin V2 receptors (truncated by the introduction of a stop codon into either the i3, i2, or e2 loop) with the wild-type receptor led to a decrease in maximum binding and agonist-stimulated cyclic adenosine monophosphate (cAMP) levels *(25)*. Similarly, an attenuation of agonist-mediated signaling efficiency was observed after co-expression of a dominant interfering mutant of the α-mating factor receptor STE2 with the wild-type receptor *(26)*. This did not result from G protein sequestration, because the effect was observed even after overexpression of G proteins *(26)*. Additionally, co-expression of an endocytosis-deficient mutant receptor with the wild-type receptor led to efficient receptor internalization, suggesting that dimerization plays a role in agonist-mediated endocytosis of the α-mating factor receptor *(26)*.

The physiological consequences of receptor dimerization have been illustrated in studies using natural variants and/or mutants of GPCRs *(27,28; see* Chapter 14). In the case of the melanocortin-4 receptor, researchers have observed that a D90N mutant receptor initially isolated in a patient with

GPCR Dimer Pharmacology

severe early onset obesity is highly expressed at the cell surface and binds agonist with the same affinity as the wild-type receptor, but it exhibits a complete loss of G_s-mediated adenylyl cyclase activation. Cotransfection of this mutant receptor with the wild-type receptor leads to suppression of agonist-mediated cAMP stimulation, and this is dependent on the amount of mutant receptor transfected *(27)*. Studies with other GPCRs have shown that co-expression of naturally occurring deletion or truncated mutants that exhibit decreased agonist-mediated signaling with wild-type receptors leads to a significant decrease in surface expression and agonist-mediated signaling of the wild-type receptor *(28; see* Chapter 13).

These functional complementation studies, which resulted in reconstitution, increased, or decreased binding or function of wild-type receptors, suggest that dimerization plays a major role in modulation of receptor activity.

2.4. Ligand-Mediated Modulation of Receptor Dimerization

The fact that ligands significantly modulate the dimerization of other membrane receptors, such as the tyrosine kinase receptors, raised the issue of whether ligands can modulate GPCR dimers. Studies have found that ligand treatment leads to no change or an increase or decrease in the level of receptor dimers.

Increases in the level of receptor dimers following agonist treatment have been observed with a few GPCRs, such as the β_2-AR *(14)*. Receptor isolation by affinity chromatography, followed by nonreducing sodium dodecyl sulfate-polyacrylamide gel electrophoresis (SDS-PAGE) and Western blotting with an antireceptor antibody, led to the isolation of both monomeric and dimeric forms of the β_2-AR *(14)* as well as the m3 muscarinic receptor *(29)*, suggesting that both monomers and dimers could bind ligand. Treatment with the agonist increased and with the inverse agonist decreased the amount of dimers of β_2AR, suggesting that agonists stabilized the dimeric species of the receptor, whereas inverse agonists stabilized the monomeric species *(14)*.

These results are consistent with data from bioluminescence resonance energy transfer (BRET) studies in which agonist treatment led to an increase in the BRET signal, which was blocked by the selective antagonist *(30)*. Agonist treatment also led to a dose- and time-dependent increase in BRET signal in the case of the thyrotrophin-releasing hormone receptor *(31,32)*. Interestingly, in the case of the melatonin MT2 receptors (but not MT1) agonists, neutral antagonists and inverse agonists were able to induce an increase in the BRET signal, suggesting that receptor occupancy is sufficient to modulate the proximity of dimers *(33)*. In the case of the bradykinin B_2 receptor, crosslinking studies showed that treatment with the agonist bradykinin, but not a selective antagonist, led to an increase in receptor dimers *(17)*. Agonist treatment was also shown to induce dimerization of the chemokine CXCR4 and CCR2b receptors *(23,34)*.

An interesting study with the gonadotropin-releasing hormone receptor used a bivalent antibody against the antagonist; treatment with this antibody in combination with the antagonist promoted receptor signaling. This led the authors to suggest that the antagonist–bivalent antibody complex induced receptor activation by bringing two receptor proteins to close proximity *(35)*. More recently, using fluourescence resonance energy transfer and BRET techniques, researchers have shown that agonists, but not antagonists, can cause increased oligomerization of this receptor *(36–38)*. Similarly, in the case of CCR2b receptors, it was shown that a bivalent agonistic antichemokine receptor monoclonal antibody induced receptor dimerization, whereas monovalent Fab fragments did not *(23)*.

In several cases, ligand treatment did not lead to alterations in the level of receptor dimers, suggesting that these were constitutive dimers. This result was observed with the m_3 muscarinic receptor, in which the level of dimers was not significantly affected by treatment with the agonist carbachol *(29)*. Similarly, agonist treatment did not significantly alter the level of dimers for the calcium sensing *(39)*, sphingosine-1-phosphate *(40)*, mGluR5 *(41)*, neuropeptide NPY1 *(42)*, chemokine CXCR2 *(24)*, and chemokine CCR5 receptors *(43)*. In the latter case, however, a bivalent antibody, which bound to the second intracellular loop, promoted micro-aggregation of preformed receptor homodimers *(43)*.

In the case of a few GPCRs, agonist treatment led to a decrease in the level of dimers, with a corresponding increase in the level of monomers. This was observed for the oxytocin *(44)*, rhesus neuropeptide Y Y4 *(45)*, and δ-opioid receptors *(46)*. In the case of δ-opioid receptors, it was observed that only agonists that induced receptor internalization caused receptor monomerization. This suggests the involvement of an agonist-mediated decrease in the level of receptor dimers in agonist-mediated endocytosis.

In summary, these studies show that in some cases, agonist treatment can modulate the level of GPCR homodimers. Note that current techniques do not allow us to distinguish between ligand-mediated association of monomers into dimers and ligand-mediated changes in the conformation of dimers (which would appear as an increase in dimer levels). However, in the majority of cases, the level of GPCR homodimers is unaffected by agonist treatment, suggesting that these receptors generally exist as constitutive dimers.

3. HETERODIMERS

A report on interactions between muscarinic m_3 and adrenergic α_{2c} receptors was among the first studies assessing the functionality of heterodimeric receptors *(47)*. Two chimeric receptor molecules, α_{2c}/m_3 and m_3/α_{2c}, were generated by exchanging the C-terminal regions (containing TM domains VI and VII) between the α_{2c} and m_3 receptors. The mutant receptors were then expressed in COS-7 cells, either alone or in combination, and radioligand-binding studies were performed to determine their ability to bind muscarinic and adrenergic antagonists. When expressed alone, the chimeric constructs were unable to bind their selective radioligands. In contrast, co-expression of α_{2c}/m_3 and m_3/α_{2c} produced significant binding of selective α_{2c} and m_3 receptor ligands, suggesting that interactions between the two receptors created specific binding sites for each ligand. Additionally, muscarinic agonist-induced phosphatidylinositol hydrolysis was demonstrated to occur in cells co-expressing the α_{2c}/m_3 and m_3/α_{2c} receptors but not in cells expressing either receptor alone *(47)*. These data were the first to suggest that two nonfunctional chimeric receptors could physically associate to create a functional heterodimeric receptor with effective ligand-binding and signaling capabilities.

In a similar set of studies, Maggio et al. *(48)* examined a chimeric m_3/m_2 receptor that contained 16 amino acids of the m_2 receptor sequence in the third cytoplasmic loop of the m_3 receptor. Although this chimeric receptor could bind muscarinic ligands, it was unable to stimulate phosphatydylinositol hydrolysis. However, when cotransfected with a truncated form of the m_3 receptor (which was also incapable of signaling on its own), phosphatydylinositol breakdown did occur, demonstrating once again that dimerization of two nonfunctional chimeric receptors could effectively produce a functional receptor complex.

Following these pivotal findings with chimeric receptors, researchers was discovered that heterodimerization occurs between many naturally occurring GPCRs (Table 2). Indeed, this phenomenon has been shown to occur between closely related GPCR types as well as between distantly related receptors. In each case, the resultant heterodimeric receptor complex has been found to differ in its pharmacological properties from either of the individual receptors (*see* refs. *1–5*). Occasionally, heterodimerization of two receptors is necessary to constitute a functional receptor *(49–57)*. In other cases, heterodimerization simply modulates GPCR pharmacology by affecting ligand binding, G protein coupling, signaling, and/or trafficking proper-

Table 2
GPCR Heterodimers

Heterodimers	References
Heterodimerization is necessary for receptor function	
$GABA_BR1-GABA_BR2$	*49–52*
T1R2–T1R3	*56,57*
T1R1–T1R3	*55*
Heterodimerization modulates receptor pharmacology	
Closely related receptors	
κ–δ opioid	*58*
μ–δ opioid	*59–61*
SSTR1–SSTR5	*19*
D2R dopamine–D3R dopamine	*62*
CCR2–CCR5	*65*
TRHR1–TRHR2	*32*
α_{1a} adrenergic–α_{1b} adrenergic	*63*
α_{2a} adrenergic–β_1 adrenergic	*64*
Distantly related receptors	
D2R dopamine–SSTR5	*66*
A1R adenosine–P2Y1R	*68*
A1R adenosine–D1R dopamine	*69*
A2AR adenosine–D2R dopamine	*70*
A1R adenosine–mGluR1α	*71*
A2AR adenosine–mGluR5	*72*
AT_1R angiotensin–B_2R bradykinin	*73*
κ-opioid–β_2 adrenergic	*75*
δ-opioid–β_2 adrenergic	*75*
μ-opioid–SSTR2A	*76*
μ-opioid–α_{2A}-AR	*77*
μ-opioid–substance P	*78*
Heterodimerization inactivates a functional receptor	
CCR2V64I–CCR5	*82*
CCR2V64I–CXCR4	*82*
SSTR2A–SSTR3	*79*
$AT_1R - AT_2R$	*81*
β_1-adrenergic–β_2-adrenergic	*80*

Abbreviations: GABA, γ aminobutyric acid; T1R, taste 1 receptor; SSTR, somatostatin receptor; CCR, chemokine receptor; TRHR, thyrotropin-releasing hormone receptor.

ties *(58–77)*. There are also some examples of heterodimerization in which one of the GPCRs in the heterodimer is inactivated *(78–80)*. Examples of these scenarios are described in the following Subheadings 3.1.–3.3.

3.1. Heterodimerization Is Necessary for Receptor Function

A fundamental role for GPCR heterodimerization in generating active receptors has been observed in naturally occurring metabotropic γ-aminobutyric acid $(GABA)_B$ receptors. In a series of seminal studies, researchers demonstrated that the co-expression of nonfunctional $GABA_B$-R1 and $GABA_B$-R2 receptors resulted in a functional receptor, as evidenced by the high-affinity GABA binding and G protein activation *(49–51)*. In fact, when expressed alone, neither receptor is capable of activating inwardly rectifying potassium channels; heterodimeric assembly of $GABA_B$-R1 and $GABA_B$-R2 is required for the efficient signaling of $GABA_B$ receptors through these channels *(49–52)*. We now know that $GABA_B$-R2 is required for the cell-surface expression of $GABA_B$-R1 *(53)*. Heterodimerization of the two receptors results in the masking of an endoplasmic reticulum retention signal on $GABA_B$-R1, allowing the proper targeting of the assembled complexes to the plasma membrane *(54)*. Therefore, the heterodimeric interaction between $GABA_B$-R1 and $GABA_B$-R2 receptors appears to be a prerequisite for the formation of a fully functional $GABA_B$ receptor.

Similarly, heterodimerization of mammalian taste receptors is required for the recognition of specific tastes. The mammalian amino acid taste receptors T1R1 and T1R3 have been shown to function as heterodimers *(55)*. These heterodimers function as L-amino acid sensors but do not respond to D-enantiomers or other compounds. The same T1R3 receptors also dimerize with T1R2 receptors to function as broadly tuned sweet sensors, recognizing a number of sweet-tasting molecules *(56,57)*.

Although the cases described here represent those for which obligatory heterodimerization has been firmly established, there are numerous other examples in which heterodimerization appears to play a crucial role in increasing pharmacological diversity by altering the various properties of individual receptors. Many of these receptor pairs are discussed in the Subheading 3.2.

3.2. Heterodimerization Modulates Receptor Pharmacology

3.2.1. Heterodimerization Between Closely Related Receptors

Heterodimerization has been shown to modulate the ligand-binding, signaling, and receptor-trafficking properties of several related GPCRs. In the case of the opioid receptor family, δ-opioid receptors have been shown to interact with both κ- and μ-opioid receptors to form heterodimers with altered pharmacological properties *(58,59)*. In the case of interactions between κ- and δ-receptors, the resultant κ–δ heterodimers were found to have greatly reduced affinities for highly selective κ- or δ-receptor ligands but had enhanced affinities for partially selective ligands *(58)*. In the presence of a δ-selective agonist, a κ-selective agonist bound to the receptors with high affinity, and, reciprocally, a κ-selective agonist increased the binding of a δ-selective agonist. Cells co-expressing κ- and δ-receptors also exhibited synergistic effects on agonist-induced signaling (as measured by cAMP and phosphorylated mitogen-activated protein kinase levels). Finally, dimerization was found to affect the trafficking properties of these receptors because etorphine-induced trafficking of the δ receptor was significantly reduced in cells expressing κ–δ heterodimers, suggesting that δ-receptors are retained at the cell surface as a result of dimerization with κ-receptors *(58)*.

Studies with μ–δ heterodimers have also demonstrated decreased binding affinity to selective synthetic agonists *(60)*. The rank order of agonist affinities for the heterodimeric receptors was different from that of the individual receptors, suggesting allosteric modulation of the binding pocket *(59,60)*. Treatment of cells expressing μ–δ heterodimers with very low doses of δ-selective ligands produced a significant increase in the binding of a μ-selective agonist; this increase was seen irrespective of the temperature at which the binding assay was performed (Fig. 2). This treatment also enhanced μ-receptor-mediated signaling *(59,61)*. It is possible that the heterodimeric μ-δ complex associates with pertussis toxin-insensitive G proteins, because treatment with pertussis toxin did not abolish the synergistic binding (Fig. 3) and signaling *(60)*. Therefore, the unique properties of μ–δ and κ–δ-receptor dimers suggest that heterodimerization may at least partially account for pharmacologically characterized opioid receptor subtypes for which genes have not been isolated.

Heterodimerization has also been shown to affect signaling and trafficking properties of somatostatin (SST) receptor types *(19)*. Although SSTR5 receptors undergo agonist-induced internalization, SSTR1 receptors do not (and are externalized by prolonged agonist treatment). Rocheville et al. *(19)* showed that co-expression of SSTR5 with SSTR1 receptors leads to agonist-induced internalization of SSTR1 receptors. Additionally, chronic treatment of SSTR1–SSTR5 co-expressing cells with an SSTR5-selective ligand induced increases in the cell-surface expression (retention) of SSTR1 receptors. These results suggest that crosstalk between SSTR1 and SSTR5 receptors may be necessary for SSTR1 receptor trafficking and an increase in cell-surface expression (retention). Also, co-expression of a mutant SSTR5

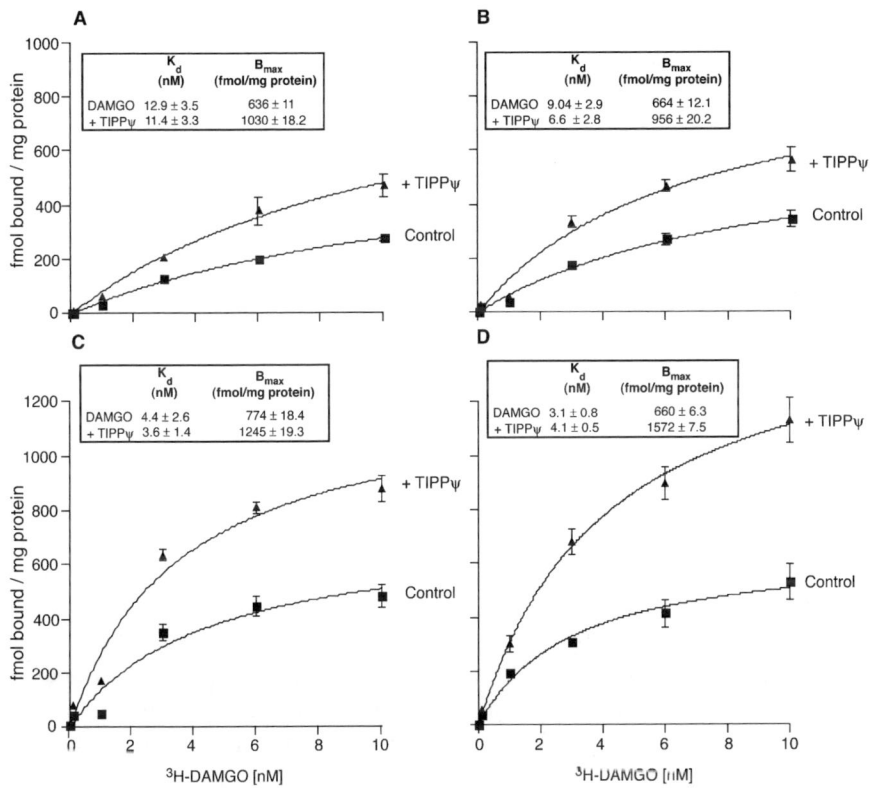

Fig. 2. Effect of temperature on potentiation of μ-agonist binding by δ-antagonist. Chinese hamster ovary (CHO) cells stably co-expressing Flag-tagged μ- and myc-tagged δ-opioid receptors were plated into 24-well plates (5×10^5 cells/well). The plates were kept at either 4°C (A), 10°C (B), room temperature (C), or 37°C (D). Cells were then incubated with [^{3}H]DAMGO (0.1–10 nM) in the presence or absence of 10 nM TIPPΨ for 2 h at the temperatures mentioned. Nonspecific binding was determined in the presence of 1 μM of DAMGO or diprenorphine. Wells were washed three times with 50 mM of ice-cold Tris-Cl, pH 7.5. Cells were lysed overnight with 1 N of NaOH and neutralized with 1 N of HCl, and radioactivity was collected and measured in a scintillation counter. Results are mean ± SEM of 3 experiments in triplicate.

receptor lacking its ligand-binding site with a mutant SSTR1 receptor containing a C-tail deletion (which is thus unable to signal) led to a significant increase in receptor signaling, as evidenced by somatostatin-induced adenylyl cyclase inhibition (19). This further supports a role for heterodimerization in modulating receptor signaling properties.

Heterodimers of closely related receptors exhibiting altered signaling and/ or trafficking properties have also been reported for the D3–D2 dopamine *(62)*, TRHR1–TRHR2 *(32)*, α_{1a}-α_{1b} adrenergic *(63)*, α_{2a}-β_1 adrenergic *(64)*, and CCR2–CCR5 chemokine receptors *(65)*. In the case of CCR2–CCR5, it was shown that heterodimers were more efficient at inducing a biological response than either receptor alone *(65)*. Additionally, although both CCR2 and CCR5 normally associate with $G_{\alpha i}$ proteins, CCR2–CCR5 heterodimers were able to recruit $G_{q/11}$ proteins and trigger a pertussis toxin-resistant calcium flux *(65)*. Taken together, these studies show that heterodimerization between specific subtypes of a GPCR can lead to alterations in pharmacological, signaling, and trafficking properties, which may account for the wide range of biological responses that follow receptor activation.

3.2.2. Heterodimerization Between Distantly Related Receptors

Heterodimeric assembly between the SSTR5 receptor and the structurally related D2 dopamine receptor generated a novel receptor that was pharmacologically distinct from either of its receptor homodimers *(66)*. It is believed that heterodimerization between these receptors may explain some of the biological interactions observed between these two neurotransmitters *(67)*. The D2–SSTR5 receptor heterodimer had a greater affinity for both dopamine and SST receptor agonists and exhibited enhanced G protein and effector coupling to adenylyl cyclase. As observed with opioid receptor dimers, synergistic binding of dopamine and SST receptor agonists occurred in D2–SSTR5 heterodimers. Additionally, the heterodimeric D2–SSTR5 receptor appeared to be most efficient when simultaneously occupied by its two agonists *(66)*. Interestingly, a dopamine receptor antagonist produced a decrease in the binding of a SSTR selective agonist. These results are in contrast to μ–δ heterodimers, in which the δ-antagonist promoted an increase in the binding and signaling by the μ-receptor selective agonist *(59,61)*.

Two distinct purinergic receptors, adenosine A1 (coupled to $G_{i/o}$) and P2Y1 (coupled to G_q), have been shown to form heterodimers with altered ligand-binding properties *(68)*. The heteromeric complex exhibited enhanced P2Y1R-like pharmacology: there was a significant reduction of A1R-agonist and antagonist binding and a 400-fold increase in the binding affinity of adenosine diphosphate-βS, a potent P2Y1R agonist *(68)*. Adenosine A1 has been shown to form heteromeric complexes with D1 dopamine receptors *(69)*, as does adenosine A2A with D2 dopamine receptors *(70)*. These phenomena may constitute the basis for adenosine–dopamine antagonism, because co-exposure of the A1R-D1R heterodimers to A1R and D1R selective agonists, but not to either agonist alone, caused a substantial reduction in D1R-mediated cAMP accumulation, which was associated with

GPCR Dimer Pharmacology

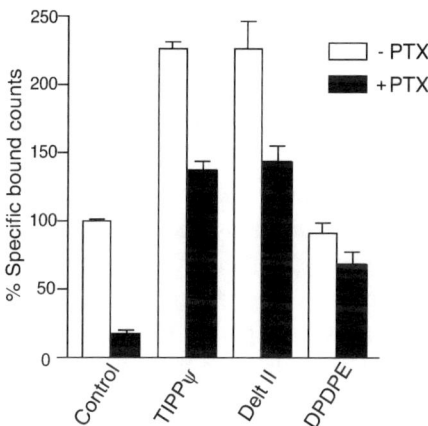

Fig. 3. Synergistic interactions between μ- and δ-opioid receptors persist after treatment with pertussis toxin. Chinese hamster ovary (CHO) cells stably co-expressing Flag-tagged μ- and myc-tagged δ-opioid receptors were plated into 24-well plates (5×10^5 cells/well). On the day of the assay, cells were untreated (control) or pretreated with 15 ng/mL of pertussis toxin (PTX) for 3 h at 37°C. Cells were then incubated with 10 nM of [^{3}H]DAMGO in the presence or absence of either TIPPΨ, Delt II, or DPDPE (final concentration: 10 nM) for 2 h at 37°C. Nonspecific binding was determined in the presence of 1 μM of DAMGO or diprenorphine. Wells were washed three times with 50 mM of ice-cold Tris-Cl, pH 7.5. Cells were lysed overnight with 1 N of NaOH and neutralized with 1 N of HCl, and radioactivity was collected and measured in a scintillation counter. Results are mean ± SEM of 3 experiments in triplicate.

receptor co-internalization *(69)*. A similar effect was observed with A2AR-D2R heterodimers *(70)*. The interactions between adenosine and dopamine receptors are believed to be responsible for the ability of adenosine agonists to inhibit and adenosine antagonists to potentiate the behavioral effects induced by dopamine agonists. This may be relevant for the development of adenosine and dopamine antagonists/agonists for the treatment of neuropsychiatric diseases in which D2R has been implicated, such as Parkinson's disease, schizophrenia, Huntington's disease, and dystonia.

In addition to modulating dopaminergic activity, adenosine also acts to inhibit glutamate neurotransmission in several brain regions. Therefore, the functional interactions between adenosine and glutamate receptors have been investigated *(71)*. In cells co-expressing A1R and mGluR1α, a glutamate/adenosine synergism was discovered at the level of calcium mobilization *(71)*. Similarly, adenosine A2AR and mGluR5 have been shown to exhibit a synergistic effect on extracellular signal-regulated kinase

(ERK1/2) phosphorylation and on the expression of the immediate-early gene c-*fos* *(72)*. A similar synergistic effect on c-*fos* expression was also observed in striatal sections after administration of A2AR and mGluR5 selective agonists to rats with intact dopaminergic innervation *(72)*. These results indicate that A2AR-mGluR5 interactions may be involved in striatal neuronal plasticity, such as long-term potentiation and depression.

Heterodimerization of two different vasoactive hormone receptors, the angiotensin II type 1 (AT1R) receptor and bradykinin B2 receptor (B2R), affects the signaling and trafficking of these receptors *(73)*. Co-expression of AT1R and B2R increased the efficacy and potency of angiotensin II but decreased the efficacy and potency of bradykinin. Additionally, angiotensin-stimulated activation of $G_{\alpha i}$ and $G_{\alpha q}$ proteins was increased by heterodimerization of the AT1R and B2R, independently of bradykinin binding. Finally, internalization of AT1R-B2R heterodimers occurred by a dynamin-dependent mechanism, as opposed to the endocytotic pathway of the individual receptors, which is dynamin- and clathrin-independent *(73)*. Interestingly, it has been shown that pre-eclamptic hypertensive women exhibit a significant increase in the levels of AT1R-B2R heterodimers, which display increased sensitivity toward angiotensin II. Thus, the hypertension in pre-eclampsia may be related to an increase in AT1R-B2R heterodimers in platelets *(74)*.

Heterodimerization between members of two distinct GPCR subfamilies has also been demonstrated using δ or κ-opioid receptors (receptors that couple to inhibitory G proteins) and $β_2$-ARs (receptors that couple to stimulatory G proteins) *(75)*. Although dimerization of opioid receptors with $β_2$-receptors did not significantly alter ligand-binding properties, it did affect receptor trafficking. When co-expressed with δ-receptors, which normally internalize rapidly, $β_2$-receptors underwent opioid-mediated endocytosis. However, when co-expressed with κ-receptors, which do not internalize rapidly, $β_2$-receptors did not undergo endocytosis in response to isoproterenol or opioids. Additionally, the loss of $β_2$-receptor internalization was accompanied by some loss of receptor signaling in κ–$β_2$ cells *(75)*. In light of these results, it is conceivable that opioid receptors and $β_2$-receptors physically associate in vivo to influence each other's function. For example, cardiac effects mediated by opioids (such as bradycardia) may, to some extent, result from dimerization between opioid receptors and $β_2$ receptors.

The μ-opioid receptor has also been shown to heterodimerize with distantly related GPCRs such as SSTR2A *(76)*, $α_{2A}$-AR *(77)*, or substance P *(78)* receptors. These receptor heterodimers exhibit properties that are quite distinct from each individual receptor, again indicating that heterodimerization may serve to increase the functional diversity of individual receptors.

3.3. Heterodimerization Inactivates a Functional Receptor

We have seen that in many cases, GPCR dimerization results in the modulation of receptor activity to exhibit enhanced function. However, heterodimerization can also result in a decrease in the activity of a fully functional receptor, in which case one of the receptors forming the heterodimer no longer signals in response to its agonist. This has been observed with SSTR2A–SSTR3 receptor heterodimers *(79)*. These heterodimers had a high affinity for an SSTR2-selective agonist but displayed a 100-fold lower affinity for an SSTR3-selective agonist. Additionally, although an SSTR2A-selective ligand stimulated strong GTPγS binding, adenylyl cyclase inhibition, and ERK1/2 activation in cells expressing the heterodimers, an SSTR3-selective agonist had no effect on signaling. These findings imply that SSTR2A-SSTR3 heterodimerization generates a receptor with a ligand-binding site and a functional profile resembling that of the SSTR2A receptor. On the other hand, the SSTR3 receptor is rendered inactive by heterodimerization. Additionally, SSTR2A–SSTR3 heterodimerization alters the desensitization rate of the receptor *(79)*.

Similarly, heterodimerization of β_1- and β_2-ARs has been shown to inhibit β_2-receptor function in HEK293 cells *(80)*. Although adenylyl cyclase activity was unaltered in cells expressing β_1-β_2 heterodimers, the ability of the β_2-receptor to activate the ERK1/2 signaling pathway was lost. Also, heterodimerization of β_1- and β_2-receptors prevented agonist-induced internalization of the β_2-receptor. Given that cardiac cells naturally express both β_1- and β_2-receptors and that β_2-receptor stimulation does result in ERK1/2 signaling in vivo, a mechanism for the sequestration of β_2-receptors from β_1-receptors may be required for β_2-receptor function.

An interesting case is that of μ-α_{2A} heterodimers, in which co-activation of both receptors leads to a decrease in signaling, whereas activation with ligands to either receptor leads to an increase in signaling. These observations were also made in primary spinal cord neurons, suggesting that these receptor interactions may play an important role in modulating pain transmission *(77)*.

Heterodimerization-mediated inactivation of a fully functional receptor can have important physiological consequences. This is illustrated in the case of AT1R-AT2R heterodimerization *(81)*. AT2R binds directly to AT1R, resulting in a decrease in receptor function *(81)*. This observation was made in cultured cells as well as human myometrial biopsies. The physiological relevance of AT1R-AT2R heterodimerization is evident during pregnancy, when the level of AT2R is seen to decrease and angiotensin II responsiveness of the myometrium increases *(78)*. Another example is provided by heterodimers between a naturally occurring mutant of the CCR2 chemokine receptor CCR2V64I (this mutation occurs at an allelic frequency of 10–25%) and chemokine CCR5 or CXCR4 receptors *(82)*. The human immunodeficiency virus (HIV) gains entry into cells via interaction with CCR5 or CXCR4 receptors. However, association of CCR2V64I with either receptor prevents HIV access to the cell *(82)* and is the likely mechanism underlying the ability of CCR2V64I to delay the progression of acquired immunodeficiency syndrome for 2 to 4 years in patients carrying this mutation. Taken together, these results suggest alterations in receptor heterodimerization in normal cell physiology and disease states. Therefore, developing drugs that selectively target receptor heterodimers would be of importance in the case of various pathologies.

4. CONCLUSIONS

Although traditional views have held that the functional GPCR comprised of a single monomeric entity, it is now evident that GPCRs exist as dimers or even oligomers. Additionally, a growing number of reports clearly demonstrate that GPCRs can heterodimerize not only with other GPCRs but also with proteins such as receptor activity-modifying proteins (Chapters 4, 9, 13, and 14), leading to a profound modulation of pharmacological, signaling, and trafficking properties. To date, the majority of studies on GPCR dimerization have been carried out in heterologous expression systems. Therefore, further studies are necessary to identify the physiological significance of receptor dimerization in vivo. Advances in the techniques used to investigate protein–protein interactions in vivo along with the development of reagents selective for the dimeric or oligomeric form of GPCRs, such as selective antibodies and/or ligands, will help explain the physiological consequences of this phenomenon. The presence of naturally occurring splice variants of some GPCRs and their effects on the pharmacology and function of the corresponding wild-type receptors indicates that in addition to generating pharmacological diversity, receptor heterodimerization may have profound influences on receptor function under normal as well as pathological conditions. This also suggests that further studies are needed to identify other naturally occurring mutant GPCRs and to analyze their role under normal and pathological conditions. The identification of heterodimers that are formed in vivo will enable the development of novel drugs and therapies for the treatment of a number of pathological conditions in which GPCR involvement has been implicated.

ACKNOWLEDGMENTS

This work was supported by grants from National Institutes of Health (DA 088360 and DA 00458) to LAD.

REFERENCES

1. Gomes I, Jordan BA, Gupta A, Rios C, Trapaidze N, Devi LA. G protein coupled receptor dimerization: implications in modulating receptor function. J Mol Med 2001;79:226–242.
2. Rios CD, Jordan BA, Gomes I, Devi LA. G-protein coupled receptor dimerization: modulation of receptor function. Pharmacol Ther 2001;92:71–87.
3. Angers S, Salahpour A, Bouvier M. Dimerization: an emerging concept for G-protein coupled receptor ontogeny and function. Annu Rev Pharmacol Toxicol 2002;42:409–435.
4. Bai M. Dimerization of G-protein coupled receptors: roles in signal transduction. Cell Signal 2004;16:175–186.
5. Kroeger KM, Pfleger KDG, Eidne KA. G-protein coupled receptor oligomerization in neuroendocrine pathways. Front Neuroendocrinol 2004;24:254–278.
6. Kunishima N, Shimada Y, Tsuji Y, et al. Structural basis of glutamate recognition by a dimeric metabotropic glutamate receptor. Nature 2000;407:971–977.
7. Palczewski K, Kumasaka T, Hori T, et al. Crystal structure of rhodopsin: a G-protein coupled receptor. Science 2000;289:739–745.
8. Fotiadis D, Liang Y, Filipek S, Saperstein DA, Engel A, Palczewski K. Rhodopsin dimers in native disc membranes. Nature 2003;241:127,128.
9. Liang Y, Fotiadis D, Filipek, S, Saperstein DA, Palczewski K, Engel A. Organization of the G-protein-coupled receptors rhodopsin and opsin in native membranes. J Biol Chem 2003;278:21,655–21,662.
10. Fotiadis D, Liang Y, Filipek S, Saperstein DA, Engel A, Palczewski K. The G-protein coupled receptor rhodopsin in the native membrane. FEBS Lett 2004;564.281–288.
11. Filipek S, Krzysko KA, Fotiadis D, et al. A concept for G-protein activation by G-protein coupled receptor dimers: the transducin/rhodopsin interface. J Photobiochem Photobiol Sci 2004;3:628–638.
12. Banères J-L, Parello J. Structure-based analysis of GPCR function: evidence for a novel pentameric assembly between the dimeric leukotriene B4 receptor BLT1 and the G-protein. J Mol Biol 2003;329:815–829.
13. Gouldson PR, Snell CR, Bywater RP, Higgs C, Reynolds CA. Domain swapping in G-protein coupled receptor dimers. Protein Eng 1998;11:1181–1193.
14. Hebert TE, Moffett S, Morello J-P, et al. A peptide derived from a β_2-adrenergic receptor transmembrane domain inhibits both receptor dimerization and activation. J Biol Chem 1996;271:16,384–16,392.
15. George SR, Lee SP, Varghese G, et al. A transmembrane domain-derived peptide inhibits D1 dopamine receptor function without affecting receptor oligomerization. J Biol Chem 1998;273:30,244–30,248.
16. Guo W, Shi L, Javitch JA. The fourth transmembrane segment forms the interface of the dopamine D2 receptor homodimer. J Biol Chem 2003;278:4385–4388.
17. AbdAlla S, Zaki E, Lother H, Quitterer U. Involvement of the amino terminus of the B2 receptor in agonist-induced receptor dimerization. J Biol Chem 1999;274:26,079–26,084.
18. Monnot C, Bihoreau C, Conchon S, Curnow KM, Corvol P, Clauser E. Polar residues in the transmembrane domains of the type I angiotensin II receptor are

required for binding and coupling: reconstitution of the binding site by coexpression of two deficient mutants. J Biol Chem 1996;271:1507–1513.
19. Rocheville M, Lange D, Kumar U, Sasi R, Patel RC, Patel YC. Subtypes of the somatostatin receptor assemble as functional homo and heterodimers. J Biol Chem 2000;275:7862–7869.
20. Carrillo JJ, Pediani J, Milligan G. Dimers of class A G-protein coupled receptors function via agonist-mediated trans-activation of associated G-proteins. J Biol Chem 2003;278:42,578–42,587.
21. Le Gouill C, Parent J- L, Caron C- A, et al. Selective modulation of wild type receptor functions by mutants of G-protein coupled receptors. J Biol Chem 1999;274:12,548–12,554.
22. Bai M, Trivedi S, Kifor O, Quinn SJ, Brown EM. Intermolecular interactions between dimeric calcium-sensing receptor monomers are important for its normal function. Proc Natl Acad Sci USA 1999;96:2834–2839.
23. Rodriguez-Frade J, Vila-Coro AJ, de Ana AM, Albar JP, Martinez-A C, Mellado M. The chemokine monocyte chemoattractant protein-1 induces functional responses through dimerization of its receptor CCR2. Proc Natl Acad Sci USA 1999;96:3628–3633.
24. Trettel F, Bartolomeo SD, Lauro C, Catalano M, Ciotti MT, Limatola C. Ligand-independent CXCR2 dimerization. J Biol Chem 2003;278:40,980–40,988.
25. Zhu X, Wess J. Truncated V2 vasopressin receptors as negative regulators of wild-type V2 receptor function. Biochemistry 1998;37:15,773–15,784.
26. Overton MC, Blumer KJ. G-protein coupled receptors function as oligomers in vivo. Curr Biol 2000;10: 341–344.
27. Biebermann H, Krude H, Elsner A, Chubanov V, Gudermann T, Gruters A. Autosomal-dominant mode of inheritance of a melanocortin-4 receptor mutation in a patient with severe early-onset obesity is due to a dominant-negative effect caused by receptor dimerization. Diabetes 2003;52:2984–2988.
28. Lee SP, O'Dwd BF, George SR. Homo- and hetero-oligomerization of G protein-coupled receptors. Life Sci 2003;74: 173–180.
29. Zeng F- Y, Wess J. Identification and molecular characterization of m3 muscarinic receptor dimers. J Biol Chem 1999;274:19,487–19,497.
30. Angers S, Salahpour A, Bouvier M. Biochemical and biophysical demonstration of GPCR oligome-rization in mammalian cells. Life Sci 2001;68:2243–2250.
31. Kroeger KM, Hanyaloglu AC, Seeber RM, Miles LEC, Eidne KA. Constitutive and agonist-dependent homo-oligomerization of the thyrotropin-releasing hormone receptor: detection in living cells using bioluminescence energy transfer. J Biol Chem 2001;276:12,736–12,743.
32. Hanyaloglu AC, Seeber RM, Kohout TA, Lefkowitz RJ, Eidne KA. Homo- and hetero-oligomerization of thyrotropin-releasing hormone (THR) receptor subtypes: differential regulation of β-arrestins 1 and 2. J Biol Chem 2002;277:50,422–50,430.

33. Ayoub MA, Couturier C, Lucas-Meunier E, et al. Monitoring of ligand-independent dimerization and ligand-induced conformational changes of melatonin receptors in living cells by bioluminescence resonance energy transfer. J Biol Chem 2002;277:21,522–21,528.
34. Vila-Coro AJ, Rodriguez-Frade JM, De Ana AN, Moreno-Ortiz MC, Martinez-A C, Mellado M. The chemokine SDF-1α triggers CXCR4 receptor dimerization and activates JAK/STAT pathway. FASEB J 1999:13:1699–1710.
35. Conn PM, Rogers DC, Steweart JM, Niedel J, Sheffield T. Conversion of a gonadotrophin-releasing hormone antagonist to an agonist. Nature 1982;296:653–655.
36. Cornea A, Janovick JA, Maya-Nunez G, Conn PM. Gonadotrophin-releasing hormone receptor microaggregation: rate monitored by fluorescence resonance energy transfer. J Biol Chem 2001;276:2153–2158.
37. Horvat RD, Roess DA, Nelson SE, Barisas BG, Clay CM. Binding of agonist but not antagonist leads to fluorescence resonance energy transfer between intrinsically fluorescent gonadotropin-releasing hormone receptors. Mol Endocrinol 2001;15:695–703.
38. Kroeger KM, Hanyaloglu AC, Seeber RM, Miles LEC, Eidne KA. Constitutive and agonist-dependent homo-oligomerization of the thyrotropin-releasing hormone receptor: detection in living cells using bioluminescence energy transfer. J Biol Chem 2001;276:12,736–12,743.
39. Bai M, Trivedi S, Brown EM. Dimerization of the extracellular calcium-sensing receptor (CaR) on the cell surface of CaR-transfected HEK293 cells. J Biol Chem 1998;273:23,605–23,610.
40. Brocklyn JRV, Behbahani B, Lee NII. Homodimerization and heterodimerization of S1P/EDG sphingosine-1-phosphate receptors. Biochim Biophys Acta 2002;1582:89–93.
41. Romano C, Yang YL, O'Malley. Metabotropic glutamate receptor 5 is a disulfide-linked dimer. J Biol Chem 1996;271:28,612–28,616.
42. Dinger MC, Bader JE, Kobor AD, Kretzschmar AK, Beck-Sickinger AG. Homodimerization of neuropeptide Y receptors investigated by fluorescence resonance energy transfer in living cells. J Biol Chem 2003;278:10,562–10,571.
43. Issafras H, Angers S, Bulenger S, et al. Constitutive agonist-independent CCR5 oligomerization and antibody-mediated clustering occurring at physiological levels of receptors J Biol Chem 2002;277:34,666–34,673.
44. Devost D, Zingg HH. Identification of dimeric and oligomeric complexes of the human oxytocin receptor by co-immunoprecipitation and bioluminescence resonance energy transfer. J Mol Endocrinol 2003;31:461–471.
45. Berglund MM, Schober DA, Esterman MA, Gehlert DR. Neuropeptide Y Y4 receptor homodimers dissociate upon agonist stimulation. J Pharmacol Exp Ther 2003;307:1120–1126.

46. Cvejic S, Devi LA. Dimerization of the δ opioid receptor: implication for a role in receptor internalization. J Biol Chem 1997;272:26,959–26,964.
47. Maggio R, Vogel Z, Wess J. Coexpression studies with mutant muscarinic/ adrenergic receptors provide evidence for intermolecular "cross-talk" between G-protein linked-receptors. Proc Natl Acad Sci USA 1993;90:3103–3107.
48. Maggio R, Barbier P, Formai F, Corsini GU. Functional role of the third cytoplasmic loop in muscarinic receptor dimerization. J Biol Chem 1996;271:31,055–31,060.
49. Jones KA, Borowski B, Tamm JA, et al. $GABA_B$ receptor function as a heterotrimeric assembly of the subunits $GABA_BR1$ and $GABA_BR2$. Nature 1998;396:674–679.
50. Kaupmann K, Malitschek B, Schuler V, et al. $GABA_B$-receptor subtypes assemble into functional heterodimeric complexes. Nature 1998;396:683–687.
51. White JH, Wise A, Main MJ, et al. Heterodimerization is required for the formation of a functional $GABA_B$ receptor. Nature 1998;396:679–682.
52. Kuner R, Köhr G, Grünewald S, Eisenhardt G, Bach A, Hans-Christian K. Role of heterodimer formation in $GABA_B$ receptor function. Science 1999;283:74–77.
53. Sullivan R, Chateauneuf A, Coulombe N, et al. Co-expression of full length γ-aminobutyric acid B ($GABA_B$) receptors with truncated receptors and metabotropic glutamate receptor 4 supports the $GABA_B$ heterodimer as the functional receptor. J Pharmacol Exp Ther 2000;293:460–467.
54. Margeta-Mitrovic M, Jan YN, Jan LY. A trafficking checkpoint controls GABA(B) receptor heterodimerization. Neuron 2000;27:97–106.
55. Nelson G, Chandrashekar J, Hoon MA, et al. An amino-acid taste receptor. Nature 2002;416:199–202.
56. Nelson G, Hoon MA, Chandrashckar J, Zhang Y, Ryba NJ, Zuker CS. Mammalian sweet taste receptors. Cell 2001;106:381–390.
57. Li X, Staszewski L, Xu H, Durick K, Zoller M, Adler E. Human receptors for sweet and umami taste. Proc Natl Acad Sci USA 2002;99:4692–4696.
58. Jordan BA, Devi LA. G-protein-coupled receptor heterodimerization modulates receptor function. Nature 1999;399:697–700.
59. Gomes I, Jordan BA, Gupta A, Trapaidze N, Nagy V, Devi LA. Heterodimerization of the mu and delta opioid receptors: a role in opiate synergy. J Neurosci 2000;20:RC110(1 5).
60. George SR, Fan T, Xie Z, et al. Oligomerization of mu and delta opioid receptors. J Biol Chem 2000;275:26,128–26,135.
61. Gomes I, Gupta A, Filipovska J, Szeto HH, Pintar JE, Devi La. A role for heterodimerization of m and d opiate receptors in enhancing morphine analgesia. Proc Natl Acad Sci USA 2004;101:5135–5139.
62. Scarselli M, Novi M, Schallmach E, et al. D_2/D_3 dopamine receptor heterodimers exhibit unique functional properties. J Biol Chem 2001;276:30,308–30,314.

63. Stanasila L, Perez J-B, Vogel H, Cotecchia S. Oligomerization of the α_{1a}- and α_{1b}-adrenergic receptor subtypes. J Biol Chem 2003;278:40,239–40,251.
64. Xu J, He J, Castleberry AM, Balasubramanian S, Lau AG, Hall RA. Heterodimerization of α_{2A}- and β_1-adrenergic receptors. J Biol Chem 2003;278:10,770–10,777
65. Mellado M, Rodriguez-Frade JM, Vila-Coro AJ, et al. Chemokine receptor homo- or heterodimerization activates distinct signaling pathways. EMBO J 2001;20:2497–2507.
66. Rocheville M, Lange DC, Kumar U, Patel SC, Patel RC, Patel YC. Receptors for dopamine and somatostatin: formation of heterodimers with enhanced functional activity. Science 2000;288:154–157.
67. Marzullo P, Ferone D, Di Somma C, et al. Efficacy of combined treatment with lanreotide and cabergoline in selected therapy-resistant acromegalic patients. Pituitary 1999;1:115–120.
68. Yoshioka K, Saitoh O, Nakata H. Heteromeric association creates a P2Y-like adenosine receptor. Proc Natl Acad Sci USA 2001;98:7617–7622.
69. Gines S, Hillion J, Torvinen M, et al. Dopamine D_1 and adenosine A_1 receptors form functionally interacting heteromeric complexes. Proc Natl Acad Sci USA 2000;97:8606–8611.
70. Hillion J, Canals M, Torvinen M, et al. Coaggregation, cointernalization, and codesensitization of adenosine A_{2A} receptors and dopamine D_2 receptors. J Biol Chem 2002;277:18,091–18,097.
71. Ciruela F, Escriche M, Burgueno J, et al. Metabotropic glutamate 1 α and adenosine A1 receptors assemble into functionally interacting complexes. J Biol Chem 2001;276:18,345–18,351.
72. Ferre S, Karcz-Kubicha M, Hope BT, et al. Synergistic interaction between adenosine A2A and glutamate mGlu5 receptors: Implications for striatal neuronal function. Proc Natl Acad Sci USA 2002;99:11,940–11,945.
73. AbdAlla S, Lother H, Quitterer U. AT_1-receptor heterodimers show enhanced G-protein activation and altered receptor sequestration. Nature 2000;407:94–98.
74. Quitterer U, Lother H, AbdAlla S. AT1 receptor heterodimers and angiotensin II responsiveness in preeclampsia. Semin Nephrol 2004;24:115–119.
75. Jordan BA, Trapaidze N, Gomes I, Nivarthi R, Devi LA. Dimerization of opioid receptors with β_2 adrenergic receptors: a role in trafficking and mitogen activated protein kinase activation. Proc Natl Acad Sci USA 2001;98:343–348.
76. Pfeiffer M, Koch T, Schröder H, Laugsch M, Höllt V, Schulz S. Heterodimerization of somatostatin and opioid receptors cross-modulates phosphorylation, internalization, and desensitization. J Biol Chem 2002;277:19,762–19,772.
77. Jordan BA, Gomes I, Rios C, Filipovska J, Devi LA. Functional interactions between µ opioid and α_{2A}-adrenergic receptors. Mol Pharmacol 2003;64:1317–1324.

78. Pfeiffer M, Kirscht S, Stumm R, et al. Heterodimerization of substance P and µ-opioid receptors regulates trafficking and resensitization. J Biol Chem 2003;278:51,630–51,637.
79. Pfeiffer M, Koch T, Schröder H, et al. Homo- and heterodimerization of somatostatin receptor subtypes. J Biol Chem 2001;276:14,027–14,036.
80. Lavoie C, Mercier JF, Salahpour A, et al. β_1/β_2-adrenergic receptor heterodimerization regulates β_2-adrengergic receptor internalization and ERK signaling efficacy. J Biol Chem 2002;277:35,402–35,410.
81. AbdAlla S, Lother H, Abdel-tawab AM, Quitterer U. The angiotensin II AT_2 receptor is an AT_1 receptor antagonist. J Biol Chem 2001; 276:39,721–39,726.
82. Mellado M, Rodriguez-Frade JM, Vila-Coro AJ, Martin de Ana A, Martinez-A C. Chemokine control of HIV-1 infection. Nature 1999; 400:723,724.

IV
RECENT DEVELOPMENTS IN DRUG DISCOVERY

16
Role of Heteromeric GPCR Interactions in Pain/Analgesia

Andrew P. Smith and Nancy M. Lee

1. INTRODUCTION

The mammalian response to pain is extremely complex, involving multiple nervous pathways in the brain and spinal cord as well as in the periphery. The chemical signaling that links individual neurons in these pathways makes use of a great variety of neurotransmitters and neuromodulators, most of which act at G protein-coupled receptors (GPCRs). In fact, most of the known GPCR types have been shown to play some role in pain processing *(1,2)*.

Pharmacological manipulations have revealed that these GPCRs interact extensively with each other. Classically, these interactions have been accounted for by synaptic connections between a population of neurons that releases one type of transmitter and a second population, which is stimulated by that transmitter, that releases a different kind of transmitter. For example, it is well-established that nociception in the spinal cord is partially controlled by descending systems from the brain that release the monoamine neurotransmitters norephinephrine, serotonin, and, perhaps, dopamine *(3–6)*. At the spinal cord level, these transmitters interact with their appropriate GPCRs on dorsal horn neurons, some of which release the transmitter enkephalin. Enkephalin, acting on opioid receptors, may then inhibit the release of transmitters, such as glutamate and substance P, from incoming sensory afferents *(7,8)*. Opioid receptor activation in the spinal cord is also associated with release of acetylcholine (ACh), a ligand for muscarinic receptors *(9)*. Activation of spinal muscarinic receptors, in turn, may be associated with feedback on α_2-adrenergic receptors (ARs) and on serotonergic receptors *(10)* as well as a reduction in release of substance P *(11)*.

From: *Contemporary Clinical Neuroscience: The G Protein-Coupled Receptors Handbook*
Edited by: L. A. Devi © Humana Press Inc., Totowa, NJ

However, not all functional interactions among GPCRs necessarily occur through multisynaptic pathways. Because a great deal of pain processing occurs in a few restricted areas of the central nervous system (CNS) (including the dorsal horn of the spinal cord, the rostroventral medulla [RVM], and the peri-aqueductal gray [PAG] area of the brain *[12,13]*) and because some GPCRs have been shown to be colocalized on certain populations of neurons *(14–17)*, the possibility exists for functional interactions of these receptors within single cells. Such interactions are commonly referred to as crosstalk and occur between GPCRs at the level of G proteins *(18,19)* and their regulated enzymes *(20,21)*.

Recently, crosstalk has also been discovered at the earliest conceivable point in signal transduction: physical association of receptors. Techniques such as sodium dodecyl sulfate gel analysis, crosslinking, immunoprecipitation, and bioluminescence resonance energy transfer have established that a growing number of GPCRs can form homomers or heteromers *(22,23)*. The formation of such complexes obviously has the potential to increase the signaling capacity of a ligand that is selective for a particular receptor. Not only may a ligand modulate the activity of a receptor it does not directly bind to but, in some cases, heteromers have been shown to have binding and/or functional properties that are different from either of their component receptors *(24–26)*. This suggests that the physical complex formed between the two receptors may associate with a different set of signal-transducing molecules from those activated by either receptor alone.

To date, there appears to be no compelling evidence that physical association of GPCRs plays a role in pain processing. Because receptor dimerization has been appreciated only recently, however, this possibility has not been thoroughly explored. Regardless, there are numerous examples of functional interactions among these receptors to which physical association might contribute. This chapter evaluates this possibility for several GPCRs that play a prominent role in pain processing.

2. ROLE OF HETEROMERIZATION IN FUNCTIONAL INTERACTIONS OF GPCRS

2.1. Interactions of Different Opioid Receptor Types

Opioid agonists, particularly morphine, are widely used as general analgesics for pain that is experienced following surgery or that is associated with terminal illnesses. The nervous pathways involved in opioid analgesia are not completely understood, but they include the PAG, RVM, and dorsal horn of the spinal cord *(27)*. Opioid agonists are active when injected into these regions, blocking transmission of pain signals on their multisynaptic journey from peripheral receptors to the brain.

Pharmacological studies initially identified three major types of opioid receptors in the mammalian CNS: μ, δ, and κ *(28,29)*. Moreover, each of these receptor types is believed to exist in two or more subtypes *(30–33)*, although to date, there is no evidence that these subtypes correspond to distinct genes or nucleotide sequences. The μ-receptor is believed to play the primary role in mediating antinociception, although δ and κ-receptors are antinociceptive when administered to certain CNS areas and antinociception assessed by certain kinds of assays. However, physiological interactions are well-established among these three receptor types. For example, ligands selective for δ-opioid receptors, such as DPDPE, can potentiate antinociception mediated by μ-selective ligands (DAMGO) in both brain *(34)* and spinal cord *(35,36)*. Many of these studies also used selective antagonists for further proof that the observed effects did not result from crossreactivity (i.e., from interaction between δ-ligand and μ-receptors). Interactions have also been reported between κ-agonists such as U-50,488H or dynorphinA *(1–17)* and morphine or other μ-agonists *(37,38)*.

Heteromer formation between μ- and δ-opioid receptors has been demonstrated both in transfected cells and in a cell line that contains both of these receptor types endogenously *(24,25)*. Heteromers of δ/κ-receptors have also been reported *(39)*, but no μ–κ heteromers. In the case of μ–δ heteromers, Gomes et al. *(25)* reported some evidence for potentiation, with occupation of one type of receptor increasing affinity of ligands for the other type; however, both agonists and antagonists had this effect. In contrast, George et al. *(24)*, found that selective μ- and δ-agonists had less affinity for heteromers than for pure receptors of the appropriate type. It should be noted that inhibitory interactions between these two kinds of ligands have also been reported in vivo *(40)*.

In summary, different types of opioid receptors manifest both functional and physical interactions. There is also some evidence for colocalization, although this evidence is not as strong. Aside from some early studies predating the development of highly selective ligands *(41,42)*, the main evidence for colocalization of different opioid receptor types has been provided by studies of neuroblastoma-derived cell lines. Not only do some of these cell lines contain more than one type of opioid receptor, but, in some cases, the receptors have been shown to interact functionally *(43–45)*. As noted earlier, heteromer formation between μ- and δ-opioid receptors was also demonstrated in one of these cell lines *(25)*.

Nevertheless, studies of endogenous cells are needed. Cheng et al. *(14)* reported colocalization of μ- and δ-opioid receptors in dorsal horn neurons, but the great majority of neurons containing one type of receptor lacked the other kind, and even in cells containing both, much of the populations were in different regions of the cell. Colocalization of two, or even all three, major opioid receptor types has also been reported in dorsal root ganglion cells *(46)*. Functional interactions among them seem likely to occur at this site, but are not as well-established as they are within the spinal cord. To our knowledge, colocalization of opioid receptor types in brain has not been reported.

2.2. Opioid–Adrenergic Interactions

As noted in Section 1., pain inhibitory systems descending from the brain to the spinal cord are partially mediated by noradrenergic processes. The latter interact intimately with opioid systems. Grabow et al. *(47)* reported that intrathecal administration of yohimbine (an α_2-antagonist) blocked antinociception induced by injection of deltorphin (a δ-opioid agonist) into the RVM, confirming the role of α_2-receptors in opioid antinociception activated at the supraspinal level. Intrathecal administration of yohimbine also blocked a synergistic effect observed when low doses of deltorphin were administered simultaneously to both RVM and spinal cord. Finally, a synergistic effect was observed when subanalgesic doses of deltorphin and an α_2-agonist were administered simultaneously to the spinal cord. Somewhat similar results were observed by Hao et al. *(48)*, who used endormorphin, a μ-opioid agonist.

Therefore, both μ- and δ-opioid receptors interact with α_2-receptors in the spinal cord. It has been conventionally believed that this functional interaction of opioid and ARs in the spinal cord results from multisynaptic mechanisms. However, the finding of a synergistic interaction between these two systems in the spinal cord suggests that the interaction is mediated not by a sequential pathway but by a parallel pathway or possibly a pathway involving action of the two transmitters on a common neuron. Colocalization of opioid and ARs has not been demonstrated in the spinal cord, but Jordan et al. *(17)* recently reported colocalization of μ-opioid receptors and α_2-receptors in hippocampal neurons. These two receptors form functional heteromers with properties that are different from the individual receptors. Moreover, dimerization of both δ- and κ-opioid receptors with β_2-ARs has been detected *(49,50)*. Therefore, a role for heteromerization in this major functional interaction in the spinal cord should not excluded.

2.3. Opioid–Substance P Interactions

The endogenous peptide substance P plays an important role in pain transmission at the initial spinal cord level. Administration of substance P to the spinal cord induces thermal nociception (as measured by the tail flick test *[51]*), and elimination of either substance P *(52)* or its receptor *(53)* in animals by homologous recombination results in altered pain response. Substance P is contained in the terminals of many primary afferent fibers to the dorsal horn of the spinal cord *(54)* and is released by noxious stimuli *(55)*. Therefore, substance P is a major neurotransmitter at the first central synapse in the nociceptive pathway.

Morphine is antinociceptive when administered to the spinal cord *(56)*, and its action is believed to at least partially involve an inhibition of the substance P-mediated nociceptive pathway *(57)*. An immunocytochemical study confirmed that opioid and substance P receptors are colocalized on dorsal horn neurons in the nociceptive pathway *(15)*. Currently, there is no study demonstrating a physical association between µ-opioid and substance P receptors; however, as discussed earlier, different opioid receptor types are known to form heteromers with each other as well as homomers. Some studies have suggested that these dimers may have altered internalization properties and that one type of receptor may promote or inhibit internalization of another *(24,58,59)*. This possibility is interesting in light of a study by Trafton et al. *(60)* on the effect of morphine on substance P receptor internalization in dorsal horn cells. Although morphine alone had little effect, the opioid decreased substance P internalization when co-administered with a dose of substance P receptor antagonist, which also was ineffective when administered alone. Thus, µ-receptor–substance P heteromerization is a possible explanation for this phenomenon.

In addition to providing insights into substance P actions, this finding may have clinical relevance. Despite the nociceptive activity of substance P, antagonists to this receptor are not antinociceptive in humans, although they are antinociceptive in animal models of neuropathic and inflammatory pain *(61)*. The study by Trafton et al. *(60)* suggests that substance P antagonist might be a more effective antinociceptive agent in the presence of morphine or other opioid agonist by blocking internalization and recycling of substance P receptors.

2.4. Opioid–Cannabinoid Interactions

The cannabinoid receptors CB1 and CB2 have recently been implicated in pain processing in both brain and spinal cord as well as in the periphery and in several pain models *(52,63)*. Although some evidence suggests they can mediate antinociception by direct interaction with cannabinoid receptors *(64)*, cannabinoid agonists such as tetrahydrocannibinol (THC) also interact with the opioidergic system. Antinociception induced by THC was potentiated by morphine and blocked by opioid antagonists *(65)*. Conversely, THC can potentiate morphine antinociception *(66)*.

Several cell lines derived from neuroblastomas have been shown to have cannabinoid and opioid receptors *(67,68)*. In NG108-15 neuroblastoma x glioma cells, cannabinoid agonists can inhibit adenylate cyclase *(69)*, similarly to δ-opioid agonists. Moreover, exposure of cells to THC reduced opioid receptor binding and opioid-mediated inhibition of adenylyl cyclase, apparently through effects on G proteins *(70)*. Recently, colocalization of CB1 and μ-opioid receptors was reported on spinal interneurons *(16)*.

Therefore, functional interactions between opioid and cannabinoid receptors are well-established, and they are likely to be colocalized at the appropriate in vivo sites. Similarly to opioid–substance P interactions, evidence of heteromer formation between these receptors is lacking. However, similarly to opioid receptors, cannabinoid receptors, are known to associate with themselves *(98)*, suggesting that studies of physical interactions between these two GPCR classes may be fruitful.

2.5. Dopamine–Adenosine Interactions

One of the most extensively documented cases of interaction between GPCRs is the interaction between receptors for dopamine and adenosine. Functional interactions were initially described in models of Parkinsonism, where adenosine agonists inhibit (whereas antagonists potentiate) the effects of dopamine agonists *(71,72)*. It was subsequently reported that dopamine D1 and adenosine A1 receptors are colocalized in cultured cortical cells *(73)*, whereas D2 and A2A receptors are colocalized in cells of the primate striatum *(74)*. Studies of both cultured neurons as well as transfected cells suggested that colocalized receptors could interact to affect binding, signal transduction, and internalization *(75–77)*. Finally, heteromer formation was demonstrated between D1/A1 and D2/A2A receptors in cultured neurons as well as in transfected cell lines *(73,77)*. Thus, it appears that many physiological interactions between dopamine and adenosine could be accounted for, in principle, by physical associations of their receptors.

The studies cited earlier are directly relevant to the role of dopamine and adenosine in motor systems. There is much less evidence of dopamine–adenosine interaction in pain processing, but both of these receptors play a significant role in pain processing in both brain and spinal cord *(6,78–80)*. A recent study reported a biphasic effect of dopamine agonist on adenosine agonist-mediated antinociception *(81)*.

2.6. Dopamine–Somatostatin Interactions

Somatostatin (SST) was first characterized as an inhibitor of pituitary secretion of growth hormone, but some evidence suggests that this peptide is nociceptive. SST is released in the spinal cord in response to thermal *(82)* or inflammatory *(83)* stimuli. Intracerebrovascular injection of SST decreased the threshold to thermal stimulus *(84)*, whereas i.t. administration of antagonist *(85)* or antibody *(86)* was antinociceptive against thermal or inflammatory pain, respectively. However, other studies have challenged the notion that SST has nociceptive effects at subtoxic doses *(87)*, and one study reported that SST administered intrathecally reduced pain in two cancer patients *(88)*.

SST and dopamine are known to play a major role in mediating motor behaviors, and numerous studies have documented a functional interaction between receptors for the two transmitter/modulators *(89,90)*. The recent report that SST5 and dopamine D2 receptors are colocalized in striatal neurons and that these receptors, when transfected at relatively low density in a stable cell line, form heteromers *(26)* suggests that physical association might account for some of the functional interactions. This possibility is further supported by studies demonstrating that dopamine agonists or antagonists modulate both SST binding and SST-mediated second messenger effects in striatal or hippocampal neurons *(91,92)*.

These motor regions are not known to be involved in pain processing, but the demonstration that dopamine and SST receptors can associate in such neurons suggests that if they are colocalized in other CNS areas, then they might also form heteromers in these motor regions. Localization studies of these receptors suggest that the dorsal horn of the spinal cord is one common area that plays a prominent role in pain processing where the receptors might be co-expressed *(93,94)*. To date, however, such studies have concentrated on SST2 and D1 and D2 receptors; mapping the distribution of other subtypes may provide further clues of possible physical and functional interactions.

3. CONCLUSIONS

The examples of GPCR interactions discussed in this chapter demonstrate that heteromer formation can, at least in principle, play an important role in mediating many major functional interactions between these receptors in pain processing. If this has not been definitively established for any particular example of functional interaction and the evidence for most cases remains rather weak, we must remember this is largely because the essential studies of colocalization and heteromer formation have not yet been carried out. Only recently have we become aware of the significance that such data could have for our understanding of how pain processing occurs.

Moreover, an even more important role may be played by homomer formation. More than a dozen different GPCR receptors have been shown to form homomers, including μ, δ-, and κ-opioid receptors *(39,58,59)*; α_2- and β_2-ARs *(49,95)*; CB *(96)*; M3 muscarinic *(97)*; D2 and D3 dopamine receptors *(98,99)*; and SST2 and SST3 receptors *(100)*. In several cases, more than one kind of subtype of the receptor can associate with itself. Obviously, in these cases, the possibility of physiological relevance is much greater than with heteromer formation, where co-expression of the two different GPCRs in endogenous systems must be demonstrated. If two receptor molecules of the same type physically associate when transfected into a cell line, then the presumption exists that they can associate within any cell in which that receptor is found—although perhaps not to the same extent if their endogenous concentrations are lower. However, for this reason, it is more difficult to demonstrate that homomerization has functional consequences (i.e., that it results in a different effect in the cell from that transduced by ligand binding to a single receptor molecule). Indeed, the growing realization that homomer formation may be common among GPCRs raises the issue of whether any effects transduced by these receptors are achieved through single molecules, as was commonly accepted only a few years ago. This issue must be resolved in future studies of GPCR association.

REFERENCES

1. Millan MJ. Descending control of pain. Prog Neurobiol 2002;66:355–374.
2. Henry JL. Concepts of pain sensation and its modulation. J Rheumatol Suppl 1989;19:104–112.
3. Yaksh TL. Pharmacology of spinal adrenergic systems which modulate spinal nociceptive processing. Pharmacol Biochem Behav 1985;22:845–858.
4. Yaksh TL, Noueihed R. The physiology and pharmacology of spinal opiates. Annu Rev Pharmacol Toxicol 1985;25:433–462.
5. Loomis CW, Jhamandas K, Milne B, Cervenko F. Monoamine and opioid interactions in spinal analgesia and tolerance. Pharmacol Biochem Behav 1987;26:445–451.
6. Weil-Fugazza J, Godefroy F. Further evidence for the involvement of the diencephalic-dopaminergic system in pain modulation: a neurochemical study on the effect of morphine in the arthritic rat. Int J Tissue React 1991;13:305–310.
7. Hylden JL, Wilcox GL. Intrathecal opioids block a spinal action of substance P in mice: functional importance of both mu- and delta-receptors. Eur J Pharmacol 1982;86:95–98.

8. Hentall ID, Fields HL. Actions of opiates, substance P, and serotonin on the excitability of primary afferent terminals and observations on interneuronal activity in the neonatal rat's dorsal horn in vitro. Neuroscience 1983;9:521–528.
9. Chen SR, Pan HL. Spinal endogenous acetylcholine contributes to the analgesic effect of systemic morphine in rats. Anesthesiology 2001;95:525–530.
10. Iwamoto ET, Marion L. Characterization of the antinociception produced by intrathecally administered muscarinic agonists in rats. J Pharmacol Exp Ther 1993;266:329–338.
11. Smith MD, Yang XH, Nha JY, Buccafusco JJ. Antinociceptive effect of spinal cholinergic stimulation: interaction with substance P. Life Sci 1989;45:1255–1261.
12. Willis WD Jr. The Pain System: The Neural Basis of Nociceptive Transmission in the Mammalian Nervous System. Basel: Karger, 1985.
13. Basbaum AI, Fields HL. Endogenous pain control systems: brainstem spinal pathways and endorphin activity. Ann Rev Neurosci 1984;7:309–338.
14. Cheng PY, Liu-Chen LY, Pickel VM. Dual ultrastructural immunocytochemical labelling of mu and delta opioid receptors in the superficial layers of the rat cervical spinal cord. Brain Res 1997;778: 367–380.
15. Aicher SA, Punnoose A, Goldberg A. mu-opioid receptors often colocalize with the substance P receptor (NK1) in the trigeminal dorsal horn. J Neurosci 2000;20:4345–4354.
16. Salio C, Fischer J, Franzoni MF, Mackie K, Kaneko T, Conrath M. CB1-cannabinoid and mu-opioid receptor co-localization on postsynaptic target in the rat dorsal horn. Neuroreport 2001;12: 3689–3692.
17. Jordan BA, Gomes I, Rios C, Filipovska J, Devi LA. Functional interactions between mu opioid and alpha2A-adrenergic receptors. Mol Pharmacol 2004;64:1317–1324.
18. Meszaros JG, Gonzalez AM, Endo-Mochizuki Y, Villegas S, Villareal F, Brunton LL. Identification of G protein-coupled signaling pathways in cardiac fibroblasts: cross talk between G(q) and G(s). Am J Physiol Cell Physiol 2000;278: C154–C162.
19. Hanke S, Nurnberg B, Groll DH, Liebmann C. Cross talk between beta-adrenergic and bradykinin B(2) receptors results in cooperative regulation of cyclic AMP accumulation and mitogen-activated protein kinase activity. Mol Cell Biol 2001;21:8452–8460.
20. Thakker DR, Standifer KM. Induction of G protein-coupled receptor kinases 2 and 3 contributes to the cross-talk between mu and ORL1 receptors following prolonged agonist exposure. Neuropharmacology 2002;43:679–690.
21. Rasolonjanahary R, Gerard C, Dufour MN, Homburger V, Enjalbert A,Guillon G. Evidence for a direct negative coupling between dopamine-D2 receptors and PLC by heterotrimeric Gi1/2 proteins in rat anterior pituitary cell membranes. Endocrinology 2002;143: 747–754.
22. Bai M. Dimerization of G -protein-coupled receptors: roles in signal transduction. Cell Signal 2004;16:175–186.
23. Kroeger KM, Pfleger KD, Eidne KA. G-protein coupled receptor oligomerization in neuroendocrine pathways. Front Neuroendocrinol 2003;24:254–278.

24. George SR, Fan T, Xie Z, et al. Oligomerization of μ and δ opioid receptors: generation of novel functional properties. J Biol Chem 2000;275(34): 26,128–26,135.
25. Gomes I, Jordan BA, Gupta A, Trapaidze N, Nagy Y,Devi LA. Heterodimerization of mu and delta opioid receptors: a role in opiate synergy. J Neurosci 2000;20:RC110(22):1–5.
26. Rocheville M, Lange DC, Kumar U, Patel SC, Patel RC, Patel YC. Receptors for dopamine and somatostatin: formation of hetero-oligomers with enhanced functional activity. Science 2000;288:154–157.
27. Fields HL. Pain modulation: expectation, opioid analgesia and virtual pain. Prog Brain Res 2000;122:245–253.
28. Martin WR, Eades CG, Thompson JA, Huppler RE, Gilbert PE. The effects of morphine and nalorphine-like drugs in the nondependent and morphine-dependent chronic spinal dog. J Pharmacol Exp Ther 1976;197(1):517–532.
29. Lord JAH, Waterfield AA, Hughes J, Kosterlitz HW. Endogenous opioid peptides: multiple agonists and receptors. Nature 1977;267(5611):495–499.
30. Gintzler AR, Pasternak GW. Multiple mu receptors: evidence for mu2 sites in the guinea pig ileum. Neurosci Lett 1983;39(1):51–56.
31. Stefano GB, Hartman A, Bilfinger TV, et al. Presence of the mu3 opiate receptor in endothelial cells. Coupling to nitric oxide production and vasodilation. J Biol Chem 1995;270(51):30,290–30,293.
32. Jiang Q, Takemori AE, Sultana M, et al. Differential antagonism of opioid delta antinociception by [D-Ala2,Leu5,Cys6]enkephalin and naltrindole 5'-isothiocyanate: evidence for delta receptor subtypes. J Pharmacol Exp Ther 1991;257(3):1069–1075.
33. Clark JA, Liu L, Price M, Hersh B, Edelson M, Pasternak GW. Kappa opiate receptor multiplicity: evidence for two U50,488-sensitive kappa 1 subtypes and a novel kappa 3 subtype. J Pharmacol Exp Ther 1989;251(2):461–468.
34. Heyman JS, Vaught JL, Mosberg HI, Haaseth RC, Porreca F. Modulation of μ-mediated antinociception by δ agonists in the mouse: selective potentiation of morphine and normorphine by DPDPE. Eur J Pharmacol 1989;165(1):1–10.
35. Malmberg AB, Yaksh TL. Isobolographic and dose–response analyses of the interaction between mu and delta agonists: effects of naltrindole and its benzofuran analog (NTB). J Pharmacol Exp Ther 1992;263(1):264–275.
36. Russell RD, Leslie JB, Su YF, Watkins WD, Chang KJ. Interaction between highly selective mu and delta opioids in vivo at the rat spinal cord. NIDA Res Monogr 1986;75:97–100.
37. Ramarao P, Jablonski HI Jr., Rehder KR,Bhargava HN. Effect of kappa-opioid receptor agonists on morphine analgesia in morphine-naive and morphine-tolerant rats. Eur J Pharmacol 1988;156(2):239–246.
38. Stachura Z, Herman ZS. The influence of the kappa agonist-spiradoline (U62066E) on the analgesic activity of some opioids at the spinal level. Pol J Pharmacol 1994;46(1-2):37–41.
39. Jordan BA, Levi LA. G-protein coupled receptor heterodimerization modulates receptor function. Nature 1999;399(6737):697–700.

40. Larson AA, Vaught JL,Takemori AE. The potentiation of spinal analgesia by leucine enkephalin. Eur J Pharmacol 1980;61(4):381–383.
41. Egan TM, North RA. Both mu and delta opiate receptors exist on the same neuron. Science 1981;214:923,924.
42. Zieglgansberger W, French ED, Mercuri N, Pelayo F,Williams J. Multiple opiate receptors on neurons of the mammalian central nervous system. Life Sci 1982;31: 2343–2346.
43. Yu VC, Richards ML, Sadee W. A human neuroblastoma cell line expresses mu and delta opioid receptor sites. J Biol Chem 1986;261:1065–1070.
44. Kazmi S, Mishra R. Comparative pharmacological properties and functional coupling of mu and delta opioid receptor sites in human neuroblastoma SH-SY5Y cells. Mol Pharmacol 1987;32:109–18.
45. Palazzi E, Ceppi E, Guglielmetti F, Catozzi L, Amoroso D, Groppetti A. Biochemical evidence of functional interaction between mu- and delta-opioid receptors in SK-N-BE neuroblastoma cell line. J Neurochem 1996;67:13–144.
46. Ji RR, Zhang Q, Law PY, Loh HH, Elde R, Hokfelt T. Expression of mu-, delta-, and kappa-opioid receptor-like immunoreactivities in rat dorsal root ganglia after carrageenan-induced inflammation. J Neurosci 1995;15:8156–8166.
47. Grabow TS, Hurley RW, Banfor PN, Hammond DL. Supraspinal and spinal delta(2) opioid receptor-mediated antinociceptive synergy is medated by spinal alpha(2) adrenoceptors. Pain 1999;83:47–55.
48. Hao S, Takahata O, Iwasaki H. Intrathecal endomorphin-1 produces antinociceptive activities modulated by alpha 2-adrenoceptors in the rat tail flick, tail pressure and formalin tests. Life Sci 2000;66:PL195–PL204.
49. McVey M, Ramsay D, Kellett E, et al.Monitoring receptor oligomerization using time-resolved fluorescence resonance energy transfer and bioluminescence resonance energy transfer. The human delta-opioid receptor displays constitutive oligomerization at the cell surface, which is not regulated by receptor occupancy. J Biol Chem 2001;276(17):14,092–14,099.
50. Jordan BA, Trapaidze N, Gomes I, Nivarthi R, Devi LA. Oligomerization of opioid receptors with beta 2-adrenergic receptors: a role in trafficking and mitogen-activated protein kinase activation. Proc Natl Acad Sci USA 2001;98(1):343–348.
51. Yasphal K, Wright DM, Henry JL. Substance P reduces tail-flick latency: implications for chronic pain syndromes. Pain 1982;14:155–167.
52. Nagahisa A, Kanai Y, Suga O, et al. Antiinflammatory and analgesic activity of a non-peptide substance P receptor antagonist. Eur J Pharmacol 1992;217:191–195.
53. Campbell EA, Gentry CT, Patel S, Panesar MS, Walpole CS, Urban L. Selective neurokinin-1 receptor antagonists are anti-hyperalgesic in a model of neuropathic pain in the guinea pig. Neuroscience 1998;87:527–532.
54. Hokfelt T, Kellerth JO, Nilsson G, Pernow B. Substance P: localization in the central nervous system and in some primary sensory neurons. Science 1975;190:889,890.

55. Duggan AW, Hendry IA, Morton CR, Hutchison WD, Zhang ZQ. Cutaneous stimuli releasing immunoreactive substance P in the dorsal horn of the cat. Brain Res 1988;451:261–273.
56. Yaksh TL, Rudy TA. Analgesia mediated by a direct spinal action of narcotics. Science 1976;192:1357–1358.
57. Besse D, Lombard MC, Zajac JM, Roques BP, Besson JM. Pre- and postsynaptic distribution of mu, delta and kappa opioid receptors in the superficial layers of the cervical dorsal horn of the rat spinal cord. Brain Res 1990;52:15–22.
58. Cvejic S, Devi LA. Dimerization of the delta opioid receptor: implication for a role in receptor internalization. J Biol Chem 1997;272(43): 26,959–26,964.
59. He L, Fong J, von Zastrow M, Whistler JL. Regulation of opioid receptor trafficking and morphine tolerance by receptor oligomerization. Cell 2002;108(2):271–282.
60. Trafton JA, Abbadie C, Marchand S, Mantyh PW, Basbaum AI. Spinal opioid analgesia: how critical is the regulation of substance P signaling. J Neurosci 1999;19:9642–9653.
61. Hill R. NK1 (substance P) receptor antagonists—why are they not analgesic in humans? Trends Pharmacol Sci 2000;21:244–246.
62. Meng ID, Manning BH, Martin WJ, Fields HL. An analgesia circuit activated by cannabinoids. Nature 1998;395:381–383.
63. Kelly S, Chapman V. Selective cannabinoid CB1 receptor activation inhibits spinal nociceptive transmission in vivo. J Neurophysiol 2001;86:3061–3064.
64. Gardell IR, Ossipov MH, Vanderah TW, Lai J, Porreca F. Dynorphin-independent spinal cannabinoid antinociception. Pain 2002;100:243–248.
65. Reche I, Fuentes JA, Ruiz-Gayo M. Potentiation of delta 9-tetrahydrocannabinol-induced analgesia by morphine in mice: involvement of mu- and kappa-opioid receptors. Eur J Pharmacol 1996;318:11–16.
66. Smith FL, Cichewicz D, Martin ZL, Welch SP. The enhancement of morphine antinociception in mice by delta9-tetrahydrocannibinol. Pharmacol Behav 1998;60:559–566.
67. Ho BY, Zhao J. Determination of the cannabinoid receptors in mouse x rat hybridoma NG108-15 cells and rat GH4C1 cells. Neurosci Lett 1996;212:123–126.
68. McIntosh HH, Song C, Howlett AC. 1 cannabinoid receptor: cellular regulation and distribution in N18TG2 neuroblastoma cells. Brain Res Mol Brain Res 1998;53:163–173.
69. Di Toro R, Campana G, Sciaretta V, Murari. Regulation of delta opioid receptors by delta9-tetrahydrocannabinol in NG108-15 hybrid cells. Life Sci 1998;63: PL197–PL204
70. Wager-Miller J, Westenbrock R, Mackie K. Dimerization of G protein-coupled receptors: CB1 cannabinoid receptors as an example. Chem Phys Lipids 2002;121:83–89.
71. Ferre S, Fredholm BB, Morelli M, Popoli P, Fuxe K. Adenosine–dopamine receptor–receptor interactions as an integrative mechanism in the basal ganglia. Trends Neurosci 1997;20:482–487.

72. Ferre S, Popoli P, Gimenez-Llort L, et al. Adenosine/dopamine interaction: implications for the treatment of Parkinson's disease. Parkinsonism Relat Disord 2001;7:235–241.
73. Gines S, Hillion J, Torvinen M, et al. Dopamine D1 and adenosine A1 receptors form functionally interacting heteromeric complexes. Proc Natl Acad Sci USA 2000;97:8606–8611.
74. Svenningsson P, Le Moine C, Aubert I, Burbaud P, Fredholm BB, Bloch B. Cellular distribution of adenosine A2A receptor mRNA in the primate striatum. J Comp Neurol 1998;399:229–240.
75. Ferre S, Torvinen M, Antonio K, et al. Adenosine A1 receptor-mediated modulation of dopamine D1 receptors in stably cotransfected fibroblast cells. J Biol Chem 1998;273:4718–4724.
76. Salim H, Ferre S, Dalal A, et al. Activation of adenosine A1 and A2A receptors modulate dopamine D2 receptor-induced responses in stably transfected human neuroblastoma cells. J Neurochem 2000;74:432–439.
77. Hillion J, Canals M, Torvinen M, et al. Coaggregation, cointernalization, and codensitization of adenosine A2A receptors and dopamine D2 receptors. J Biol Chem 2002;277:18,091–18,097.
78. Altier N, Stewart J. Dopamine receptor antagonists in the nucleus accumbens attenuate analgesia induced by ventral tegmental substance P or morphine and by nucleus accumbens amphetamine. J Pharmacol Exp Ther 1998;285:208–215.
79. Tanase D, Baghdoyan HA, Lydic R. Microinjection of an adenosine A1 agonist intothe medial pontine reticular formation increases tail flick latency to thermal stimulation. Anesthesiology 2002;97:1597–1601.
80. Nakamura I, Ohta Y, Kemmotsu O. Characterization of adenosine receptors mediating spinal sensory transmission related to nociceptive information in the rat. Anesthesiology 1997;87:577–584.
81. Malhotra J, Chaudhary G, Gupta YK. Dopaminergic involvement in adenosine A1 receptor-mediated antinociception in the tail flick model in mice. Methods Find Exp Clin Pharmacol 2000;22:37–41.
82. Kuraishi Y, Hirota N, Sato Y, Hino Y, Satoh M, Takagi H. Evidence that substance P and somatostatin transmit separate information related to pain in the spinal dorsal horn. Brain Res 1985;325:294–298.
83. Ohno H, Kuraishi Y, Nanayama T, Minami M, Kawamura M, Satoh M. Somatostatin is increased in the dorsal root ganglia of adjuvant-inflamed rat. Neurosci Res 1990;8:179–188.
84. Capasso A. Effect of somatostatin and its antagonist on morphine analgesia in mice. Zhongguo Yao Li Xue Bao 1999;20:1079–1082.
85. Ohkubo T, Shibata M, Takahashi H, Inoki R. Role of substance P and somatostatin on transmission of nociceptive information induced by formalin in spinal cord. J Pharmacol Exp Ther 1990;252:1261–1268.
86. Ohno H, Kuraishi Y, Minami M, Satoh M. Modality-specific antinociception produced by intrathecal injection of anti-somatostatin antiserum in rats. Brain Res 1988;474:197–200.

87. Gaumann DM, Yaksh TL, Post C, Wilcox GL, Rodriguez M. Intrathecal somatostatin in cat and mouse studies on pain, motor behavior, and histopathology. Anesth Analg 1989;68:623–632.
88. Meynadier J, Chrubasik J, Dubar M, Wunsch E. Intrathecal somatostatin in terminally ill patients. A report of two cases. Pain 1985;23:9–12.
89. Martin-Iverson MT, Radke JM, Vincent SR. The effects of cysteamine on dopamine-mediated behaviors: evidence for dopamine–somatostatin interactions in the striatum. Pharmacol Biochem Behav 1986;24:1707–1714.
90. Lee N, Radke JM, Vincent SR. Intra-cerebral cysteamine infusions attenuate the motor response to dopaminergic agonists. Behav Brain Res 1988;29:179–183.
91. Izquierdo-Claros RM, Boyano-Adanez MC, Larsson C, Gustavsson L, Arilla E. Acute effects of D1- and D2-receptor agonist and antagonist drugs on somatostatin binding, inhibition of adenylyl cyclase activity and accumulation of inositol 1,4,5-trisphosphate in the rat striatum. Brain Res Mol Brain Res 1997;47:99–107.
92. Rodriguez-Sanchez MN, Puebla L, Lopez-Sanudo S, et al. Dopamine enhances somatostatin receptor-mediated inhibition of adenylate cyclase in rat striatum and hippocampus. J Neurosci Res 1997;48:238–248.
93. Wamsley JK, Gehlert DR, Filloux FM, Dawson TM. Comparison of the distribution of D-1 and D-2 dopamine receptors in the rat brain. J Chem Neuroanat 1989;2:119–137.
94. Schulz S, Schreff M, Schmidt H, Handel M, Przewlocki R, Hollt V. Immunocytochemical localization of somatostatin receptor sst2A in the rat spinal cord and dorsal root ganglia. Eur J Neurosci 1998;10:3700–3708.
95. Carillo JJ, Pediani J, Milligan G. Dimers of class A G protein-coupled receptors function via agonist-mediated trans-activation. J Biol Chem 2003;278: 42,578–42,587.
96. Wager-Miller J, Westenbroek R, Mackie K. Dimerization of G protein-coupled receptors: CB1 cannabinoid receptors as an example. Chem Phys Lipids 2002;121:83–89.
97. Zeng FY, Wess J. Identification and molecular characterization of m3 muscarinic receptor dimers. J Biol Chem 1999;274(27):19,487–19,497.
98. Ng GY, O'Dowd BF, Lee SP, et al. Dopamine D2 receptor dimers and receptor-blocking peptides. Biochem Biophys Res Commun 1996;227(1):200–204.
99. Nimchinsky EA, Hof PR, Janssen WG, Morrison JH, Schmauss C. Expression of dopamine D3 receptor dimers and tetramers in brain and in transfected cells. J Biol Chem 1997;272(46):29,229–29,237.
100. Pfeiffer M, Koch T, Schroeder H, et al. Homo- and heterodimerization of somatostatin receptor subtypes. Inactivation of sst(3) receptor function by heterodimerization with sst(2A). J Biol Chem 2001;276:14,027-14,036.

ID # 17

Conformational Plasticity of GPCR Binding Sites

Structural Basis for Evolutionary Diversity in Ligand Recognition

Xavier Deupi, Cedric Govaerts, Lei Shi, Jonathan A. Javitch, Leonardo Pardo, and Juan Ballesteros

1. LINKING G PROTEIN ACTIVATION TO LIGAND DIVERSITY: EVOLUTION OF THE G PROTEIN-COUPLED RECEPTOR FAMILY

1.1. G Protein-Coupled Receptors Transduce Signals of Exceptionally Different Chemical Nature

Genome sequencing projects have identified the G protein-coupled receptor (GPCR) superfamily as one of the largest classes of proteins in mammalian genomes *(1)*. For example, preliminary analyses of the human genome have revealed up to 600 GPCRs *(2,3)*. Additionally, GPCRs are scored as the most common family in the human proteome at the Proteome Analysis Database of the European Bioinformatics Institute(see http://www.ebi.ac.uk/proteome/HUMAN/interpro/top15f.html), with more than 800 sequences. Based on phylogenetic analyses of the human genome, these receptors have been classified into five main families: glutamate, rhodopsin, adhesion, frizzled/taste2, and secretin *(4)*.

This vast family of proteins carries out two complementary functions: (a) transduction and (b) amplification of extracellular chemical signals across

From: *Contemporary Clinical Neuroscience: The G Protein-Coupled Receptors Handbook*
Edited by: L. A. Devi © Humana Press Inc., Totowa, NJ

the cell membrane. To perform these functions, extracellular ligand binding to the receptor is translated into regulation of the activity of intracellular proteins—primarily heterotrimeric guanine–nucleotide exchange proteins *(5)*. In turn, these G proteins modulate several cellular signaling pathways, such as adenylyl cyclase, phopholipase C, or potassium and calcium channels.

One of the most striking properties of GPCRs is the variety in the chemical nature of their cognate extracellular ligands. Different subfamilies respond to signals of great structural diversity, such as hormones, peptides, nucleotides, amino acids, neurotransmitters, lipids, ions, or light *(6)*, all of which interact with GPCRs in distinctive binding modes *(7)*. Despite such agonist diversity, activation of a GPCR invariably leads to G protein activation. Therefore, the question arises regarding how this protein family, stemming from a common ancestor, has evolved such diversity in ligand recognition while preserving a common scaffold composed of seven-transmembrane helices (TMHs) and a set of shared intracellular partners.

1.2. Activation of GPCRs Requires Conformational Changes in the Transmembrane Bundle

Because of the difficulty in producing and purifying a recombinant receptor, only one GPCR, bovine rhodopsin, has been crystallized thus far. Different analyses of X-ray diffraction data of the crystals have led to four structural models of bovine rhodopsin, which are available at the Protein Data Bank (PDB) at resolutions of 2.8 Å (PDB identifiers 1F88 and 1HZX), 2.6 Å (PDB identifier 1L9H [8]), and 2.65 Å (1GZM) *(9)*. These crystal structures show that bovine rhodopsin consists of seven-membrane-spanning α-helices joined by three cytoplasmic and three extracellular loops, with an extracellular N-terminus and a cytoplasmic C-terminus. An additional intracellular helix lies approximately parallel to the plane of the membrane, pointing away from the helical bundle. The presence of highly conserved patterns in the sequences of rhodopsin-like GPCRs and a plethora of data from biochemical and biophysical experiments indicate that the overall folding of the transmembrane bundle is conserved throughout the family.

Although little is known about the details of the molecular mechanisms underlying GPCR activation, various experiments have suggested that motions of transmembrane helices are associated with this process. For example, changes in the relative orientations of the cytoplasmic ends of TMH3 and TMH6 *(10,11)* or the positions of TMH7 *(12)* and TMH2 *(13)* (for a review, *see* ref. *14*) have been associated with receptor activation. These motions have been observed in different receptor subtypes and are likely to be a common feature of GPCR activation, at least for rhodopsin-like receptors.

1.3. Family-Specific and Conserved Steps in the GPCR Activation Mechanism

Specific interactions between a receptor and its cognate agonist typically involve the extracellular domain and/or the extracellular part of the transmembrane region of the receptor. As expected from the ligand diversity, these areas are poorly conserved throughout the GPCR family. On the other hand, there is a higher degree of sequence conservation toward the intracellular part of the transmembrane domain *(9)* in a region where helix movements have been observed during activation. Interestingly, little sequence conservation is observed in the intracellular loops, although these domains are known to be important for binding the G protein. Evolutionary pressure to diversify or modify the coupling specificity has probably favored this variability, whereas convergent evolution has most likely allowed distant receptors to couple with the same G protein *(15,16)*. Such a sequence conservation pattern suggests a mechanistic similarity in the activation process, where the sequence conservation would parallel the functional conservation. Each receptor subfamily must have evolved specific motifs to adapt to the structural characteristics of its cognate ligands; therefore, domains involved in ligand binding must also be subfamily specific. On the other hand, with a common intracellular partner, the conformational rearrangement of the helices close to the G protein are expected to be preserved to a certain extent. Thus, conceptually, the signal transduction process can be seen as a progress from the extracellular side to the intracellular domain through conformational changes in the transmembrane bundle, which are propagated through interactions between specific residues *(17)*. These conformational changes are more conserved as the signal progresses toward the cytoplasm and the G protein (Fig. 1).

However, with the helical bundle embedded in the lipid bilayer, the sequence of the transmembrane domain is highly constrained. Therefore, it is challenging to understand how evolution has selected and adapted the similar architecture of GPCRs to recognize such a large variety of ligands while preserving common mechanisms of activation. It appears that GPCRs have accomplished this tour de force by achieving a significant degree of conformational plasticity on an otherwise preserved scaffold. We use this term to describe the structural differences within the binding site crevices among different receptor subfamilies responsible for recognition of diverse ligands. Additionally, individual receptors also feature some degree of flexibility, denoting the ability of a receptor to dynamically adapt to agonist binding and allow the conformational changes related to mechanisms of activation.

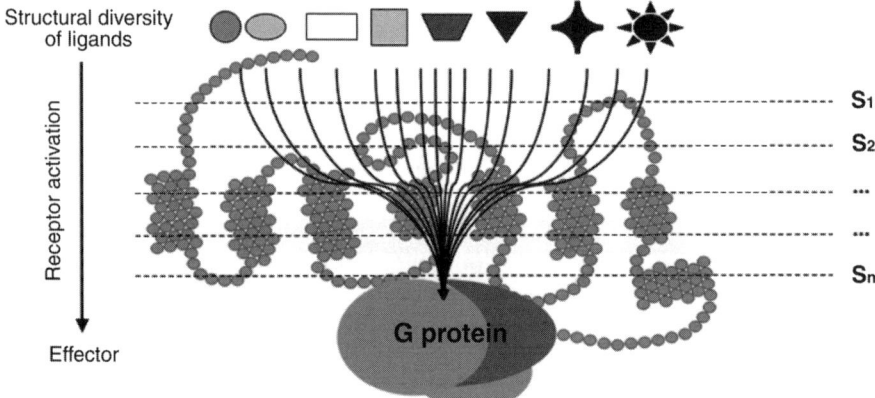

Fig. 1. The highly diverse extracellular signals converge toward structurally conserved mechanisms of effector activation. Whereas the first steps of these mechanisms (S_1, S_2) are specific for each subfamily, the last steps (Sn-1, Sn) share many common structural features (e.g., specific movements of the intracellular region of the TMHs).

We are beginning to understand the molecular actors responsible for these properties. Careful analysis of sequences of receptor subfamilies allows us to detect motifs that could potentially be implicated in this conformational plasticity and/or flexibility. In a second step, these motifs are studied in silico using simulation techniques that reveal their structural properties. Finally, the motifs can be tested experimentally (e.g., using site-directed mutagenesis and bioassays) to identify their biological roles. This chapter focuses on work done on class A GPCRs *(18)* and refers to positions in the receptor sequences using the general notation of Ballesteros and Weinstein *(19)*.

2. SEQUENCE PATTERNS INVOLVED IN RECEPTOR PLASTICITY

2.1. Prolines: The Strongest Disruptors of α-Helical Geometry

Proline (Pro) residues have long been known to disrupt the structure of α-helices *(20)*. The proline side-chain is bonded back to its own amino group, forming a bulky pyrrolidine ring that restricts the conformation of adjacent residues. In an α-helix, the proline ring moves away from the fourth preceding residue (i-4) to avoid a steric clash with the backbone carbonyl, leading to a global kink and to a local disruption in the intrahelical hydrogen bond network that normally stabilizes the α-helix *(21,22)*. The average bend angle of an helix resulting from the presence of Pro averages 20° *(20,23,24)*. Moreover, the presence of the pyrrolidine ring induces a local opening in the helix at the Pro-kinked turn *(25)*, which is translated in a change of direction of the kinked helix *(24,25)*.

Although Pro residues are infrequent within helical segments in globular proteins, they are regularly found in TMHs *(22,24)*. For instance, TMH5, -6, and -7 of class A GPCRs almost invariably contain a Pro, whereas the other helices (with the exception of TMH3) contain Pro residues at specific positions, ranging from 5 to 60% of conservation *(25)*. These Pro residues tend to be conserved within subfamilies; for example, in TMH4, Pro4.59 (60% of conservation) is present in peptide, amine, and opsin receptors, whereas Pro4.60 (34% of conservation) is characteristic of hormone, cannabinoid, and prostanoid receptors but is also found in opsins and some amine receptors. As a result, vertebrate opsins and some amine receptor subfamilies feature a highly conserved Pro4.59 Pro4.60 motif.

Proline residues present in α-helical stretches are proposed to have structural or functional importance, for example, acting as helix breakers *(26)* or facilitating the packing of helical structures *(27,28)*. Beyond these static roles, they are also dynamically important for protein function. Specifically, there is wide experimental *(29–33)* and theoretical *(34–38)* evidence that Pro residues are involved in the regulation of the structure of TMHs in relation to biological function. Because of its enhanced local dynamic flexibility, the Pro-containing helix can adopt different conformations that would correspond to open/closed states in channels or active/inactive conformations in receptors (for reviews, *see* refs. *39* and *40*).

2.2. Ser and Thr Also Deform the Structure of α-Helices

Ser and Thr, unlike other polar or charged residues, do not destabilize TMHs *(41)*. As a result, they are the most frequently occurring polar residues in TMHs: Whereas Leu, Ile, Val, Ala, Phe, and Gly account for two-thirds of the composition of α-helices, the next most common amino acids are Ser and Thr *(42)*. In class A GPCRs, these Ser and Thr transmembranes are often conserved in specific positions of the sequence (Table 1).

In known protein structures, the side-chain hydroxyl of Ser and Thr located within α-helices is often hydrogen bonded to the backbone carbonyls of the preceding turns, as shown in Fig. 2 *(43)*. This interaction could mimic the known interaction of OH moieties of water molecules that hydrogen bond the backbone carbonyls of α-helices, which seems responsible for the concave shape of solvent-exposed α-helices in soluble globular proteins. The hydrophobic environment of the lipid membranes enhances the strength

Table 1
Degree of Conservation of Selected Transmembrane Ser, Thr, and Cys Residues in the Amine Receptor Subfamily of GPCRs

Position	2.45	3.37	3.39	4.57	5.42	5.43	5.46	7.46
%Ser	88	3	99	74	64	51	51	100
%Thr	0	81	0	2	14	19	6	0
%Cys	0	8	1	4	1	2	1	0
Total	88	92	100	80	79	72	58	100

All the sequence analyses were performed on the multiple sequence alignments deposited at the GPCRDB (v6.1) *(69)*.

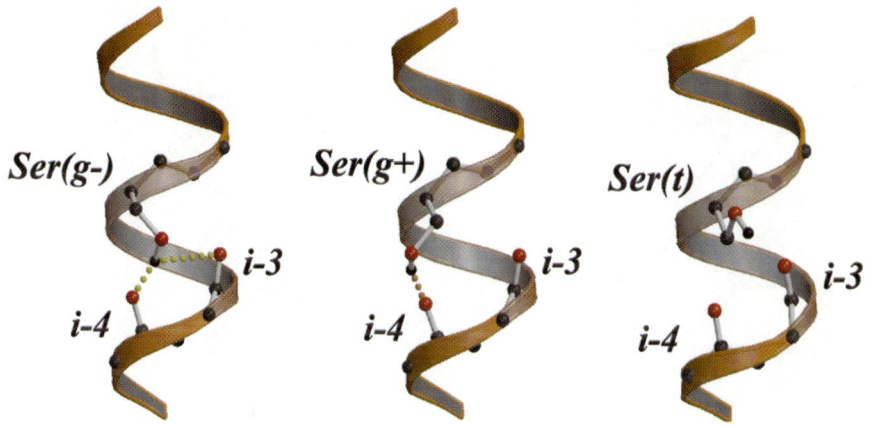

Color Plate 1, Fig. 2. (*see* full caption and discussion in Ch. 17, p. 369). The gauche-conformation of Ser and Thr can disrupt the hydrogen bond network of TMHs. All the figures of molecular models have been created using MolScript v2.1.1 (70) and Raster 3D v2.5 (71).

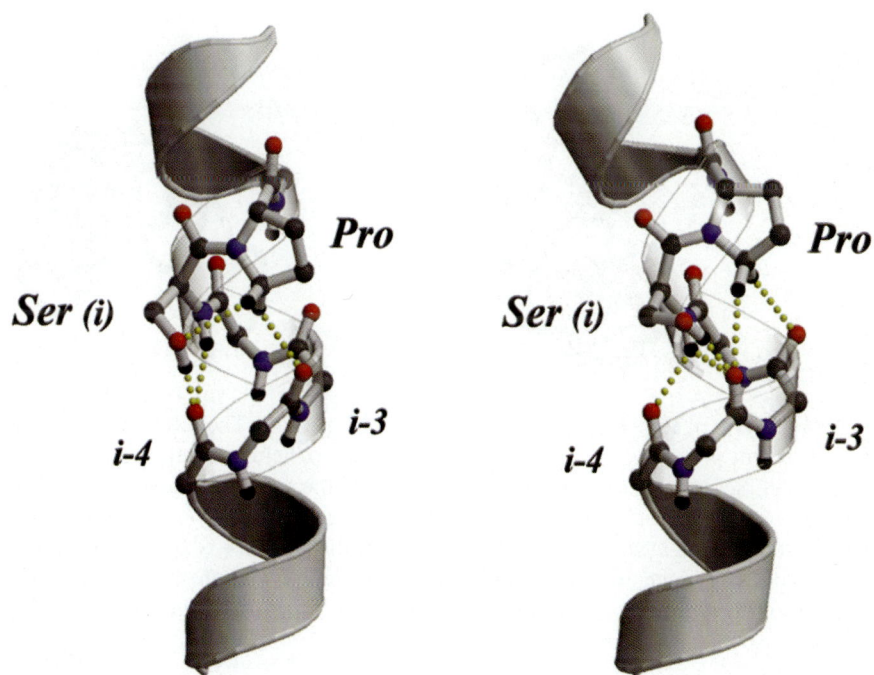

Color Plate 2, Fig. 3. (*see* full caption and discussion in Ch. 17, p. 373). Influence of the side-chain conformation of Ser in the geometry of an α-helix featuring a Ser-X-Pro motif. These local changes induce or stabilize strong distortions of the overall helical structure.

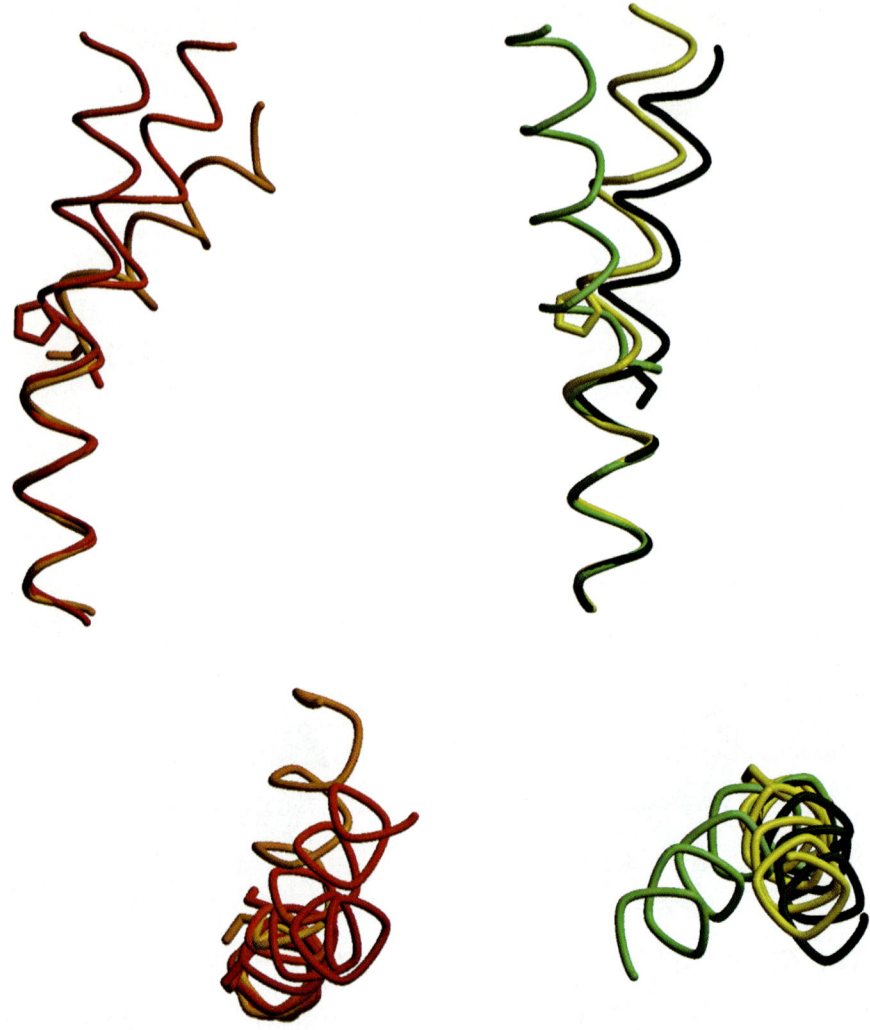

Color Plate 3, Fig. 4. (*see* discussion in Ch. 17, p. 374). The left panels display two different views of representative structures of α-helices featuring a Ser-Pro motif, with Ser in the gauche+ rotamer (red), and in the gauche- rotamer interacting with two different backbone carbonyls (dark orange and light orange). The right panels show helices with a Ser-Ala-Pro motif in the gauche+ (dark green), trans- (yellow), and gauche- rotamers (light green) of Ser. Variation in the side-chain rotamer or the hydrogen bond pattern covers a wide structural range in the α-helix.

Color Plate 4, Fig. 7. (*see* discussion and full caption in Ch. 17, p. 378). Structure of TMH3 in the 5-HT1A receptor (yellow) compared with rhodopsin (grey).

Color Plate 5, Fig. 8. (*see* full caption and discussion in Ch. 17, p. 379). Rotation of Ser and Thr side-chain. In this figure, the red helix represents an ideal (non bent) α-helix, taken as a reference.

Color Plate 6, Fig. 9. (*see* discussion in Ch. 17, p. 382). Structure of TMH2 in the CCR5 receptor (orange) compared with rhodopsin (grey). The left panel shows how the presence of the Thr82 2.56-X-Pro84 2.58 (highly conserved in some peptide receptor families) forces the helix to point in a different direction. The magnitude of this effect relative to the transmembrane bundle, viewed from the extracellular side, is shown in the right panel.

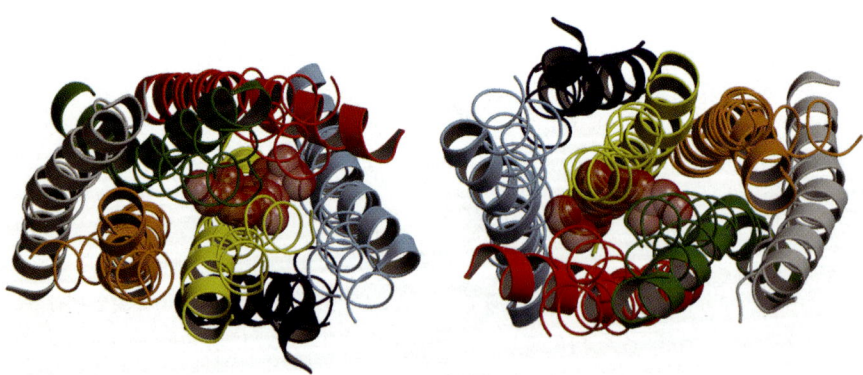

Color Plate 7, Fig. 10. (*see* full caption and discussion in Ch. 17, p. 383). Conformational plasticity of GPCRs. The figures represent two different orientations (rotated 180°) of the transmembrane bundle viewed from the extracellular side.
Color code: TMH1, grey; TMH2, orange; TMH3, yellow; TMH4, dark blue; TMH5, light blue; TMH6, red; TMH7, green. 1X is any nonpolar residue, with the exception of Pro. In the simulations, Ala was used.

Conformational Plasticity of GPCR Binding Sites

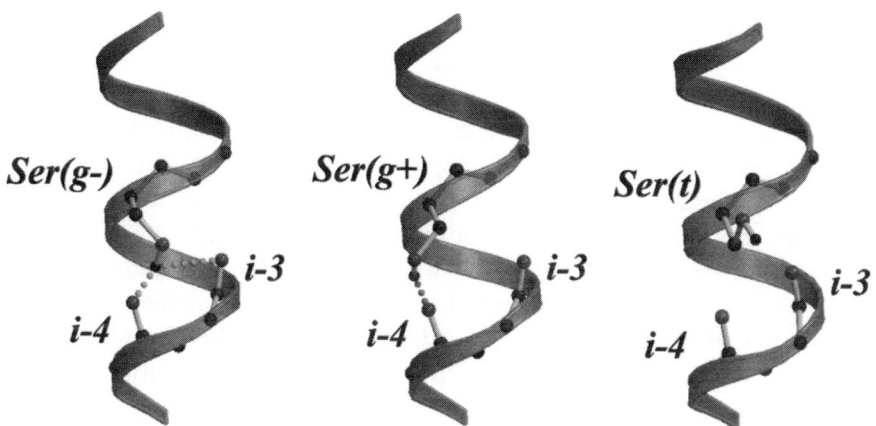

Fig. 2. The gauche-conformation of Ser and Thr can disrupt the hydrogen bond network of TMHs. Whereas, in principle, this conformation allows two possible hydrogen bonds, with the carbonyls at i-3 and i-4 positions (relative to the Ser/Thr) the gauche+ conformation only permits the interaction with the carbonyl one turn away at i-4. Finally, the trans-conformation excludes the possibility of intrahelical hydrogen bonds. All the figures of molecular models have been created using MolScript v2.1.1 *(70)* and Raster 3D v2.5 *(71)*.

of these intrahelical hydrogen bonding interactions by Ser and Thr residues. Therefore, the enhanced force applied on the helical structure may induce larger conformational changes in TMHs than are observed in water-soluble proteins.

The additional hydrogen bond formed between the hydroxyl group of Ser and Thr side-chains in its gauche- conformation and the peptide carbonyls in the previous turn of the helix (Fig. 2) disrupts the hydrogen bond network that stabilizes TMHs, inducing or stabilizing a bend or kink in the helix *(44)*. This seemingly small distortion results in a significant displacement of the residues located a few turns away in the helix. Moreover, the effect of other nearby polar residues—either consecutive or located on the same face of the helix (i.e., three/four residues apart)—can increase the magnitude of this structural effect.

2.3. Ser and Thr Can Modulate the Structure of Pro-Kinked Transmembrane α-Helices

Larger conformational changes can be induced in α-helical structures when the two helix-disrupting motifs reviewed earlier occur together. In these cases, Ser or Thr residues can significantly modulate the structure of

Pro-kinked TMHs as a result of changes induced in the hydrogen bond network of the disrupted turn; this results in a synergistic distortion of the helical structure *(25)*. Importantly, sequence motifs of Pro with a nearby Ser or Thr are common in transmembrane helical segments—particularly within the transmembrane bundle of rhodopsin-like GPCRs. Statistical analyses show that there is a clear tendency for Ser and Thr residues to be found at positions one to two residues preceding Pro, with the exception of the Ser-Pro motif (Tables 2 and 3). Analysis of molecular dynamics simulations of Pro-kinked α-helices containing Ser or Thr residues has revealed possible structural roles for these sequence motifs. Ser and Thr residues in the gauche-side-chain conformation can hydrogen bond either of the two carbonyls from the previous turn (three or four positions before in the sequence) *(43)*. Our simulations demonstrate how this different pattern of interaction results in a different rearrangement of the hydrogen bond network in the helical turn (Fig. 3). These differences can be translated into a local change in the opening of the helix and into an overall change of the helical bend angle. The precise modulation of the structure depends on the relative position of Ser/Thr and Pro residues, the side-chain conformation of the polar residues, and the nature of the intrahelical hydrogen bond between these residues and the backbone carbonyls in the preceding turn of the helix. The changes in the local structure are diverse and can either increase or decrease the bend of the helix compared to the standard Pro-kink. In (Ser/Thr)-X1-Pro and (Ser/Thr)-Pro motifs, an increase of the bend is measured, which apparently is caused by the additional hydrogen bond formed between the side-chain of Ser/Thr and the backbone carbonyl oxygen. In contrast, a decrease of the helical bend angle is observed in (Ser/Thr)-X-X-Pro, Pro-X-(Ser/Thr), and Pro-X-X-(Ser/Thr) motifs, either because of reducing the steric clash between the pyrrolidine ring of Pro and the helical backbone or because of the addition of a constraint in the form of a hydrogen bond in the curved-in face of the helix. In the case of Ser-Pro and Ser-X-Pro motifs, a change in the direction of the helix is observed when the Ser is in the gauche-rotamer (i.e., when the hydroxyl group of Ser hydrogen bonds the backbone carbonyl three positions before in the sequence), which appears to induce a strongly distorted helical turn.

Interestingly, a follow-up of these findings shows that changes in the local hydrogen bond network of the helix, triggered by a change in the Ser or Thr side-chain conformation and amplified by the presence of a nearby Pro residue, can lead to conformational modification of the entire helix in a dynamic fashion. Figure 4 shows the conformations of helices with different Ser/Thr and Pro combinations, in different side-chain conformations of the polar

Table 2
Observed and Expected Number of Occurrences of the (S/T)xxP, (S/T)xP, (S/T)P, P(S/T), Px(S/T), and Pxx(S/T) Motifs in a Nonhomologous Database of Transmembrane Helices

Pair	Observed	Expected	Odds ratio	Significance
SxxP	292	294.9	0.99	0.88
TxxP	267	268.7	0.99	0.95
SxP	353	314.6	1.12	0.02
TxP	316	286.6	1.10	0.07
SP	291	334.2	0.87	0.01
TP	325	304.6	1.07	0.21
PS	280	334.2	0.84	0.001
PT	308	304.6	1.01	0.83
PxS	297	314.6	0.94	0.31
PxT	283	286.6	0.99	0.85
PxxS	249	294.9	0.84	0.004
PxxT	250	268.7	0.93	0.23

Calculated with the TMSTAT formalism (see http://bioinfo.mbb.yale.edu/tmstat/; ref. 42). We used a p value cutoff of 0.10 to select statistically significant overrepresented (odds ratio: <1) and under-represented (odds ratio: >1) patterns. Both TxP and SxP motifs are overrepresented pairs, suggesting that (S/T)xP is a common pattern in transmembrane helices. In contrast, the SP, PS, and PxxS are underrepresented pairs.

residues, and with different hydrogen bond patterns. Clearly, variation of the side-chain rotamer or in the hydrogen bond network generates a wide structural diversity in the α-helix. This change in rotamer conformation, triggered by interactions with an external partner (i.e., bound ligand) or with nearby residues, may have an effect on the structure of the helix, with possible functional importance. In other words, by changing conformation from a gauche+ or gauche- rotamer (i.e., hydrogen bonding the backbone) to trans (not hydrogen bonding to backbone), Thr or Ser could act as conformational switches during receptor activation, especially if present within a Pro-kink motif. This switch of hydrogen bonding partners for the Ser or Thr side-chain hydroxyl moiety could be induced, for example, by direct hydrogen bonding with the ligand, resulting in receptor activation (agonist) or inactivation (inverse agonist).

In summary, simulations on model peptides combined with analysis of known protein structures offer molecular explanations for the observed high association and conservation of Ser and Thr residues with Pro in TMHs regarding specific structural and/or functional roles. These residues modulate the structural deformations caused by the pyrrolidine ring of Pro through

Table 3
Percentage of Occurrences of Ser/Thr and Pro Combinations in Class A GPCRs.

Position of Pro	Number of Occurrence S of Pro	SxxP	TxxP	SxP	TxP	SP	TP	PS	PT	PxS	PxT	PxxS	PxxT	Total
1.36	378	2	4	0	1	5	0	6	12	2	1	3	0	35
1.48	203	0	13	9	0	0	12	2	1	1	0	3	2	44
2.58	546	6	1	10	71	0	1	0	1	—	0	5	2	100
2.59	875	6	16	3	15	5	6	2	3	16	10	1	3	86
2.60	103	5	0	3	0	0	0	1	1	0	1	8	28	47
4.59	1139	2	1	42	11	6	4	3	3	0	2	2	0	76
4.60	634	34	12	3	2	1	3	1	1	1	2	5	0	66
5.50	1588	0	1	3	3	0	3	2	4	11	9	1	5	40
6.50	1603	8	3	1	0	3	15	1	0	4	4	7	4	51
6.59	97	0	6	1	4	6	0	0	1	0	2	0	0	21
7.38	235	3	8	9	53	0	0	18	0	0	0	0	3	94
7.45	101	13	4	0	13	0	0	3	0	2	4	0	0	39
7.46	325	1	10	0	0	2	40	4	4	0	0	1	0	62
7.50	1888	5	4	1	1	0	1	0	0	0	0	0	0	13
	9715	6	5	7	9	2	6	2	2	4	3	2	2	52

Partly or highly conserved prolines (>5% of the sequences) can be found in TMH1 (positions 1.36 and 1.48), TMH2 (2.58, 2.59, and 2.60), TMH4 (4.59 and 4.60), TMH5 (highly conserved 5.50), TMH6 (highly conserved at position 6.50 and occasionally at 6.59), and TMH7 (7.38, 7.45, 7.46, and 7.50). Table 3 also shows the probability of finding a Ser or Thr residue in the direct vicinity of the Pro. Although 52% of the Pro-containing TMHs contain either Ser or Thr in the vicinity of the proline, its distribution is not uniform over the different Pro-kinks.

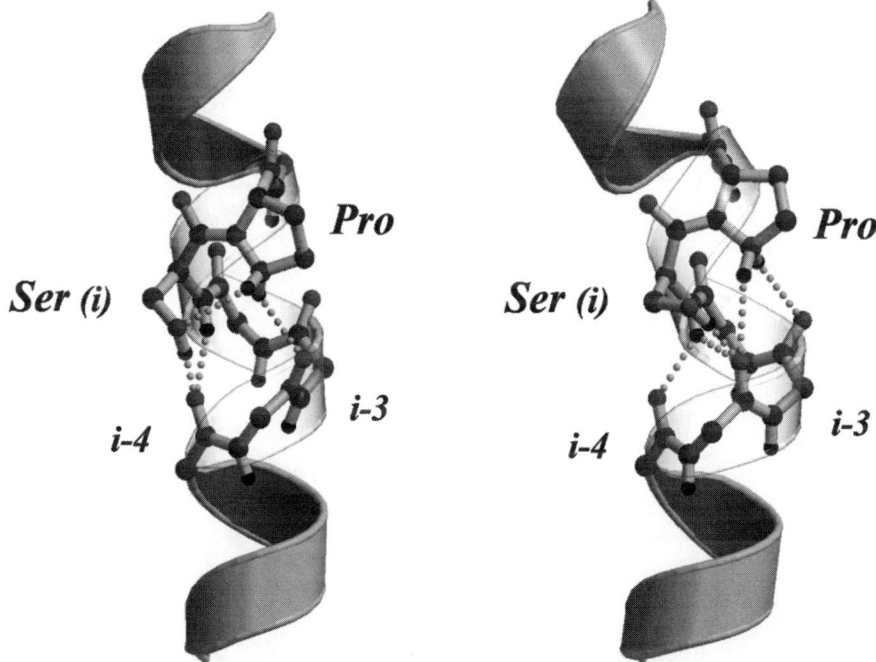

Fig. 3. Influence of the side-chain conformation of Ser in the geometry of an α-helix featuring a Ser-X-Pro motif. The local hydrogen bond network in the helical turn (yellow dots) is different when Ser is in *gauche+* (left panel) or *gauche-* (right panel). These local changes induce or stabilize strong distortions of the overall helical structure.

specific interactions of their polar side-chains. These significant conformational changes may have important functional roles, such as modifying the three-dimensional arrangements of the receptor's binding site crevice, what we refer to as "plasticity" of the binding site. These conformational variations within the binding sites, which allow or prevent specific interactions to take place, may be responsible for divergent recognition of different ligands through evolution. Therefore, combinations of Ser and Thr residues with Pro appear as a likely mechanism for structural adaptation of membrane proteins through evolution.

2.4. Glycine and Cysteine

Residues other than Pro, Ser, and Thr are involved in modulation of the α-helical structure. For example, Glycine (Gly) residues, which have been shown to promote helix-helix association *(42,45,46)*, are also responsible

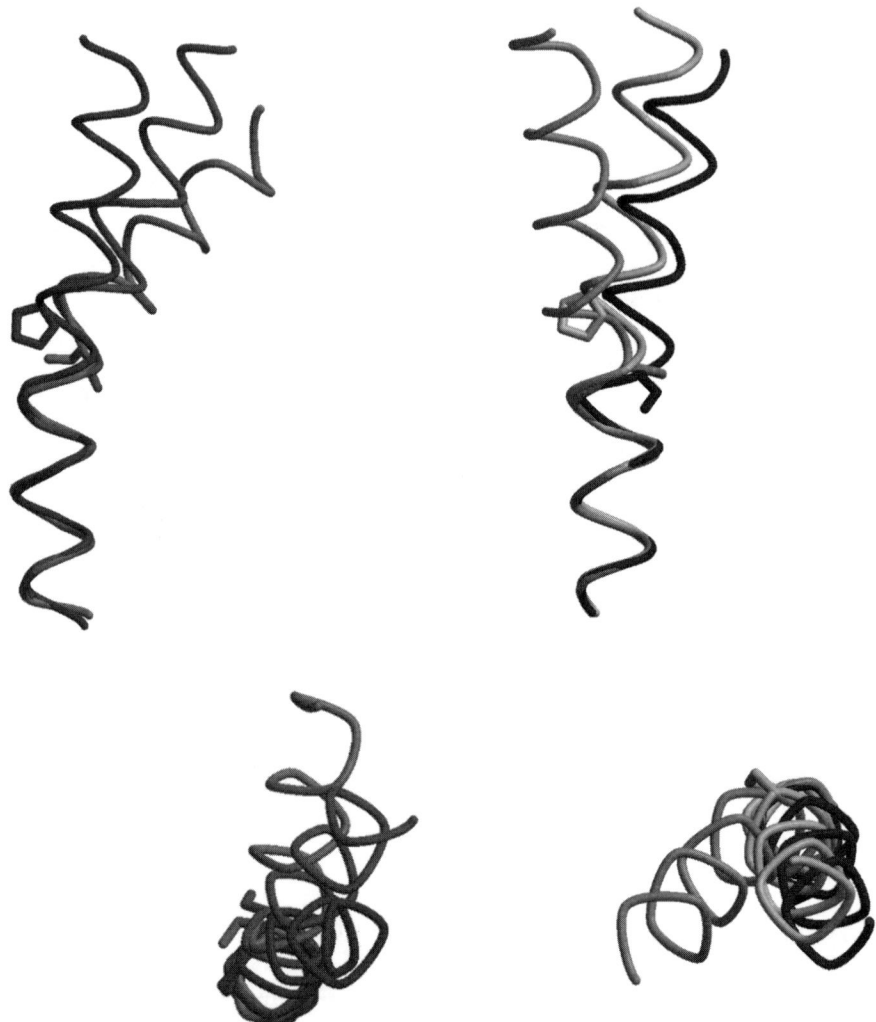

Fig. 4. The left panels display two different views of representative structures of α-helices featuring a Ser-Pro motif, with Ser in the gauche+ rotamer (red), and in the gauche- rotamer interacting with two different backbone carbonyls (dark orange and light orange). The right panels show helices with a Ser-Ala-Pro motif in the gauche+ (dark green), trans- (yellow), and gauche- rotamers (light green) of Ser. Variation in the side-chain rotamer or the hydrogen bond pattern covers a wide structural range in the α-helix.

for enhanced local flexibility in helices because of the lack of a side-chain *(47)*. It is remarkable that although Gly residues are rarely found in α-heli-

Conformational Plasticity of GPCR Binding Sites 375

ces of soluble proteins (where they are thought to destabilize this secondary structure), they are much more common in TMHs. The reason may be the lack (or decreased presence) of water molecules within transmembrane domains. The β-methyl group that is absent in Gly residues that is present in all other side-chains shields the carbonyl of the preceding turn from potential hydrogen bonding partners such as water molecules. Therefore, it might be possible that Gly residues in TMHs do not destabilize α-helices and can thus be used more often to confer the desired structural and/or functional roles. For example, the crystal structures of bovine rhodopsin show a strong bend in TMH2 at the level of Gly89 2.56 and Gly90 2.57, stabilized in part by Thr residues in the next turn hydrogen bonding their preceding carbonyls as described earlier *(48)*. Interestingly, Gly residues are also often found near transmembrane Pro residues, increasing the local flexibility of Pro-kinked α-helices *(24,49)*. In fact, a Gly-Pro motif has been proposed to be responsible for the divergent agonist recognition between the cannabinoid (CB)2 (Phe-Pro) and CB1 receptors (Gly-Pro) *(50)*.

Finally, Cys residues contain a sulfhydryl group in their side-chain that is also able to establish an hydrogen bond with the backbone carbonyl of the α-helix *(43)*. Although this weak hydrogen bond can not promote a structural distortion in the helix on its own *(44)*, it is likely that similarly to Ser and Thr, its presence synergistically influences the effect of nearby Pro *(51)* or polar residues *(52)*, leading to substantial conformational changes in TMHs. This would explain the conservation of several Cysteine (Cys) residues within GPCRs, such as the conserved Cys6.47 that recently has been shown to modulate the Pro-kink of TMH6 in the β_2-adrenergic receptor (AR) *(51)*. It has been proposed that this Cys unusually hydrogen bonds its own carbonyl in the inactive state of the receptor in the trans-rotamer configuration, thereby exploiting the flexibility of the Pro-kink motif, whereas it hydrogen bonds the backbone carbonyl of the fourth preceding residues (like Ser or Thr residues) in the active state *(51)*.

3. EXAMPLES OF PLASTICITY

3.1. Amine Receptors:
From Ligand Recognition to Activation Mechanism

Conserved Ser and Thr residues can play different roles in maintaining the structure or regulating the activity of GPCRs. We hypothesize that these functional roles are related to the capability of these residues to affect the geometry of TMHs. To illustrate this point, this section discusses the example of amine receptors.

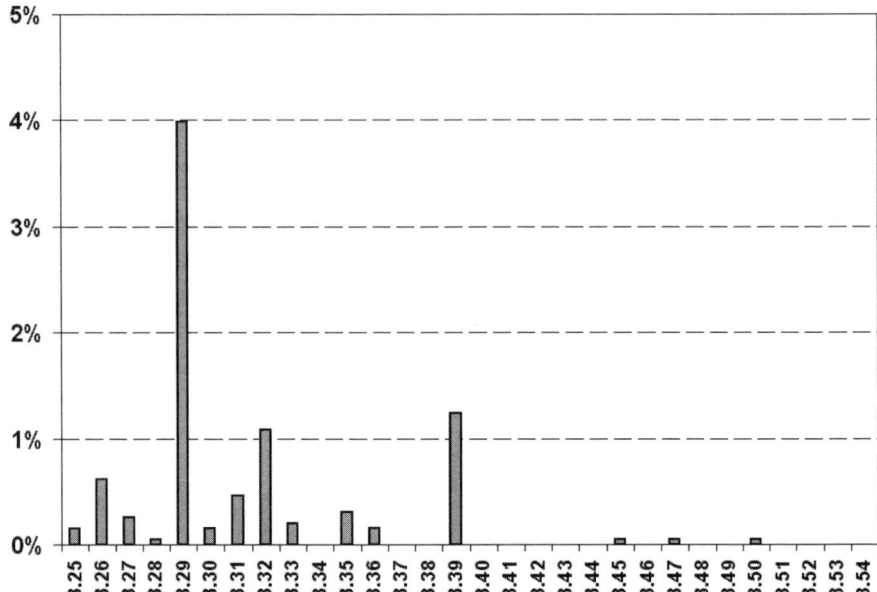

Fig. 5. Presence and degree of conservation of Pro residues in TMH3 of class A GPCRs. Only a few and very specific subfamilies feature this residue-for example, some peptide receptors at 3.29 (4% of class A), gonadotropin-releasing hormone and prostanoid receptors in 3.39 (1% of class A), or tachykinin receptors at 3.32 (1% of class A).

Sequence analysis of the amine receptor family reveals several conserved Ser or Thr residues in TMH3 (18), the helix with the fewest conserved Pro residues in the rhodopsin-like family of GPCRs (25) (Fig. 5). Whereas 3.37 and 3.39 positions are highly conserved in the entire amine receptor family (forming the Thr3.37-(Ala/Ser)3.38-Ser3.39 signature of amine receptors) and 3.30 and 3.47 are also conserved (to a lesser extent), other positions are conserved only within certain subfamilies (Fig. 6). Located in the center of the helical bundle, TMH3 interacts with nearly all the other helices, and, therefore, these polar residues could facilitate specific interhelical contacts. Our structural studies suggest an alternative role for these residues: Whereas the lack of Pro in this helix suggests that it will not feature strong hinges, these Ser and Thr can provide a certain degree of flexibility and plasticity to the helix, leading to various alternative TMH3 conformations that might represent different functional states of the receptor, as has been proposed for the D2 dopamine receptor (18). The actual distortions depend on the precise positions of Ser and Thr residues in TMH3. Because this helix is known to

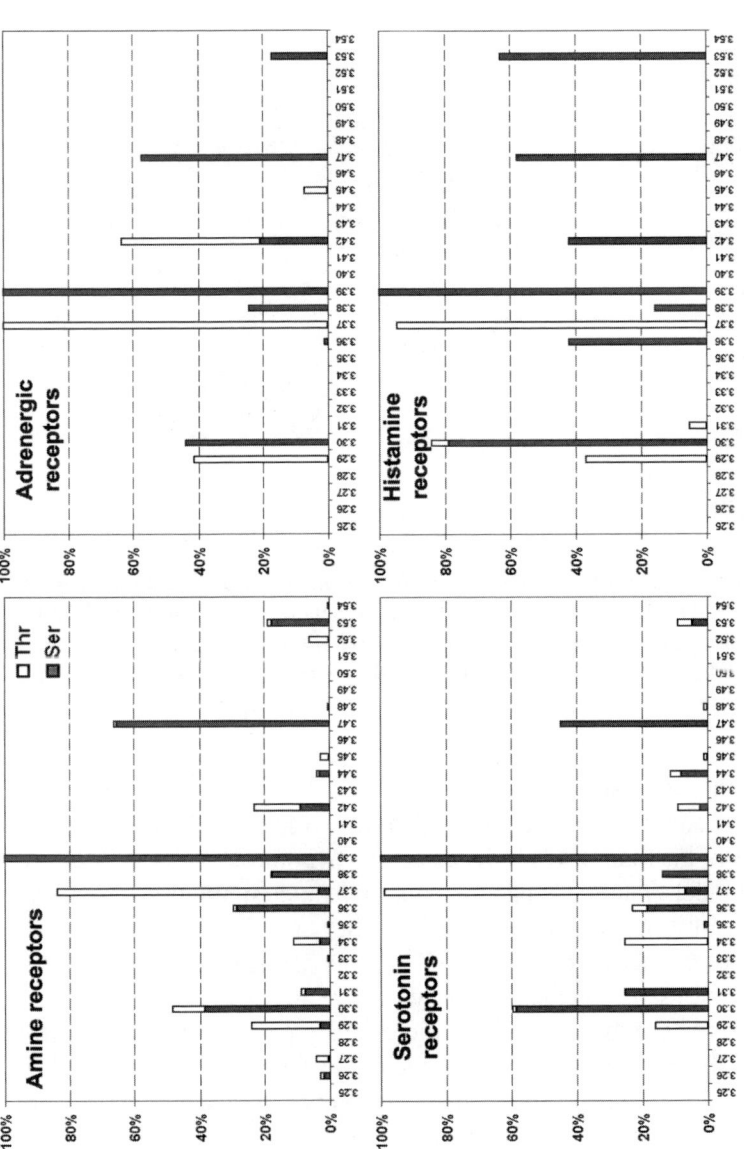

Fig. 6. Presence and degree of conservation of Ser and Thr residues in TMH3 of amine receptors and in the serotonin, adrenergic, and histamine receptors subfamilies. The similarity of these residues regarding the region of sequence can be observed. Although some positions present highly conserved Ser or Thr, there is a certain degree of variability. For example, whereas Thr3.37 and Ser3.39 are highly conserved and 3.42 is either a Ser or a Thr in more than 60% of ARs, it is exclusively a Ser for approx 40% of histamine receptors, and none of these residues in virtually all the serotonin receptors

377

Fig. 7. Structure of TMH3 in the 5-HT1A receptor (yellow) compared with rhodopsin (grey). The left panel shows how the presence of Cys3.36 and Thr3.37 conserved in the amine receptor family of class A GPCRs (59 and 80%, respectively) induces a bend in the helix. The effect of this bend relative to the transmembrane bundle, viewed from the extracellular side, is shown in the right panel.

interact with the agonists in most GPCRs, variability in the number and locations of Ser and Thr offers a mechanism for structural adaptation to the characteristics of the ligand. We have tested this hypothesis with the 5-HT1A receptor as a model system. Using molecular dynamics simulations, we have shown that because of its specific patterns of Ser and Thr, TMH3 of the 5-HT1A receptor tends to bend toward TMH5. This deformation would facilitate the interactions between the agonist and residues Asp3.32, Ser5.42, Ser5.43, and Ser5.46, which have been identified experimentally. A statistical analysis of the molecular dynamics trajectories led to the conclusion that these structural divergences result from the difference in sequence at the 3.36 and 3.37 positions; whereas rhodopsin vertebrate type 1 receptors possess a conserved (85%) Gly3.36-Glu3.37 motif, amine receptors feature either a completely conserved Cys3.36-Thr3.37 or a Cys3.35-X-Thr3.37 motif. The predicted relocation of TMH3 in the 5-HT1A receptor (Fig. 7) substantially alters the structure of the binding pocket. These changes allowed us to explain the experimentally determined pattern of binding affinities of synthetic

Conformational Plasticity of GPCR Binding Sites

Fig. 8. Rotation of Ser and Thr side-chain can induce different conformation of the helix. In this figure, the red helix represents an ideal (non bent) α-helix, taken as a reference. Rotation of Ser side-chain from gauche- (blue) to the trans- (green) conformation induces a change in the direction of the helix toward different locations.

ligands (52). Interestingly, amine receptors also have a highly conserved Cys residue at position 3.44; the structural and/or functional role of this Cys residue has not been elucidated but may be involved in selecting specific helical conformations, as proposed for Ser and Thr residues.

Additionally, there is indirect experimental evidence for a structural role of Ser residues in TMH5. Conserved Ser5.42 (53), Ser5.43 (53), and Ser5.46 (54) have been shown to be involved in ligand recognition, interacting with the hydroxyl groups of the catechol moiety in catecholamines. Also, Ser2045.42 and Ser2055.43 of the β_2-AR modulates the equilibrium between the active and inactive form of the receptor (55). We suggest that these Ser adopt the gauche- conformation in the absence of the extracellular ligand, thus satisfying its hydrogen bonding potential through interactions with the backbone carbonyls. During ligand binding, Ser must adopt the trans-conformation to optimally interact with the hydroxyl moieties of the ligand. This transition ultimately leads to an alteration of the orientation of the helix toward different positions in space (Fig. 8), which can be related to the transition toward an active form of the receptor. Notably, the presence of several Ser or Thr residues conserved within the same turn of TMH5 at 5.42, 5.43, and 5.46 may result in a much larger degree of conformational plasticity because of synergistic helical distortions, as shown for the D2 dopamine receptor *(18)*.

Although the role of other conserved Ser of the amine receptor family is not clear, some studies point to a role in the structure or function of these receptors. Substituted Cys accessibility method (SCAM) experiments in the D2 dopamine receptor *(56)* show that Ser1213.39 faces a water-accessible crevice. According to models based on the crystal structure of rhodopsin *(31,52)*, this residue would be facing Asn7.49 and Asp2.50, which are highly conserved in the whole class A family and are involved in receptor activation *(57)*. In this region, the most recent crystal structures of bovine rhodopsin (PDB access codes 1L9H and 1GZM) show the presence of water molecules *(58)*. Interestingly, bovine rhodopsin features an Ala residue at position 3.39. Therefore, one can suggest that in the amine receptor family, the polar side-chain of Ser3.39 replaces the function of one of these water molecules. Mutations of Ser1213.39 in the D2 dopamine receptor *(59)* and of Ser1093.39 in acetylcholine receptor *(60)* appear to destabilize the active state of the receptors, likely by disruption of this hydrogen bond network that results from the loss of the polar Ser3.39. A similar role has been proposed for Ser7.46 based on mutagenesis studies on the acetylcholine receptor. This position, which is likely to be near Asn7.49 and Asp2.50, has been proposed to be involved in a network of interactions stabilizing the active state of the receptor *(61)*.

Finally, some of these conserved Ser and Thr may be involved in maintaining the local structure of certain regions in the transmembrane bundle through specific helix-helix interactions. For example, Ser2.45 of amine receptors is believed to interact with the nearby Trp4.50 (which is highly conserved in all class A GPCRs) through a hydrogen bond interaction; a similar pattern is present in rhodopsin, where position 2.45 holds an Asn residue. However, single Ser and Thr residues are not able to promote helix-helix association. Motifs of multiple Ser and/or Thr are needed to create a network of hydrogen bonds that is strong enough to promote association *(62)*.

In summary, we hypothesize that Ser and Thr residues can affect the structure of the transmembrane bundle, conferring some degree of plasticity to the receptor. Dynamically, these motifs also provide flexibility to the structure, because changes in the conformation of the side-chain can modulate the deformation of TMHs. Such changes in side-chain conformation could be triggered either by changes in the protonation state or conformation of nearby side-chains or by ligand binding.

3.2. Combinations of Ser/Thr and Pro Can Account for Structural Specificities of Peptide Receptors

Combinations of Ser/Thr and Pro residues found in the transmembrane bundle of GPCRs tend to be conserved within specific subfamilies (see Table 3). For example, the Thr2.56-X-Pro2.58 motif in TMH2 is present in leukotriene B4, platelet activating factor, and viral and peptide receptors. In the latter subfamily, it is highly conserved in the APJ-like, angiotensin, chemokine, Fmet-leu-phe, interleukin-8 and opioid receptors subfamilies, accounting for almost 50% of the entire peptide receptor subgroup. Interestingly, in the corresponding position of the sequence, opsin receptors possess Gly89 2.56-Gly90 2.57, a highly conserved motif (90%) in rhodopsin vertebrate type 1 receptors. Although both the Thr2.56-X-Pro2.58 and the Gly2.56-Gly2.57 motifs are able to induce a distortion in the α-helix, these motifs are likely to bend TMH2 in different directions, thereby producing a different local structure in this region (Fig. 9). Hence, the structural consequences of this sequence motif are expected to be a structural determinant of at least some of the listed families, thus accounting for specificities in the structure and/or function among GPCRs.

This putative functional role has been experimentally tested in the Thr82 2.56-X-Pro84 2.58 motif of TMH2 by site-directed mutagenesis and functional assays of the chemokine receptor CCR5 *(31,63)*. In summary, the results show that although mutation of Thr82 2.56 to either Ser, Cys, Ala, or Val does not affect chemokine binding, it strongly influences the functional response. The functional impairment is highly dependent on the specific side-chain substituted for the Thr, and the rank order parallels the structural deformation of the α-helix that was observed in the molecular dynamics simulations *(25)*. These simulations showed that the presence of Thr, Ser, and Cys side-chains two positions before the Pro residue increases the average bend angle of the α-helix. The observation that the polar side-chains form hydrogen bonds with the helix backbone during the simulations suggests that the observed effect on the bending angle arises from local deformations in helix geometry induced by these bonds. As a result of this effect, chemokine receptors require a re-arrangement of the transmembrane bundle interactions (relative to rhodopsin), because the presence of Pro84 2.58 orients the extracellular part of TMH2 toward TMH 3 and would not close to TMH1 (as observed in the crystal structure of rhodopsin), whereas the additional presence of Thr82 2.56 accentuates this effect (Fig. 9). Our modeling study also suggests that in chemokine receptors, the extracellular region of TMH2 interacts with TMH3, which is not feasible in the rhodopsin struc-

Fig. 9. Structure of TMH2 in the CCR5 receptor (orange) compared with rhodopsin (grey). The left panel shows how the presence of the Thr822.56-X-Pro842.58 (highly conserved in some peptide receptor families) forces the helix to point in a different direction. The magnitude of this effect relative to the transmembrane bundle, viewed from the extracellular side, is shown in the right panel.

ture. This pattern of interaction has been further experimentally tested for the CCR5 receptor (63). These results lead us to suggest that the Thr822.56-X-Pro842.58 motif in TMH2 is a structural determinant in the chemokine- and possibly other peptide-receptor family by virtue of significant local effects on the helix conformation, which propagate through a lever action to the extracellular parts of the transmembrane bundle.

3.3. Ser/Thr/Pro-Induced Plasticity: A Common Attribute of Membrane Proteins

Because our findings are based on general principles of protein structure, it is conceivable that Ser and Thr residues in α-helices of other integral membrane proteins may also participate in conformational changes. For example, mutations of Thr86 in the second transmembrane segment of connexin32 (part of a Thr86-Pro87 motif) shift the conductance-voltage relationship of the wild-type channel, leading to the proposal that the hydrogen bonding potential of Thr86, together with the structural effect of the nearby Pro87, mediates the conformational changes between open- and closed-channel

Conformational Plasticity of GPCR Binding Sites 383

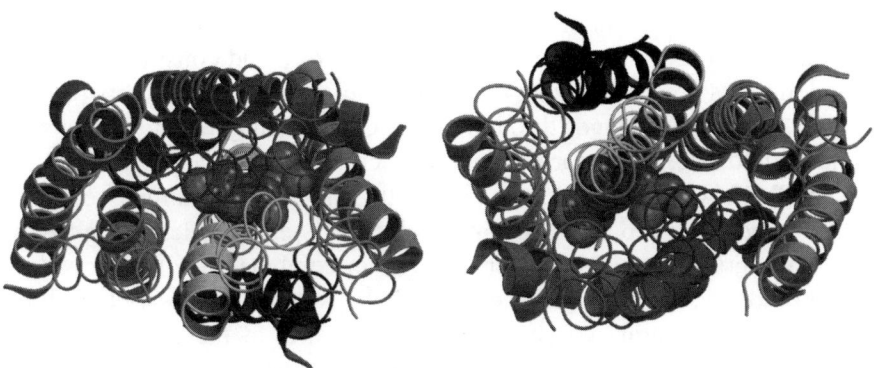

Fig. 10. Conformational plasticity of GPCRs. The figures represent two different orientations (rotated 180°) of the transmembrane bundle viewed from the extracellular side. The rhodopsin template is represented by thick ribbons, whereas possible alternative conformations for each helix are represented by thin ribbons. This alternative conformations arise from different conformations of selected sequence patterns: a Ser1.46-X-Pro1.48 motif (conserved in adenosine type 1 receptors); a Ser2.56-X-Pro2.58 motif, with Ser in gauche- and gauche+ conformations (conserved in some peptide receptor families); Ser3.30 (conserved in histamine, some serotonin and β_2-ARs) and a Cys3.36-Thr3.37 motif (conserved in amine receptors); a Ser4.59-X-Pro4.61 motif, with Ser in gauche- and gauche+ conformations (conserved in some peptide receptor families); Ser5.43 and Ser5.46 (conserved in amine receptors); and different conformations of the Pro-kink for Pro6.50 and Pro 7.50. It can be observed how these sequence motifs are able to account for either static structural plasticity or dynamic flexibility of the transmembrane bundle, which can be related, respectively, to ligand recognition and activation mechanisms. To help visualize the location of the binding crevice, retinal is represented in space-filling spheres.

Color code: TMH1, grey; TMH2, orange; TMH3, yellow; TMH4, dark blue; TMH5, light blue; TMH6, red; TMH7, green.1X is any nonpolar residue, with the exception of Pro. In the simulations, Ala was used.

states *(30)*. In bacteriorhodopsin, mutation of Thr90 to Ala in the Thr90-Pro91 motif present in TMH C alters the proton pumping efficiency, suggesting that Thr90 also plays an important structural role in the proton pumping mechanism *(64)*.

4. CONCLUSIONS

Numerous studies have provided growing evidence that GPCRs can coexist in different conformations, strongly suggesting an inherent flexibility. For example, in the β_2-AR, it has been demonstrated that there are different states of the native form of the receptor *(65,66)*, different ligands can stabilize distinct conformations of the active state *(67)*, and constitutively activated mutants present a higher degree of conformational flexibility *(68)*.

The diversity of ligands throughout the GPCR family implies that a very significant degree of conformational plasticity has been achieved within the seven-TMHs. Our work and the work of others indicate that simple sequence motifs, involving Pro, Ser and Thr residues, are responsible for the necessary structural variability. Either alone or in combinations, these residues can provide the structural diversity in α-helices that is necessary to accommodate the structural and chemical characteristics of the cognate ligands (Fig. 10). The remarkable abundance of these motifs in class A GPCRs shows that evolution has selected such motifs as a generic way to achieve plasticity. As a corollary, this indicates that the seven-TMH architecture is suitable to allow local structural variations while preserving global structure and activation mechanism. Therefore, we suggest that the evolutionary success of GPCRs (numerous and diverse, as are their associated biological functions) results from their ability to allow for structural plasticity using limited sequence variation, thanks to the structural motifs reviewed in this chapter. Finally, it appears that not only can these motifs confer plasticity to the protein family, but they can also provide flexibility. In several cases, Ser and Thr could act as conformational switches, leading to conformational changes upon ligand binding. Therefore, understanding the action of these motifs is key in understanding the mechanisms of receptor activation at a molecular level.

REFERENCES

1. Takeda S, Kadowaki S, Haga T, Takaesu H, Mitaku S. Identification of G protein-coupled receptor genes from the human genome sequence. FEBS Lett 2002;520:97–101.
2. Lander ES, Linton LM, Birren B, et al. Initial sequencing and analysis of the human genome. Nature 2001;409:860–921.
3. Venter JC, Adams MD, Myers EW, et al. The sequence of the human genome. Science 2001;291:1304–1351.
4. Fredriksson R, Lagerstrom MC, Lundin LG, Schioth HB. The G protein-coupled receptors in the human genome form five main families. Phylogenetic analysis, paralogon groups, and fingerprints. Mol Pharmacol 2003;63:1256–1272.
5. Bourne HR. How receptors talk to trimeric G proteins. Curr Opin Cell Biol 1997; 9(2):134–142.
6. Gether U. Uncovering molecular mechanisms involved in activation of G protein- coupled receptors. Endocr Rev 2000;21:90–113.
7. Ji TH, Grossmann M, Ji I. G protein-coupled receptors. I. Diversity of receptor-ligand interactions. J Biol Chem 1998;273:17,299–17,302.

8. Okada T, Ernst OP, Palczewski K, Hofmann KP. Activation of rhodopsin: new insights from structural and biochemical studies. Trends Biochem Sci 2001;26:318–324.
9. Mirzadegan T, Benko G, Filipek S, Palczewski K. Sequence analyses of G protein-coupled receptors: similarities to rhodopsin. Biochemistry 2003;42:2759–2767.
10. Farrens DL, Altenbach C, Yang K, Hubbell WL, Khorana HG. Requirement of rigid-body motion of transmembrane helices for light activation of rhodopsin. Science 1996;274:768–770.
11. Altenbach C, Yang K, Farrens DL, Farahbakhsh ZT, Khorana HG, Hubbell WL. Structural features and light-dependent changes in the cytoplasmic interhelical E-F loop region of rhodopsin: a site-directed spin-labeling study. Biochemistry 1996;35:12,470–12,478.
12. Abdulaev NG, Ridge KD. Light-induced exposure of the cytoplasmic end of transmembrane helix seven in rhodopsin. Proc Natl Acad Sci USA 1998;95: 12,854–12,859.
13. Miura S, Karnik SS. Constitutive activation of angiotensin II type 1 receptor alters the orientation of transmembrane helix-2. J Biol Chem 2002;277(27): 24,299–24,305.
14. Visiers I, Ballesteros JA, Weinstein H. Three-dimensional representations of G protein-coupled receptor structures and mechanisms. Methods Enzymol 2002;343:329–371.
15. Donnelly D, Findlay JB, Blundell TL. The evolution and structure of aminergic G protein-coupled receptors. Receptors Channels 1994;2(1):61–78.
16. Horn F, van der Wenden EM, Oliveira L, IJzerman AP, Vriend G. Receptors coupling to G proteins: is there a signal behind the sequence? Proteins 2000;41:448–459.
17. Madabushi S, Gross AK, Philippi A, Meng EC, Wensel TG, Lichtarge O. Evolutionary trace of G protein-coupled receptors reveals clusters of residues that determine global and class-specific functions. J Biol Chem 2004;279(9): 8126–8132.
18. Ballesteros JA, Shi L, Javitch JA. Structural mimicry in G protein-coupled receptors: implications of the high-resolution structure of rhodopsin for structure-function analysis of rhodopsin-like receptors. Mol Pharmacol 2001;60:1–19.
19. Ballesteros JA, Weinstein H. Integrated methods for the construction of three dimensional models and computational probing of structure-function relations in G-protein coupled receptors. Methods Neurosci 1995;25:366–428.
20. Barlow DJ, Thornton JM. Helix geometry in proteins. J Mol Biol 1988;201:601–619.
21. MacArthur MW, Thornton JM. Influence of proline residues on protein conformation. J Mol Biol 1991;218(2):397–412.
22. von Heijne G. Proline kinks in transmembrane alpha-helices. J Mol Biol 1991;218:499–503.
23. Sankararamakrishnan R, Vishveshwara S. Geometry of proline-containing alpha-helices in proteins. Int J Pept Protein Res 1992;39:356–363.

24. Cordes FS, Bright JN, Sansom MS. Proline-induced distortions of transmembrane helices. J Mol Biol 2002;323:951–960.
25. Deupi X, Olivella M, Govaerts C, Ballesteros JA, Campillo M, Pardo L. Ser and Thr residues modulate the conformation of Pro-kinked transmembrane alpha-helices. Biophys J 2004;86:105–115.
26. Richardson JS, Richardson DC. Amino acid preferences for specific locations at the ends of alpha helices. Science 1988;240(4859):1648–1652.
27. Woolfson DN, Williams DH. The influence of proline residues on alpha-helical structure. FEBS Lett 1990;277:185–188.
28. Orzaez M, Salgado J, Gimenez-Giner A, Perez-Paya E, Mingarro I. Influence of proline residues in transmembrane helix packing. J Mol Biol 2004;335(2):631–640.
29. Hong S, Ryu KS, Oh MS, Ji I, Ji TH. Roles of transmembrane prolines and proline-induced kinks of the lutropin/choriogonadotropin receptor. J Biol Chem 1997;272:4166–4171.
30. Ri Y, Ballesteros JA, Abrams CK, et al. The role of a conserved proline residue in mediating conformational changes associated with voltage gating of Cx32 gap junctions. Biophys J 1999;76:2887–2898.
31. Govaerts C, Blanpain C, Deupi X, et al. The TXP motif in the second transmembrane helix of CCR5. a structural determinant of chemokine-induced activation. J Biol Chem 2001;276:13,217–13,225.
32. Stitham J, Martin KA, Hwa J. The critical role of transmembrane prolines in human prostacyclin receptor activation. Mol Pharmacol 2002;61:1202–1210.
33. Slepkov ER, Chow S, Lemieux MJ, Fliegel L. Proline residues in transmembrane segment IV are critical for activity, expression and targeting of the Na+/H+ exchanger isoform 1. Biochem J 2004;379(pt 1):31–38.
34. Yun RH, Anderson A, Hermans J. Proline in alpha-helix: stability and conformation studied by dynamics simulation. Proteins 1991;10:219–228.
35. Sankararamakrishnan R, Vishveshwara S. Characterization of proline-containing alpha-helix (helix F model of bacteriorhodopsin) by molecular dynamics studies. Proteins 1993;15(1):26–41.
36. Biggin PC, Breed J, Son HS, Sansom MS. Simulation studies of alamethicin-bilayer interactions. Biophys J 1997;72(2 pt 1):627–636.
37. Tieleman DP, Shrivastava IH, Ulmschneider MR, Sansom MS. Proline-induced hinges in transmembrane helices: possible roles in ion channel gating. Proteins 2001;44(2):63–72.
38. Bright JN, Shrivastava IH, Cordes FS, Sansom MS. Conformational dynamics of helix S6 from Shaker potassium channel: simulation studies. Biopolymers 2002;64(6):303–313.
39. Sansom MS, Weinstein H. Hinges, swivels and switches: the role of prolines in signalling via transmembrane alpha-helices. Trends Pharmacol Sci 2000;21:445–451.
40. Reiersen H, Rees AR. The hunchback and its neighbours: proline as an environmental modulator. Trends Biochem Sci 2001;26(11):679–684.
41. Monne M, Hermansson M, von Heijne G. A turn propensity scale for transmembrane helices. J Mol Biol 1999;288:141–145.

42. Senes A, Gerstein M, Engelman DM. Statistical analysis of amino acid patterns in transmembrane helices: the GxxxG motif occurs frequently and in association with beta-branched residues at neighboring positions. J Mol Biol 2000;296:921–936.
43. Gray TM, Matthews BW. Intrahelical hydrogen bonding of serine, threonine and cysteine residues within alpha-helices and its relevance to membrane-bound proteins. J Mol Biol 1984;175(1):75–81.
44. Ballesteros JA, Deupi X, Olivella M, Haaksma EE, Pardo L. Serine and threonine residues bend alpha-helices in the chi(1) = g(-) conformation. Biophys J 2000;79:2754–2760.
45. Lemmon MA, Treutlein HR, Adams PD, Brunger AT, Engelman DM. A dimerization motif for transmembrane alpha-helices. Nat Struct Biol 1994;1(3):157–163.
46. Brosig B, Langosch D. The dimerization motif of the glycophorin A transmembrane segment in membranes: importance of glycine residues. Protein Sci 1998;7(4):1052–1056.
47. Li SC, Deber CM. Influence of glycine residues on peptide conformation in membrane environments. Int J Pept Protein Res 1992;40(3–4):243–248.
48. Palczewski K, Kumasaka T, Hori T, et al. Crystal structure of rhodopsin: a G protein-coupled receptor. Science 2000;289:739–745.
49. Jacob J, Duclohier H, Cafiso DS. The role of proline and glycine in determining the backbone flexibility of a channel-forming peptide. Biophys J 1999;76(3):1367–1376.
50. Barnett-Norris J, Hurst DP, Buehner K, Ballesteros J, Guarnieri F, Reggio PH. Agonist alkyl tail interaction with cannabinoid CB1 receptor V6.43/I6.46 groove induces a helix 6 active conformation. Intl J Quantum Chem 2002;88:76–86.
51. Shi L, Liapakis G, Xu R, Guarnieri F, Ballesteros JA, Javitch JA. Beta2 adrenergic receptor activation. Modulation of the proline kink in transmembrane 6 by a rotamer toggle switch. J Biol Chem 2002;277:40,989–40,996.
52. Lopez-Rodriguez ML, Vicente B, Deupi X, al. Design, synthesis and pharmacological evaluation of 5- hydroxytryptamine 1a receptor ligands to explore the three-dimensional structure of the receptor. Mol Pharmacol 2002;62:15–21.
53. Liapakis G, Ballesteros JA, Papachristou S, Chan WC, Chen X, Javitch JA. The forgotten serine. A critical role for Ser-2035.42 in ligand binding to and activation of the beta 2-adrenergic receptor. J Biol Chem 2000;275:37,779–37,788.
54. van Rhee AM, Jacobson KA. Molecular architecture of G protein-coupled receptors. Drug Dev Res 1996;37:1–38.
55. Ambrosio C, Molinari P, Cotecchia S, Costa T. Catechol-binding serines of beta 2 adrenergic receptors control the equilibrium between active and inactive receptor states. Mol Pharmacol 2000;57:198–210.
56. Javitch JA, Fu D, Chen J, Karlin A. Mapping the binding-site crevice of the dopamine D2 receptor by the substituted-cysteine accessibility method. Neuron 1995; 14:825–831.

57. Govaerts C, Lefort A, Costagliola S, et al. A conserved Asn in transmembrane helix 7 is an on/off switch in the activation of the thyrotropin receptor. J Biol Chem 2001;276:22,991–22,999.
58. Okada T, Fujiyoshi Y, Silow M, Navarro J, Landau EM, Shichida Y. Functional role of internal water molecules in rhodopsin revealed by X- ray crystallography. Proc Natl Acad Sci USA 2002;99:5982–5987.
59. Neve KA, Cumbay MG, Thompson KR, et al. Modeling and mutational analysis of a putative sodium-binding pocket on the dopamine D2 receptor. Mol Pharmacol 2001;60(2):373–381.
60. Lu ZL, Hulme EC. The functional topography of transmembrane domain 3 of the M1 muscarinic acetylcholine receptor, revealed by scanning mutagenesis. J Biol Chem 1999;274(11):7309–7315.
61. Lu ZL, Saldanha JW, Hulme EC. Transmembrane domains 4 and 7 of the M(1) muscarinic acetylcholine receptor are critical for ligand binding and the receptor activation switch. J Biol Chem 2001;276(36):34,098–34,104.
62. Dawson JP, Weinger JS, Engelman DM. Motifs of serine and threonine can drive association of transmembrane helices. J Mol Biol 2002;316:799–805.
63. Govaerts C, Bondue A, Springael JY, et al. Activation of CCR5 by chemokines involves an aromatic cluster between transmembrane helices 2 and 3. J Biol Chem 2003;278:1892–1903.
64. Peralvarez A, Barnadas R, Sabes M, Querol E, Padros E. Thr90 is a key residue of the bacteriorhodopsin proton pumping mechanism. FEBS Lett 2001;508:399–402.
65. Peleg G, Ghanouni P, Kobilka BK, Zare RN. Single-molecule spectroscopy of the beta 2 adrenergic receptor: observation of conformational substates in a membrane protein. Proc Natl Acad Sci USA 2001;98(15):8469–8474.
66. Swaminath G, Xiang Y, Lee TW, Steenhuis J, Parnot C, Kobilka BK. Sequential binding of agonists to the beta 2 adrenoceptor. Kinetic evidence for intermediate conformational states. J Biol Chem 2004;279(1):686–691.
67. Ghanouni P, Gryczynski Z, Steenhuis JJ, et al. Functionally different agonists induce distinct conformations in the G protein coupling domain of the beta 2 adrenergic receptor. J Biol Chem 2001;276(27):24,433–24,436.
68. Gether U, Ballesteros JA, Seifert R, Sanders-Bush E, Weinstein H, Kobilka BK. Structural instability of a constitutively active G protein-coupled receptor: Agonist-independent activation due to conformational flexibility. J Biol Chem 1997;272:2587–2590.
69. Horn F, Bettler E, Oliveira L, Campagne F, Cohen FE, Vriend G. GPCRDB information system for G protein-coupled receptors. Nucleic Acids Res 2003;31:294–297.
70. Kraulis J. MOLSCRIPT: a program to produce both detailed and schematic plots of protein structure. J Appl Crystallogr 1991; 24:946–950.
71. Merritt EA, Bacon DJ. Raster3D: photorealistic molecular graphics. Methods Enzymol 1997;277:505–524.

18
De-Orphanizing GPCRs and Drug Development

Rainer K. Reinscheid and Olivier Civelli

1. INTRODUCTION

Traditionally, the family of G protein-coupled receptors (GPCRs) has been divided into "classical" GPCRs and olfactory (or gustatory) GPCRs. The classical GPCRs comprise receptors for all neurotransmitters and hormones, displaying a vast chemical diversity among their natural ligands. Odorant receptors are believed to bind volatile molecules, but the mechanisms of ligand binding in this group are less well-understood.

Our current knowledge about GPCRs and their natural ligands is the combined work of several decades of research. However, we have only come half the distance, or even less. Cloning of the human genome has revealed about 1000 GPCRs—400 for neurotransmitters and hormones (transmitters, as a general term) and about 600 presumed odorant and gustatory receptors (sensory receptors). Currently, we know the natural ligands for approx 180 GPCRs and virtually none of the odorant ligands *(1)*. Therefore, the work has just begun.

By definition, orphan GPCRs are GPCRs that are not matched to any know natural ligand; in this area, the olfactory GPCRs constitute by far the largest group of orphan receptors. However, odorant receptors pose specific technical challenges and are not currently viewed as targets of drug development. Therefore, we focus on the progress made over recent years regarding the classical GPCRs (transmitter GPCRs) and their ligands. Orphan GPCRs and the quest for their ligands clearly is a product of the postgenomic era. Technologies are being developed at the frontline of biomedical research, and the benefits for future drug development are emerging.

From: *Contemporary Clinical Neuroscience: The G Protein-Coupled Receptors Handbook*
Edited by: L. A. Devi © Humana Press Inc., Totowa, NJ

As members of the most successful family of current drug targets, the orphan GPCRs are certainly the best candidates for drugs of the future and hold special promise for treatment of brain and mental disorders *(2)*.

2. The Family of Orphan GPCRs

By definition, orphan GPCRs are members of the superfamily of GPCRs that await identification of their natural ligand(s). In this sense, every GPCR that was cloned through homology was an orphan receptor because its discovery was based on sequence data. The first reported orphan GPCRs were the 5-HT 1A and the D2 dopamine receptors, which were both cloned in 1988 using the β_2-adrenergic receptor complementary DNA (cDNA) as a probe *(3,4)*. Similarly, the muscarinic m3 receptor *(5)* and the cannabinoid CB1 receptor *(6)* can be considered examples of early orphan receptors that were soon paired with their respective ligands.

With application of polymerase chain reaction technology, large groups of orphan GPCRs were soon discovered in the genome or cDNA libraries *(7)*. Ensuing research quickly showed that the total number of GPCRs in the genome was larger than anyone had anticipated. Shortly after, another groundbreaking paper provided compelling evidence that GPCRs also lie at the basis of olfaction and that the number of olfactory receptors even exceeds the number of "classical" GPCRs *(8)*.

Progress in identification of olfactory ligands has been slow for three primary reasons. First, there is an almost endless number of odorants that humans or animals can smell, and the search for a specific molecule activating an olfactory receptor can be endless. Second, technical problems have hampered the expression of olfactory receptors in non-neuronal cellular environments *(9)*. Finally, there is obviously a smaller commercial interest in this group of GPCRs compared to the so-called "druggable" orphan receptors. At first sight, the search for natural ligands of orphan receptors appears similar to a classical academic field of research: high-risk, uncertain results and unpredictable timing. However, the biggest progress and successes in the hunt for new transmitter molecules were made in the pharmaceutical industry *(10)*. Academia has largely avoided the field for three major reasons: (a) until recently, orphan receptor research was considered too speculative to receive sufficient public funding; (b) high-tech equipment mandates high costs; and (c) the projects generally require long durations.

This chapter summarizes the status quo of research in the field of orphan GPCRs, the advancement of knowledge about physiological functions, and the impact of de-orphanized receptors for pharmaceutical research of the future.

3. TECHNOLOGICAL APPROACHES TO IDENTIFY LIGANDS OF ORPHAN GPCRS

Three fundamentally different strategies have been employed for ligand identification: (a) isolation of natural molecules from tissue extracts; (b) matching of large collections of known chemicals; and (c) prediction of peptide ligands through bio-informatic approaches, followed by matching them to cloned orphan GPCRs. As a common principle, these approaches use cellular assays of receptor activation that monitor changes in second messengers, such as Ca^{2+}, cyclic adenosine monophosphate, and arachidonic acid, or quantify increases in γ-^{35}S GTP binding as a measure of receptor activation. Because most of the search for new ligands was performed in the pharmaceutical industry, the matching approach was completed quickly and efficiently. In a few years, more than 30 orphan GPCRs were matched with known molecules. Some of these known ligands were neuropeptides that had been missing a corresponding receptor, such as melanin-concentrating hormone (MCH), urotensin II (UII), neuromedin U (NMU), or neuropeptides FF and AF. In the cases of MCH, UII, and NMU, the same ligands were also simultaneously isolated from tissue extracts by using the expressed orphan GPCR as a cellular detector system (*see* Table 1). Together, these results confirmed that the peptides were indeed bona fide ligands.

Matching has been most successful in the groups of nucleotide and lipid receptors. In most cases, the key for success has been to find the right cellular environment that allows detection of receptor activation. Both nucleotide and lipid receptors are almost ubiquitous, and finding a background-free cell system has been one of the most difficult tasks in this field of orphan GPCR research.

Isolation of novel bio-active molecules from tissue extracts has been slow and tedious. To date, all newly discovered natural ligands have been peptides. This is not surprising because quantities of natural ligands are usually infinitesimally low, and the available methods of analytical chemistry and structure elucidation are most sophisticated for peptides. For example, peptide sequencing by Edman degradation or mass spectrometry requires only femto- to low picomolar amounts of material, whereas the structural analysis of a novel lipid or nucleotide molecule by nuclear magnetic resonance might require 1000 times more isolated material to obtain the full structure.

In 1995, our group and the team of Jean-Claude Meunier isolated and independently identified the first ligand of an orphan GPCR, orphanin FQ/Nociceptin (OFQ/N), *(11,12)*. Since then, only 10 truly novel ligands have been discovered, although more than 150 orphan GPCRs are available for screening. At this rate, the discovery of the natural ligands for all orphan

Table 1
New GPCR Ligands Identified by Isolation From Tissue Extracts

Ligand	Year	Physiological Functions	Synthetic Agonist/Antagonist	Reference
OFQ/N	1995	Pain, stress	Yes	11,12
Hypocretins/Orexins	1998	Sleep, feeding	Yes	36,37
Apelin	1998	Immune modulation		70
PrRP	1998	Appetite, stress, sleep		46
MCH	1999	Feeding	Yes	53–55
Ghrelin	1999	Feeding	Yes	62
Urotensin II	1999	Cardiovascular	Yes	71–73
Neuromedin U	2000	Feeding, stress		74
Metastin	2001	Cell proliferation, Development		75
PK1 and PK2	2002	Circadian rhythm		76,77
NPB and NPW	2002	Pain, feeding		78–80

Abbreviations: PrRP, prolactin-releasing hormone; MCH, melanin-concentrating hormone; NPB, neuropeptide B; NPW, neuropeptide W; PK, prokineticin.

GPCRs—excluding olfactory receptors—may require years of work. Therefore, development of new technologies for this task is highly anticipated.

Recently, a bio-informatic approach for finding novel peptide ligands has been used in several cases. Secretory proteins are commonly predicted from databases using sophisticated software that scans open reading frames for structural hallmarks of peptide precursors, such as a hydrophobic signal peptide, endoprotease cleavage sites, and glycine residues that can be converted into a C-terminal amide. This strategy has been most successfully employed to discover new members in the family of the RFamide peptides, which all terminate in the sequence Arg-Phe-amide. The predicted peptides were synthesized and then matched with orphan GPCRs *(13–15)*.

By definition, matching of known compounds to a receptor can never identify totally new molecules because it relies on a collection of chemicals that are usually commercially available. Therefore, only the isolation of bio-active molecules from tissue extracts or prediction of novel secretory proteins has the potential to discover new molecules.

4. EXAMPLES OF LIGAND DISCOVERIES AND THEIR IMPACT ON PHARMACEUTICAL RESEARCH

This section discusses several examples of newly identified ligands of orphan GPCRs and novel therapeutic concepts that have emerged from studies of their physiological functions. In many cases, our understanding of a novel ligand–receptor system is still in an embryonic and continuously evolving stage, and we have selected only those cases in which sufficient experimental evidence or preclinical studies with synthetic small molecules are available. A list of molecules that have been isolated as ligands of orphan GPCRs and their primary physiological functions (based on current knowledge) is provided in Table 1. Nomenclature of both ligands and receptors is particularly confusing in this field because ligands have occasionally been identified by independent groups and receptors have been named inconsistently. We have tried to use the nomenclature of the original publications.

4.1. Orphanin FQ/Nociceptin

OFQ/N was identified as the endogenous ligand of an orphan GPCR that showed high homology to the opiate receptors (ORL-1, now called NOP) *(11,12)*. The primary structure of the peptide showed resemblance to opioid peptides—particularly dynorphin A; therefore, it was no surprise that this novel ligand was quickly embraced by the opiate community, and a large number of physiological functions were described. OFQ/N can modulate nociception, although in a different way than common opioid peptides

(11,12,16–18). The peptide was demonstrated to reverse stress-induced analgesia *(19)*, modulate transmitter and hormone release *(20–22)*, influence cardiac function *(23)* and feeding *(24)*, modulate learning and memory *(25,26)*, and suppress coughing *(27)*. OFQ/N may also be involved in the pathophysiology of seizures and stroke *(28)*. The role of OFQ/N in analgesia is not clearly understood, but it appears that, at least in the spinal chord, OFQ/N antagonists could have therapeutic potential as analgesics in models of chronic or inflammatory pain *(29,30)*. One of the major functions of central OFQ/N appears to be the attenuation of stress responses *(31,32)*. Converging evidence from both pharmacological and genetic studies showed that OFQ/N or small molecule agonists *(33)* produced anxiolytic effects, and mice lacking the peptide displayed increased responsiveness to acute and chronic stress *(34)*. Several studies have been published regarding small-molecule agonists and antagonists for the NOP receptor *(35)*. Generally, OFQ/N antagonists are being developed as analgesics, whereas OFQ/N agonists may be a new class of anxiolytic drugs in the future.

4.2. Hypocretins/Orexins

Hypocretin (Hcrt) 1 and 2 (also termed Orexin A and B, Hcrt/Ox) were identified as hypothalamus-specific peptides *(36,37)*. Because of their expression in the lateral hypothalamus, Hcrt/Ox were expected and demonstrated to regulate feeding behavior. However, when mice were engineered to be devoid of Hcrt/Ox, they exhibited pronounced narcoleptic behavior *(38)*. This conclusion was paralleled by the results of positional cloning analyses performed on an autosomal recessive mutation that was responsible for narcolepsy in dogs *(39)* that showed a mis-sense mutation in one of the orexin receptors. Moreover, human patients with narcolepsy were found to have a selective ablation of Hcrt/Ox-producing neurons *(40)*. This phenotype is now believed to be the result of an auto-immune disease. Consequently, although Hcrt/Ox have a range of other effects *(41,42)*, they are now seen as important modulators of sleep and wakefulness. To date, one selective antagonist (SB-334867-A) has been characterized in detail and was found to modulate wakefulness but also displayed anorectic effects *(43,44*; for a discussion of the link between feeding and vigilance states, *see* ref. *45)*.

4.3. Prolactin-Releasing Peptide

Prolactin-releasing peptide (PrRP) was first characterized as a potent stimulator of prolactin release in in vitro experiments *(46)*. However, later studies showed no evidence for a similar role of endogenous PrRP *(47)*. Since its discovery, PrRP has ben found to produce anorectic effects that may be mediated via activation of corticotropin-releasing hormone *(48)*.

Additionally, PrRP appears to modulate stress responses *(49)* and wakefulness *(50)*, making it an interesting target for drug development in the field of anorexia nervosa or similar stress disorders. To date, no synthetic molecules modulating PrRP activity in vivo have been published, but it can be assumed that the interest in this system will rise as more physiological functions of PrRP are described.

4.4. Melanin-Concentrating Hormone

MCH is another high-profile player in the control of feeding and energy balances. Before identification of its receptor, MCH was demonstrated to potently stimulate food intake *(51)*, and MCH knockout mice were found to be lean *(52)*. The first MCH receptor (MCH1R) was identified both by matching and isolation of the natural peptide from brain extracts *(53–55)*. Soon after, a second MCH receptor (MCH2R) was described whose function is less clear *(56)*. Synthetic antagonists of MCH1R were found to suppress MCH-induced food intake and to reduce body weight after chronic administration *(57–59)*. Unexpectedly, one MCH antagonist (SNAP-7941) was also reported to have antidepressant- and anxiolytic-like effects in rodent models *(58)*. Obviously, MCH receptor antagonists are prime candidates for novel therapeutic concepts in control of body weight.

4.5. Ghrelin

The history of ghrelin and its receptor is a story with surprises and serendipity. At the beginning, there was a synthetic compound identified and even clinically tested as a growth hormone secretagogue (GHS) *(60)*. This compound was also used to clone a GPCR (termed GHS-R) that was found to be predominantly expressed in the brain *(61)*. Therefore, it was surprising when ghrelin was isolated from stomach extracts *(62)*. Ghrelin was found to be profoundly involved in energy homeostasis and the control of food intake, likely as a hunger signal from the gut to the brain *(63)*. Ghrelin enhances fat-mass deposition and food intake by activating hypothalamic feeding centers and promotion of NPY expression *(64)*. Under normal conditions, plasma ghrelin levels peak before food intake. Interestingly, it was discovered that morbidly obese patients undergoing gastric bypass surgery lack this ghrelin peak, indicating that the disruption of ghrelin signaling may at least partially contribute to the weight-reducing effects of this surgical procedure *(65)*. There are currently no reports about small-molecule ghrelin antagonists, but it can be assumed that preclinical development is intensely progressing on this target. The usefulness of ghrelin antagonists in the treatment of obesity was recently questioned by the finding that ghrelin knockout mice had completely normal body weight and displayed no gross phenotypical abnormalities *(66)*. Future studies and the availability of synthetic ghrelin antagonists will help to clarify the function of ghrelin in feeding responses and control of body weight.

4.6. ADP as the Ligand of P2Y12

The identification of the long-sought platelet P2Y receptor (now termed P2Y12) and its endogenous ligand ADP was achieved in two different ways. First, ADP was purified from tissue extracts as a natural compound activating an orphan GPCR that was clearly a member of the purinergic subfamily of receptors, making it the first nonpeptide ligand isolated using the orphan receptor strategy *(67)*. However, the physiological identity of the receptor clone was unclear. Soon after, expression cloning of the platelet ADP receptor was reported, verifying the results of the former study *(68)*. The P2Y12 receptor had been a major focus of pharmaceutical development, because the antithrombotic drugs ticlopidine and clopidogrel were known to antagonize the platelet ADP receptor *(69)*. However, these drugs were developed before the molecular structure of P2Y12 was available. Blocking of P2Y12 receptors can potently reduce blood coagulation and is clinically used to prevent thrombosis. The identification of P2Y12 and its endogenous ligand ADP will enable pharmaceutical companies to develop more selective antagonists with an improved side effect profile.

5. CONCLUSIONS

The identification of ligands for orphan GPCRs is a very juvenile, but promising, field of postgenomic research. Both academia and the pharmaceutical industry have readily understood the great potential for unraveling basic neurochemical mechanisms and development of truly innovative drugs. At this early stage, several small-molecule agonists and antagonists to newly identified ligand–receptor systems are already in preclinical development, and hopefully, we will soon witness the first clinical trials based on orphan GPCR research. Because of the successful history of GPCRs as drug targets, these newly identified GPCRs and their natural ligands will certainly become the focus of future preclinical and clinical development.

A large amount of work lies ahead, and development of new technologies might be necessary to identify all natural ligands of orphan receptors. The knowledge of all transmitters and their cognate receptors involved in cell–cell communication will eventually lead to fundamentally new views of brain functions and associated diseases. Because GPCRs usually have a modulatory effects on neurotransmission, it can be assumed that the results from orphan receptor research will have their greatest impact on the understanding and treatment of psychiatric disorders, where current evidence often suggests a lack of fine-tuning in neuronal communication. In this respect, work on orphan GPCRs holds tremendous potential—not only for the development of drugs that will fill unmet therapeutic demands but also for the neurosciences in general.

REFERENCES

1. Civelli O, Nothacker HP, Saito Y, Wang Z, Lin S, Reinscheid RK. Discovery of novel neurotransmitters as natural ligands of orphan G protein-coupled receptors. Trends Neuro Sci 2001;24:230–237.
2. Drews J. Drug discovery: a historical perspective. Science 2000;287:1960–1964.
3. Fargin A, Raymond JR, Lohse MJ, Kobilka BK, Caron MG, Lefkowitz RJ. The genomic clone G-21 which resembles a beta-adrenergic receptor sequence encodes the 5-HT1A receptor. Nature 1988;335:358–360.
4. Bunzow JR, Van Tol HH, Grandy DK, et al. Cloning and expression of a rat D2 dopamine receptor cDNA. Nature 1988;336:783–787
5. Bonner TI, Buckley NJ, Young AC, Brann MR. Identification of a family of muscarinic acetylcholine receptor genes. Science 1987;237:527–532.
6. Matsuda LA, Lolait SJ, Brownstein MJ, Young AC, Bonner TI. Structure of a cannabinoid receptor and functional expression of the cloned cDNA. Nature 1990;346:561–564.
7. Libert F, Parmentier M, Lefort A, et al. Selective amplification and cloning of four new members of the G protein-coupled receptor family. Science 1989;244:569–572.
8. Buck L, Axel R. A novel multigene family may encode odorant receptors: a molecular basis for odor recognition. Cell 1991;65:175–187.
9. Zhao H, Ivic L, Otaki JM, Hashimoto M, Mikoshiba K, Firestein S. Functional expression of a mammalian odorant receptor. Science 1998;279:237–242.
10. Wilson S, Bergsma DJ, Chambers JK, et al. Orphan G-protein-coupled receptors: the next generation of drug targets? Br J Pharmacol 1998;125:1387–1392.
11. Reinscheid RK, Nothacker HP, Bourson A, et al. Orphanin FQ: a neuropeptide that activates an opioidlike G protein-coupled receptor. Science 1995;270:792–794.
12. Meunier JC, Mollereau C, Toll L, et al. Isolation and structure of the endogenous agonist of opioid receptor-like ORL1 receptor. Nature 1995;377:532–535.
13. Hinuma S, Shintani Y, Fukusumi S, et al. New neuropeptides containing carboxy-terminal RFamide and their receptor in mammals. Nat Cell Biol 2000;2:703–708.
14. Jiang Y, Luo L, Gustafson EL, et al. Identification and characterization of a novel RF-amide peptide ligand for orphan G-protein-coupled receptor SP9155. J Biol Chem 2003;278:27,652–27,657.
15. Fukusumi S, Yoshida H, Fujii R, et al. A new peptidic ligand and its receptor regulating adrenal function in rats. J Biol Chem 2003;278:46,387–46,395.

16. Flores CA, Wang XM, Zhang KM, Mokha SS. Orphanin FQ produces gender-specific modulation of trigeminal nociception: behavioral and electrophysiological observations. Neuroscience 2001;105:489–498.
17. Erb K, Liebel JT, Tegeder I, Zeilhofer HU, Brune K, Geisslinger. Spinally delivered nociceptin/orphanin FQ reduces flinching behaviour in the rat formalin test. Neuroreport 1997;8:1967–1970.
18. Zeilhofer HU, Calo G. Nociceptin/orphanin FQ and its receptor—potential targets for pain therapy? J Pharmacol Exp Ther 2003;306:423–429.
19. Mogil JS, Grisel JE, Reinscheid RK, Civelli O, Belknap JK, Grandy DK. Orphanin FQ is a functional anti-opioid peptide. Neuroscience 1996;75:333–337.
20. Murphy NP, Ly HT, Maidment NT. Intracerebroventricular orphanin FQ/nociceptin suppresses dopamine release in the nucleus accumbens of anaesthetized rats. Neuroscience 1996;75:1–4.
21. Rominger A, Forster S, Zentner J, et al. Comparison of the ORL1 receptor-mediated inhibition of noradrenaline release in human and rat neocortical slices. Br J Pharmacol 2002;135:800–806.
22. Bryant W, Janik J, Baumann M, Callahan P. Orphanin FQ stimulates prolactin and growth hormone release in male and female rats. Brain Res 1998;807:228–233.
23. Armstead WM. Role of Nociceptin/Orphanin FQ in the physiologic and pathologic control of the cerebral circulation. Exp Biol Med 2002;227:957–968.
24. Polidori C, de Caro G, Massi M. The hyperphagic effect of nociceptin/orphanin FQ in rats. Peptides 2000;21:1051–1062.
25. Sandin J, Georgieva J, Schott PA, Ogren SO, Terenius L. Nociceptin/orphanin FQ microinjected into hippocampus impairs spatial learning in rats. Eur J Neurosci 1997;9:194–197.
26. Mamiya T, Yamada K, Miyamoto Y, et al. Neuronal mechanism of nociceptin-induced modulation of learning and memory: involvement of N-methyl-D-aspartate receptors. Mol Psychiatry 2003;8:752–765.
27. McLeod RL, Parra LE, Mutter JC, et al. Nociceptin inhibits cough in the guinea-pig by activation of ORL(1) receptors. Br J Pharmacol 2001;132:1175–1178.
28. Bregola G, Zucchini S, Rodi D, et al. Involvement of the neuropeptide nociceptin/orphanin FQ in kainate seizures. J Neurosci 2002;22:10,030–10,038.
29. Hao JX, Xu IS, Wiesenfeld-Hallin Z, Xu XJ. Anti-hyperalgesic and anti-allodynic effects of intrathecal nociceptin/orphanin FQ in rats after spinal cord injury, peripheral nerve injury and inflammation. Pain 1998;76:385–393.
30. Ito S, Okuda-Ashitaka E, Minami T. Central and peripheral roles of prostaglandins in pain and their interactions with novel neuropeptides nociceptin and nocistatin. Neurosci Res 2001;41:299–332.
31. Jenck F, Moreau JL, Martin JR, et al. Orphanin FQ acts as an anxiolytic to attenuate behavioral responses to stress. Proc Natl Acad Sci USA 1997;94:14,854–14,858.
32. Griebel G, Perrault G, Sanger DJ. Orphanin FQ, a novel neuropeptide with anti-stress-like activity. Brain Res 1999;836:221–224.

33. Jenck F, Wichmann J, Dautzenberg FM, et al. A synthetic agonist at the orphanin FQ/nociceptin receptor ORL1: anxiolytic profile in the rat. Proc Natl Acad Sci USA 2000;97:4938–4943.
34. Köster A, Montkowski A, Schulz S, et al. Targeted disruption of the orphanin FQ/nociceptin gene increases stress susceptibility and impairs stress adaptation in mice. Proc Natl Acad Sci USA 1999;96:10,444–10,449.
35. Meunier JC. Utilizing functional genomics to identify new pain treatments : the example of nociceptin. Am J Pharmacogenomics 2003;3:117–130.
36. de Lecea L, Kilduff TS, Peyron C, et al. The hypocretins: hypothalamus-specific peptides with neuroexcitatory activity. Proc Natl Acad Sci USA 1998;95:322–327.
37. Sakurai T, Amemiya A, Ishii M, et al. Orexins and orexin receptors: a family of hypothalamic neuropeptides and G protein-coupled receptors that regulate feeding behavior. Cell 1998;92:573–585.
38. Chemelli RM, Willie JT, Sinton CM, et al. Narcolepsy in orexin knockout mice: molecular genetics of sleep regulation. Cell 1999;98:437–451.
39. Lin L, Faraco J, Li R, et al. The sleep disorder canine narcolepsy is caused by a mutation in the hypocretin (orexin) receptor 2 gene. Cell 1999;98:365–376.
40. Peyron C, Faraco J, Rogers W, et al. A mutation in a case of early onset narcolepsy and a generalized absence of hypocretin peptides in human narcoleptic brains. Nat Med 2000;6:991–997.
41. Bingham S, Davey PT, Babbs AJ, et al. Orexin-A, an hypothalamic peptide with analgesic properties. Pain 2001;92:81–90.
42. Hirota K, Kushikata T, Kudo M, Kudo T, Smart D, Matsuki A. Effects of central hypocretin-1 administration on hemodynamic responses in young-adult and middle-aged rats. Brain Res 2003;981:143–150.
43. Rodgers RJ, Halford JC, Nunes de Souza RL, et al. SB-334867, a selective orexin-1 receptor antagonist, enhances behavioural satiety and blocks the hyperphagic effect of orexin-A in rats. Eur J Neurosci 2001;13:1444–1452.
44. Haynes AC, Chapman H, Taylor C, et al. Anorectic, thermogenic and anti-obesity activity of a selective orexin-1 receptor antagonist in ob/ob mice. Regul Pept 2002;104:153–159.
45. Sutcliffe JG, de Lecea L. The hypocretins: setting the arousal threshold. Nat Rev Neurosci 2002;3:339–349.
46. Hinuma S, Habata Y, Fujii R, et al. A prolactin-releasing peptide in the brain. Nature 1998;393:272–276.
47. Skinner DC, Caraty A. Prolactin release during the estradiol-induced LH surge in ewes: modulation by progesterone but no evidence for prolactin-releasing peptide involvement. J Endocrinol 2003;177:453–460.
48. Lawrence CB, Celsi F, Brennand J, Luckman SM. Alternative role for prolactin-releasing peptide in the regulation of food intake. Nat Neurosci 2000;3:645,646.
49. Seal LJ, Small CJ, Dhillo WS, Kennedy AR, Ghatei MA, Bloom SR. Prolactin-releasing peptide releases corticotropin-releasing hormone and increases

plasma adrenocorticotropin via the paraventricular nucleus of the hypothalamus. Neuroendocrinology 2002;76:70–78.
50. Lin SH, Arai AC, Espana RA, et al. Prolactin-releasing peptide (PrRP) promotes awakening and suppresses absence seizures. Neuroscience 2002;114:229–238.
51. Qu D, Ludwig DS, Gammeltoft S, et al. A role for melanin-concentrating hormone in the central regulation of feeding behaviour. Nature 1996;380:243–247.
52. Shimada M, Tritos NA, Lowell BB, Flier JS, Maratos-Flier E. Mice lacking melanin-concentrating hormone are hypophagic and lean. Nature 1998;396:670–674.
53. Chambers J, Ames RS, Bergsma D, et al. Melanin-concentrating hormone is the cognate ligand for the orphan G-protein-coupled receptor SLC-1. Nature 1999;400:261–265.
54. Saito Y, Nothacker HP, Wang Z, Lin SH, Leslie F, Civelli O. Molecular characterization of the melanin-concentrating-hormone receptor. Nature 1999;400:265–269.
55. Lembo PM, Grazzini E, Cao J, et al. The receptor for the orexigenic peptide melanin-concentrating hormone is a G-protein-coupled receptor. Nat Cell Biol 1999;1:267–271.
56. Hill J, Duckworth M, Murdock P, et al. Molecular cloning and functional characterization of MCH2, a novel human MCH receptor. J Biol Chem 2001;276:20,125–20,129.
57. Takekawa S, Asami A, Ishihara Y, et al. T-226296: a novel, orally active and selective melanin-concentrating hormone receptor antagonist. Eur J Pharmacol 2002;438:129–135.
58. Borowsky B, Durkin MM, Ogozalek K, et al. Antidepressant, anxiolytic and anorectic effects of a melanin-concentrating hormone-1 receptor antagonist. Nat Med 2002;8:825–830.
59. Shearman LP, Camacho RE, Sloan Stribling D, et al. Chronic MCH-1 receptor modulation alters appetite, body weight and adiposity in rats. Eur J Pharmacol 2003;475:37–47.
60. Smith RG, Cheng K, Schoen WR, et al. A nonpeptidyl growth hormone secretagogue. Science 1993;260:1640–1643.
61. Howard AD, Feighner SD, Cully DF, et al. A receptor in pituitary and hypothalamus that functions in growth hormone release. Science 1996;273:974–977.
62. Kojima M, Hosoda H, Date Y, Nakazato M, Matsuo H, Kangawa K. Ghrelin is a growth-hormone-releasing acylated peptide from stomach. Nature 1999;402:656–660.
63. Tschop M, Smiley DL, Heiman ML. Ghrelin induces adiposity in rodents. Nature 2000;407:908–913.
64. Ghrelin, an endogenous growth hormone secretagogue, is a novel orexigenic peptide that antagonizes leptin action through the activation of hypothalamic neuropeptide Y/Y1 receptor pathway. Diabetes 2001;50:227–232.
65. Cummings DE, Weigle DS, Frayo RS, et al. Plasma ghrelin levels after diet-induced weight loss or gastric bypass surgery. N Engl J Med 2002;346:1623–1630.

66. Sun Y, Ahmed S, Smith RG. Deletion of ghrelin impairs neither growth nor appetite. Mol Cell Biol 2003;23:7973–7981.
67. Zhang FL, Luo L, Gustafson E, et al. ADP is the cognate ligand for the orphan G protein-coupled receptor SP1999. J Biol Chem 2001;276:8608–8615.
68. Hollopeter G, Jantzen HM, Vincent D, et al. Identification of the platelet ADP receptor targeted by antithrombotic drugs. Nature 2001;409:202–207.
69. Storey F. The P2Y12 receptor as a therapeutic target in cardiovascular disease. Platelets 2001;12:197–209.
70. Tatemoto K, Hosoya M, Habata Y, et al. Isolation and characterization of a novel endogenous peptide ligand for the human APJ receptor. Biochem Biophys Res Commun 1998;251:471–476.
71. Ames RS, Sarau HM, Chambers JK, et al. Human urotensin-II is a potent vasoconstrictor and agonist for the orphan receptor GPR14. Nature 1999;401:282–286.
72. Nothacker HP, Wang Z, McNeill AM, et al. Identification of the natural ligand of an orphan G-protein-coupled receptor involved in the regulation of vasoconstriction. Nat Cell Biol 1999;1:383–385.
73. Mori M, Sugo T, Abe M, et al. Urotensin II is the endogenous ligand of a G-protein-coupled orphan receptor, SENR (GPR14). Biochem Biophys Res Commun 1999;265:123–129.
74. Kojima M, Haruno R, Nakazato M, et al. Purification and identification of neuromedin U as an endogenous ligand for an orphan receptor GPR66 (FM3). Biochem Biophys Res Commun 2000;276:435–438.
75. Ohtaki T, Shintani Y, Honda S, et al. Metastasis suppressor gene KiSS-1 encodes peptide ligand of a G-protein-coupled receptor. Nature 2001;411:613–617.
76. Lin DC, Bullock CM, Ehlert FJ, Chen JL, Tian H, Zhou QY. Identification and molecular characterization of two closely related G protein-coupled receptors activated by prokineticins/endocrine gland vascular endothelial growth factor. J Biol Chem 2002;277:19,276–19,280.
77. Masuda Y, Takatsu Y, Terao Y, et al. Isolation and identification of EG-VEGF/ prokineticins as cognate ligands for two orphan G-protein-coupled receptors. Biochem Biophys Res Commun 2002;293:396–402.
78. Shimomura Y, Harada M, Goto M, et al. Identification of neuropeptide W as the endogenous ligand for orphan G-protein-coupled receptors GPR7 and GPR8. J Biol Chem 2002;277:35,826–35,832.
79. Fujii R, Yoshida H, Fukusumi S, et al. Identification of a neuropeptide modified with bromine as an endogenous ligand for GPR7. J Biol Chem 2002;277:34,010–34,016.
80. Tanaka H, Yoshida T, Miyamoto N, et al. Characterization of a family of endogenous neuropeptide ligands for the G protein-coupled receptors GPR7 and GPR8. Proc Natl Acad Sci USA 2003;100:6251–6256.

INDEX

A

AC, see Adenylyl cyclase
Activation, G protein-coupled receptors, see also Constitutively active mutant, conformational plasticity, see Rhodopsin; Structure, G protein-coupled receptors
  conserved mechanisms, 365, 366
  palmitoylation modulation, 17–19
  rhodopsin, see Rhodopsin
  ternary complex models, 11–13
  transmembrane bundle conformational changes, 364
Adenylyl cyclase (AC), signal transduction, 121, 122
ADP, identification as P2Y12 ligand, 396
AFM, see Atomic force microscopy
Akt, arrestin interactions, 185, 186
Analgesia, see Opioid receptors
Arrestins,
  G protein-coupled receptor, desensitization,
    heterologous vs homologous desensitization, 165, 167, 168
    homologous desensitization in vivo and β-arrestins, 168, 169
    overview, 159, 160
    sequestration and β-arrestins, 169–171
  downregulation, resensitization, and recycling role, 173, 174
  interactions, 174, 175
  scaffolds for signaling,
    Akt, 185, 186
    extracellular signal-regulating kinase, 181–184
    Jun N-terminal kinase, 180
    mitogen-activated protein kinases, 179, 180, 184
    overview, 175, 176
    phosphodiesterases, 184, 185
    Ral, 185
    receptor tyrosine kinases, 177–179
    ubiquitin ligase, 184
  isoforms and functional specialization, 171, 172
  non-G protein signaling mediation, 186–188
  posttranslational modifications of β-arrestins, 172, 173
  structure–function relationships, 161, 163, 165
  types, 160, 161
Atomic force microscopy (AFM), G protein-coupled receptor dimerization studies, 324

B

Bioluminescence resonance energy transfer (BRET), receptor oligomerization studies, 228–233

BRET, see Bioluminescence
    resonance energy transfer
C
Calmodulin, regulation of G
    protein-coupled receptor
    kinases, 155, 156
CAM, see Constitutively active
    mutant
Cdc42, G protein regulation, 126
Chaperones, see Folding, G protein-
    coupled receptors
Classification, G protein-coupled
    receptors, 4, 5, 112, 113, 363
Conformational change, see
    Activation, G protein-coupled
    receptors; Rhodopsin;
    Structure, G protein-coupled
    receptors
Constitutively active mutant
    (CAM),
    6.34(X_3) site mutations, 45, 46
    conformational flexibility, 384
    cysteine accessibility studies, 55,
        56
    D/E3.49 protonation and
        mutation studies, 41–43, 45
    D/E6.30(X_1) site protonation
        and mutation studies, 46,
        47, 49, 50
    multiple activated state
        evidence,
        constitutive internalization,
            57, 58
        distinct signaling pathways,
            57
        opioid receptor random
            mutagenesis studies, 58–60
    overview, 13, 14, 40, 41
    R3.50
        constraint, 41–43
        mutants, 50, 51

CTC, see Cubic ternary complex
    model
Cubic ternary complex model
    (CTC), receptor activation, 13

D

Desensitization,
    endocytosis, 97
    G protein-coupled receptor
        kinase role, 167, 168
    heterologous vs homologous
        desensitization, 165, 167,
        168
    homologous desensitization in
        vivo and β-arrestins, 168,
        169
    overview, 159, 160
    receptor dimerization role,
        311–313
    sequestration and β-arrestins,
        169–171
Dimerization, see Oligomerization,
    G protein-coupled receptors
Dopamine receptors,
    adenosine receptor interactions,
        336, 354
    DRIP78, cell surface trafficking
        role, 298, 299
    somatostatin receptor
        interactions, 355
DRIP78, receptor cell surface
    trafficking role, 298, 299

E

ECM, see Extended ternary
    complex model
Electron paramagnetic resonance
    (EPR), rhodopsin activation
    studies, 38, 39
Endocytosis, see Desensitization;
    Trafficking, G protein-
    coupled receptors

Index

EPR, *see* Electron paramagnetic resonance
ERK, *see* Extracellular signal-regulating kinase
Expression cloning, protein–protein interactions, 201
Extended ternary complex model (ECM), receptor activation, 11–13
Extracellular signal-regulated kinase (ERK),
  arrestin interactions, 181–184
  regulation of G protein-coupled receptor kinases, 154, 155

F

Fluorescence resonance energy transfer (FRET), receptor oligomerization studies,
  photobleaching, 227, 228
  principles, 224
  probes, 224, 225
  time-resolved studies, 225, 227
Folding, G protein-coupled receptors,
  chaperones in oligomerization, 292–294
  chemical chaperones, 77, 78
  endoplasmic reticulum quality control,
    associated degradation, 76, 77
    chaperones and folding factors, 74–76
  glycosylation, 73, 74
  opioid receptor, 73
  overview, 71, 72
  pharmacological chaperones,
    clinical implications, 84, 85
    mechanisms of action, 83
    rescue studies, 78–82
  two-stage model, 72

FRET, *see* Fluorescence resonance energy transfer
Functional complementation, *see* Oligomerization, G protein-coupled receptors

G

gClq-R, receptor cell surface trafficking role, 299
G protein,
  activation cycle, 3, 135
  receptor phosphorylation and uncoupling from G proteins, 95, 96
  subunits,
    $G\alpha_{12/13}$ family, 118–120
    $G\alpha_{i/o}$ family, 116, 117
    $G\alpha_{q/11}$ family, 117, 118
    $G\alpha_s$ family, 114, 115
    $G\beta\gamma$ complex, 120
    heterotrimeric complex, 113
  signaling,
    adenylyl cyclase, 121, 122
    GTPase regulators, 125, 125
    ion channels, 123–125
    phospholipase C, 123
    protein kinase A, 121
  neuronal regulation, 126–128
G protein-coupled receptor (GPCR), *see also specific receptors*,
  abundance, 3, 112, 153, 389
  activation, *see* Constitutively active mutant; Rhodopsin; Activation, G protein-coupled receptors
  classification, 4, 5, 112, 113, 363
  difficulty of study, 4
  endocytosis, *see* Desensitization; Trafficking, G protein-coupled receptors

folding, see Folding, G protein-coupled receptors
interacting proteins, see Protein–protein interactions; specific proteins,
ligand diversity, 364, 384
oligomerization, see Oligomerization, G protein-coupled receptors
orphan receptors, see Orphan G protein-coupled receptors
regulators, see Arrestins; G protein-coupled receptor kinases
structure, see Structure, G protein-coupled receptors
G protein-coupled receptor kinases (GRKs),
degradation, 156
desensitization role, 167, 168
expression regulation, 156
functional overview, 135, 149, 151
receptor phosphorylation and uncoupling from G proteins, 95, 96
regulators of localization and activity,
calcium-binding proteins, 155, 156
extracellular signal-regulated protein kinase, 154, 155
G proteins, 154
phospholipids, 154
protein kinase A, 155
protein kinase C, 155
c-Src, 155
structure, 150
substrate specificity, 151–153
types, 95, 96, 149, 168
G protein inwardly-rectifying potassium channel (GIRK),
classification, 124

Ghrelin, identification as orphan receptor ligand, 395, 396
GIRK, see G protein inwardly-rectifying potassium channel
Glycosylation, G protein-coupled receptors,
folding role, 73, 74
oligomer formation role, 291, 292
GPCR, see G protein-coupled receptor
GRKs, see G protein-coupled receptor kinases

H

HIV, see Human immunodeficiency virus
Homer proteins, receptor cell surface trafficking role, 298
Human immunodeficiency virus (HIV), CCR5 coreceptor dimerization and trafficking, 315, 340

I

Immunoprecipitation, receptor oligomerization studies, 219, 222, 223, 268
Insulin-like growth factor-1 receptor, arrestins in signaling, 186, 187

J, K

Jun N-terminal kinase (JNK), arrestin interactions, 180
Kir, see G protein inwardly-rectifying potassium channel
KNK, see Jun N-terminal kinase

L

Ligand-binding sites, see Activation, G protein-coupled

Index

receptors; Structure, G protein-coupled receptors
Ligand discovery, *see* Orphan G protein-coupled receptors

M

MAPK, *see* Mitogen-activated protein kinase
Maturation, *see* Oligomerization, G protein-coupled receptors
MCH, *see* Melanin-concentrating hormone
Melanin-concentrating hormone (MCH), identification as orphan receptor ligand, 395
Mitogen-activated protein kinase (MAPK), arrestin interactions, 179, 180, 184
Morphine, *see* Opioid receptors

N

Neurite outgrowth, G protein regulation, 126–128
NinaA, receptor cell surface trafficking role, 297
Nociceptin, *see* Orphanin FQ

O

Odorant response mutant proteins (ODRs), receptor cell surface trafficking role, 297, 298
ODRs, *see* Odorant response mutant proteins
Oligomerization, G protein-coupled receptors,
  activation role, 243, 244
  analysis techniques,
    bioluminescence resonance energy transfer, 228–233
    comparison of techniques, 232–234
    computational analysis, 244, 251–258
    fluorescence resonance energy transfer,
      photobleaching, 227, 228
      principles, 224
      probes, 224, 225
      time-resolved studies, 225, 227
    immunoprecipitation, 219, 222, 223, 268
    overview, 219–221
  dimers vs oligomers, 230, 231
  domain interactions,
    contact dimerization, 249, 250
    domain swapping, 249
    extracellular amino terminus domain, 245, 246
    intracellular C-terminal domain, 246, 247
    transmembrane domains, 247, 248
  functional complementation studies,
    mutant receptor pairs, 268–270
    overview, 267, 268
    prospects, 280, 281
    receptor–G protein fusion pairs,
      cell membrane studies, 273–279
      fusion protein constructs, 270–273
      intact cell studies, 279, 280
      trans-complementation, 218, 219
  heterodimers, *see also* Opioid receptors,
    examples, 309, 310, 316, 331–333
    functional receptor inactivation, 339, 340

obligatory activation, 333
pharmacology studies,
  closely-related receptor
    dimerization, 333–336
  distantly-related receptor
    dimerization, 336–338
history of study, 217, 218, 309
homodimer studies,
  atomic force microscopy, 324
  examples, 325
  functional complementation,
    326–329
  ligand modulation, 329, 330
  peptide competition studies,
    324, 326
  X-ray crystallography, 324
ligand induction studies, 231,
  232, 329, 330
maturation,
  endoplasmic reticulum
    chaperone mediation,
    292–294
  glycosylation role, 291, 292
  intracellular formation of
    oligomers, 290, 291
  overview, 287–290
rhodopsin oligomerization
  interface,
  computational analysis, 244,
    251–258
  cysteine crosslinking studies,
    258–260
trafficking role,
  accessory protein regulation,
    DRIP78, 298, 299
  gClq-R, 299
  Homer proteins, 298
  ninaA, 297
  odorant response mutant
    proteins, 297, 298
  receptor activity-
    modifying proteins,
    296

α_1-adrenergic receptor
  subtype hetero-
    oligomerization, 295
  desensitization and
    endocytosis, 311–313
$GABA_B$ receptor subtype
  hetero-oligomerization,
    295
postendocyte trafficking, 313,
  314
secretion, 310, 311
therapeutic implications,
  human immunodeficiency
    virus coreceptor, 315
  morphine tolerance and
    addiction, 315, 316
Opioid receptors,
  dimerization and morphine
    tolerance and addiction,
    315, 316
  folding, 73
  heteromerization,
    adrenergic receptors, 338, 352
    cannabinoid receptors,
      353, 354
    opioid receptor subtypes, 333,
      334, 350–352
    somatostatin receptor, 338
  random mutagenesis and
    constitutively active
    mutants, 58–60
  substance P interactions,
    352, 353
Orexin, identification as orphan
  receptor ligand, 394
Orphan G protein-coupled
  receptors,
  definition, 389, 390
  discovery, 390
  ligand identification,
    bioinformatic prediction of
      peptide ligands, 393
    examples of discovery,

ADP as ligand of P2Y12, 396
ghrelin, 395, 396
melanin-concentrating hormone, 395
orexins, 394
orphanin FQ, 393, 394
prolactin-releasing peptide, 394, 395
matching large collections of known chemicals, 391, 393
pharmaceutical implications, 396, 397
tissue extract isolates, 391, 392
Orphanin FQ, identification as orphan receptor ligand, 393, 394

P

Pain, *see also* Opioid receptors,
G protein-coupled receptors in signaling, 349, 350
neurotransmission, 349
Palmitoylation, receptor activation modulation, 17–19
Phosphodiesterase, arrestin interactions, 184, 185
Phospholipase C (PLC), signal transduction, 123
Phospholipase D2 (PLD2), G protein-coupled receptor interactions, 209
PKA, *see* Protein kinase A
PKC, *see* Protein kinase C
PLC, *see* Phospholipase C
PLD2, *see* Phospholipase D2
Prolactin-releasing peptide (PrRP), identification as orphan receptor ligand, 394, 395
Protein kinase A (PKA), regulation of G protein-coupled receptor kinases, 155

signal transduction, 121
Protein kinase C (PKC),
interacting protein, 208, 209
regulation of G protein-coupled receptor kinases, 155
Protein–protein interactions, expression cloning, 201
G protein-coupled receptors, *see also specific proteins*,
ligand recognition modification, 205–207
non-PDZ domain interactions, 204, 205
oligomerization, see Oligomerization, G protein-coupled receptors
PDZ domain interactions, 203, 204
trafficking modification, 207, 208
proteomics, 201, 202
yeast two-hybrid system, 200

R

Rac, G protein regulation, 125, 126
Raf-1, G protein regulation, 125
Ral, arrestin interactions, 185
RAMPs, *see* Receptor activity-modifying proteins
Rap, G protein regulation, 126
Ras, G protein regulation, 125
Receptor activity-modifying proteins (RAMPs), receptor cell surface trafficking role, 296
Receptor tyrosine kinases, arrestin interactions, 177–179
regulators of G protein signaling linkage with G protein-coupled receptors, 143
Recoverin, regulation of G protein-coupled receptor kinases, 155

Regulators of G protein signaling
(RGS),
bridge signaling between
heterotrimeric and
monomeric G proteins, 142
classification, 136, 138
expression regulation, 139
functional overview, 135, 136
Gβ interactions, 141, 142
ion channel modification in
neurons, 143, 144
splice variants, 139
structure, 136, 140
subcellular localization, 139, 140
tissue distribution, 137, 139
tyrosine kinase signaling
linkage, 143
RGS, *see* Regulators of G protein
signaling
Rhodopsin,
activation,
3.32–7.43 interactions, 54, 55
arginine cage hypothesis,
51, 52
crosslinking of TM3 and TM6
for inactive
conformation
stabilization, 37, 38
mechanisms, 14–17
polar pocket hypothesis, 51, 52
rearrangement of TM3, TM6,
and TM5
disulfide crosslinking
studies, 39
electron paramagnetic
resonance studies,
38, 39
fluorescence spectroscopy,
39
reversal, 159, 160
TM3–TM6 interactions, 53, 54
TM6 aromatic cluster, 52, 53
TM6–TM7 interactions, 55
TM7 movement, 55
Y7.53–Y7.60 interactions, 54
constitutively active mutant
studies,
6.34(X3) site mutations, 45, 46
cysteine accessibility, 55, 56
D/E3.49 protonation and
mutation studies,
41–43, 45
D/E6.30(X_1) site protonation
and mutation studies,
46, 47, 49, 50
overview, 13, 14, 40, 41
R3.50
constraint, 41–43
mutants, 50, 51
crystal structures, 33–35
ninaA role in trafficking, 297
numbering schemes for
sequences, 35
oligomerization interface
analysis,
computational analysis, 244,
251–258
cysteine crosslinking studies,
258–260
phosphorylation and
uncoupling from G
proteins, 95, 96
Somatostatin receptors,
heteromerization,
dopamine receptor
interactions, 336, 355
inactivation, 339
somatostatin receptor
subtype interactions,
334, 335
opioid receptor interactions,
338
Src,
arrestin interactions, 177–179

Index

c-Src regulation of G protein-coupled receptor kinases, 155
Structure, G protein-coupled receptors, *see also* Rhodopsin,
  class A receptors,
    ligand-binding sites,
      distribution of agonist- and antagonist-binding sites, 11
      extracellular loop 2 and conserved disulfide bridge, 10, 11
      extracellular regions, 9, 10
      transmembrane regions, 7, 9
    topography, 6
  multiple conformations and G protein coupling, 19–21
  plasticity examples, ligand recognition and activation mechanism,
    amine receptors, 375, 376, 378–380
    peptide receptors, 381, 382
    serine, threonine, and proline-induced plasticity, 382, 383
  sequence patterns in plasticity, alpha-helix disruption,
    proline, 366, 367
    serine, 367, 369
    threonine, 367, 369
  cysteine modulation of alpha-helix structure, 373–375
  glycine modulation of alpha-helix structure, 373–375
  serine/threonine modulation of proline-kinked alpha-helices, 369

T

TCM, *see* Ternary complex model

Ternary complex model (TCM), receptor activation, 11–13
Trafficking, G protein-coupled receptors,
  cell surface trafficking role of oligomerization,
    accessory protein regulation,
      DRIP78, 298, 299
      gClq-R, 299
      Homer proteins, 298
      ninaA, 297
      odorant response mutant proteins, 297, 298
      receptor activity-modifying proteins, 296
    α_1-adrenergic receptor subtype hetero-oligomerization, 295
    $GABA_B$ receptor subtype hetero-oligomerization, 295
  dimerization effects,
  desensitization and endocytosis, 311–313
  postendocyte trafficking, 313, 314
  secretion, 310, 311
  therapeutic implications,
    human immunodeficiency virus coreceptor, 315
    morphine tolerance and addiction, 315, 316
  endocytosis,
    agonist induction, 96, 97
    roles,
      desensitization, 97
      resensitization, 99
      proteolytic downregulation mediation, 99
    intracellular retainment significance, 299

lysosomal sorting mechanisms, 100
modification by interacting proteins, 207, 208
phosphorylation and uncoupling from G proteins, 95, 96
prospects for study, 101, 102
recycling mechanisms, 99, 100
regulation,
  endoplasmic reticulum folding and export, 100, 101
  traffic after exit from Golgi apparatus, 101

Transforming growth factor-β receptor, arrestins in signaling, 187

U–Y

Ubiquitin ligase, arrestin interactions, 184
VDCC, *see* Voltage-gated calcium channel
Voltage-gated calcium channel (VDCC), G protein activation, 124, 125
Wnt, arrestins in signaling, 187, 188
Yeast two-hybrid system, protein–protein interactions, 200